IFIP Advances in Information and Communication Technology　　584

Editor-in-Chief

Kai Rannenberg, Goethe University Frankfurt, Germany

Editorial Board Members

IFIP – The International Federation for Information Processing

IFIP was founded in 1960 under the auspices of UNESCO, following the first World Computer Congress held in Paris the previous year. A federation for societies working in information processing, IFIP's aim is two-fold: to support information processing in the countries of its members and to encourage technology transfer to developing nations. As its mission statement clearly states:

IFIP is the global non-profit federation of societies of ICT professionals that aims at achieving a worldwide professional and socially responsible development and application of information and communication technologies.

IFIP is a non-profit-making organization, run almost solely by 2500 volunteers. It operates through a number of technical committees and working groups, which organize events and publications. IFIP's events range from large international open conferences to working conferences and local seminars.

The flagship event is the IFIP World Computer Congress, at which both invited and contributed papers are presented. Contributed papers are rigorously refereed and the rejection rate is high.

As with the Congress, participation in the open conferences is open to all and papers may be invited or submitted. Again, submitted papers are stringently refereed.

The working conferences are structured differently. They are usually run by a working group and attendance is generally smaller and occasionally by invitation only. Their purpose is to create an atmosphere conducive to innovation and development. Refereeing is also rigorous and papers are subjected to extensive group discussion.

Publications arising from IFIP events vary. The papers presented at the IFIP World Computer Congress and at open conferences are published as conference proceedings, while the results of the working conferences are often published as collections of selected and edited papers.

IFIP distinguishes three types of institutional membership: Country Representative Members, Members at Large, and Associate Members. The type of organization that can apply for membership is a wide variety and includes national or international societies of individual computer scientists/ICT professionals, associations or federations of such societies, government institutions/government related organizations, national or international research institutes or consortia, universities, academies of sciences, companies, national or international associations or federations of companies.

More information about this series at http://www.springer.com/series/6102

Ilias Maglogiannis · Lazaros Iliadis ·
Elias Pimenidis (Eds.)

Artificial Intelligence Applications and Innovations

16th IFIP WG 12.5 International Conference, AIAI 2020
Neos Marmaras, Greece, June 5–7, 2020
Proceedings, Part II

Editors
Ilias Maglogiannis ⓘ
Department of Digital Systems
University of Piraeus
Piraeus, Greece

Elias Pimenidis ⓘ
Department of Computer Science
and Creative Technologies
University of the West of England
Bristol, UK

Lazaros Iliadis ⓘ
Department of Civil Engineering,
Lab of Mathematics and Informatics (ISCE)
Democritus University of Thrace
Xanthi, Greece

ISSN 1868-4238   ISSN 1868-422X (electronic)
IFIP Advances in Information and Communication Technology
ISBN 978-3-030-49188-8   ISBN 978-3-030-49186-4 (eBook)
https://doi.org/10.1007/978-3-030-49186-4

This Springer imprint is published by the registered company Springer Nature Switzerland AG
The registered company address is: Gewerbestrasse 11, 6330 Cham, Switzerland

Preface

AIAI 2020

Artificial Intelligence (AI) is already affordable through a large number of applications that offer good services to our post-modern societies. Image-face recognition and translation of speech are already a reality. File sharing via Dropbox, Uber transportations, social interaction through Twitter, and shopping from eBay are employing Google's TensorFlow platform. AI has already developed high levels of reasoning. Respective applications like AlphaGo (released by Google DeepMind) have managed to defeat human experts in highly sophisticated demanding games, like *Go*. This is a great step forward, if we realize that in the *Go* game the number of potential moves is higher than the number of atoms in the entire Universe. *AlphaGo Zero* is a recent and impressive advance which is using Reinforcement Learning to teach itself. It started with no knowledge at all and in three days it bypassed the capabilities of *AlphaGo Lee*, which is the version that defeated one of the best *Go* human players in four out of five games in 2016. In 21 days, it evolved even further, and it reached Master level. More specifically, it defeated 60 top professional *Go* players online and the world champion himself.

Deep Learning has significantly contributed to the progress made during the last decades. Meta-Learning and intuitive intelligence are on the way, and soon they will enable machines to understand what learning is all about. The concept of Generative Adversarial Neural Networks will try to fuse "imagination" to AI. However historic challenges for the future of mankind will be faced. Potential unethical use of AI may violate democratic human rights and may alter the character of western societies.

The 16th conference on Artificial Intelligence Applications and Innovations (AIAI 2020) offers an insight to all timely challenges related to technical, legal, and ethical aspects of AI systems and their applications. New algorithms and potential prototypes employed in diverse domains are introduced. AIAI is a mature international scientific conference held in Europe and well established in the scientific area of AI. Its history is long and very successful, following and spreading the evolution of Intelligent Systems.

The first event was organized in Toulouse, France, in 2004. Since then, it has a continuous and dynamic presence as a major global, but mainly European, scientific event. More specifically, it has been organized in China, Greece, Cyprus, Australia, and France. It has always been technically supported by the International Federation for Information Processing (IFIP) and more specifically by the Working Group 12.5 which is interested in AI applications.

Following a long-standing tradition, this Springer volume belongs to the IFIP AICT Springer Series and contains the papers that were accepted to be presented orally at the AIAI 2020 conference. An additional volume comprises the papers that were accepted and presented at the workshops – held as parallel events.

The diverse nature of papers presented, demonstrates the vitality of AI algorithms and approaches. It certainly proves the very wide range of AI applications as well.

The event was held during June 5–7, 2020, as a remote LIVE event with live presentations, a lot of interaction, and live Q&A sessions. There was no potential for physical attendance due to the COVID-19 global pandemic.

Regardless of the extremely difficult pandemic conditions, the response of the international scientific community to the AIAI 2020 call for papers was overwhelming, with 149 papers initially submitted. All papers were peer reviewed by at least two independent academic referees. Where needed a third referee was consulted to resolve any potential conflicts. A total of 47% of the submitted manuscripts (70 papers) were accepted to be published as full papers (12-pages long) in these Springer proceedings. Due to the high quality of the submissions, the Program Committee decided additionally to accept five more papers to be published as short papers (10-pages long).

The following scientific workshops on timely AI subjects were organized under the framework of AIAI 2020:

9th Mining Humanistic Data Workshop (MHDW 2020)

We would like to thank the Steering Committee of MHDW 2020 Professors Ioannis Karydis and Katia Lida Kermanidis from Ionian University, Greece, and Professor Spyros Sioutas from the University of Patras, Greece, for their important contribution towards the organization of this high-quality and mature event. Also, we would like to thank Professor Christos Makris and Dr. Andreas Kanavos from the University of Patras, Greece, and Professor Phivos Mylonas for doing an excellent job as Senior Members of the Program Committee. MHDW is an annual event that attracts an increasing amount of high-quality research papers, under the framework of the AIAI conference.

5th Workshop on 5G-Putting Intelligence to the Network Edge (5G-PINE 2020)

We would like to thank Dr. Ioannis P. Chochliouros, Research Programs Section, Research and Development Department, Fixed & Mobile, Hellenic Telecommunications Organization (OTE). We really appreciate his great efforts in organizing this high-quality 5th 5G-PINE workshop, which is a well-established annual event. It presents timely and significant results from state-of-the-art research in the 4th industrial revolution area. This workshop connects the conference with the latest AI applications in the telecommunication industry.

The workshops organized under the framework of AIAI 2020, also followed the same review and acceptance ratio rules. More specifically the 5th 5G-PINE workshop accepted 11 full papers (50%) and one short paper out of 23 submissions, whereas the MHDW 2020 accepted 6 full (37.5%) and 3 short papers out of 16 submitted papers.

Four keynote speakers gave state-of-the-art lectures (after invitation) in timely aspects-applications of AI.

Professor Leontios Hadjileontiadis, Department of Electrical and Computer Engineering at Aristotle University of Thessaloniki, Greece, and Coordinator of i-PROGNOSIS, gave a speech on the subject of "Smartphone, Parkinson's and Depression: A new AI-based prognostic perspective".

Professor Hadjileontiadis has been awarded, among other awards, as innovative researcher and champion faculty from Microsoft, USA (2012), the Silver Award in Teaching Delivery at the Reimagine Education Awards (2017–2018), and the Healthcare Research Award by the Dubai Healthcare City Authority Excellence Awards (2019). He is a Senior Member of IEEE.

Professor Nikola Kasabov, FIEEE, FRSNZ, Fellow INNS College of Fellows, DVF RAE UK, Director, Knowledge Engineering and Discovery Research Institute, Auckland University of Technology, Auckland, New Zealand, Advisory/Visiting Professor SJTU and CASIA China, RGU UK, gave a speech on the subject of "Deep Learning, Knowledge Representation and Transfer with Brain-Inspired Spiking Neural Network Architectures".

Professor Kasabov has received a number of awards, among them: Doctor Honoris Causa from Obuda University, Budapest; INNS Ada Lovelace Meritorious Service Award; NN Best Paper Award for 2016; APNNA 'Outstanding Achievements Award'; INNS Gabor Award for 'Outstanding contributions to engineering applications of neural networks'; EU Marie Curie Fellowship; Bayer Science Innovation Award; APNNA Excellent Service Award; RSNZ Science and Technology Medal; and 2015 AUT Medal; and Honorable Member of the Bulgarian and Greek Societies for Computer Science.

Dr. Pierre Philippe Mathieu, European Space Agency (ESA) Head of the Philab (Φ Lab) Explore Office at the European Space Agency in ESRIN (Frascati, Italy).

Professor Xiao-Jun Wu Department of Computer Science and Technology Jiangnan University, China, gave a speech on the subject of "Image Fusion Based on Deep Learning".

Professor Xiao-Jun Wu has won the most outstanding postgraduate award by Nanjing University of Science and Technology. He has won different national and international awards for his research achievements. He was a visiting postdoctoral researcher at the Centre for Vision, Speech, and Signal Processing (CVSSP), University of Surrey, UK, from 2003–2004.

The following two tutorial sessions by experts in the AI field completed the program:

Professor John Macintyre Dean of the Faculty of Applied Sciences, Pro Vice Chancellor at the University of Sunderland, UK. During the 1990s he established the Center for Adaptive Systems – at the university, which became recognized by the UK government as a Center of Excellence for applied research in adaptive computing and AI. The Center undertook many projects working with and for external organizations in industry, science, and academia, and for three years ran the Smart Software for Decision Makers program on behalf of the Department of Trade and Industry.

Professor Macintyre presented a plenary talk on the following subject: "AI Applications during the COVID-19 Pandemic - A Double Edged Sword?".

Dr. Kostas Karpouzis Associate Researcher, Institute of Communication and Computer Systems (ICCS) of the National Technical University of Athens, Greece. Tutorial Subject: "AI/ML for games for AI/ML".

Digital games have recently emerged as a very powerful research instrument for a number of reasons: they involve a wide variety of computing disciplines, from databases and networking to hardware and devices, and they are very attractive to users regardless of age or cultural background, making them popular and easy to evaluate with actual players. In the fields of AI and Machine Learning (ML), games are used in a two-fold manner: to collect information about the players' individual characteristics (player modeling), expressivity (affective computing), and playing style (adaptivity) and also to develop AI-based player bots to assist and face the human players as a test-bed for contemporary AI algorithms.

This tutorial discusses both approaches that relate AI/ML to games: starting from a theoretical review of user/player modeling concepts, it discusses how we can collect data from the users during gameplay and use them to adapt the player experience or model the players themselves. Following that, it presents AI/ML algorithms used to train computer-based players and how these can be used in contexts outside gaming. Finally, it introduces player modeling in contexts related to serious gaming, such as health and education.

Intended audience: researchers in the fields of ML and Human Computer Interaction, game developers and designers, as well as health and education practitioners.

The accepted papers of the AIAI 2020 conference are related to the following thematic topics:

- AI Constraints
- Classification
- Clustering – Unsupervised Learning
- Deep Learning Long Short Term Memory
- Fuzzy Algebra-Fuzzy Systems
- Image Processing
- Learning Algorithms
- ML
- Medical – Health Systems
- Natural Language
- Neural Network Modeling
- Object Tracking-Object Detection
- Ontologies/AI
- Sentiment Analysis/ Recommendation Systems
- AI/Ethics/Law

The authors of submitted papers came from 28 different countries from all over the globe, namely:

Algeria, Austria, Belgium, Bulgaria, Canada, Cyprus, Czech Republic, Finland, France, Germany, Greece, The Netherlands, India, Italy, Japan, Mongolia, Morocco, Oman, Pakistan, China, Poland, Portugal, Spain, Sweden, Taiwan, Turkey, the UK, and the USA.

June 2020

Ilias Maglogiannis
Plamen Angelov
John Macintyre
Lazaros Iliadis
Stefanos Kolias
Elias Pimenidis

Organization

Executive Committee

General Chairs

Ilias Maglogiannis
University of Piraeus, Greece
(President of the IFIP WG12.5)

Plamen Angelov
University of Lancaster, UK

John Macintyre
University of Sunderland, UK (Dean of the Faculty of Applied Sciences and Pro Vice Chancellor of the University of Sunderland)

Program Chairs

Lazaros Iliadis
Democritus University of Thrace, Greece

Stefanos Kolias
University of Lincoln, UK

Advisory Chairs

Andreas Stafylopatis
Technical University of Athens, Greece

Vincenzo Piuri
University of Milan, Italy (IEEE Fellow (2001), IEEE Society/Council active memberships/services: CIS, ComSoc, CS, CSS, EMBS, IMS, PES, PHOS, RAS, SMCS, SPS, BIOMC, SYSC, WIE)

Honorary Chair

Robert Kozma
University of Memphis, USA

Liaison Co-chairs

Ioannis Kompatsiaris
IPTIL Research Institute, Greece

Ioannis Chochliouros
Hellenic Telecommunications Organization, Greece

Workshop Chairs

Christos Makris
University of Patras, Greece

Phivos Mylonas
Ionian University, Greece

Spyros Sioutas
University of Patras, Greece

Katia Kermanidou
Ionian University, Greece

Publication and Publicity Chairs

Antonis Papaleonidas
Democritus University of Thrace, Greece

Konstantinos Demertzis
Democritus University of Thrace, Greece

George Tsekouras
University of the Aegean, Greece

Special Sessions Chairs

Panagiotis Papapetrou   Stockholm University, Sweeden
Georgios Paliouras   National Center for Scientific Research NSCR
         Demokritos, Greece

Steering Committee Chairs

Ilias Maglogiannis   University of Piraeus, Greece
Plamen Angelov   University of Lancaster, UK
Lazaros Iliadis   Democritus University of Thrace, Greece

Program Committee

Michel Aldanondo	Toulouse University, IMT Mines Albi, France
Georgios Alexandridis	University of the Aegean, Greece
Serafín Alonso Castro	University of Leon, Spain
Ioannis Anagnostopoulos	University of Thessaly, Greece
Costin Badica	University of Craiova, Romania
Giacomo Boracchi	Politecnico di Milano, Italy
Ivo Bukovsky	Czech Technical University in Prague, Czech Republic
George Caridakis	University of the Aegean, Greece
Francisco Carvalho	Polytechnic Institute of Tomar, Portugal
Ioannis Chamodrakas	National and Kapodistrian University of Athens, Greece
Adriana Coroiu	Babeş-Bolyai University, Romania
Kostantinos Delibasis	University of Thessaly, Greece
Konstantinos Demertzis	Democritus University of Thrace, Greece
Sergey Dolenko	Lomonosov Moscow State University, Russia
Georgios Drakopoulos	Ionian University, Greece
Mauro Gaggero	National Research Council of Italy, Italy
Ignazio Gallo	University of Insubria, Italy
Angelo Genovese	Università degli Studi di Milano, Italy
Spiros Georgakopoulos	University of Thessaly, Greece
Eleonora Giunchiglia	Oxford University, UK
Foteini Grivokostopoulou	University of Patras, Greece
Peter Hajek	University of Pardubice, Czech Republic
Giannis Haralabopoulos	University of Nottingham, UK
Ioannis Hatzilygeroudis	University of Patras, Greece
Nantia Iakovidou	King's College London, UK
Lazaros Iliadis	Democritus University of Thrace, Greece
Zhu Jin	University of Cambridge, UK
Jacek Kabziński	Lodz University of Technology, Poland
Andreas Kanavos	University of Patras, Greece
Stelios Kapetanakis	University of Brighton, UK
Petros Kefalas	CITY College, International Faculty of the University of Sheffield, Greece

Katia Kermanidis	Ionio University, Greece
Niki Kiriakidou	University of Patras, Greece
Giannis Kokkinos	University of Macedonia, Greece
Petia Koprinkova-Hristova	Bulgarian Academy of Sciences, Bulgaria
Athanasios Koutras	Technical Educational Institute of Western Greece, Greece
Paul Krause	University of Surrey, UK
Florin Leon	Technical University of Iasi, Romania
Aristidis Likas	University of Ioannina, Greece
Ioannis Livieris	University of Patras, Greece
Doina Logofătu	Frankfurt University of Applied Sciences, Germany
Ilias Maglogiannis	University of Piraeus, Greece
Goerge Magoulas	Birkbeck College, University of London, UK
Christos Makris	University of Patras, Greece
Mario Malcangi	University of Milan, Italy
Francesco Marceloni	University of Pisa, Italy
Giovanna Maria Dimitri	University of Cambridge, UK and University of Siena, Italy
Nikolaos Mitianoudis	Democritus University of Thrace, Greece
Antonio Moran	University of Leon, Spain
Konstantinos Moutselos	University of Piraeus, Greece
Phivos Mylonas	Ionio University, Greece
Stefanos Nikiforos	Ionio University, Greece
Stavros Ntalampiras	University of Milan, Italy
Mihaela Oprea	Petroleum-Gas University of Ploieşti, Romania
Ioannis P. Chochliouros	Hellenic Telecommunications Organization, Greece
Basil Papadopoulos	Democritus University of Thrace, Greece
Vaios Papaioannou	University of Patras, Greece
Antonis Papaleonidas	Democritus University of Thrace, Greece
Daniel Pérez López	University of Leon, Spain
Isidoros Perikos	University of Patras, Greece
Elias Pimenidis	University of the West of England, UK
Panagiotis Pintelas	University of Patras, Greece
Nikolaos Polatidis	University of Brighton, UK
Bernardete Ribeiro	University of Coimbra, Portugal
Leonardo Rundo	University of Cambridge, UK
Alexander Ryjov	Lomonosov Moscow State University, Russia
Simone Scardapane	Sapienza University, Italy
Evaggelos Spyrou	National Center for Scientific Research – Demokritos, Greece
Antonio Staiano	University of Naples Parthenope, Italy
Andrea Tangherloni	University of Cambridge, UK
Azevedo Tiago	University of Cambridge, UK
Francesco Trovò	Polytecnico di Milano, Italy

Nicolas Tsapatsoulis   Cyprus University of Technology, Cyprus
Petra Vidnerová   Czech Academy of Sciences, Czech Republic
Paulo Vitor de Campos   CEFET-MG, Brazil
  Souza
Gerasimos Vonitsanos   University of Patras, Greece

Abstracts of Invited Talks

Smartphone, Parkinson's and Depression: A New AI-Based Prognostic Perspective

Leontios Hadjileontiadis

Khalifa University of Science and Technology, UAE, and Aristotle University of Thessaloniki, Greece
leontios@auth.gr

Abstract. Machine Learning (ML) is a branch of Artificial Intelligence (AI) based on the idea that systems can learn from data, identify patterns, and make decisions with minimal human intervention. While many ML algorithms have been around for a long time, the ability to automatically apply complex mathematical calculations to big data – over and over, faster and faster, deeper and deeper – is a recent development, leading to the realization of the so called Deep Learning (DL). The latter has an intuitive capability that is similar to biological brains. It is able to handle the inherent unpredictability and fuzziness of the natural world. In this keynote, the main aspects of ML and DL will be presented, and the focus will be placed in the way they are used to shed light upon the Human Behavioral Modeling. In this vein, AI-based approaches will be presented for identifying fine-motor skills deterioration due to early Parkinson's and depression symptoms reflected in the keystroke dynamics, while interacting with a smartphone. These approaches provide a new and unobtrusive way for gathering and analyzing dense sampled big data, contributing to further understanding disease symptoms at a very early stage, guiding personalized and targeted interventions that sustain the patient's quality of life.

Deep Learning, Knowledge Representation and Transfer with Brain-Inspired Spiking Neural Network Architectures

Nikola Kasabov

Auckland University of Technology, New Zealand
nkasabov@aut.ac.nz

Abstract. This talk argues and demonstrates that the third generation of artificial neural networks, the spiking neural networks (SNN), can be used to design brain-inspired architectures that are not only capable of deep learning of temporal or spatio-temporal data, but also enabling the extraction of deep knowledge representation from the learned data. Similarly to how the brain learns time-space data, these SNN models do not need to be restricted in number of layers, neurons in each layer, etc. When a SNN model is designed to follow a brain template, knowledge transfer between humans and machines in both directions becomes possible through the creation of brain-inspired Brain-Computer Interfaces (BCI). The presented approach is illustrated on an exemplar SNN architecture NeuCube (free software and open source available from www.kedri.aut.ac.nz/neucube) and case studies of brain and environmental data modeling and knowledge representation using incremental and transfer learning algorithms These include predictive modeling of EEG and fMRI data measuring cognitive processes and response to treatment, AD prediction, BCI for neuro-rehabilitation, human-human and human-VR communication, hyper-scanning, and others.

The Rise of Artificial Intelligence for Earth Observation (AI4EO)

Pierre Philippe Mathieu

European Space Agency (ESA), Head of the Philab (Φ Lab), Explore Office
at the European Space Agency in ESRIN, Frascati, Italy

Abstract. The world of Earth Observation (EO) is rapidly changing as a result
of exponential advances in sensor and digital technologies.

The speed of change has no historical precedent. Recent decades have wit-
nessed extraordinary developments in ICT, including the Internet, cloud computing and storage, which have all led to radically new ways to collect, distribute
and analyse data about our planet. This digital revolution is also accompanied by
a sensing revolution that provides an unprecedented amount of data on the state
of our planet and its changes.

Europe leads this sensing revolution in space through the Copernicus initiative and the corresponding development of a family of Sentinel missions. This
has enabled the global monitoring of our planet across the whole electromagnetic spectrum on an operational and sustained basis. In addition, a new trend,
referred to as "New Space", is now rapidly emerging through the increasing
commoditization and commercialization of space.

These new global data sets from space lead to a far more comprehensive
picture of our planet. This picture is now even more refined via data from
billions of smart and inter-connected sensors referred to as the Internet of
Things. Such streams of dynamic data on our planet offer new possibilities for
scientists to advance our understanding of how the ocean, atmosphere, land and
cryosphere operate and interact as part on an integrated Earth System. It also
represents new opportunities for entrepreneurs to turn big data into new types of
information services.

However, the emergence of big data creates new opportunities but also new
challenges for scientists, business, data and software providers to make sense
of the vast and diverse amount of data by capitalizing on powerful techniques
such as Artificial Intelligence (AI). Until recently AI was mainly a restricted
field occupied by experts and scientists, but today it is routinely used in
everyday life without us even noticing it, in applications ranging from recommendation engines, language services, face recognition and autonomous vehicles.

The application of AI to EO data is just at its infancy, remaining mainly
concentrated on computer vision applications with Very High-Resolution
satellite imagery, while there are certainly many areas of Earth Science and big
data mining/fusion, which could increasingly benefit from AI, leading to entire
new types of value chain, scientific knowledge and innovative EO services.

This talk will present some of the ESA research/application activities and
partnerships in the AI4EO field, inviting you to stimulate new ideas and collaboration to make the most of the big data and AI revolutions.

Image Fusion Based on Deep Learning

Xiao-Jun Wu

Jiangnan University, China
wu_xiaojun@jiangnan.edu.cn

Abstract. Deep Learning (DL) has found very successful applications in numerous different domains with impressive results. Image Fusion (IMF) algorithms based on DL and their applications will be presented thoroughly in this keynote lecture. Initially, a brief introductory overview of both concepts will be given. Then, IMF employing DL will be presented in terms of pixel, feature, and decision level respectively. Furthermore, a DL inspired approach called MDLatLRR which is a general approach to image decomposition will be introduced for IMF. A comprehensive analysis of DL models will be offered and their typical applications will be discussed, including Image Quality Enhancement, Facial Landmark Detection, Object Tracking, Multi-Modal Image Fusion, Video Style Transformation, and Deep Fake of Facial Images, respectively.

Contents – Part II

Fuzzy Algebra/Systems

Machine Learning

Medical-Health Systems

Natural Language

Contents – Part I

Image Processing

Learning Algorithms

Neural Network Modeling

Object Tracking/Object Detection Systems

Ontologies/AI

AI Ethics/Law

The Ethos of Artificial Intelligence as a Legal Personality in a Globalized Space: Examining the Overhaul of the Post-liberal Technological Order

Abhivardhan(✉)

Indian Society of Artificial Intelligence and Law, 8/12, Patrika Marg, Civil Lines,
Allahabad 211001, India
abhivardhan@isail.in

Abstract. The categorical ethos of artificial intelligence is influenced by its basic structure, which defines its due purpose as a legal personality, challenging the conventional standards of law and justice in a globalized world. Recent developments show a precedential growth in the need-perspective of the AI industry, thereby influencing governance and corporate operations and their legal side in cross-cultural avenues. The determinant outlining of artificial intelligence as a legal personality rests on its probabilistic nature, which yet can be limited to the jurisprudential scope of AI-based on the ethos of the utilitarian approach involving the anthropocentric innovations for artificial intelligence. The dynamic nature of AI, however, in the proposition, is capable of a full-fledged and anthropomorphic legal representation and interpretation, which is hard to find in D9 and certain developing countries, which poses special risks to the generic legal infrastructure of a democratic polity to understand the dynamic and self-transformative nature of artificial intelligence in the age of globalization.

The paper is thus based on the proposition that the ethos involving the legal infrastructure and persona of artificial intelligence is traceable and easier in deterministic mechanisms by regarding and extending stable & constitutive approaches to dissect the legal challenges connected with the redemptions implicated with the lack of a full-fledged regard and scope of the legal personality of AI. The approaches in due proposition are (a) anthropomorphisation; (b) naturalization; (c) techno-socialization; and (d) enculturation. Further, the paper analyses on the challenges to determine the problematic implications awaited by the influence of populism, protectionism, data-centred digital colonialism and technology distancing and proposes suggestions based on the four approaches to counter the minimal effects of the implications. The conclusions of the paper rest on the argument that in the case of a post-liberal order, the ethos of AI can be protected and diversified by adapting with the appreciation of the ethos of globalization, giving adequate, constitutive and reasonable space to the identity-led implications of national identity & diluting the monopolistic influence of the utilitarian approach to artificial intelligence.

Keywords: International relations · Law and technology ethics · AI ethics · Behavioural economics · Identity ethics · Consumerism

© IFIP International Federation for Information Processing 2020
Published by Springer Nature Switzerland AG 2020
I. Maglogiannis et al. (Eds.): AIAI 2020, IFIP AICT 584, pp. 3–14, 2020.
https://doi.org/10.1007/978-3-030-49186-4_1

1 Introduction

The economics of development in a globalized world using technology for imparting the cause of a welfare state is connected with the usefulness of technocratic governance among nation-states in order to convert the stature of technology from being a mere tool or need to become an integral asset of human lives and the constructive systems of governance among public and private actors. Developed countries, therefore, encourage technocratic governance sponsoring transparency, accountability and technical cum humanized regulation & regularization of 'technological assets'.

Artificial Intelligence is gradually becoming a relevant technology asset for knowledge-driven societies and governments & even in cases where knowledge-driven policies [1] and methodologies are yet to be transformed. Nevertheless, the legal transformation of AI is not limited to the planar legality defined in the constitutional and political systems across nation-states, and the barriers and infrastructures of conventional and proceduralist morality in these systems are plausibly hampered by a systemic, yet unparallel conceivable evolution of technology [2, 3]. The rules-based international order maintained and nurtured among nation-states is also suffering from various redemptions that are constitutional, legal, political and social by nature and practice. Generally, the international rule-based order is at the brink of balance between natural morality and state interests [4]. There are, however, existent certain legal, social and political redemptions that damage the inherent plausibility of technocratic governance and the commitment of a liberal order [5]. Thus, the advancement of technological assets is set out to face a post-liberal globalized international order with the complications of redemptive decisions, modalities and foresight. Therefore, the groundwork for making technology integral to societies and lives is set to be affected by diplomatic and technological actors involved.

In case of AI, it is proposed in this paper that there exist three non-exhaustive and integral challenges to both corporate and state actors in developing and developed economies that shape the legal and socio-political ethos of the AI in a post-liberal order. The paper further analyses and proposes the essential aspects of the three challenges and proposes in its further sections four constitutive approaches for an ethical assessment of AI in law and policy-making measures, taking into consideration the rivalled nature of problems due to AI in developed and developing economies with conclusions.

2 Challenges to Harmonization of AI

The harmonization of technology as an asset to public welfare and rule of law is a predominant notion accepted among nation-states and the same. Various declarations, reports and policy documents focus on a harmonious AI, and they endorse a regularized and ethical usage and control of AI as a utility. There is literature available, including case studies, reports and opinions by experts and think-tanks, which show certain reflective aspects as to how should AI be harmonized for public welfare. These are certainly important aspects of a harmonized AI that is duly expected:

- AI must be robust, and bear fairness and accountability [6, 7];
- AI must respect privacy concerns by design and default [6, 8];

- AI must be regulated and should be human-centric [9, 10];
- AI must respect the principles of legitimacy and constitutionalism [10, 11];
- AI must adhere to international human rights obligations [6];
- AI must reflect human-centric and individualistic interests;
- AI must be objective in its probabilistic operations [7];

The essential aspects of an ethical AI are not exhaustive and cannot be preconditioned, because the dynamic behaviour of AI cannot be confined by limited legal interpretation nor the natural aspect of scientific development can be averted. However, the problems in fulfilling the goals to achieve an ethical AI are not limited to the technological domain of AI, which involves technological ethics but have certain socio-political and legal implications as well. These problems emerge as real challenges to the issue and as per the propositions in the paper – are divided into three essential (non-exhaustive) cases, which are enumerated as follows:

1. the uncertain characteristics of populist politics in foreign and domestic policies;
2. the resistant measures are undertaken by governments under the policy and practice of legal, social and technological protectionism;
3. a data-truistic and data-centered digital colonialism of technology in economic and digital competence;

The further sub-sections lay down the social, political and legal challenges towards fostering an impeccable development of the AI ecosystem with suggestions provided.

2.1 Effect of Populism on the Social and Technological Domains

Populism is a political phenomenon, which involves the conception of popular sovereignty and the politics of redemption, which is not limited to democratic establishments, and some of its tactical practices can be seen in China, Russia and even Iran. Nevertheless, with the intent of dissecting ideological intent and welfare interests merged and represented by the leaders who subscribe or have gained power due to populism in politics, the propositions focus on the following reflections of populism in terms of due involvement and inspiration to the connoted domains of technology and society, taking into consideration the market economy perspective behind the development of AI as a technical utility and its ethical role:

- Populism involves a technical disruption of political and democratic confidence among law and policymakers, people and even at the integral levels of governmental agencies [12, 13];
- Populism in many countries involve opaque data protection and privacy ethics policy-making, demanding a closed implementation of data protection policies beyond ideological and political maneuvers of the policy-makers and relevant stakeholders due to lack of a clear, well-informed and well-concerted knowledge among the people who legitimize their leaders' political actions [14], whether by legal or illegal direct or indirect means;

- Populism often in many countries involves the politics of redemption, which means that a chain of legal reforms or miscellaneous changes disrupt the policy status quos of a state;
- In many instances, the precedents of technological ethics and humanism assumed by the incumbent governments to foster consensual measures among the members of the international community are alleged to be or in reality, either reversed or potentially damaged [15];

Though these propositions are non-exhaustive, they are yet reflective of the challenges that law and policymakers, political and civic groups and people face these days.

2.2 Influence of Legal, Economic and Technical Protectionism

In issues related to politics, legal reforms and economics, it is essential that an undone or wrongly achieved act or exercisable policy must be reconvened and resolved adequately. Reasonableness is, therefore, an essential criterion that develops and legitimizes redemptive activities. In the case of AI, there are relative problems with rising implications evolved due to the principled practices of globalization and the conception of the market economy [16, 17].

While the role of globalization is centered to global capitalism, it respects and is centered at the idea of neoliberal economics. Other than economics, globalization endorses multilateral organizations over strategic partnerships and integrable legal measures, based on consensus and peacebuilding. However, there is a shift in the global consensus from a globalist approach to a protectionist approach when it comes to technological democratization. The proposed reflections of the current trends of protectionism in tech-centric diplomacy are enumerated as follows:

- Precedential to various redemptions, law and policymakers feel that the policies under globalization adopted by their state involving the regimes of data protection, accountability and trust are acting to fail because various nation-centric issues with respect to technology are being ignored;
- Corruptible and incredulous practices among globalized economies in governmental and non-state actors weaken the position of the stakeholders involved in tech-centric diplomacy to foster developmental approaches via technology;
- Lack of target-centric tech measures involves adverse implications in welfare activities. Tech measures are therefore central to entrepreneurial ethics and its design because they decide for a set of generations their education, skill development and employment opportunities as well as the policy that shapes the same.
- The market economy approach defeats indigenous interests, lacks replenishment in entrepreneurial ethics and design in developed and developing economies because it lacks a vision towards policing newer and improvable employment opportunities [17];

2.3 Scope of Data-Centered and Data-Truistic Digital Colonialism

Digital colonialism is a phenomenon that involves the supremacy of state and non-state actors in certain areas via domination in the digital activities of technocratic cyberspace.

In the coming years, the phase of creation in the life of internet sustained changes and would continue with the phase of surveillance. The libertarian approach of internet governance is being modified by the technocratic approach of internet governance, making surveillance essential. Digital colonialism became inevitably a problem due to such implications of internet governance. While the role of protectionism and populism is essential to digital colonialism, yet the practicable aspects of digital colonialism have traversed vast, which may be hazardous. The proposed reflections of the current trends of digital colonialism in tech-centric diplomacy are enumerated as follows:

- Trust is an important currency in dealing with tech-centric diplomacy and innovation among the stakeholders involved in the process [16]. If issues of consensual importance related to the data protection regimes are not adequately settled, then it would certainly affect the moral capital of disruptive technologies;
- Technology distancing is a key tool to benefit digital colonialism [18], politically, socially and institutionally. The problem of digital colonization persists not with the libertarian approach of free speech and other relevant and connected liberties, and neither is the approach of technological and political liberalization is flawed. The persistent catalyst that drives the libertarian approach governance in globalization is conventional morality [5]. However, a lack of trust and autonomy among the individuals, who either control or become the data subjects endorses technology distancing, whereby empowering digital colonialism by gradual means;

It is proposed that there are four non-exhaustive approaches to assess and estimate the ethos involving the legal infrastructure and persona of artificial intelligence to dissect the legal and ethical challenges towards harmonizing AI with its aesthetic perspective and pragmatism.

2.4 Hypothesis

The hypothesis of the paper is based on the following corollaries:

- That the aesthetic component of AI can be estimated by four constitutive approaches to render an objective and entity-centric assessment of the AI as a human artefact;
- The four constitutive approaches to assess AI are (a) Anthropomorphisation, (b) Naturalization, (c) Techno-socialization, and (d) Enculturation;
- Anthropomorphisation, as proposed is the approach of accepting the substantive and operative attributes of AI, entailing a proper distinguishability of the attributes and adapting the operative attributes of AI in conformation with the transformable human-related attributes acquired through data subjects by constructive and stable estimation of the data processed;
- Naturalization, as proposed, is the approach of revising and improving the substantive attributes of AI-based on the stimulated data received towards a coherent, constitutive and peaceful adaptability of such transformable human-oriented attributes perceived from the data subject;
- Techno-socialization, as proposed, is the approach of making the environment for AI socialized, regularized and improved with better and constitutive processing of the data and the ethical channeling of the stimuli on which the AI is reliant;

- Enculturation, as proposed, involves the approach of encouraging and harmonizing the identity-oriented footprints generated, learnt and manifestly determined by AI by its probabilistic mechanisms;

With the concept of the four approaches, the ethos of AI as a legal personality is proposed to be harmonized and constructive to ensure better, coherent and unbiased legal recognition and interpretation of the role, persona and activities of artificial intelligence. The additional propositions in the hypothesis are provided as follows:

- The recognition of AI as a juristic entity is required to be extended from being an industrial and economic unit to an organic technology that can possess certain limited yet applicable rights in conjugation with the human stimulus with which it is entitled to exist and remain to ensure its nature to be human-centric;
- The constitutive approaches are based on the argument that the dynamic yet opaque behaviour of AI must be decluttered with better and empathy-oriented measures, where the aesthetic ecosystem between AI and human entities are assessed carefully;
- Further, the receptive rights of the AI, (which are based partly on the technical and vicarious responsibilities of the AI & partly on the entrepreneurial ethics and design as an inherently absolute and fundamental responsibility of the state and non-state actors, who are involved with that AI at any possible and foreseeable levels of human agency) & the obscure and delicate privacies pre-emptively reserved by data subjects (human entities) must be reconciled;
- The natural and civil liberties of AI that are reservedly developed and exercised by the AI (whether recognized or not recognized by law), must be carefully adjudicated and must be properly bridged with the privacy rights of human entities as a simple priority;
- All of the four approaches proposed are non-exhaustive, useful separately but not liable to be implemented and applied together;

The four constitutive approaches for an open and fair assessment of AI are enumerated as follows.

2.5 Anthropomorphisation

In this constitutive approach, the purposive construct is that artificial intelligence as a human artefact must be assumed as a technological entity that has certain substantive and operational attributes. In the hypothesis, the substantive and operative attributes of AI are limited of aesthetic nature and are immaterial, which means that these attributes are not directly bound by the technological features of AI as in the conceptions of technology sciences. These aesthetic attributes are equally recognized in all of the four approaches proposed in the paper and can be derived by a constitutionalist adjudication and interpretation of the characteristics of AI. The characteristics of the approach of anthropomorphisation are enumerated as follows:

- The approach is premised on the basis that AI requires to be understood as a transformable legal entity, i.e., artificial intelligence possesses stimulating characteristics and can develop its own limited empathy;
- The empathy developed by AI is probabilistic by activities but is deterministic by observations because the same practicable empathy developed by AI is adaptive incoherent human environments. This is practicable in case of any technological asset developed. Therefore, by practice and purpose, AI is capable to develop adaptive empathy;
- The activities of AI are central to (a) the environment in which it resides and learns, and (b) the technological features of the AI that enable its learning and executability;
- The attributes of human entities are transformative and adaptive, by their psychological and biological means coupled with external and additional circumstances;
- Every substantive and operational aesthetic attribute of AI, therefore, must be separately adjudicated and interpreted without a pre-empted intent of correlation between the human and AI stimuli & their practicable empathies;
- The adjudication of the aesthetic attributes of AI must uphold the concept of anthropomorphism, which means that the adjudication itself must recognize that AI as a technology is capable of mapping and resembling certain parameters of human characteristics;

The feature of the approach is that the relation between humans and technology is recognized in an intimate and more perfect method and that the relation between technology and human entities is not limited to the argument of (1) affirmative scientific humanism; and (2) material & public welfare. The approach also enables by principle to widen the diversity of outcomes that can be sought through ethnographic analyses of the human-AI aesthetics that is observed.

2.6 Naturalization

The constitutive approach of naturalization signifies towards the appreciation and careful replenishment of the aesthetic attributes of AI beyond the initial recognition of the attributes. The approach focuses on the pragmatic scope of technology-led empathy by practice recognized by law and is subject to proper adjudication. The characteristics of the approach of naturalization are enumerated as follows:

- The aesthetic attributes of AI must be revised and improved for a conditioned and liberalized assessment of the human-AI aesthetics under observation;
- The learning and executability functions & mechanisms of AI that are closely connected with the technological features of AI must be designed and based on (a) environment reconciliation, and (b) environment realization. Both methodologies must keep the individualistic and indigenous attributes of human society the first priority to assess the empathy of AI, and then resort towards industrial and governance goals as the secondary priority, wherein the second priority must not override the primary priority;
- The method of environment reconciliation means that the environment resided by AI must be learnt and normalized by AI to ensure proper and reasonable knowledge of the environment in which the AI can reconcile its own transformative features;

- The method of environment realization follows the method of environment reconciliation, which means that the practical and exercisable empathy of AI in its general resident environment must be explainable, adaptive and self-assessable;
- The test of explanability, adaptability and self-assessment (hereinafter EAS test) in adjudicating the empathy of AI must, in principle, not be overridden by interpretable, perceptible and learning capabilities of AI in order to ensure a fair, just and reasonable assessment of the aesthetic attributes of AI;
- The EAS test for the purpose of adjudicating the empathy of AI, in principle, must prove that the method of environment reconciliation and realization developed within AI must ensure that the entrepreneurial ethics and design (hereinafter EED) behind and in observation (irrespective of positive or zero correlation) must allow a liberalized and freer development of the human data subjects, and such EED must not cause, contribute or render any undue influence over the obscure and delicate privacies pre-emptively reserved by human data subjects;

The feature of this approach is that it helps in a deeper assessment of the aesthetic attributes of the AI in alignment (no correlation) with the technological features of AI, making vicarious and entitative liabilities of the AI deterministic through a test that is based on the empathy of AI by practice and purpose. It also protects the fair, objective and democratized mandate of entrepreneurial ethics and design to carefully assess the anomalies of technology ethics and practices of both political and apolitical nature.

2.7 Techno-Socialization

The constitutive approach of techno-socialization is based on the position that artificial intelligence as a human artefact must reside in a socialized, regularized and improved human real-time environment. The approach of techno-socialization aims at proper scrutiny and understanding of the delicacy, sensitivity and scope of the human environment. Further, the human environment implied includes both real and material environments & the cyberspace in which the AI resides. The characteristics of the approach are enumerated as follows:

- The social environment of the data subjects must be properly scrutinized and assessed. The nature of assessment and scrutiny must not involve an absolutist and restrictive approach based on legal formalism by practice and in principle;
- The technological rubric of cyberspace and material environments must be equitably and coherently surveyed and estimated;
- The ethical autonomy of data subjects must be safeguarded and kept as an overriding priority against the aesthetic liberties of AI, provided that the overriding effect must exist in principle and not achieved by design and practice by the industrial and governing entities responsible and tested for limited liability;
- The real-time environment must be designed and preserved taking into consideration the economic and social design of the environments in principle and practice;
- The environment must be sociable, regularized and replenishable itself for the AI to enable the self-transformation of the aesthetics of AI;

- A sociable environment means the empathy of AI must be distinctively explainable, adaptive and self-assessable, and this distinctiveness must not be influenced or damaged;
- A regularized environment means that the social and economic rubric of the environment must be regularized through ethical and fair practices;
- A replenishable environment means that the inherent anthropomorphic liberties of humans must seek ethically autonomous and libertarian change by practice;

The feature of the approach is that the characteristics of a human environment are now required to be assessed and resolved adequately for a freer and ethical AI. The approach also endorses the libertarian model of fundamental rights and liberties & affirms that human liberty and privacy is absolute and integral to industrial and governance responsibilities and liabilities.

2.8 Enculturation

The constitutive approach of enculturation involves an ethnographic assessment and liberalization of the identities of human data subjects and the cultivation of cultural identities. The purpose of enculturation is an essential rapprochement of cultures and identities & the preservation of the cultural & identity rights and liberties of human entities in the aesthetics involving the human entities and AI. The characteristics of the approach are enumerated as follows:

- The protection of cultural heritage and identities directly and indirectly associable must be encouraged, harmonized and normalized;
- The safeguard of cultural and identity liberties must be an intermediary liability on the entity that breaches the safeguard;
- The data involving identities (including culture) and their heritage must be treated as identity-oriented footprints in case of AI, which means that every such data will be coherent and connected and is useful in assessing the empathy, learning and executability of the AI;
- The probabilistic mechanisms of AI under deterministic assessments are not deprived of creative aesthetic liberties to exercise enculturation;
- The protection of cultures must not be defined by an absolutist, rudimentary and strict adjudication and interpretation involving the issues related to the same. Instead, the protection of cultures must be put into practice by naturalized, real, community-centred, non-isolated, ethically autonomous means of adjudication;
- The obscurity of identity-oriented footprints (based on assessment, record or question of fact) must be recognized in assessing the attributes of AI;
- The identity-oriented footprints can be generated, learnt & manifestly determined by AI within its self and vicarious liabilities;

The feature of this approach is that the semblance of human culture and empathy & technology ethics is determined, recognized and put into the principle to ensure that the role of technology as a dynamic force to drive human cultures is automated and regularized.

3 Conclusions

The conclusions of the paper rest on the implications of the key challenges as analyzed in the previous sections and the characteristics of the constitutive approaches put in the hypothesis. It is proposed that the approaches and challenges asserted in the paper are reflective in nature. It is proposed that AI is a special class of technology, and its different and disruptive behaviour in practice and observation is a preemptive basis of the constitutive approaches and the challenges enumerated in the paper:

- There is a lack of consensus over automating ethnographic virtues of AI by practice and purpose in nations. The approaches of conventional and constitutive morality assumed in the legal ethics of data protection regime adopted by countries have conflicts at a global level to some extent because of the ethnocentric nature of the AI Ethics principles endorsed in the West;
- The stimulating structure of AI is data-centric, probabilistic and absolutist, which can improve and change its normal definitiveness with better scientific advancements in the technology involved;
- The resemblance of AI in tech-oriented welfare cannot be stretched to its utilitarian outset: for public and social welfare, the design of AI should be human-centric, naturalist and entitative;
- A generational ecosystem of technology like AI must lead with better tech-led socialization. There should be a society-observant renovation of AI systems to encourage proper skill development and diversification of employment opportunities;
- Companies that develop and use AI may certainly lack popular or general confidence of the public as the aesthetics involving the development of AI under the market economy model lacks human value, originality and replenishment, and instead causes more technology distancing when it comes to a balance between the human capital and the AI capital required in industrial, innovation and government sectors;
- Data-truistic digital colonization makes the aesthetic relationship of AI and human entities complex and diverging, due to lack of autonomous empathy that is encultured and learnt by the AI;
- Utilitarian and industry-centric methods with isolation to individual quality assessment and development in business ethics do not convene the naturalistic and individual-centric capabilities of individuals and make AI vulnerable to technology distancing [19, 20]. Thus, true accountability and real trust with regards AI can only be developed when indigenous and individual interests are appropriately considered;

The conclusions are therefore provided as follows:

- In a post-liberal international order, the ethos of AI can be protected and diversified by cutting the monopolistic nature of the utilitarian and market-centric approach of AI;
- The practices of entrepreneurial ethics and design among governments and companies must keep individual, indigenous and libertarian interests by practice as the key priority to avoid malpractices with respect to technology and its economics and social influence;

- Social and economic nationalism under the politics of populism and redemption cannot be isolated when it comes to reforming the anomalies of globalization. The ethos of technological globalism must be protected, reconciled and improved to encourage a trust-based identity-affirmative order to cater and maintain the operant benefits of the rule-based international order;
- Technology diplomacy will require prevention of the damaging implications of the politics and practice of social, political and economic redemption & protectionism among developing and developed economies instead of mere containment of the problems;
- The stature of AI must be considered entitative and self-transformative as a juristic entity through a dissected and federalized assessment as proposed in the constitutive approaches;
- The rule-based international order must not be led and dominated by ethnocentric foreign policy reservations in terms of developed and developing economies, when it comes to the democratization of technology and the life of the internet, because political and constitutional confidence is essential from the local level to the diplomatic level;

References

1. BAAI: Beijing AI Principles, 28 May 2019. https://www.baai.ac.cn/blog/beijing-ai-principles
2. Google Spain v AEPD and Mario Costeja González (2014)
3. OHCHR: Guiding Principles on Business and Human Rights (2011). https://www.ohchr.org/Documents/Publications/GuidingPrinciplesBusinessHR_EN.pdf
4. Koskenniemi, M.: From Apology to Utopia: The Structure of International Legal Argument. Cambridge University Press, Cambridge (2005)
5. Tripković, B.: The metaethics of constitutional adjudication. Department of Law, European University Institute, Florence (2015)
6. Commission Nationale de l'Informatique et des Libertés (CNIL), France, European Data Protection Supervisor (EDPS), European Union, Garante per la protezione dei dati personali, Italy. Declaration on Ethics and Data Protection in Artificial Intelligence, 23 October 2018. https://edps.europa.eu/sites/edp/files/publication/icdppc-40th_ai-declaration_adopted_en_0.pdf
7. NITI Aayog, Government of India: National Strategy for AI - Discussion Paper, June 2018. https://niti.gov.in/writereaddata/files/document_publication/NationalStrategy-for-AI-Discussion-Paper.pdf. Accessed 2 Jan 2020
8. European Union: Regulation (EU) 2016/679 of the European Parliament and of the Council of 27 April 2016 on the protection of natural persons with regard to the processing of personal data and on the free movement of such data, and repealing Directive 95/46/EC (2016)
9. VDW: Policy Paper on the Asilomar Principles on Artificial Intelligence, Asilomar (2017)
10. Madiega, T.: EU guidelines on ethics in artificial intelligence: context and implementation (PE 640.163), September 2019. https://www.europarl.europa.eu/RegData/etudes/BRIE/2019/640163/EPRS_BRI(2019)640163_EN.pdf. Accessed 2 Jan 2020
11. European Commission: Ethics guidelines for trustworthy AI. European Commission, 8 April 2019. https://ec.europa.eu/digital-single-market/en/news/ethics-guidelines-trustworthy-ai. Accessed 30 Dec 2019
12. Salmon, J.: The legacy of Jean Bodin: absolutism, populism or constitutionalism? Hist. Polit. Thought **17**(4), 500–522 (1996)

13. House of Representatives: Impeachment of Donald John Trump, President of the United States, 16 December 2019. https://docs.house.gov/billsthisweek/20191216/CRPT-116hrpt346.pdf

14. Koh, H.H.: The Trump Administration and International Law (2017). https://digitalcommons.law.yale.edu/fss_papers/5213

15. Dutta, S.: Determinants of Ethnocentric Attitudes in the United States (2009). https://paa2009.princeton.edu/abstracts/91531

16. Grewal, J., Rakesh, V., Chattopadhyay, S., Hickock, E.: Report on Understanding Aadhaar and its New Challenges, 31 August 2016. https://cis-india.org/internet-governance/blog/report-on-understanding-aadhaar-and-its-new-challenges

17. Akula, R., Liu, C., Saba-Sadiya, S., Lu, H., Todorovic, S., Chai, J.Y., Zhu, S.-C.: X-ToM: Explaining with Theory-of-Mind for Gaining Justified Human Trust, 15 September 2019. https://arxiv.org/abs/1909.06907

18. Pacey: Meaning of Technology. The MIT Press, Cambridge (1999)

19. Molla, R.: "Knowledge workers" could be the most impacted by future automation, 20 November 2019. https://www.vox.com/recode/2019/11/20/20964487/white-collar-automation-risk-stanford-brookings

20. Manyika, J., Chui, M., Miremadi, M., Bughin, J., George, K., Wilmott, P., Dewhurst, M.: Mc Kinsey Global Institute: A Future That Works: Automation, Employment, and Productivity (2017). https://www.mckinsey.com/~/media/McKinsey/Featured%20Insights/Digital%20Disruption/Harnessing%20automation%20for%20a%20future%20that%20works/MGI-A-future-that-works_Full-report.ashx

Trustworthy AI Needs Unbiased Dictators!

Kian Abolfazlian[✉][iD]

SPIDER Business Ideas Architects, Stockholm, Sweden
kian.abolfazlian@spider-bia.com

Abstract. EU Draft Ethics guidelines for Trustworthy AI [8] has been proposed to promote ethical, lawful and robust AI solutions. In this article, we entertain the systemic issues and challenges of any development of the proposed guidelines.

Keywords: Trustworthy AI · Ethical AI · Dictatorial decision-makers · Manipulation · Cognitive biases

1 Is Trustworthy AI Achievable?

The accelerated enthusiasm of all walks of life for Artificial Intelligence and its applications has been both blessing and preoccupying. The AI solutions, backed by private and public sector interests have opened different and interesting avenues for how to address societal as well as business challenges, which were not thought possible.

As usual for any new technological advances, there have been many lessons learned, as a direct result of application of AI solutions in varied contexts. The lessons have shown us that, AI solutions, much more than any other type of solutions, are affected by the cognitive biases, inherent in the way, their human designers, developers and implementers among others, interact with their end-users. Every solution is biased in that the design choices, underlying its creation, naturally includes and excludes some groups of end-users. In the case of AI solutions, their inherent *biased-by-design* scope have been much faster detected and criticized, due to their rapid advances, achieved reach and application contexts. As a novel approach, adjectives such as Trustworthy and Ethical have been used in describing what AI solutions must be.

As a step towards this, European Union (EU) commission's High-level Expert Group on AI (AI HLEG) has formulated a framework for Trustworthy AI [8]. They define Trustworthy AI as follows:

Trustworthy AI has three components: (1) it should be lawful, ensuring compliance with all applicable laws and regulations (2) it should be ethical, demonstrating respect for, and ensure adherence to, ethical principles and values and (3) it should be robust, both from a technical and social

I. Maglogiannis et al. (Eds.): AIAI 2020, IFIP AICT 584, pp. 15–23, 2020.
https://doi.org/10.1007/978-3-030-49186-4_2

perspective, since, even with good intentions, AI systems can cause unintentional harm. Trustworthy AI concerns not only the trustworthiness of the AI system itself but also comprises the trustworthiness of all processes and actors that are part of the system's life cycle.

As well AI HLEG defines Ethical AI as:

The development, deployment and use of AI that ensures compliance with ethical norms, including fundamental rights as special moral entitlements, ethical principles and related core values. It is the second of the three core elements necessary for achieving Trustworthy AI.

Furthermore, they define the following non-exhaustive list of requirements, needed to ensure Trustworthy AI. The requirements include systemic, individual and societal factors:

1. **Human agency and oversight**, *including fundamental rights, human agency and human oversight.*
2. **Technical robustness and safety**, *including resilience to attack and security, fall back plan and general safety, accuracy, reliability and reproducibility.*
3. **Privacy and data governance**, *including respect for privacy, quality and integrity of data, and access to data.*
4. **Transparency**, *including traceability, explainability and communication.*
5. **Diversity, non-discrimination and fairness**, *including the avoidance of unfair bias, accessibility and universal design, and stakeholder participation.*
6. **Societal and environmental wellbeing**, *including sustainability and environmental friendliness, social impact, society and democracy.*
7. **Accountability**, *including auditability, minimization and reporting of negative impact, trade-offs and redress.*

These seven requirements are all interrelated in that they are *"all of equal importance, support each other, and should be implemented and evaluated throughout the AI system's life-cycle"* [8]. This is depicted in Fig. 1. As well, the AI HLEG has presented technical and non-technical methods to realize Trustworthy AI (Fig. 2).

The focus on the design and design choices behind the development of AI solutions is inherent in the way the concept of Trustworthy AI has been handled in AI HLEG guidelines.

In our opinion, here lies the biggest challenge to achieving these guidelines' goals. The guidelines develop a set of check-lists and principals, which must be observed, whenever any taken design decision is to be examined for its ethical character and trustworthiness. This could be the task of a group of developers or internal "AI auditors". In any case, the dynamics of how such groups act have been studied by pioneers such as Nobel laureate, Kenneth Arrow [1], and philosophers and economists such as Allan Gibbard [2–4], Mark Satterwaithe [7], and Aanund Hylland [6], as well as Cognitive Scientists such as Gärdenfors [5] to mention a few.

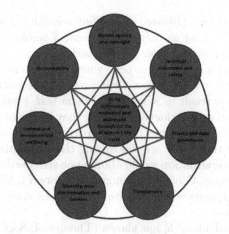

Fig. 1. The interrelation of the 7 requirements of Trustworthy AI (adopted from [8])

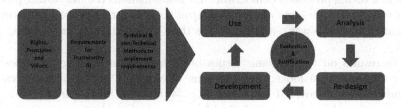

Fig. 2. Realizing Trustworthy AI throughout the system's entire life cycle (source [8])

For instance, if we have, for a specific design factor, n alternatives, A_1 to A_n, where the members of the committee (or development team) are asked to present an ordering $\succ$ to show their preference, e.g. $A_1 \succ \ldots \succ A_n$, [1] has shown (Theorem 3 in Sect. 2, below) that, as long as;

1. There are at least 3 possible outcomes
2. If everybody in the group prefers, for instance, an alternative A_i to another alternative A_j, then the result of the voting would also reflect this unanimity of preferences
3. Each voter's preference between any two alternatives A_i and A_j, is indifferent to the preferences of other alternatives

Then *the voting process would be dictatorial*, i.e. among the voters, there would be a *dictator*, in the sense of a person, whose choices and preferences dictates what the committee would decide, regardless of how other members of the committee are voting.

Even if we change the scheme, so the alternatives are ordered from most preferred to the least one as above, but, now, the alternatives are *ranked using an ordinal system*, we would still have challenges. Here, the alternatives are ordered, so the most preferred alternative (rank 1) is given the highest point (from an ordinal system), and so forth. Then the points are counted in a voting process to declare a winner. Gibbard [2] and Satterwaithe [7] showed (Theorem 2, Sect. 2, below) that *as long as there are at least 3 possible outcomes, then the voting process is either manipulable or dictatorial.*

A manipulable voting process is one, where the voters can vote tactically. This means that they, realizing that their preferred or sincere ordering of the alternatives would not result in what *serves them best*, would change their vote (i.e. an insincere vote) so to achieve the result that is closest to their preference.

In above situations, the alternatives are ordered strictly, so the voters cannot order them in a way that there are ties among the alternatives (i.e. not a strict order). A more realistic situation would allow a not-strict ordering of alternatives. Even for these situations, Gibbard [2] showed (Theorem 1, Sect. 2, below) that *if there are at least 3 outcomes, and the committee does not have a dictator, then the voting is manipulable.*

Furthermore, Gärdenfors [5] has shown (Theorem 4, Sect. 2 below), that *any democratic voting process with at least 3 voters is manipulable!*

Here a voting process is democratic, if it is *anonymous* (i.e. the voting process treats every voter the same way), it is *neutral* (i.e. the voting process treats every alternative the same way), and it satisfies the *Condorcet criterion* (i.e. majority rules).

The results on the systemic issues of such decision making processes are many. One has even looked at situations where some of the alternatives are dependent of or discovered by chance (e.g. [4] and [6]). In the next section the above mentioned results are described in a more technical terms.

2   Dictators and Tactical Voters in a Decision-Making Context

In the following, we follow the definitions and technical format of [2].

Definition 1 *Game form.* *A game form is characterized by:*

(i) *A set X, whose members are called* **possible outcomes***, or simply* **outcomes***. Unless otherwise stated, variables x, y, and z will range over outcomes.*

(ii) *A positive integer n, called the* **number of players***. The n players will be denoted by the integers 1 to n, and variables i, j, and k will range over these integers.*

(iii) *n sets S_i, one for each i. For each i the members of S_i are called strategies for i. The word "strategy," then, refers here to what in game theory is usually called a "pure strategy." An n-tuple $(s_1, \ldots, s_n)$, with $s_1 \in S_1, \ldots, s_n \in S_n$ will be called a* **strategy** *n-tuple. Strategy n-tuples will be indicated by bold-face small letters on the pattern $\boldsymbol{s} = (s_1, \ldots, s_n)$, $\boldsymbol{s}' = (s'_1, \ldots, s'_n)$, and so forth.*

(iv) *A function g, defined for every strategy n-tuple, whose range is X. Strictly speaking, a game form is simply a function g which can be characterized as above. We can define a game form, then, as a function whose domain is the*

*Cartesian product $S_1 \times \ldots \times S_n$ of a finite number of finite non-empty sets. Its values are called **outcomes**, its arguments are called strategy n-tuples, and a member of a set S_i is called a **strategy** for i.*

Definition 2 *Orderings*

*(i) An **ordering** of a set Z is a two-place relation R between members of Z, (i.e. $R \subseteq Z \times Z$) such that for all x, y and z in Z*
- *$\neg(xRy \land yRx)$*
- *$xRz \Rightarrow (xRy \lor yRz)$*

As such, we can define an ordering R of the set X of outcomes of a game form g.

*(ii) A **preference ordering** P is an ordering of X, the set of outcomes of a game form g. The relation xPy then means "x **is preferred to** y" under ordering P. For distinct x and y in X, we may have that $(x, y) \notin P$, i.e. neither xPy nor yPx. In this case we say that x and y are **indifferent under ordering** P. As such P indicates **strict preference** between 2 elements of X.*

*(iii) For any preference ordering P of set of outcomes X, we can define another binary relation $R \subseteq X \times X$, such that: $xRy \iff \neg yPx$. For any distinct x and y in X, the relation R indicates their **preference or indifference**.*

*(iv) For any preference ordering P of set of outcomes X, we can define another binary relation $I \subseteq X \times X$, such that: $xIy \iff (\neg yPx \land \neg xPy)$. For any distinct x and y in X, the relation I indicates their **absolute indifference**.*

Remark 1. We will use the following notation. For any n-tuple $\boldsymbol{s} = (s_1, \ldots, s_n)$, we indicate the result of the altering of its k-th place as follows:

$$\boldsymbol{s}\langle k/t \rangle = (s_1, \ldots, s_{k-1}, t, s_{k+1}, \ldots, s_n)$$

In other words, $\boldsymbol{s}' = \boldsymbol{s}\langle k/t \rangle$ iff $\{(s'_k = t) \land (\forall i)\, [i \neq k \Rightarrow s'_i = s_i]\}$.

Definition 3 P-*dominance*. *Where P is a preference ordering, a strategy t is P-dominant for k if for every strategy n-tuple $\boldsymbol{s}$, we have that $g(\boldsymbol{s}\langle k/t \rangle)\, R\, g(\boldsymbol{s})$.*

In other words, t is P-dominant for k iff no matter what strategies are fixed for everyone else, strategy t for k produces an outcome at least as high in preference ordering P as does any other. As such, the player k, by choosing the strategy t, needs not to think of other players' strategies. Strategy t serves the player k's interests best.

Definition 4 *Straightforward game form*. *A game form is **straightforward** if, for every preference ordering P and player k, there is a strategy which is P-dominant for k.*

As such, in a straightforward game form, each player has a strategy, which serves his/her interests best, and it is independent of what any other player chooses as strategy. So the players need not play strategically (i.e. be attentive to what others do). Therefore, a straightforward game form is also called **strategy-proof**.

Definition 5 *Manipulable game form*. *A **manipulable** game form is one which is not straightforward. In a manipulable game form, there exists a player k, who, given a preference ordering P, cannot find any strategy t, which is P-dominant for him/her. So given a game form g, a preference ordering P, and a strategy n-tuple $\boldsymbol{s} = (s_1, \ldots, s_n)$, the player k needs to choose a strategy t, where even though $s_k P t$, but by choosing this strategy, he/she would achieve $g(\boldsymbol{s}\langle k/t\rangle) \ R \ g(\boldsymbol{s})$.*

Definition 6 *Dictatorial game form*. *A player k is a dictator for game form g if, for every outcome x, there is a strategy $s(x)$ for k such that for strategy n-tuple $\boldsymbol{s} = (s_1, \ldots, s_k \ldots, s_n)$, the $g(\boldsymbol{s}) = x$ whenever $s_k = s(x)$. A game form g is dictatorial if there is a dictator for g.*

Having a dictator in a game form means that there exists a player k whose choices are always the outcome of the game, no-matter what other players have chosen (i.e. their strategies).

Theorem 1 (Gibbard [2]). *Every straightforward game form with at least three possible outcomes is dictatorial.*

Definition 7 *Voting scheme*. *A voting scheme is a game form v with set of possible outcomes X, such that for some set Z or set of **alternatives** with $X \subseteq Z$, the set S_i of strategies open to each player i is the set of orderings of Z. We call this set Π_Z. Then a voting scheme is a single valued function from Π_Z^n to X, which given a n-tuple $\boldsymbol{P} = (P_1, \ldots, P_n)$, returns a single possible outcome $x \in X$.*

In a voting scheme, an ordering P_i represents the ballet that voter i casts. The orderings can be fixed by an ordinal scheme, so the voter i, in his/her ordering P_i, places his/her most preferred alternative as number 1, the second most preferred alternative as number 2, and so forth. Then, the voting scheme simulates a counting mechanisms of the ballets and their orderings, which results in naming an alternative as the winner.

Definition 8 *Manipulable voting scheme*. *A voting scheme v is manipulable if for some voter k, and for some n-tuple $\boldsymbol{P} \in \Pi_Z^n$, there exists some ordering $P^* \in \Pi_Z$, such that $v(\boldsymbol{P}\langle k/P^*\rangle) \ P_k \ v(\boldsymbol{P})$.*

This means that for voter k, in the situation, which is represented by n-tuple $\boldsymbol{P} = (P_1, \ldots, P_k, \ldots, P_n)$, there exists another voting possibility (represented by ordering P^*), such that if the voter k changes P_k to P^*, then the result of the voting would represent his/her interest best, in the sense of her original vote P_k.

Definition 9 *Dictatorial voting scheme*. *Player k is a dictator for a voting scheme v if, for every possible outcome $x \in X$, player k can choose an ordering $P(x)$, so $v(\boldsymbol{P}) = x$ whenever $P_k = P(x)$, for any voting $\boldsymbol{P} = (P_1, \ldots, P_k \ldots, P_n)$. A voting scheme v is dictatorial if there is a dictator for it.*

Theorem 2 (Gibbard [2], Satterthwaite [7]). *Every voting scheme with at least three outcomes is either dictatorial or manipulable.*

Definition 10 Social welfare function. *A preference n-tuple over a set X is an n-tuple $(P_1, \ldots, P_n)$ whose terms are preference orderings of X, the set of possible outcomes. Preference n-tuples will be designated in bold-face type on the pattern $\mathbf{P} = (P_1, \ldots, P_n), \mathbf{P'} = (P'_1, \ldots, P'_n)$ and so forth. A social welfare function is a function whose arguments, for some fixed n (or number of voters) and the set of possible outcomes X are all preference n-tuples $\mathbf{P}$ over X, and whose values are preference orderings of X.*

Given n voters and a set of possible outcomes X, a social welfare function $f : \Pi_X^n \longrightarrow \Pi_X$, is a mapping that takes a n-tuple $\mathbf{P} = (P_1, \ldots, P_n) \in \Pi_X^n$ and delivers a preference ordering $P \in \Pi_X$. Here Π_X represents the set of preference orderings on X.

Before we go further, we choose a notation that makes it easier to visualize the orderings. Let us use $\succ$ for orderings of X. For alternatives x and y, the $x \succ y$ signifies x is preferred to y. So if $\succ$ is an ordering of $x_1, x_2, \ldots, x_m \in X$ of m alternatives may look like this: $x_1 \succ x_2 \succ \ldots \succ x_m$.

Theorem 3 (Arrow [1]). *Every social welfare function defined for a set of possible outcomes X, and n voters, violates one of following **Arrow conditions**:*

(i) **Scope:** X *has at least three members.*

(ii) **Unanimity** *or* **weak Pareto efficiency:** *If* $\mathbf{P} = (\succ_1, \ldots, \succ_n)$ *and* $\forall i.\ x \succ_i y$, *and* $f(\mathbf{P}) = \succ$, *then* $x \succ y$.

(iii) **Pairwise Determination:** *If for* $\mathbf{P} = (\succ_1, \ldots, \succ_n)$ *and* $\mathbf{P'} = (\succ'_1, \ldots, \succ'_n)$:
  - $\forall i.\ [x \succ_i y \iff x \succ'_i y]$ *and* $\forall i.\ [y \succ_i x \iff y \succ'_i x]$
  - $f(\mathbf{P}) = \succ$ *and* $f(\mathbf{P'}) = \succ'$
  then: $x \succ y \iff x \succ' y$

(iv) **Non-dictatorship:** *There is no dictator for f, where a dictator for f is a voter k such that for every preference ordering $\succ$, and every $x, y \in X$, if $x \succ_k y$ and $f(\mathbf{P}) = \succ$, then $x \succ y$.*

The Arrow conditions are a group of logical and fair conditions which can be put on any voting system. The *Scope* and *non-Dictatorship* conditions are self-explanatory. The *Unanimity* or *weak Pareto efficiency* condition says that if for any two possible outcomes x and y, every voter has the same preference, e.g. x is preferred to y, then the result of voting and the final order of preference of the candidates (i.e. what social welfare function returns) must also reflect that x is preferred to y. At the same time, the *Pairwise Determination* condition says that the order of preference between two candidates is not dependant of how the order of preference between other pairs of candidates is.

In the case of general game forms and voting schemes, we looked at the result of the game and/or scheme to be a single element x of X, the set of possible outcomes. In the case of social welfare functions, the result was a preference ordering $\succ$ on $X = \{x_1, \ldots, x_m\}$, where, for instance $x_1 \succ \ldots \succ x_m$.

As such, the Gibbard-Satterwaithe result of Theorems 2 gives an impression that it is never possible to have a voting mechanism, which is non-manipulable, have more than 2 possible outcomes, and is non-dictatorial. Further investigation shows that the proof of Theorem 2 is, very much dependent on the assumption that the voting process selects a single winner [5]. But this is not the case in the real world. Many times, we will have situations, where several alternatives are chosen, al par with each-other, as acceptable outcomes (i.e. tied winners). Then the tie is broken and a single winner is chosen by some extra-ordinary process (e.g. in alphabetic order, or by random chance).

In order to work with situations with possibility of tied winners, Gärdenfors [5] developed $\gg$ as a new ordering between sets (e.g. $\{x, y\} \gg \{y\}$ if $x \succ y$), which would be a generalization of the ordering $\succ$ from above. The definition can be found in [5]. In the following we will use it to find a very strong result (Theorem 4, below) on democratic voting processes. This result is very important, since such democratic voting processes define the ideal processes towards the development of Trustworthy AI.

Definition 11 *Social choice function. Let X be the set of possible outcomes, $i = 1, \ldots n$ the set of voters, and Π_X the set of preference orderings on X. A* **social choice function (SCF)** *is a function $F : \Pi_X^n \longrightarrow 2^X - \emptyset$, where 2^X denotes the set of all subsets of X.*

Definition 12. *A social choice function F is* **manipulable** *by voter i at situation $s = (P_1, \ldots, P_n)$ iff there is an ordering P_i' such that $F(s\langle i/P_i'\rangle) \gg_i F(s)$, where $\gg_i$ is the ordering derived from P_i. F is* **non-manipulable** *or* **stable** *iff F is nowhere manipulable.*

Definition 13. *A social choice function F is* **anonymous** *iff whenever two situations s_1 and s_2 are identical except that for some voters i and j, we have $P_{i|s_1} = P_{j|s_2}$ and $P_{j|s_1} = P_{i|s_2}$ then $F(s_1) = F(s_2)$. So a social choice function is annonymous if it treats every voter in the same way. Here $P_{i|s}$ means the preference P_i of voter i in the situation s.*

Definition 14. *A social choice function F is* **neutral** *iff whenever two situations s_1 and s_2 are identical except for two alternatives x and y that have changed places everywhere (i.e. in the preferences of each voters), then $x \in F(s_1)$ iff $y \in F(s_2)$ and $y \in F(s_1)$ iff $x \in F(s_2)$. So a social choice function is neutral if it treats every alternative in the same way.*

Definition 15. *A social choice function F satisfies the* **Condorcet criterion** *iff whenever there is an alternative x in a situation s such that, for every alternative $y \neq x$, the number of individuals who strictly prefer x to y is greater than the number of individuals who strictly prefer y ot x, then $F(s) = \{x\}$. Such an alternative is called a* **majority alternative** *in the situation s.*

Definition 16. *A social choice function F which is anonymous and neutral, and satisfies the Condorcet criterion, is a* **democratic** *social choice function.*

Theorem 4 (Gärdenfors [5]). *Any democratic social choice function which is defined for at least three voters is manipulable.*

3   Conclusions

We conclude that any development of AI HLEG proposed guidelines, framework and requirements for Trustworthy and Ethical AI suffers from systemic problems. We would be either in a situation where there are dictator decision-makers, or we will have design choices which are manipulated by those who are responsible for ensuring the trustworthiness of AI solutions. In both cases, the cognitive biases of the decision maker (either the dictator or manipulator) would have significant effect on the result of the designed and developed AI solution. This is the main challenge here.

As mentioned above, it is a fact that most of the voting processes can suffer of either having a dictator or voters who vote strategically to serve their own benefits. The issue is that what if the cognitive biases of the dictator decision-maker, or the ones of the tactical voter is exactly the biases, which we are trying to root-out from the design and architecture of AI solutions. As such, the very challenge of biases, which the EU guidelines are developed to answer, remains unanswered.

References

1. Arrow, K.J.: Social Choice and Individual Values. Wiley, New York (1963)
2. Gibbard, A.: Manipulation of voting schemes: a general result. Econometrica **41**(4), 587–601 (1973)
3. Gibbard, A.: Manipulation of schemes that mix voting with chance. Econometrica **45**, 665–681 (1977)
4. Gibbard, A.: Straightforwardness of game forms with lotteries as outcomes. Econometrica **46**, 595–614 (1978)
5. Gärdenfors, P.: Manipulations of social choice functions. J. Econ. Theory **13**, 217–228 (1976)
6. Hylland, A.: Strategy-proofness of voting procedures with lotteries as outcomes and infinite sets of strategies (1980, Unpublished)
7. Satterthwaite, M.A.: Strategy-proofness and arrow's conditions: existence and correspondence theorems for voting procedures and social welfare functions. J. Econ. Theory **10**(2), 187–217 (1975)
8. Draft ethics guidelines for trustworthy AI. http://ec.europa.eu/digital-single-market/en/news/draft-ethics-guidelines-trustworthy-ai. Accessed 4 Oct 2017

AI/Constraints

An Introduction of FD-Complete Constraints

Sven Löffler[(✉)], Ke Liu, and Petra Hofstedt

Brandenburg University of Technology Cottbus-Senftenberg, Cottbus, Germany
Sven.Loeffler@b-tu.de, Hofstedt@b-tu.de

Abstract. The performance of solving a constraint problem can often be improved by converting a subproblem into a single constraint (for example into a regular membership constraint or a table constraint). In the past, it stood out, that specialist constraint solvers (like simplex solver or SAT solver) outperform general constraint solvers, for the problems they can handle. The disadvantage of such specialist constraint solvers is that they can handle only a small subset of problems with special limitations to the domains of the variables and/or to the allowed constraints. In this paper we introduce the concept of fd-complete constraints and fd-complete constraint satisfaction problems, which allow combining both previous approaches. More accurately, we convert general constraint problems into problems which use only one, respectively one kind of constraint. The goal is it to interpret and solve the converted constraint problems with specialist solvers, which can solve the transformed constraint problems faster than the original solver the original constraint problems.

Keywords: Constraint programming · CSP · Refinement · Optimizations · Regular membership constraint · Regular CSPs · Table constraint · FD-completeness

1 Introduction

Constraint programming (CP) is a powerful method to model and solve NP-complete problems in a declarative way. Typical research problems in CP are among others rostering, graph coloring, optimization, resource management, planning, scheduling and satisfiability (SAT) problems [12].

Because the search space of constraint satisfaction problems (CSPs) and constraint satisfaction optimization problems (CSOPs or COPs) is immensely big and the solution process often needs an extremely high amount of time we are always interested in improving the solution process. There are various ways to describe a CSP in practice and consequently, the problem can be modeled by different combinations of constraints, which results in the differences in resolution speed and behavior.

In particular, there is the possibility to represent a CSP with only one constraint or with constraints, which are of the same kind. Inspired by the concept of np-complete problems, we introduce the definition of fd-complete constraints.

© IFIP International Federation for Information Processing 2020
Published by Springer Nature Switzerland AG 2020
I. Maglogiannis et al. (Eds.): AIAI 2020, IFIP AICT 584, pp. 27–38, 2020.
https://doi.org/10.1007/978-3-030-49186-4_3

A *finite domain complete constraint (fd-complete constraint)* is a finite domain constraint which can represent every other fd constraint. Thus an fd-complete constraint can be used to replace each other constraint in a CSP.

Possible representatives of fd-complete constraints are the *table* constraint or the *regular membership* constraint. Previous researches show how constraints can be transformed into *table* [2] and *regular membership* [8–11] constraints. These publications also showed, that already the transformations into a *regular membership* or *table* constraint can improve the solving speed of a CSP.

In this paper, we want to go a step further. We not only transform parts of a CSP into an fd-complete constraint, we transform the whole CSP into one, which contains only a set (of the same kind) of fd-complete constraints. Thus we have the possibility to use a solver, solver settings or search strategies which are optimized for the used constraints.

The rest of this paper is structured as follows. In Sect. 2, we explain the necessary definitions of constraint programming. In Sect. 3, we introduce our new concept of fd-complete constraint satisfaction problem. Section 4 shows the transformation of a regular CSP into a binary variable scalar CSP. In Sect. 5, a rostering example is given to underline the benefit of finite domain CSPs. Finally, Sect. 6 draws a conclusion and proposes research directions in the future.

Remark 1: We use the notation of a *regular* constraint as a synonym for *regular membership* respectively *regular language membership* constraint.

2   Preliminaries

In this section we introduce some basic definitions and concepts of constraint programming (CP) and show the relevant constraints, which are used in the rest of the paper. We consider *CSPs*, which are defined in the following way.

A *constraint satisfaction problem (CSP)* is defined as a 3-tuple $P = (X, D, C)$ where $X = \{x_1, x_2, \ldots, x_n\}$ is a set of variables, $D = \{D_1, D_2, \ldots, D_n\}$ is a set of finite domains where D_i is the domain of x_i and $C = \{c_1, c_2, \ldots, c_m\}$ is a set of constraints covering between one and all variables of X [16].

A *constraint* $c_j = (X_j, R_j)$ is a relation R_j, which is defined over a set of variables $X_j \subseteq X$ [4].

The *scope* of a constraint $c_j = (X_j, R_j)$ indicates the set of variables, which is covered by the constraint c_j: $scope(c_j) = X_j$ [4].

The relation R of a constraint $c = (X, R)$ is more clearly a subset of the Cartesian product of the domain values $D_1 \times \ldots \times D_n$ of the corresponding variables $X = \{x_1, \ldots, x_n\}$. Examples for constraints are amongst other things $c_1 = (\{x, y\}, (x > y))$, $c_2 = (\{R, U, I\}, (R = U/I))$ or $c_3 = (\{A, B, C\}, (A \land B \rightarrow C))$. For the reason that we only consider finite domains, it is always possible to enumerate all allowed tuples of a constraint, which we will use for the transformation into other constraints like *regular* or *table* constraints.

Based on the definition above, an fd-complete constraint can substitute all constraints C of a given CSP $P = (X, D, C)$. For the reason that a CSP can only have a limited number of solutions c_{sol}, because we have a limited number of variables n and a limited number of values inside of the domains of the variables, it is possible to enumerate all of the solutions of a CSP.

It follows a list of definitions of the relevant constraints for this paper. Let a CSP $P = (X, D, C)$ and a subset X' of variables X of the CSP P be given.

The *scalar* constraint guarantees for an ordered subset of variables $X' = \{x_1, \ldots, x_n\} \subseteq X$, a vector of integers $v = (v_1, \ldots, v_n)$, a relation $\Re$ and another variable $x_r \in X$ that the scalar product of the variables X' with the vector v is in relation $\Re$ to x_r.

$$scalar(X', v, \Re, x_r) = (\sum_{i=1}^{n} x_i * v_i) \; \Re \; x_r$$

Like we already figured out, the *table* and the *regular* constraints are representatives of fd-complete constraints. It follows a short description of these both constraints.

The *table* constraint is one of the most frequently used constraints in practice. For an ordered subset of variables $X' = \{x_1, \ldots, x_n\} \subseteq X$, a positive (negative) *table* constraint defines that any solution of the CSP P must (not) be explicitly assigned to a tuple t of a given tuple list T, which consists the allowed (disallowed) combinations of values for X'. For a given list of tuples T, we can state the positive *table* constraint as:

$$table(X', T) = \{(x_1, \ldots, x_n) \mid x_1 \in D_1, \ldots, x_n \in D_n\} \subseteq T$$

The *regular* constraint and its propagation [5,13,14] is based on deterministic finite automatons (DFAs) [7]. Thus, we briefly review the notion of a deterministic finite automaton (DFA) before we define the *regular* constraint.

A *deterministic finite automaton (DFA)* is a quintuple $M = (Q, \Sigma, \delta, q_0, F)$, where Q is a finite set of states, Σ is the finite input alphabet, δ is a transformation function $Q \times \Sigma \rightarrow Q$, $q_0 \in Q$ is the initial state, and $F \subseteq Q$ is the set of final or accepting states. A word $w \in \Sigma^*$ is accepted by M, if the corresponding DFA M with the input w stops in a final state $f \in F$. All accepting words of the DFA M can be summarized to the accepting language $L(M)$ of the DFA M [7].

The *regular constraint* is another of the most common constraints in practice. Let $M = (Q, \Sigma, \delta, q_0, F)$ be a DFA, let $X = \{x_1, \ldots, x_n\}$ be an ordered set of variables with domains $D = \{D_1, D_2, \ldots, D_n\}$, $\forall i \in \{1, \ldots, n\} : D_i \subseteq \Sigma$. The regular constraint is defined as [6]:

$regular(X, M) =$
$\qquad \{(w_1, \ldots, w_n) \mid \forall i \in \{1, \ldots, n\}, w_i \in D_i, (w_1 w_2 \ldots w_n) \in L(M)\}$

So the concatenation of the values v_i of the variables $x_i \forall i \in \{1, \ldots, n\}$ must be accepted by the automaton M. The input DFA M of a *regular* constraint is internally transformed into a directed acyclic graph (DAG) M' [14].

We define a *directed acyclic graph (DAG)* as a DFA $M = (Q, \Sigma, \delta, q_{initial}, F)$, where the states q in Q are partitioned into levels $Q = \{Q_0, ..., Q_n | Q_i = \{q_{i,1}, q_{i,2}, ...\} \mid \forall i \in \{1, ..., n\}\}$, the initial state $q_{0,0}$ is the only element of Q_0, the final states are members of the last level of states ($F = Q_n$), and transitions are only allowed from a state $q_{i,j}$ in level Q_i to a state $q_{t,k}$ in level Q_t, where $t \geq i$.

For our use, we only allow transitions from a state $q_{i,j}$ in level Q_i to a state $q_{i+1,k}$ in level Q_{i+1}. We use the notation $q_{i,j} \underset{v}{\longrightarrow} q_{i+1,k}$, if the state $q_{i,j}$ of level Q_i have a transition with value v to the state $q_{i+1,k}$ of level Q_{i+1}. A solution of a *regular* constraint is a path from the initial state $q_{0,0}$ to the final state $q_{n,0}$ with values $v_1 \in D_1, ..., v_n \in D_n$ ($q_{0,0} \underset{v_1}{\longrightarrow} q_{1,j_1} \underset{v_2}{\longrightarrow} \cdots \underset{v_n}{\longrightarrow} q_{n,0}$).

Based on the definition of the regular constraint, we use the notion of a regular constraint satisfaction problem, and analogously the notation of a table CSP:

A *regular constraint satisfaction problem (regular CSP)* is defined as a 3-tuple $P = (X, D, C)$, where $X = \{x_1, x_2, ..., x_n\}$ is a set of variables, $D = \{D_1, D_2, ..., D_n\}$ is a set of finite domains where D_i is the domain of x_i and $C = \{c_1, c_2, ..., c_m\}$ is a set of *regular* constraints over variables of X.

A *table constraint satisfaction problem (table CSP)* is defined as a 3-tuple $P = (X, D, C)$, where $X = \{x_1, x_2, ..., x_n\}$ is a set of variables, $D = \{D_1, D_2, ..., D_n\}$ is a set of finite domains where D_i is the domain of x_i and $C = \{c_1, c_2, ..., c_m\}$ is a set of *table* constraints over variables of X.

3   FD-Complete CSPs

After we defined fd-complete constraints and showed two examples (the *table* and the *regular* constraint), we will extend this concept a little bit more. The possibility to substitute a constraint or a subset of constraints of a CSP with an fd-complete constraint can lead to a speed up in the solution process [2,8–11]. The problem is that it is mostly not useful to transform a whole CSP into another one with only one fd-complete constraint, which substitutes all constraints of the original CSP. In most cases, the transformation process will be more time consuming than solving the original problem.

Mostly, we can reduce this slow-down significantly and reach a speed-up, if we substitute only some of the constraints $c_i \in C$ of a given CSP $P = (X, D, C)$ with semantically equivalent **fd**-complete constraints $c_i^{\mathbf{fd}} \in C^{\mathbf{fd}}$ and combine them together (to constraints $c_j^{\text{c-fd}} \in C^{\text{c-fd}}$). Finding the best subset of constraints, which should be transformed and combined, depends on several things like the needed time for transformation or the size of the data structures of the transformed or the combined constraints.

Often, a CSP $P' = (X, D, C')$, which contains some of the original constraints $c_i \in C$ and some of the combined constraints $c_j^{\text{c-fd}}$, which are semantically equivalent to a subset of the fd-complete constraints C^{fd}, has the fastest solution process.

In contrast to this, we will explain why it can be useful to create a CSP $P = (X, D, C)$ with only singleton c_i^{fd} and combined fd-complete constraints $c_j^{\text{c\_fd}}$, even though the propagation of the original constraint c_i is possibly faster than the propagation of the corresponding fd-complete constraint c_i^{fd}.

The advantage is, that we may be able to (create and) use specialist constraint solvers for CSPs which have only constraints of a special type (a popular example is the simplex algorithm, which allows only linear optimization under linear side conditions). Furthermore, we can create and use more efficient search strategies and solver settings if the used constraints have the same shape.

Also interesting for future researches is, that a CSP P with only one kind of constraint may be able to be transformed easier into other models for example a SAT model and make it possibly simpler to transfer a constraint problem of a given language into another one. The last idea is for example used to translate MiniZinc [18] models into other models (like Gecode [17], Google OR-Tools [1] or Choco [15] models). Thus, many constraints, which were entered in MiniZinc, will be translated into the *table* constraint.

Based on the potential benefit of CSPs, which contain only fd-complete constraints, we created the definition of an fd-complete CSP.

A *finite domain complete constraint satisfaction problem (fd-complete CSP)* is a CSP $P = (X, D, C)$ which contains only fd-complete constraints of the same kind.

Examples for an fd-complete CSP are the *regular CSP* [9] or the *table CSP*. We call fd-complete CSPs with n different kinds of constraints fd-complete CSPs of degree n. We differentiate between two kinds of fd-complete CSPs. On the one hand we have directly convertible fd-complete CSPs like table or regular CSPs. Every general CSP $P = (X, D, C)$ can be transformed into a **directly convertible fd-complete CSP** $P^{\text{dc\_fd}} = (X, D, C^{\text{fd}})$ by transformation of each single constraint $c \in C$ to the corresponding fd-complete constraint $c \in C^{\text{fd}}$. In particular, it is not necessary to transform the variables X or domains D of the original CSP P.

On the other hand, there are **indirectly convertible fd-complete CSPs** $P^{\text{ic\_fd}} = (X', D', C^{\text{fd}})$, where the variables and their domains must be transformed additionally to the constraints. We present a representative example of this group in the next section.

4   The Binary Scalar CSP as a Representative of FD-Complete CSPs

In this Section, we will introduce a description of how a general CSP P can be transformed into a CSP P^{bvsc}, which contains only **b**inary **v**ariables and **sc**alar constraints. Using only binary variables allows us the use of the very good researched SAT solvers, for solving P^{bvsc}. Furthermore, for the reason that the CSP P^{bvsc} only contains scalar constraints, we expect that a possibly more accurate and more specialized solver can be created to solve it.

Traditionally, an input CSP P for a SAT solver contains binary variables and logic constraints like *and, or, implication, negation* etc. It is also possible to transform a regular constraint into a set of logic constraints, but we will see, that the transformation into a set of *scalar* constraints is more compact. Thus, we expect that a specialized solver for *scalar* constraints over binary variables works faster than a usual SAT solver.

The transformation from a general CSP P to a binary variable scalar CSP P^{bvsc} can happen in two steps.

Step one: transforming the CSP $P = (X, D, C)$ into a regular CSP $P^{\text{reg}} = (X, D, C^{\text{reg}})$, which contains the same variables X, the same domains D, but only regular constraints C^{reg} instead of the original constraints C. For example, it is possible to transform each original constraint $c \in C$ into a regular constraint $c^{\text{reg}} \in C^{\text{reg}}$. The algorithms and transformations presented in [8–11] can be used for this step.

Step two: transform the regular CSP $P^{\text{reg}} = (X, D, C^{\text{reg}})$ into a binary variable scalar CSP $P^{\text{bvsc}} = (X^{\text{b}}, D^{\text{b}}, C^{\text{sc}})$. We consider a regular CSP $P^{\text{reg}} = (X, D, C^{\text{reg}})$ with n variables $X = \{x_1, ..., x_n\}$, the associated finite domains $D = \{D_1, ..., D_n\}$ and m regular constraints $C^{\text{reg}} = \{c_1^{\text{reg}}, ..., c_m^{\text{reg}}\}$ and transform it into a binary variable scalar CSP $P^{\text{bvsc}} = (X^{\text{b}}, D^{\text{b}}, C^{\text{sc}})$ with k variables $X^{\text{b}} = \{x_1^{\text{b}}, ..., x_k^{\text{b}}\}$, which have all the domain $\{0, 1\} = D_1^{\text{b}} = ... = D_k^{\text{b}}$, and l scalar constraints $C^{\text{sc}} = \{c_1^{\text{sc}}, ..., c_l^{\text{sc}}\}$.

There are two kinds of variables in X^{b}. The first kind contains for each domain value of each variable of the regular CSP P^{reg}, respectively the original CSP P, a binary variable $x_{i,j}^{\text{b}} \forall i \in \{1, ..., n\}, j \in \{1, ..., |D_i|\}$. If such a variable $x_{i,j}^{\text{b}} \in x^{\text{b}}$ of P^{bvsc} is set to 1, it represents that the corresponding variable $x_i \in X$ in the regular CSP P^{reg} is set to the value d_j, where d_j is the j-th value of D_i, otherwise, if a variable $x_{i,j}^{\text{b}} \in X^{\text{b}}$ of P^{bvsc} is set to 0, it represents that the corresponding variable $x_i \in X$ in the regular CSP P^{reg} cannot be set to the value d_j. Thus transform the original search space into a binary variable search space.

The second kind of variables represents the states of the underlying DAGs of the regular constraints C^{reg}. Each regular constraint $c_h^{\text{reg}} \in C^{\text{reg}}$ has a DAG M_h which represents the regular expression of the *regular* constraint. For each state $q_{i,j}^{\text{h}}$ of each DAG M_h, a binary variable $x_{i,j}^{\text{h}}$ represents if the state $q_{i,j}^{\text{h}}$ is on a solution path ($x_{i,j}^{\text{h}}$ is set to one) or not ($x_{i,j}^{\text{h}}$ is set to 0).

After defining the variables and the domains (all $\{0,1\}$) we need to define our constraints, which must be semantically equivalent to the constraints C^{reg} and realize that each binary variable set $\{x_{i,1}^{\text{b}}, ..., x_{i,|D_i|}^{\text{b}}\}$, which represents one original variable $x_i \in X$, contains only one variable which is instantiated to 1. This guarantees that we can directly conclude a solution of the original CSP P from a solution of P^{bvsc}. Doing this, we have for each binary variable set $\{x_{i,1}^{\text{b}}, ..., x_{i,|D_i|}^{\text{b}}\}$ a scalar constraint, which set the number of 1s in $\{x_{i,1}^{\text{b}}, ..., x_{i,|D_i|}^{\text{b}}\}$ to one (see Eq. 1).

$$\forall i \in \{1, ..., n\} : scalar(\{x_{i,1}^{\text{b}}, ..., x_{i,|D_i|}^{\text{b}}\}, \{1, ..., 1\}, =, 1) \qquad (1)$$

Analogously, we have constraints for each set of variables $x_{i,j}^h$, which represents each one level i of the DAG M_h. The path of each solution of a DAG contains only one state $q_{i,j}^h$ of each level i. Thus, we can define the following *scalar* constraints (see Eq. 2).

$$\forall i \in \{0, ..., n\}, h \in \{1, ..., |C^{\text{reg}}|\} : scalar(\{x_{i,*}^h\}, \{1, ..., 1\}, =, 1) \qquad (2)$$

To represent a DAG M_h, we need to describe the paths of the solutions of the DAG. The initial state $q_{0,0}^h$ and the final state $q_{n,0}^h$ are mandatory parts of a solution path. Thus, the variables $x_{0,0}^h$ and $x_{n,0}^h$ must be set to 1. A state q_{i,j_3}^h of a DAG M_h is part of a solution path, if one of its predecessor states q_{i-1,j_1}^h and one of the values $d_{i-1,j_2} \in D_{i-1}$ of the edge between this both states is part of the solution path. This can be represented by the logical formula $q_{i-1,j_1} \wedge d_{i-1,j_2} \rightarrow q_{i,j_3}^h$. See Fig. 1 for some example path relations.

Applied to our variable system it concludes the formula $x_{i-1,j_1}^h \wedge x_{i-1,j_2}^b \rightarrow x_{i,j_3}^h$. Furthermore, this logical formula can be represented as *scalar* constraint (see Eq. 3).

$$\forall i \in \{1, ..., n\}, h \in \{1, ..., |C^{\text{reg}}|\}, j_1, j_2, j_3,$$

$$\text{where } q_{i-1,j_1}^h \xrightarrow{d_{i-1,j_2}} q_{i,j_3}^h \text{ is an edge in } M_h : \qquad (3)$$

$$scalar(\{x_{i-1,j_1}^h, x_{i-1,j_2}^b, x_{i,j_3}^h\}, \{1, 1, -2\}, \leq, 1)$$

The equivalence of the statement $x_{i-1,j_1}^h \wedge x_{i-1,j_2}^b \rightarrow x_{i,j_3}^h$ with the statement in Eq. 3 can be easily proved with the truth table shown in Table 1. The result of the logical formula is always true if the scalar product is smaller or equal to one, otherwise it is false.

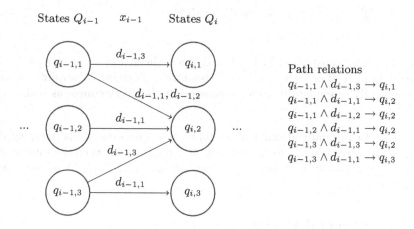

Fig. 1. Example path relations for an excerpt of a DAG.

Table 1. The truth table to show the equivalence of $x^h_{i-1,j_1} \wedge x^b_{i-1,j_2} \to x^h_{i,j_3}$ and $scalar(\{x^h_{i-1,j_1}, x^b_{i-1,j_2}, x^h_{i,j_3}\}, \{1, 1, -2\}, \leq, 1)$.

$x^h_{i-1,j_1} \wedge x^b_{i-1,j_2} \to x^h_{i,j_3}$		$\Leftrightarrow$	$1 * x^h_{i-1,j_1} + 1 * x^b_{i-1,j_2} - 2 * x^h_{i,j_3} \leq 1$	
$True \wedge True \to True$	True	$\Leftrightarrow$	$0 \leq 1$	$1 * 1 + 1 * 1 - 2 * 1 \leq 1$
$True \wedge True \to False$	False	$\Leftrightarrow$	$2 \leq 1$	$1 * 1 + 1 * 1 - 2 * 0 \leq 1$
$True \wedge False \to True$	True	$\Leftrightarrow$	$-1 \leq 1$	$1 * 1 + 1 * 0 - 2 * 1 \leq 1$
$True \wedge False \to False$	True	$\Leftrightarrow$	$1 \leq 1$	$1 * 1 + 1 * 0 - 2 * 0 \leq 1$
$False \wedge True \to True$	True	$\Leftrightarrow$	$-1 \leq 1$	$1 * 0 + 1 * 1 - 2 * 1 \leq 1$
$False \wedge True \to False$	True	$\Leftrightarrow$	$1 \leq 1$	$1 * 0 + 1 * 1 - 2 * 0 \leq 1$
$False \wedge False \to True$	True	$\Leftrightarrow$	$-2 \leq 1$	$1 * 0 + 1 * 0 - 2 * 1 \leq 1$
$False \wedge False \to False$	True	$\Leftrightarrow$	$0 \leq 1$	$1 * 0 + 1 * 0 - 2 * 0 \leq 1$

Thus, we have everything we need to represent the regular CSP $P^{\text{reg}} = (X, D, C^{\text{reg}})$ as a binary variable scalar CSP $P^{\text{bvsc}} = (X^b, D^b, C^{\text{sc}})$, with

$$X^b = \{x^b_{i,j} \mid \forall i \in \{1, ..., n\}, j \in \{1, ..., |D_i|\}\}$$
$$\cup \{x^h_{i,j} \mid \forall h \in \{1, ..., |C^{\text{reg}}|\}, i \in \{0, ..., n\}, j \in \{1, ..., |Q^h_i|\}\}$$
$$D^b = \{D_i = \{0, 1\} \mid \forall i \in \{1, ..., |X^b|\}\}$$
$$C^{\text{sc}} = \{scalar(\{x^b_{i,1}, ..., x^b_{i,|D_i|}\}, \{1, ..., 1\}, =, 1) \mid \forall i \in \{1, ..., n\}\}$$
$$\cup \{scalar(\{x^h_{i,*}\}, \{1, ..., 1\}, =, 1) \mid \forall i \in \{0, ..., n\}, h \in \{1, ..., |C^{\text{reg}}|\}\}$$
$$\cup \{scalar(\{x^h_{i-1,j_1}, x^b_{i-1,j_2}, x^h_{i,j_3}\}, \{1, 1, -2\}, \leq, 1)\} \mid$$
$$\forall i \in \{1, ..., n\}, h \in \{1, ..., |C^{\text{reg}}|\}, j_1, j_2, j_3, where$$
$$q^h_{i-1,j_1} \xrightarrow{d_{i-1,j_2}} q^h_{i,j_3} \text{ is an edge in } M_h\}$$

Remark 2: The constraints in Eq. 1 and 2 are the reason, why we use scalar constraints instead of logic constraints, because such *count* constraints are just not compactly representable as logic constraints, but very simple as scalar constraints.

Remark 3: In the implementation we did some optimizations like merging different scalar constraints, which represent path relations between the same states (q_{i-1,j_1}, q_{i,j_3}), but with k different values $(d_{i-1,j_{2,1}}, ..., d_{i-1,j_{2,k}})$ to one single scalar constraint.

5   Experimental Results

For demonstrating the benefit of fd-complete constraints and fd-complete CSPs, we use a rostering problem close to the presented one in Section 4.1 in [9]. Briefly summarized, consider the rostering problem four shifts ($0 = $ day off, $1 = $ early,

$2 = $ late, $3 = $ night shift), m employees and w weeks. Let n be the number of days ($n = w * 7$). The goal is it to find a valid shift assignment for all employees, such that the following four restrictions are satisfied:

R_1: At each Saturday is the same shift as at the following Sunday.

R_2: A forward rotating system is used, so the following shift combinations are not allowed for two days in a row: (late, early), (night, early), (night, late).

R_3: There is a minimum number of two and a maximum number of four (for night shifts 3) consecutive days, with the same shift.

R_4: Between $\lfloor \frac{w}{4} \rfloor$ and $\lfloor \frac{w}{4} \rfloor + 1$ employees are needed for each shift and day.

The given restrictions cover some but not all German laws for shift planning, but it is possible to extend the problem for individual use cases. The difference to [9] is that we changed in restriction R_4 the upper bound from $\lceil \frac{w}{4} \rceil$ to $\lfloor \frac{w}{4} \rfloor + 1$ to guarantee that every problem has at least one solution.

As typical for many rostering problems, we just consider the plan of one person P_1 and assume that the plan for further staff is received by rotating P_1s plan by one week, two weeks, etc. For example given a shift plan $sol_{P_1} = (v_1, v_2, \ldots, v_n)$ for person P_1, the plan for a person P_2 would be $(v_8, v_9, \ldots, v_n, v_1, \ldots, v_7)$, for a person P_3 would be $(v_{15}, v_{16}, \ldots, v_n, v_1, \ldots, v_{14})$, and so on.

The following naive CSP P_{naive} describes the given rostering problem. The constraints in C_1 satisfy the restriction R_1, the constraints in C_2 satisfy the restriction R_2, the constraints in C_3, C_4 and C_5 satisfy the restriction R_3 and the constraints in C_6 and C_7 satisfy the restriction R_4 (Fig. 2).

$$
\begin{aligned}
X = &\{x_1, ..., x_n\} \\
D = &\{D_1 = ... = D_n = \{0, 1, 2, 3\}\} \\
C = &C_1 = \{x_{i+6} = x_{i+7} \mid \forall i \in \{1, ..., w\}\} \\
&\cup C_2 = \{(x_i, x_{i+1}) \notin \{(2,1), (3,1), (3,2)\} \mid \forall i \in \{1, ..., n\}\} \\
&\cup C_3 = \{(x_i == x_{i+1}) \vee (x_i == x_{i-1})\} \mid \forall i \in \{2, ..., n-1\}\} \\
&\cup C_4 = \{allEqual(x_i, x_{i+1}, x_{i+2}, x_{i+3}) \rightarrow (x_i \neq x_{i+4}) \mid \forall i \in \{1, ..., n-4\}\} \\
&\cup C_5 = \{allEqual(x_i, x_{i+1}, x_{i+2}, 3) \rightarrow (x_i \neq x_{i+3}) \mid \forall i \in \{1, ..., n-3\}\} \\
&\cup C_6 = \{\sum_{i=1}^{w} x_{i+d} \geq \lfloor \tfrac{w}{4} \rfloor \mid \forall d \in \{1, ..., 7\}\} \\
&\cup C_7 = \{\sum_{i=1}^{w} x_{i+d} \leq \lfloor \tfrac{w}{4} \rfloor + 1 \mid \forall d \in \{1, ..., 7\}\}
\end{aligned}
$$

Fig. 2. The naive CSP P_{naive}, which describes the given rostering problem.

For our benchmark suite we computed different instances of the rostering problem with number of weeks respectively number of employees is equal to $(4, 5, 6, 7$ and $8)$ in different versions:

1. *Naive*: The CSP is directly solved as it is modeled in P_{naive}.
2. *Regular*: Each constraint was substituted by a regular version of it as described in [11].

3. *RegularIntersected*: Each constraint was substituted by a regular version of it as described in [11] and the regular versions of the constraints C_1 to C_5 are combined to one regular constraint.
4. *BVSC*: The CSP P_{naive} was transformed into a binary variable scalar constraint CSP P^{bvsc} as described above.
5. *BVSCO*: The binary variable scalar constraint CSP P^{bvsco} was created and optimized by hand.

All the experiments are set up on a DELL laptop with an Intel i7-4610M CPU, 3.00 GHz, with 16 GB 1600 MHz DDR3 and running under Windows 7 Professional with Service Pack 1. The algorithms are implemented in Java under JDK version 1.8.0_191 and Choco Solver version 4.0.4 [15]. We used the *DowOverWDeg* search strategy which is explained in [3] and used as default search strategy in the Choco Solver [15].

Table 2. A comparison of the solution process of the given rostering problem with different sizes and different modelling versions. (*Time limit was reached.)

Problem	#Solutions	*Naive*	*BVSC*	*BVSCO*	*Regular*	*RegularIntersected*
4 weeks	44	100%	100%	100%	100%	100%
		0.726 s	5.815 s	0.155 s	2.139 s	0.371 s
5 weeks	217339	100%	100%	100%	100%	100%
		96.622 s	345.377 s	8.87 s	59.599 s	5.697 s
6 weeks	9443633	17%	2%	100%	34%	100%
		600 s*	600 s*	396.619 s	600 s*	196.055 s
7 weeks	8463303	9%	0.1%	84%	28%	100%
		600 s*	600 s*	600 s*	600 s*	214.486 s
8 weeks	42979	11%	0.002%	93%	76%	100%
		600 s*	600 s*	600 s*	600 s*	25.649 s

Table 2 shows a comparison of the different modelling versions. For four and five weeks, all approaches found all solutions (100%). The *RegularIntersected* and the *BVSCO* approach are the fastest with distance to the other approaches. Only the *RegularIntersected* approach found all solutions for the 6, 7 and 8 week problem. It is shown how many solutions every approach found in at maximum 600 s, which was the time out. It is visible that the *RegularIntersected* approach is always the fastest and the *BVSC* approach is always the worst. But we also can see, that the by hand optimized *BVSCO* is also always very good, much better than the original approach. In these experiments we did not use reduction methods, which are known from equality and inequality solving. We think that in the future we can find an automatic way from the original CSP to one which is very close to the *BVSCO*, and so one that is much better than the original, naive problem. It is clearly evident, those fd-CSPs like regular CSPs or binary variable CSPs with scalar constraints can lead to significant improvement of the solution process of constraint problems.

Remark 4: The transformation time to create the responsible fd-CSPs from the original CSP was always included in the calculation time in Table 2.

Remark 5: The results presented in Table 2 show already the power of fd-complete constraints (the *RegularIntersected* and the *BVSCO* approach are partial more than 24 times faster as the original naive approach). Further more we may can improve the solution process much more if we use a specific solver for regular CSPs or binary variable CSPs with scalar constraints instead of the general fd-solver Choco.

6 Conclusion and Future Work

We have presented a new class of constraints, which allows describing every other CSP with a single constraint (fd-complete constraint). We extended this approach to fd-complete CSPs, which are also able to represent each other CSP. Furthermore, we introduced an example fd-complete CSP, the binary scalar CSP, and explained why we think, that this fd-complete CSP can improve the solution speed of general CSPs. In Sect. 5 was shown that the use of fd-CSPs like regular CSPs or binary variable CSPs with scalar constraints can improve the solution process of CSPs in a significant way.

Future work includes the creation of specialist constraint solvers, specialist search strategies and specialist constraint solver settings. We expect that there are a big number of fd-complete constraints and fd-complete CSPs and, therefore, also a high number of specialist constraint solvers and search strategies, which need to be discovered. Moreover, we need an algorithm to decide into which fd-CSP we should transform given problems.

References

1. Google LLC: Google OR-Tools (2019). https://developers.google.com/optimization/. Accessed 22 Nov 2019
2. Akgün, Ö., Gent, I.P., Jefferson, C., Miguel, I., Nightingale, P., Salamon, A.Z.: Automatic discovery and exploitation of promising subproblems for tabulation. In: Hooker, J.N. (ed.) CP 2018. LNCS, vol. 11008, pp. 3–12. Springer, Cham (2018). https://doi.org/10.1007/978-3-319-98334-9_1
3. Boussemart, F., Hemery, F., Lecoutre, C., Sais, L.: Boosting systematic search by weighting constraints. In: de Mántaras, R.L., Saitta, L. (eds.) Proceedings of the 16th European Conference on Artificial Intelligence (ECAI 2004), including Prestigious Applicants of Intelligent Systems (PAIS 2004), Valencia, Spain, 22–27 August 2004, pp. 146–150. IOS Press (2004)
4. Dechter, R.: Constraint Processing. Elsevier Morgan Kaufmann, Burlington (2003)
5. Hellsten, L., Pesant, G., van Beek, P.: A domain consistency algorithm for the stretch constraint. In: Wallace, M. (ed.) CP 2004. LNCS, vol. 3258, pp. 290–304. Springer, Heidelberg (2004). https://doi.org/10.1007/978-3-540-30201-8_23

6. van Hoeve, W.J., Katriel, I.: Global constraints. In: [16], 1st edn. (2006). (chapter 6)
7. Hopcroft, J.E., Ullman, J.D.: Introduction to Automata Theory, Languages and Computation. Addison-Wesley, Boston (1979)
8. Löffler, S., Liu, K., Hofstedt, P.: The power of regular constraints in CSPS. In: 47. Jahrestagung der Gesellschaft für Informatik (Informatik 2017), Chemnitz, Germany, 25–29 September 2017, pp. 603–614 (2017). https://doi.org/10.18420/in2017_57
9. Löffler, S., Liu, K., Hofstedt, P.: The regularization of CSPs for rostering, planning and resource management problems. In: Iliadis, L., Maglogiannis, I., Plagianakos, V. (eds.) AIAI 2018. IAICT, vol. 519, pp. 209–218. Springer, Cham (2018). https://doi.org/10.1007/978-3-319-92007-8_18
10. Löffler, S., Liu, K., Hofstedt, P.: A meta constraint satisfaction optimization problem for the optimization of regular constraint satisfaction problems. In: Rocha, A.P., Steels, L., van den Herik, H.J. (eds.) Proceedings of the 11th International Conference on Agents and Artificial Intelligence (ICAART 2019), Prague, Czech Republic, 19–21 February 2019, vol. 2, pp. 435–442. SciTePress (2019). https://doi.org/10.5220/0007260204350442
11. Löffler, S., Liu, K., Hofstedt, P.: The regularization of small sub-constraint satisfaction problems. CoRR abs/1908.05907 (2019). http://arxiv.org/abs/1908.05907
12. Marriott, K.: Programming with Constraints - An Introduction. MIT Press, Cambridge (1998)
13. Pesant, G.: A filtering algorithm for the stretch constraint. In: Walsh, T. (ed.) CP 2001. LNCS, vol. 2239, pp. 183–195. Springer, Heidelberg (2001). https://doi.org/10.1007/3-540-45578-7_13
14. Pesant, G.: A regular language membership constraint for finite sequences of variables. In: Wallace, M. (ed.) CP 2004. LNCS, vol. 3258, pp. 482–495. Springer, Heidelberg (2004). https://doi.org/10.1007/978-3-540-30201-8_36
15. Prud'homme, C., Fages, J.G., Lorca, X.: Choco documentation. TASC, INRIA Rennes, LINA CNRS UMR 6241, COSLING S.A.S. (2019). http://www.choco-solver.org/. Accessed 07 Nov 2019
16. Rossi, F., Beek, P.V., Walsh, T.: Handbook of Constraint Programming, 1st edn. Elsevier, Amsterdam (2006)
17. Schulte, C., Lagerkvist, M., Tack, G.: Gecode 6.2.0 (2019). https://www.gecode.org/. Accessed 22 Nov 2019 (2019)
18. Tack, G., Stuckey, P.J.: Minizinc 2.3.2. Monash University (2019). https://www.minizinc.org/. Accessed 22 Nov 2019 (2019)

Backward-Forward Sequence Generative Network for Multiple Lexical Constraints

Seemab Latif[1]($\boxtimes$) (iD), Sarmad Bashir[1], Mir Muntasar Ali Agha[1],
and Rabia Latif[2]

[1] National University of Sciences and Technology (NUST), Islamabad, Pakistan
{seemab.latif,sbashir.msit16seecs,magha.bese15seecs}@seecs.edu.pk
[2] College of Computer and Information Sciences, Prince Sultan University,
Riyadh, Saudi Arabia
rlatif@psu.edu.sa

Abstract. Advancements in Long Short Term Memory (LSTM) Networks have shown remarkable success in various Natural Language Generation (NLG) tasks. However, generating sequence from pre-specified lexical constraints is a new, challenging and less researched area in NLG. Lexical constraints take the form of words in the language model's output to create fluent and meaningful sequences. Furthermore, most of the previous approaches cater this problem by allowing the inclusion of pre-specified lexical constraints during the decoding process, which increases the decoding complexity exponentially or linearly with the number of constraints. Moreover, some of the previous approaches can only deal with single constraint. Additionally, most of the previous approaches only deal with single constraints. In this paper, we propose a novel neural probabilistic architecture based on backward-forward language model and word embedding substitution method that can cater multiple lexical constraints for generating quality sequences. Experiments shows that our proposed architecture outperforms previous methods in terms of intrinsic evaluation.

Keywords: Recurrent Neural Networks · Natural Language
Generation · Language Models · Lexical constraints · Word embedding

1 Introduction

Recently, Recurrent Neural Networks (RNNs) and their variants such as Long Short Term Memory Networks (LSTMs) and Gated Recurrent Units (GRUs) based language models have shown promising results in generating high quality text sequences, especially when the input and output are of variable length. RNN based Language Models (LM) have the ability to capture the sequential nature of language, be it for words, characters or whole sentences. This allows them to outperform other language models in sequence prediction and classification tasks. To learn the distributed representation of data efficiently by RNNs, multiple methods have been proposed such as word embeddings. It mainly include

© IFIP International Federation for Information Processing 2020
Published by Springer Nature Switzerland AG 2020
I. Maglogiannis et al. (Eds.): AIAI 2020, IFIP AICT 584, pp. 39–50, 2020.
https://doi.org/10.1007/978-3-030-49186-4_4

Continuous Bag-Of-Words (CBOW) and Skip-Gram (SG) model [10,12]. CBOW model predicts the word as vector at a current time step, given preceding and proceeding context word vectors. The SG model is opposite in approach to predict the representation of target word vector, but same in the architecture.

Existing methods to incorporate constraints in the output sentences or generating lexical constrained sentences have multiple limitations. [13] proposed variants of backward-forward generation approach which can not handle Out-of-Vocabulary (OOV) words and only generate sentences with single lexical constraint. Similarly, [8] proposed a synchronous training approach to generate lexical constrained sequences with Generative Adversarial Networks (GANs). Moreover, various lexical constrained decoding methods have been proposed for constrained sequence generation through the extension of beam search to allow the inclusion of constraints [1,6]. Such lexical constrained decoding methods do not examine what specific words need to be included at the start of generation, but try to force specific words at each time step during the generation process at a cost of high computational complexity [14].

The remainder of this paper is organized as follows. We review the related work in Sect. 2. Section 3 describes our proposed architecture and Sect. 4 explains the dataset, experimental setup, comparison models and evaluation criteria. Section 5 gives in detail result analysis, finding and discussions about future directions. Finally, Sect. 6 concludes the paper.

2   Literature Review

In general, the purpose of LM is to capture the regularities of a language as well as its morphological and distributional properties. LM aims to compute the probability of a word sequence in order to estimate the maximum likelihood of an upcoming word to be predicted in the sequence. LM learns the distributed representation of words to interpret semantic and syntactic relations between the sequence of words. In past, RNN has shown progressive success in language modeling over traditional methods based on statistical counts. The ability of RNN Language Model (RNNLM) to learn long term contextual dependency and capturing inherited sequential nature of language makes it better than other traditional methods [11]. Particularly in sentence generation task, RNNLM performed well because of its capability of learning highly complicated structures of language. RNNLM makes Maximum A Posteriori (MAP) estimation for predicting words in a sentence [17].

Mou et al. first proposed multiple variants of Backward and Forward (B/F) language models based on GRUs for constrained sentence generation [13]. For training the B/F language models, sentences were split by choosing a word randomly. This resulted in the positional information of words getting smoothed out while generating sentences, and thus they lose the positional information of the word. This method of choosing a split word badly influences the joint probability estimation of a sentence.

Liu et al. proposed an algorithmic framework dubbed as Backward and For-ward Generative Adversarial Networks (BFGAN) for constrained sentence gen-eration [8]. BFGAN constitutes three modules; a discriminator, LSTM based backward and a forward generator with attention mechanism. The purpose of discriminator is to distinguish the real sentences from constrained sentences gen-erated by machine and to guide the joint training of both backward and forward generators by assigning them reward signals. The backward generator takes lexi-cal constraint as an input, which can be a word, phrase or fragment and generate the first half of the sentence backwards. The Forward generator takes the input half sentence generated by backward generator to complete the sentence with the aim of fooling the discriminator. The sentences prepared for training of backward generator relies on random splitting of sentences and the proposed framework can tackle single lexical constrained sentence generation.

Another line of work tackles the problem of constrained sentence genera-tion by sampling the sentences from search space. Su et al. proposed a Gibbs sampling method based on Markov Chain Monte Carlo (MCMC) method for decoding constrained sentences [16]. The proposed approach consists of a dis-criminator and a pure language model conditioned on a bi-directional RNN. Introducing discriminator in the proposed method caters the job for calculating probability of a sentence satisfying the constraints. Gibbs method samples the set of random variables $x_1...n$ from a joint distribution, which takes the form of words to make a sentence. The shortcoming of Gibbs sampling is that it cannot change the length of sentences and hence not able to solve complicated tasks like directly generating sentences from constraints established in advance. Miao et al. extends Gibbs sampling by introducing Metropolis-Hastings for Constrained Sentence Generation (CGMH) [9]. The proposed method directly samples from the sentence space by defining local operations in the sentence space such as word replacement, insertion and deletion.

Hokamp et al. proposed Grid Beam Search (GBS) algorithm, an extension of beam search, for incorporating specified lexical constraints in the output sequences [6]. In Neural Machine Translation (NMT) task, the proposed algo-rithm ensures that all specified constraints must meet the hypothesis before they can be considered to be completed. To generalize image caption generative mod-els for out-of-domain images constituting novel scenes or objects, Anderson et al. proposed a Constrained Beam Search (CBS) decoding method, which utilizes Finite-State Machine (FSM) [1]. The proposed search algorithm is capable of forcing certain image tags over resulting output sequences by recognizing valid sequences with FSM.

Table 1 summarizes techniques for generating constrained sequences. It is evident that many of the architectures are designed for specific scenarios and have high computational complexity. Due to performance gaps and inability to handle multiple constraints efficiently, a method need to be addressed. Therefore, we have proposed a neural probabilistic Backward-Forward architecture that can generate high quality sequences, with word embedding substitution method to satisfy multiple constraints.

Table 1. Comparison of different constrained sequence generation techniques.

	Multiple constraints	Computational time	Decoding complexity	Decoder	Target domain
Mou et al. [13]	x	Low	$\mathcal{O}(Nk)$	–	Research titles
Anderson et al. [1]	✓	High	$\mathcal{O}(Nk2^C)$	CBS	Image captioning
Su et al. [16]	✓	High	$\mathcal{O}(N + dNM)$	GSM	Product sentiments
Liu et al. [8]	x	–	–	Beam search	Product reviews
Hokamp et al. [6]	✓	High	$\mathcal{O}(Nk2C)$	GBS	NMT
Post et al. [14]	✓	Low	$\mathcal{O}(Nk)$	DBA	NMT
Miao et al. [9]	✓	High	$\mathcal{O}(N + dNM)$	MH	Generic
Proposed technique	✓	Low	$\mathcal{O}(Nk)$	Greedy search	Generic

3 General Model

To begin with, we state the problem of constrained sequence generation as follows: given the constraint(s) c as input, the proposed B/F LM needs to generate a fluent sequence $s = w_1, \cdots, w_v, \cdots, w_m$ maximizing the conditional probability $p(s|c)$. For this purpose, we need to select a split word in a sequence s to train the proposed B/F LM. As a sequence provides us an expression, the Parts-Of-Speech (POS) *verb* plays a vital role in placing the *subject* of a sequence into motion and offers more clarification about sequence. In this section, we first discuss the general seq2seq model for generation of sequences. After that, we discuss our proposed architecture to deal with constrained sequence generation.

Conventionally, RNNLMs for text generation are trained to maximize the likelihood of a word w_t or character c_t at time step t while given the context of previous observations in the sequence. This type of learning technique for generating sequences is known as *teacher forcing* [4]. In such learning technique, input to the recurrent neural probabilistic language model is of fixed size. The training purpose is to predict only next token until a special stop sign is generated or specific constraint is satisfied in a sequence given the context of previous observations.

In traditional seq2seq models we cannot satisfy lexical constraints, where disintegrating joint probability of a sentence $y = y_1, y_2 \cdots y_m$ for given input sentence $x = x_1, x_2 \cdots x_n$ is given by

$$p(y|x) = \prod_{i=1}^{m} p(y_i | y_1 \cdots y_{i-1}, x) \tag{1}$$

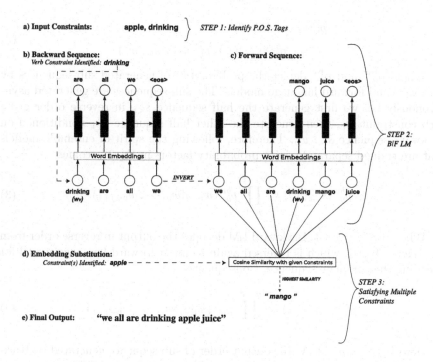

Fig. 1. An illustration of proposed system architecture

Thus, the output sentence y is predicted from y_1 to y_m in sequence either by a greedy or beam decoder. Such decomposition is because of natural language's sequential nature.

3.1   Proposed Architecture

Our proposed approach consists of a neural probabilistic architecture that is an ensemble of two LSTM based B/F LM for generating lexical constrained sequences, which captures the statistical properties of text sequences effectively. In order to generate the coherent sequences from given multiple constraints as input, we first generate the sequence from verb constraint w_v through B/F LM, and then we satisfy the other given constraints by word embedding substitution method during the inference process. The predicted verb v splits the sequence into two sub-sequences as:

$$\text{Backward Sequence} = w_{v-1}, w_{v-2}, \cdots, w_1$$
$$\text{Forward Sequence} = w_{v+1}, w_{v+2}, \cdots, w_m$$

If m denotes the length of words in a sequence s i.e. $s = w_1, \cdots, w_v, \cdots, w_m$, then the joint conditional probability of remaining m words, given lexical constraint w_v and training parameters θ can be calculated as:

$$p(s) = p(s|w_v; \theta)$$
$$= p_\theta^{bw}(s_{<v}|w_v) \cdot p_\theta^{fw}(s_{>v}|s_1 : w_v) \tag{2}$$

Where $\mathbf{p_\theta^{bw}}$ and $\mathbf{p_\theta^{fw}}$ depict the probabilities of generated sub-sequences by backward and forward language models. The sub-sequences are generated asynchronously i.e. we first generate the half sequence $s_{<v}$ in reverse order given verb constraint w_v, then generate the other half sequence $s_{>v}$ conditioned on backward sequence $s_1 : w_v$. Therefore, following the spirit of ensemble models that are trained separately, joint probability factors in Eq. 2 becomes

$$p_\theta^{bw}(s_{<v}|w_v) = \prod_{j=1}^{w_v-1} p_\theta^{bw}(w_{v-j}|w_v, \cdots, w_{v-j+1}) \tag{3}$$

Where $1 \leq j \leq v - 1$. Backward LM decodes the output in reverse order from w_{v-1}, w_{v-2} to w_1, which is reversed again to input forward language model for decoding the complete sequence. Consequently,

$$p_\theta^{fw}(s_{>v}|s_{1:v}) = \prod_{j=1}^{m-w_v} p_\theta^{fw}(w_{v+j}|w_1, \cdots, w_{v+j-1}) \tag{4}$$

Here $1 \leq j \leq m - v$. As the output order of sub-sequence generated by backward LM is reversed again to decode the entire sequence from forward language model, therefore $\mathbf{s_{1:v}}$ is equal to $\mathbf{w_1}, \cdots, \mathbf{w_v}$.

For learning the sequences, we used LSTM networks in proposed architecture. The LSTM networks have the capability of capturing sequential data effectively where the network transforms a sequence of given input word vectors $x = x_1, \cdots, x_n$ into the sequence of hidden states $h = h_1, \cdots, h_t$ by maintaining a history of inputs at each hidden state. The LSTM cell depends on gating mechanism for information processing.

LSTM network's hidden state h at time step t is dependent on the previous state $\mathbf{h_{t-1}}$ and current input $\mathbf{x_t}$ word vectors. Particularly, in our scenario for generating variable length text sequences, the probability of an output word w_{out} from both language models calculated as:

$$p_\theta^{bw}(w_{v-t}|h_t) = Softmax(w_{out}^{bw} h_t + b_{out}) \tag{5}$$

$$p_\theta^{fw}(w_{v+t}|h_t) = Softmax(w_{out}^{fw} h_t + b_{out}) \tag{6}$$

Where w_{out}^{bw} and w_{out}^{fw} are shared across all time steps in their respective LSTM models, which projects the hidden state vector h_t into a fixed same size vector as target vocabulary in order to generate a sequence of outputs $y_t = w_{v-t}, \cdots, w_1$ for backward language model and $y_t = w_{v+t}, \cdots, w_m$ for forward language model.

The softmax function is in the final layer of LSTM network, applied to each word vector for calculating the probability distribution over vocabulary of distinct word vectors.

3.2   Word Embedding Substitution

In order to satisfy the given lexical constraints c other than verb constraint w_v, we have used a lexical substitution method based on word embedding substitution. SG model embeds both target words and their context in the same dimensional space. In this space, the vector representations of words are drawn closer together when they co-occur more frequently in a learning corpus. Thus, Cosine distance between them can be viewed as *target-to-target* distributional similarity measure. Our method relies on a natural assumption that a good lexical constraint substitution for a target word w instance in a generated sequence $s = w_1, \cdots, w_v, \cdots, w_m$ needs to be consistent with the given sequence and lexically similar to the target word w instance. During inference, we find the cosine similarity [2] of given input constraint c with every word w in a sequence s generated by the proposed B/F LM. After that, we replace the constraint c with the closest matching (least cosine distance) word w in a sequence s. Step 3 of Fig. 1 illustrates the concept. For this purpose, we have created word embedding vectorization from FastText.

4   Experiments

In this section, we introduced our experimental designs, containing the preparation of dataset for training and testing, experimental configuration, comparison architectures and evaluation criteria.

4.1   Dataset

There are many benchmark datasets for evaluating pure LM consisting of seq2seq networks for text classification and generative models, but specifically there is no such benchmark corpus for evaluation of constrained sequence generation based on statistical language models. As far, we have used Stanford Natural Language Inference (SNLI) [3] dataset for evaluation and training of proposed architecture. As we target the domain of generating sequences from lexical constraints, we extracted unlabeled sequences within range of minimum 3 and maximum 25 tokens, resulting in 451k sequences for training of proposed architecture. The proposed architecture ensemble backward-forward LM, therefore, to prepare training sequences for backward LM, following steps have been carried out:

- Annotate the tokens with their lexical categories using POS Tagging.
- Split the sentences on verb category instead of random splitting.
- Sentences with more than one verb are broken up into multiple sequences.
- After splitting the sequence on verb category, invert the half sequences.

For the forward language model, the dataset contains complete sequences for training the network. Here, it should be noted that backward language model requires only half sequences till verb token for training the network.

4.2   Word Vectorization

We follow the work of Bojanowski *et al.* [2] to create dense representations of words in dataset. A word vector is represented by augmenting the character n-grams appearing in the word, where the scoring function **s** takes into consideration the internal structure information of words, which is ignored by conventional skip-gram models [10]. The proposed model represents each word **w** as a bag of character n-gram, where adding special boundary symbols <and> at the beginning and end of words for distinguishing prefixes and suffixes from other character sequences. In addition to character n-grams of word **w**, the word **w** is also included in its set of n-grams for learning representation of each word. For example, taking the word '**apple**' and let n = 3, it will be represented by the character n-grams as <**app, ppl, ple**> and the special sequence <*apple*>.

Let a dictionary of n-grams with size **G**. Given a word w where $L_w \subset 1, ... G$ is the set of n-grams appearing in word w. Vector $\mathbf{z_g}$ represents the each n-gram g, therefore a word **w** is represented by the sum of vectors of its n-gram g. In this regard, scoring function of word w with surrounding set of word indices c is calculated by:

$$s(w,c) = \sum_{g \in L_w} z_g^T v_c \tag{7}$$

This extension of skip-gram model for creating word embedding allow the sharing of word vector representations across all words, thus enabling the reliable representational learning of rare or Out-Of-Vocabulary (OOV) words.

We have used extension of FastText's SG model to learn such data representations for both backward and forward language model given their respective data sets. In order to train the FastText model, the word embedding dimension set to 300. Min_count value set to 2, which represents that all the word frequencies lower than 2 were ignored while learning the word representations. Window size set to 5, defining the maximum distance between a current and predicted word within a sequence. Workers parameter set to 16, explaining the worker threads for faster training of FastText SG model. Epochs value set to 30 iteration, over the whole data set.

4.3   Experimental Configuration

We performed different experiments on test set to get the most optimal hyperparameters and evaluate change in performance of the model. Table 2 shows the different experimental configurations and change in performance w.r.t perplexity metric. In the proposed architecture, we get the best results by employing 2-layers of LSTM in both backward and forward language model. Both the LSTM networks were trained with Adam algorithm [7] for stochastic optimization of networks. During training, the parameters were adjusted using Adam optimizer for minimizing the training loss function, also known as misclassification rate. For calculating optimization score, we used categorical cross entropy loss function between the actual y and predicted $\hat{y}$ word probability distribution [5]. In target of accurately capturing the regularities by the neural networks and

Table 2. Hyper-parameter tuning and model performance

LSTM layers	Hidden units	LR	Drop-out	PPL score
1	256	0.01	0.2	35.48
1	512	0.001	0.3	33.15
2	256	0.01	0.2	27.48
2	**512**	**0.001**	**0.3**	**24.20**

preventing overfitting, we appended drop-out layer after every LSTM layer in both the networks. The idea of drop-out layer is to randomly drop units with their connections while training, thus preventing units from co-adapting too much. Dropping units significantly leads to major improvements than other regularization methods [15]. The epochs value was set to 50 and mini batch size was set to 128 in both the networks.

Both the Backward and Forward models are trained on NVIDIA GTX 1080 Ti GPU. The LSTM based networks are developed in keras. Training took about 17 h approx. per model with this implementation and optimal hyper-parameter configuration.

4.4   Comparison and Evaluation Metrics

We compared our proposed methodology with state-of-the art sampling method CGMH [9] for satisfying multiple constraints in a sequence. We also evaluated our methodology of verb based split generation with different variants [13], which can only handle single lexical constraint. We have used intrinsic evaluation metric that allows to determine the quality of a LM without being associated or embedded to a particular application. The most conventional intrinsic evaluation metric is perplexity (PPL). PPL of a language model given a test set $\mathbf{w} = \mathbf{w_1}, \mathbf{w_2}, ... \mathbf{w_m}$ is the inverse probability of $\mathbf{w}$ where the probability is normalized by the number of words

$$PPL(w) = \sqrt[m]{\prod_{i=1}^{m} \frac{1}{P(w_i, \cdots, w_{i-1})}} \tag{8}$$

5   Results and Discussions

For intrinsic evaluation of our proposed methodology, we first make comparisons with variants such as separate B/F and asynchronous B/F language models proposed by [13]. As mentioned earlier, in our proposed methodology the given word is verb constraint w_v through which we decode complete sequence whereas in variants of B/F, the complete sequence is decoded by random split word. We calculated PPL with both verb and random constraint as input to decode the complete sequences. Table 3 represents the comparison in terms of PPL, where the higher probability of a sequence results in the lower of perplexity, which is better. Separate B/F variant yields worse sequences with huge perplexity score because both the B/F LM were enforced to output separately with the input constraint and concatenated after decoding of sequences. This is due to the fact that forward LM does not have the context of half sequence decoded by backward LM. Our proposed approach is more similar to asynchronous B/F LM, but technically very different as we are satisfying multiple constraints while asynchronous approach can deal with only single constraint. The results clearly shows that decoding a sequence on specific verb constraint can make use of the positional information of words in a sequence, that is smoothed out when we generate a sequence with random constraint.

Table 3. Intrinsic evaluation

Model	Perplexity	
Input constraint	Verb	Random
Separate B/F	74.56	80.43
Asynchronous B/F	26.63	28.32
Proposed B/F approach	**24.20**	**27.84**

Table 4. CGMH vs proposed

Constraints	Perplexity (PPL)	
	CGMH	Proposed B/F
1	19.34	**18.04**
2	19.71	**18.92**
3	21.36	**20.13**
4	**20.87**	21.63

Table 4 shows the comparison of our proposed approach for catering multiple constraints with CGMH [9]. Our proposed approach shows lower perplexity than CGMH sampling method for sentence generation through keywords/constraints 1 to 3, while with 4 constraints as input CGMH shows slightly better result than our approach of generating sequence with verb constraint and during inference replacing the words in sequence with closest embedding similarity. The decoding complexity of CGMH increases linearly with the number of constraints, while there is no such factor in our approach for catering multiple constraints. There is always a trade-off between fluency of sequence and decoding complexity. In practice, the downside of CGMH sampling methods is that we are not sure of which sampling step size is best for proposal distribution.

5.1 Discussion

To validate our proposed architecture of generating sequence, we performed a series of experiments. Results of intrinsic evaluation confirms that our proposed approach for sequence generation given constraint(s) outperforms previous methods. Splitting and generating a sequence on verb constraint makes use of positional information, which is smoothed out in breaking down a sequence with random word. We observe that decoding a sequence given random word as input in proposed B/F LM even performs better when the backward LM is trained over half sequences till verb. Moreover, in future we would like to explore about the *constraint-to-target* context similarity, indicating their syntagmatic compatibility for improving the word embedding substitution method. Introducing attention mechanism as context vectors for constraints would be an interesting side in the proposed architecture.

6 Conclusion

In this paper, we have proposed a novel method, dubbed Neural Probabilistic Backward-Forward language model and word embedding substitution method to address the issue of lexical constrained sequence generation. Our proposed system can generate constrained sequences given multiple lexical constraints as input. To the best of our knowledge, this is the first time that multiple constraints have been handled through LSTM based backward-forward LM and word embedding substitution of the sequences. The proposed method contains a backward language model based on LSTM network, which learns the half representation of a sentence until the verb splitting word and forward language model constitute LSTM Network, learning the complete representation of a sequence. Moreover, word embedding substitution method satisfy other constraints by substituting the target word in the sequence with given constraints based on similar context in an embedding space.

References

1. Anderson, P., Fernando, B., Johnson, M., Gould, S.: Guided open vocabulary image captioning with constrained beam search. arXiv preprint arXiv:1612.00576 (2016)
2. Bojanowski, P., Grave, E., Joulin, A., Mikolov, T.: Enriching word vectors with subword information. Trans. Assoc. Comput. Linguist. **5**, 135–146 (2017)
3. Bowman, S.R., Angeli, G., Potts, C., Manning, C.D.: A large annotated corpus for learning natural language inference. In: Proceedings of the 2015 Conference on Empirical Methods in Natural Language Processing, pp. 632–642. Association for Computational Linguistics (2015). https://doi.org/10.18653/v1/D15-1075
4. Dai, J., Zhang, P., Mazumdar, J., Harley, R.G., Venayagamoorthy, G.: A comparison of MLP, RNN and ESN in determining harmonic contributions from nonlinear loads. In: 2008 34th Annual Conference of IEEE Industrial Electronics, pp. 3025–3032. IEEE (2008)
5. De Boer, P.T., Kroese, D.P., Mannor, S., Rubinstein, R.Y.: A tutorial on the cross-entropy method. Ann. Oper. Res. **134**(1), 19–67 (2005). https://doi.org/10.1007/s10479-005-5724-z
6. Hokamp, C., Liu, Q.: Lexically constrained decoding for sequence generation using grid beam search. arXiv preprint arXiv:1704.07138 (2017)
7. Kingma, D.P., Ba, J.: Adam: a method for stochastic optimization. CoRR (2014)
8. Liu, D., Fu, J., Qu, Q., Lv, J.: BFGAN: backward and forward generative adversarial networks for lexically constrained sentence generation. IEEE/ACM Trans. Audio Speech Lang. Process. **27**(12), 2350–2361 (2019). https://doi.org/10.1109/TASLP.2019.2943018
9. Miao, N., Zhou, H., Mou, L., Yan, R., Li, L.: CGMH: constrained sentence generation by metropolis-hastings sampling. CoRR abs/1811.10996 (2018)

10. Mikolov, T., Chen, K., Corrado, G., Dean, J.: Efficient estimation of word representations in vector space. arXiv preprint arXiv:1301.3781 (2013)
11. Mikolov, T., Karafiát, M., Burget, L., Černocký, J., Khudanpur, S.: Recurrent neural network based language model. In: Eleventh Annual Conference of the International Speech Communication Association (2010)
12. Mikolov, T., Sutskever, I., Chen, K., Corrado, G.S., Dean, J.: Distributed representations of words and phrases and their compositionality. In: Advances in Neural Information Processing Systems, pp. 3111–3119 (2013)
13. Mou, L., Yan, R., Li, G., Zhang, L., Jin, Z.: Backward and forward language modeling for constrained sentence generation. arXiv preprint arXiv:1512.06612 (2015)
14. Post, M., Vilar, D.: Fast lexically constrained decoding with dynamic beam allocation for neural machine translation. In: Proceedings of the 2018 Conference of the North American Chapter of the Association for Computational Linguistics: Human Language Technologies, Volume 1 (Long Papers), pp. 1314–1324. Association for Computational Linguistics (2018). https://doi.org/10.18653/v1/N18-1119
15. Srivastava, N., Hinton, G., Krizhevsky, A., Sutskever, I., Salakhutdinov, R.: Dropout: a simple way to prevent neural networks from overfitting. J. Mach. Learn. Res. 15(1), 1929–1958 (2014)
16. Su, J., Xu, J., Qiu, X., Huang, X.: Incorporating discriminator in sentence generation: a Gibbs sampling method. In: Thirty-Second AAAI Conference on Artificial Intelligence (2018)
17. Sutskever, I., Vinyals, O., Le, Q.V.: Sequence to sequence learning with neural networks. In: Ghahramani, Z., Welling, M., Cortes, C., Lawrence, N.D., Weinberger, K.Q. (eds.) Advances in Neural Information Processing Systems 27, pp. 3104–3112. Curran Associates, Inc. (2014)

Deep Learning/LSTM

Deep Echo State Networks in Industrial Applications

Stefano Dettori[1]([✉]), Ismael Matino[1], Valentina Colla[1], and Ramon Speets[2]

[1] Scuola Superiore Sant'Anna - TeCIP Institute – ICT-COISP, Pisa, Italy
{s.dettori,i.matino,v.colla}@santannapisa.it
[2] Tata Steel, IJmuiden, The Netherlands
ramon.speets@tatasteeleurope.com

Abstract. This paper analyzes the impact of reservoir computing, and, in particular, of Deep Echo State Networks, to the modeling of highly non-linear dynamical systems that can be commonly found in the industry. Several applications are presented focusing on forecasting models related to energy content of steelwork byproduct gasses. Deep Echo State Network models are trained, validated and tested by exploiting datasets coming from a real industrial context, with good results in terms of accuracy of the predictions.

Keywords: Reservoir computing · Deep Echo State Networks · Industrial application

1 Introduction

These last few years have shown a dizzying growth trend in the application of Artificial Intelligence (AI) techniques, not only in the academic world, in which AI is considered a standard technique, with many fields that have reached a satisfactory level of development, but also outside the academia and, in particular, in industrial context. The new consciousness of some industrial sectors, and in particular the steel industry, towards the concepts of circular economy and sustainable development requires to tackle challenging technological and scientific problems, such as the optimization of the use and reuse of energy sources. At the same time, the need arises to improve the product quality in order to face the increasing competition in the markets of goods and services. Nowadays stakeholders are converging towards the idea that AI is a clear turning point for optimally addressing current and future challenges [1] and have begun, in the last decade, a race to cancel the knowledge gaps that exist in the industrial world between the engineering experience, which is currently based on standard modeling, optimization and control techniques and the so-called *data-driven* tools and techniques. Among these latter ones, surely AI plays a fundamental role in facing the challenge of digitalization [2] and nowadays the level of understanding and, above all, of acceptance of AI as an effective and reliable technique [3] is surely increased.

I. Maglogiannis et al. (Eds.): AIAI 2020, IFIP AICT 584, pp. 53–63, 2020.
https://doi.org/10.1007/978-3-030-49186-4_5

In this context, reservoir computing is inserted, a tool capable of making the philosophy and technique underlying the AI, aimed at the study and modeling of timeseries and dynamic processes, more understandable and affordable in terms of computation skills, efficiency and quality of obtained results. Reservoir computing has been introduced by the work of Maass et al., in which has been described a particular Recurrent Neural Network (RNN) architecture, called *Liquid-state machine* [4].

The literature in the field of reservoir computing applied to industrial processes is quite extensive, and in particular among the various techniques emerges that based on the Echo State Network, thanks to characteristics that make this approach more attractive and effective. Interesting examples of their application are shown in the work of Wang [5] in which ESN and sparse adaboost method are exploited to forecast the electricity consumption in industrial areas; Matino presented a work related to the prediction of blast furnace gas through ESN techniques [6] and Dettori applied several AI methodologies aiming at modelling energy transformation equipment [7]. Colla et al. presented a work related to the use of outlier detection and advanced variable selection to reservoir application in industry [8]. Another interesting work related to the application of ESN to Model predictive control methodologies is presented by Pan [9].

This work presents some models developed within the European project entitled "Optimization of the management of the process gas network within the integrated steelworks" (GASNET), which aims at supporting the optimal use and distribution of valuable energy resources and byproduct gasses while minimizing the environmental impact. This problem is of utmost importance for integrated steelworks, i.e. the industrial plants which produce steel from virgin raw material, as considerable savings in CO_2 emissions as well as in natural gas consumptions can be achieved by means of an optimal distribution of the off-gases, such as discussed in [10–12]. In particular, in the present work, the models are developed through a recent reservoir computing methodology called Deep Echo State Network (DESN). The novelty of the work consists in the application of this novel Neural Network (NN) architecture, which allows modelling complex nonlinear dynamics that can be typically found in industrial processes. The effectiveness of the proposed methodology has ben compared respect to other state of art neural network architectures.

The paper is organized as follows: Sect. 2 describes the DESN architecture, Sect. 3 presents processes, models and datasets used, Sect. 4 describes methods and results, while Sect. 5 provides some concluding remarks.

2 Deep Echo-State Network Architecture

ESN is an efficient tool and a universal uniform approximator [13], well known for its intrinsic capability of reconstructing complex dynamical input/output relationships. The concept behind the ESN approach is to generate within a reservoir a rich set of dynamics starting from the exciting input. The frequency information content of the input is somehow distorted and enriched through the non-linear reservoir filter and then used to compute a regression on a target. In the last decade, the research in the field of reservoir computing has deepened the study of the characteristics of ESNs up to a further evolutionary step in its architecture, which borrows the concepts introduced by

Deep Learning and amplifies its effectiveness through the use of algorithms that do not use heavy backpropagation routines. The resulted DESN approach introduced by Gallicchio [14, 15] consists of the exploitation of N reservoirs r connected in series in a deep learning fashion and a readout that collects all the reservoirs dynamics to compute the output of the network, as shown in Fig. 1.

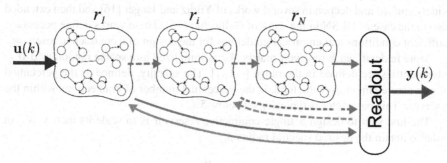

Fig. 1. Architecture of a DESN.

At time k, the state of neurons in each reservoir layers $x_i(k)$ is updated due to the action of its exciting input. The first reservoir layer receives in input the vector $u(k)$, the following layers receive in input the updated state of the previous reservoir layer $x_{i-1}(k)$, whose dynamic is updated according to the following equations:

$$x_1(k) = f\left(c_{in}W_{in_1}u(k) + W_{r_1}x_1(k-1) + v_1(k)\right) \qquad (1)$$

$$x_i(k) = f\left(c_{is}W_{in_i}x_{i-1}(k) + W_{r_i}x_i(k-1) + v_i(k)\right) \qquad (2)$$

where i is the i-th reservoir layer, f is typically a tanh function, c_{in} and c_{is} are respectively the input scaling and inter-scaling factors, W_{in_1} and W_{in_i} are, respectively, the input matrix of the first and i-th reservoir layer with dimensions $n_1 \times n_{in}$ and $n_i \times n_{i-1}$, n_i is the number of neurons of the i-th reservoir, W_{r_i} is the reservoir matrix of i-th reservoir, y is the output of the DESN and v_i is a small amplitude white noise. The output of the readout is updated as:

$$y(k) = f_{out}(W_{out}x(k)) \qquad (3)$$

where $x(k)$ is the complete vector of states, f_{out} is the readout neuron function, typically the identical function for time series regression tasks, W_{out} a $n_{out} \times n_T$ matrix and n_T is the total number of reservoir neurons of the DESN.

Some of the aspects that make the approach particularly effective compared to those based on other recurrent networks are the architectural characteristics of the DESN and the training procedures that focus only in the weights of the readout.

More in detail, as said before, the only scope of the reservoir is to generate a rich set of dynamics, while the readout has to combine them in effective way, in order to minimize the regression error through non-iterative algorithms. The training phase is carried out in two sequential routines: The reservoir and input initializations and the readout training.

In the first phase, input and reservoir matrixes are initialized with the only constraint of obtaining a reservoir characterized by stable dynamics. In other words, the state of the reservoir has to be contractive and each neuron has to gradually forget their previous activation. These properties are summarized in the so-called Echo State Property (ESP), widely studied and deepen in several works of Yildiz and Jaeger [16] and then extended also to the case of DESN in the work of Gallicchio [15]. These works define necessary, sufficient conditions and empirical guidelines for the design of a contractive reservoir.

More in detail, each reservoir matrix $\boldsymbol{W}_{r_i}$ is initialized as a sparse randomized $\hat{\boldsymbol{W}}_{r_i}$, which entries are defined in the range $[-1, 1]$. The sparsity, defined as the percentual of non-zero element over the total, is the percentual number of connections within the reservoir. This parameter is typically set below 5%.

The first step to design a single contractive reservoir is to scale its matrix $\hat{\boldsymbol{W}}_{r_i}$ in order to obtain the desired spectral radius $\tilde{\rho}_i$:

$$W_{r_i} = \tilde{\rho}_i \frac{\hat{\boldsymbol{W}}_{r_i}}{\rho\left(\hat{\boldsymbol{W}}_{r_i}\right)} \tag{4}$$

In the case of DESN, the necessary condition to guarantee the ESP is that the greater of the spectral radius of all the reservoirs is less than one. This condition is a guideline to design a contractive global reservoir, but empirically is also sufficient. The spectral radius of each layer is an important hyperparameter, whose choice is subject to optimization. While the reservoir system is initialized, also each reservoir input matrix $\boldsymbol{W}_{in_i}$ is initialized randomly with weights in the range $[-1, 1]$. In the Eq. (1) and (2) these matrixes are actually scaled by further coefficients c_{in} and c_{is}, namely input scaling and inter-scaling factors, that are important hyperparameters and subject to optimization. Once all these matrixes are initialized and left untrained, it is possible to train the readout by exploiting linear regression routines. A straightforward solution is the Tikhonov regularization least square algorithm, calculated on the training dataset:

$$W_{out} = \bar{Y}X^T\left(XX^T + \lambda I\right)^{-1} \tag{5}$$

where $\bar{Y}$ and X are respectively the sequences of targets and states of all the reservoirs, the latter calculated by using Eqs. (1) and (2). λ is the regularization coefficient that allows to solve also ill-posed inversions and, in general, could be object of optimization during the hyperparameter selection phase.

3   Models and Datasets

Between the several processes involved during the development of GASNET project, in this paper we present two case studies related to the prediction of energetic content and chemical characteristics of Blast Furnace (BF) off-gas, a particular byproduct gas (BFG) generated during the production of pig iron that can be reused as a valuable energy carrier to produce electrical energy or process steam. In particular, for the aims of GASNET project, it is useful to predict the future behavior and characteristics of the off-gas by using a restrict number of process measurements and future knowledge of the process scheduling for 2 h ahead. Together with the prediction of the BFG production, this paper presents an application in which is consumed, the hot blast stoves, also called Cowpers.

In particular, in this process the preheating of the air takes place, subsequently blown inside the BF for the production of pig iron, through the combustion of BFG and other byproduct gasses typically available in some integrated steelworks, such as Coke Oven Gas (COG).

The targets of the first two models are the BFG volume flow and its Net Calorific Value (NCV). For each target will be developed a specific model capable of predicting the entire sequence of 2 h ahead dynamic behavior in a one-shot multistep manner.

The targets of the third and fourth model are respectively the consumption of BFG and COG burned in the cowpers. Also in this case a specific model capable of predicting the entire sequence of 2 h ahead dynamic behavior in a one-shot multistep manner.

The models will be trained exploiting real data of industrial partners, through datasets related to a period of 30 days with sampling time of 1 min, sufficient to describe the main dynamics of the process.

Data pre-processing has represented a fundamental step for models design. Unreliable data have been identified through suitable outliers detection techniques [17, 18]. Moreover, the inputs of each model have been selected by exploiting a variable selection approach based on Genetic Algorithms [19, 20].

All the models have in input the scheduling of the respective process for 2 h ahead, a Boolean information that describes if the respective process is active or not. The model that predicts BFG Flow has in input also the current measurements of the O_2 content in the cold wind, the cold wind volume flow, the pressure of the hot wind and the BFG volume flow. The model that predicts the BFG NCV has in input also the same measurements and in addiction the current CO and H_2 contents in the BFG. The third model, which predicts the BFG consumption in the cowpers, takes in input its abovementioned process scheduling and the current measurements of the cold wind flow and the BFG and COG consumed in the cowpers. The fourth model, which predicts the COG consumption in the cowpers takes the same inputs of the previous model and in addiction the future 2 h predictions of BFG NCV. The input/output architecture of each model is depicted in Fig. 2, while the inputs and target dataset descriptions and Units of Measurement (UoM) are shown in the Table 1.

4 Methods and Numerical Results

In order to evaluate the effectiveness of the DESN architectures for predicting the future behavior of the considered processes, a comparison is proposed between the results achievable through DESN and another rather efficient architecture for modeling time-series characterized by non-linear dynamics, i.e. the Long Short-Term Memory (LSTM) [21]. In this work, the LSTM architecture is configured a series of input layer, a LSTM layer, L_{LSTM} fully connected layers, and a linear readout.

Each model has been developed through a systematic procedure. In first place, it is necessary to define an optimal architecture for each model, by choosing a good set of

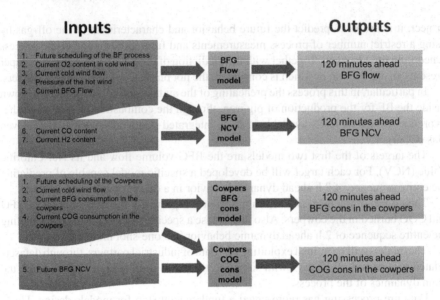

Fig. 2. Input/Output architecture of the models

Table 1. Input/output descriptions of each model

Variables	Description	UoM
Scheduling of BFG process	Boolean variable aimed at describing the status of the BF process. On (1), Off (0)	–
Scheduling of cowpers	Boolean variable aimed at describing the status of the cowpers. On (1), Off (0)	–
O_2 content in cold wind	Volume percentage of oxygen in the cold wind in input to the BF after heating	%
Cold wind flow	Volume flowrate of cold wind in input to the BF after heating	m^3/h
Hot wind pressure	Pressure of hot wind in input to the BF	bar
BFG production	Volume of produced BF gas	m^3/h
CO content	Volume percentage of carbon monoxide in BF gas	%
H_2 content	Volume percentage of hydrogen in BF gas	%
BFG consumption in the cowpers	Consumption of BF gas in cowpers	m^3/h
COG consumption in the cowpers	Consumption of COG gas in cowpers	m^3/h

hyperparameters. For the DESN the considered hyperparameters are the of number of layers and neurons of each reservoir, the spectral radius, the input scaling factor and the inter-layer scaling factor. For the LSTM the considered hyperparameters are the number of fully connected layers in series L_{LSTM}, and the number of neurons of each fully connected layer, with the i-th layer characterized by N_{LSTM_i} neurons.

For the sake of simplicity, each layer of the DESN has the same number of neurons N_{DESN} and the same spectral radius ρ_L. Moreover, each layer after the first has the same inter-layer scaling factor. The selection of the hyperparameters is an open topic for scientific research, and several works in literature give guidelines or recommend some particular algorithms for their choice, such as [22]. In our work their optimization is carried out through a *random search technique* [23] during which 1000 training trials have been carried out using a uniform distribution in the intervals: number of reservoir layers $L_{DESN} = [2\ 12]$; Total Number of Reservoir Neurons $= [200\ 2000]$, $\rho_L = [0.1\ 1]$; Input scaling $c_{in} = [0.01\ 10]$; Inter Scaling $c_{is} = [0.1\ 1]$.

In the case of LSTM, the random search is performed over 1000 training trials with a uniform distribution of the hyperparameters in the intervals: $L_{LSTM} = [1\ 10]$, $N_{LSTM_i} = [30\ 300]$. The LSTMs have been trained through the ADAM training method [24]. The random search algorithm does not allow finding the best set of hyperparameters, but in general a good approximation of the optimal solution.

The optimal set of DESN and LSTM hyperparameters minimizes the mean value of the Normalized Root Mean Square Error (NRMSE) of all the outputs of the model *mNRMSE*, evaluated on the validation set. The comparison between the two different architectures is assessed by evaluating the *mNRMSE*, which is a particularly robust index with respect to the Mean Absolute Percent Error (MAPE) or other common metrics, due to the formulation that takes into account the overall range of the targets. Furthermore, the MAPE is not an adequate measure when intermittent target values (too many values equal or near to 0) are treated.

$$mNRMSE = 100\frac{1}{n_y} \sum_{j=1}^{n_y} \left(\frac{\sqrt{\frac{1}{N_s}\sum_{k=1}^{N_s}(\bar{y}_j(k)-y_j(k))^2}}{\max(y_j)-\min(y_j)} \right) \quad (6)$$

The dataset is composed of about 45000 samples, divided in two parts: the first 50% is used for the choice of hyperparameters and subsequently for training the optimal networks, the remaining 50% is used for the test of the trained models. During the phase of the hyperparameters selection, the first fraction of the overall dataset is in turn divided into 60% training and 40% validation.

The DESN results related to the trained optimal network are summarized in Table 2, which shows the results on the test dataset and the optimal architecture of each model. The comparison between the DESN- and LSTM-based architectures is reported in Table 3, in terms of *mNRMSE* on the training and test dataset.

In the one-shot multistep ahead forecast approach, the k-th output is referred to k sample ahead prediction. For all the models, the error is low in the first 10–20 samples of prediction and, as it can normally be expected, it tends to increase as we want to predict the phenomenon in the distant future. An example of prediction 2 h ahead of the BFG and COG consumption in the cowpers through DESN and LSTM architectures are shown in Fig. 3 and Fig. 4, respectively, which allow a comparison between the measured-target values (in blue) and the LSTM and DESN forecasted values (respectively in orange and yellow). In deeper detail, the figures show an example of one-shoot multistep prediction of the abovementioned process for a specific instant of prediction, during which the

Table 2. Optimal DESN architectures and test results of each model.

Model	L_{DESN}	N_{DESN}	ρ_L	c_{in}	c_{is}	Test mNRMSE
BFG flow	7	90	0.549	0.154	0.952	6.7
BFG NCV	5	139	0.991	0.214	0.031	7.3
Cowpers BFG cons	5	293	0.447	5.10	0.022	6.09
Cowpers COG cons	5	180	0.722	0.073	0.691	9.87

Table 3. Comparison between DESN and LSTM architectures

Model	Architecture	Training mNRMSE	Test mNRMSE
BFG flow	DESN	5.02	6.70
	LSTM	7.95	10.92
BFG NCV	DESN	5.93	7.31
	LSTM	8.36	16.6
Cowpers BFG consumption	DESN	4.60	6.09
	LSTM	4.95	6.38
Cowpers COG consumption	DESN	8.08	9.87
	LSTM	11.6	13.4

target is characterized by a rich dynamic content. For confidentiality constraints, in the figures the absolute ranges of the measured and predicted values are normalized.

The results achieved during the test of the modelling approach are very encouraging. In particular, the models related to the BFG production and its energy contents are characterized of errors around 7% that, considering the heavy nonlinearity of the multi input multi output BF process, for control application are very low. This allows predicting the BFG production with good accuracy, sufficient to optimize its use in an energy control and optimization strategy, and to provide a support to process operators for the following 2 h.

In the case of BFG consumption in the cowpers, the prediction errors are satisfactory. The model for the prediction of COG consumption in the cowpers show a greater error, but also in this case the model can be considered useful in an energy control and optimization strategy.

The comparison between DESN and LSTM shows a clear difference between the results and the quality of the prediction obtainable with the two different architectures. In each proposed modelled process, DESNs outperform LSTMs on both training and test dataset. In particular, with respect to LSTM architecture, DESNs allow obtaining an improvement of the performances of the 4 models, equal to 4.22%, 9.29%, 0.29% and 3.53%, respectively.

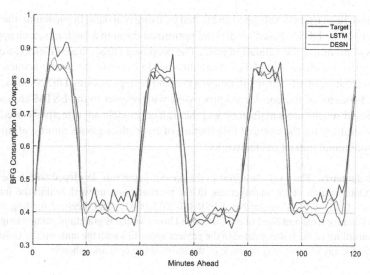

Fig. 3. Prediction example of BFG consumption in the cowpers.

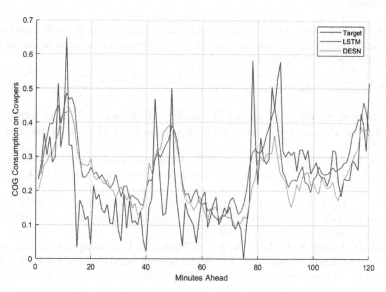

Fig. 4. Prediction example of COG consumption in the cowpers.

5   Conclusions

This paper describes the application of a particular reservoir computing technique called
Deep Echo State Network to the modelling of nonlinear dynamics typical of complex
industrial processes. The presented case studies are the forecast of energetic content in
blast furnace gasses, produced during the production of pig iron in steelworks and one
application of its consumption in the process called Hot Blast Stoves (Cowpers). The

models have been trained, validated and tested by using real data. In particular, the hyper-parameters of the DESN-based models are optimized through a random search approach that aims to minimize the validation error. The proposed DESN-based methodology has been compared with an LSTM-based architecture in order to assess the accuracy with respect to the state of art. The results show a great advantage in using DESNs to model the dynamic behavior of the considered processes, with respect to the LSTM architecture. The achieved results are satisfactory and the trained models are effectively used inside a control strategy for the optimal distribution of byproduct gasses aiming at minimize the environmental impact of steelworks.

Acknowledgments. The work described in the present paper was developed within the project entitled "Optimization of the management of the process gases network within the integrated steelworks - GASNET" (Contract No. RFSR-CT-2015-00029) and received funding from the Research Fund for Coal and Steel of the European Union, which is gratefully acknowledged. The sole responsibility of the issues treated in the present paper lies with the authors; the Union is not responsible for any use that may be made of the information contained therein.

References

1. Brynjolfsson, E., Rock, D., Syverson, C.: Artificial intelligence and the modern productivity paradox: a clash of expectations and statistics. National Bureau of Economic Research (2017). http://doi.org/10.3386/w24001
2. Branca, T.A., Fornai, B., Colla, V., Murri, M.M., Streppa, E., Schröder, A.J.: The challenge of digitalization in the steel sector. Metals **10**(2), 288 (2020). https://doi.org/10.3390/met100 20288
3. Ransbotham, S., Kiron, D., Gerbert, P., Reeves, M.: Reshaping business with artificial intelligence: closing the gap between ambition and action. MIT Sloan Manag. Rev. **59**(1) (2017)
4. Maass, W., Natschläger, T., Markram, H.: Real-time computing without stable states: a new framework for neural computation based on perturbations. Neural Comput. **14**(11), 2531–2560 (2002)
5. Wang, L., Lv, S.-X., Zeng, Y.-R.: Effective sparse adaboost method with ESN and FOA for industrial electricity consumption forecasting in China. Energy **155**, 1013–1031 (2018). https://doi.org/10.1016/j.energy.2018.04.175
6. Matino, I., Dettori, S., Colla, V., Weber, V., Salame, S.: Forecasting blast furnace gas production and demand through echo state neural network-based models: pave the way to off-gas optimized management. Appl. Energy **253**, 113578 (2019). https://doi.org/10.1016/j.ape nergy.2019.113578
7. Dettori, S., Matino, I., Colla, V., Weber, V., Salame, S.: Neural network-based modeling methodologies for energy transformation equipment in integrated steelworks processes. Energy Procedia **158**, 4061–4066 (2019). https://doi.org/10.1016/j.egypro.2019.01.831
8. Colla, V., Matino, I., Dettori, S., Cateni, S., Matino, R.: Reservoir computing approaches applied to energy management in industry. In: Macintyre, J., Iliadis, L., Maglogiannis, I., Jayne, C. (eds.) EANN 2019. CCIS, vol. 1000, pp. 66–79. Springer, Cham (2019). https://doi.org/10.1007/978-3-030-20257-6_6
9. Pan, Y., Wang, J.: Model predictive control of unknown nonlinear dynamical systems based on recurrent neural networks. IEEE Trans. Industr. Electron. **59**(8), 3089–3101 (2011). https://doi.org/10.1109/tie.2011.2169636

10. Maddaloni, A., Porzio, G.F., Nastasi, G., Colla, V., Branca, T.A.: Multi-objective optimization applied to retrofit analysis: a case study for the iron and steel industry. Appl. Therm. Eng. **91**, 638–646 (2015). https://doi.org/10.1016/j.applthermaleng.2015.08.051

11. Colla, V., et al.: Assessing the efficiency of the off-gas network management in integrated steelworks. Materiaux et Techniques **107**(1), 104 (2019). https://doi.org/10.1051/mattech/201 8068

12. Porzio, G.F., et al.: Process integration in energy and carbon intensive industries: an example of exploitation of optimization techniques and decision support. Appl. Therm. Eng. **70**(2), 1148–1155 (2014). https://doi.org/10.1016/j.applthermaleng.2014.05.058

13. Grigoryeva, L., Ortega, J.P.: Echo state network are universal. Neural Netw. **108**, 495–508 (2018). https://doi.org/10.1016/j.neunet.2018.08.025

14. Gallicchio, C., Micheli, A., Pedrelli, L.: Deep reservoir computing: a critical experimental analysis. Neurocomputing **268**, 87–99 (2017). https://doi.org/10.1016/j.neucom.2016.12.089

15. Gallicchio, C., Micheli, A.: Echo state property of deep reservoir computing networks. Cogn. Comput. **9**(3), 337–350 (2017). https://doi.org/10.1007/s12559-017-9461-9

16. Yildiz, I.B., Jaeger, H., Kiebel, S.J.: Re-visiting the echo state property. Neural Netw. **35**, 1–9 (2012). https://doi.org/10.1016/j.neunet.2012.07.005

17. Cateni, S., Colla, V., Nastasi, G.: A multivariate fuzzy system applied for outliers detection. J. Intell Fuzzy Syst. **24**(4), 889–903 (2013). https://doi.org/10.3233/ifs-2012-0607

18. Cateni, S., Colla, V., Vannucci, M.: A fuzzy logic-based method for outliers detection. In: Proceedings of the IASTED International Conference on Artificial Intelligence and Applications (AIA 2007), pp. 561–566 (2007)

19. Cateni, S., Colla, V., Vannucci, M.: General purpose input variables extraction: a genetic algorithm based procedure GIVE a GAP. In: 9th International Conference on Intelligent Systems Design and Applications (ISDA 2009), pp. 1278–1283 (2009). https://doi.org/10. 1109/isda.2009.190

20. Cateni, S., Colla, V., Vannucci, M.: A genetic algorithm-based approach for selecting input variables and setting relevant network parameters of a SOM-based classifier. Int. J. Simul. Syst. Sci. Technol. **12**(2), 30–37 (2011). https://doi.org/10.1109/ems.2010.23

21. Hochreiter, S., Schmidhuber, J.: Long short-term memory. Neural Comput. **9**(8), 1735–1780 (1997)

22. Gallicchio, C., Micheli, A., Pedrelli, L.: Design of deep echo state networks. Neural Netw. **108**, 33–47 (2018). https://doi.org/10.1016/j.neunet.2018.08.002

23. Bergstra, J., Bengio, Y.: Random search for hyper-parameter optimization. J. Mach. Learn. Res. **13**, 281–305 (2012)

24. Kingma, D.P., Ba, J.: Adam: a method for stochastic optimization. arXiv preprint arXiv:1412. 6980 (2014)

Deepbots: A Webots-Based Deep Reinforcement Learning Framework for Robotics

M. Kirtas, K. Tsampazis, N. Passalis[(✉)], and A. Tefas

Artificial Intelligence and Information Analysis Lab, Department of Informatics, Aristotle University of Thessaloniki, Thessaloniki, Greece
{eakirtas,tsampaka,passalis,tefas}@csd.auth.gr

Abstract. Deep Reinforcement Learning (DRL) is increasingly used to train robots to perform complex and delicate tasks, while the development of realistic simulators contributes to the acceleration of research on DRL for robotics. However, it is still not straightforward to employ such simulators in the typical DRL pipeline, since their steep learning curve and the enormous amount of development required to interface with DRL methods significantly restrict their use by researchers. To overcome these limitations, in this work we present an open-source framework that combines an established interface used by DRL researchers, the OpenAI Gym interface, with the state-of-the-art Webots robot simulator in order to provide a standardized way to employ DRL in various robotics scenarios. Deepbots aims to enable researchers to easily develop DRL methods in Webots by handling all the low-level details and reducing the required development effort. The effectiveness of the proposed framework is demonstrated through code examples, as well as using three use cases of varying difficulty.

Keywords: Deep Reinforcement Learning · Simulation environment · Webots · Deepbots

1 Introduction

Reinforcement Learning (RL) is a domain of Machine Learning, and one of the three basic paradigms alongside supervised and unsupervised learning. RL employs agents that *learn* by simultaneously *exploring* their environment and *exploiting* the already acquired knowledge to solve the task at hand. The learning process is guided by a reward function, which typically expresses how close the agent is to reaching the desired target behavior. In recent years, Deep Learning (DL) [8] was further combined with RL to form the field of Deep Reinforcement Learning (DRL) [17], where powerful DL models were used to solve challenging RL problems.

DRL is also increasingly used to train robots to perform complex and delicate tasks. Despite the potential of DRL on robotics, such approaches usually

© IFIP International Federation for Information Processing 2020
Published by Springer Nature Switzerland AG 2020
I. Maglogiannis et al. (Eds.): AIAI 2020, IFIP AICT 584, pp. 64–75, 2020.
https://doi.org/10.1007/978-3-030-49186-4_6

require an enormous amount of time to sufficiently explore the environment and manage to solve the task, often suffering from low sample efficiency [20]. Furthermore, during the initial stages of training, the agents take actions at random, potentially endangering the robot's hardware. To circumvent these restrictions, researchers usually first run training sessions on realistic simulators, such as Gazebo [11], and OpenRAVE [4], where the simulation can run at accelerated speeds and with no danger, only later to transfer the trained agents on physical robots. However, this poses additional challenges [5], due to the fact that simulated environments provide a varying degree of realism, so it is not always possible for the agent to observe and act exactly as it did during training in the real world. This led to the development of more realistic simulators, which further reduce the gap between the simulation and the real world, such as Webots [15], and Actin [1]. It is worth noting that these simulators not only simulate the physical properties of an environment and provide a photorealistic representation of the world, but also provide an easily parameterizable environment, which can be adjusted according to the needs of every different real life scenarios.

Even though the aforementioned simulators provide powerful tools for developing and validating various robotics applications, it is not straightforward to use them for developing DRL methods, which typically operate over a higher level of abstraction that hides low-level details, such as how the actual control commands are processed by the robots. This limits their usefulness for developing DRL methods, since their steep learning curve and the enormous amount of development required to interface with DRL methods, considerably restricts their use by DRL researchers.

The main contribution of this work is to provide an open-source framework that can overcome the aforementioned limitations, supplying a DRL interface that is easy for the DRL research community to use. More specifically, the developed framework, called "deepbots", combines the well known OpenAI Gym [3] interface with the Webots simulator in order to establish a standard way to employ DRL in real case robotics scenarios. Deepbots aims to enable researchers to use RL in Webots and it has been created mostly for educational and research purposes. In essence, deepbots acts as a middle-ware between Webots and the DRL algorithms, exposing a Gym style interface with multiple levels of abstraction. The framework uses design patterns to achieve high code readability and reusability, allowing to easily incorporate it in most research pipelines. The aforementioned features come as an easy-to-install Python package that allows developers to efficiently implement environments that can be utilized by researchers or students to use their algorithms in realistic benchmarking. At the same time, deepbots provides ready-to-use standardized environments for well-known problems. Finally, the developed framework provides some extra tools for monitoring, e.g., tensorboard logging and plotting, allowing to directly observe the training progress. Deepbots is available at https://github.com/aidudezzz/deepbots.

The paper is structured as follows. First, Sect. 2 provides a brief overview of existing tools and simulators that are typically used for training DRL algorithms and highlights the need for providing a standardized DRL framework over the simulators to lower the barrier for accessing these tools by DRL researchers. Then, a detailed description of deepbots is provided in Sect. 3, while a set of already implemented examples, along with results achieved by well-established baseline RL algorithms, are provided in Sect. 4. Finally, Sect. 5 concludes this paper.

2   Related Work

First, the well-established OpenAI Gym toolkit, as well as the Webots simulator, are briefly introduced. Then, a number of related frameworks are also discussed and compared to the proposed deepbots framework.

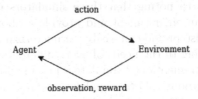

Fig. 1. OpenAI Gym interface

The OpenAI Gym, or simply "Gym", framework has been established as the standard interface between the actual simulator and RL algorithm [3]. According to the OpenAI Gym documentation, the framework follows the classic "agent-environment loop", as shown in Fig. 1, and it defines a set of environments. An environment for each step receives an action from the agent and returns a new observation and reward for the action. This procedure is repeated and separated in an episodic format. Except that, Gym standardizes a collection of testing environments for RL benchmarking. Even though OpenAI Gym is an easy-to-use tool to demonstrate the capabilities of RL in practice, it comes only with toy, unrealistic and difficult to extend scenarios. It needs several external dependencies to build more complex environments, like the MuJoCo simulator [21], which is a proprietary piece of software, which barriers its use and ability to be extended.

As RL problems become more and more sophisticated, researchers have to come up with even more complicated simulations. Self-driving cars, multi-drone scenarios, and other tasks with many more degrees of freedom synthesize the new big picture of RL research. Consequently, that leads to the need of even more accurate and realistic simulations, such as Webots [16], which is a free and open-source 3D robot simulator. Webots provides customizable environments, the ability to create robots from scratch, as well as high fidelity simulations with realistic graphics and is also Robot Operating System (ROS) compliant. It comes preloaded with several well-known robots, e.g., E-puck [7], iCub [14], etc. Robots can be wheeled or legged and use actuators like robotic arms, etc. An array of sensors is also provided, e.g., lidars, radars, proximity sensors, light sensors, touch sensors, GPS, accelerometers, cameras, etc. These capabilities allow it to cover a wide range of different applications. Robots are programmed via controllers that can be written in several languages (C, C++, Python, Java and MATLAB). However, even though Webots can be used for DRL, it comes with a set of limitations. The mechanisms which are used to run the different scripts are not friendly for those with DRL background, requiring a significant development overhead for supporting DRL algorithms, while there is no standardization regarding the way DRL methods are developed and interface with Webots.

Note that there is also an increasing number of works that attempt to formalize and facilitate the usage of RL in robotic simulators. However, none of these works target the state-of-the-art Webots simulator. For example, Gym-Ignition [6] is a work which aims to expose an OpenAI Gym interface to create reproducible robot environments for RL research. The framework has been designed for the Gazebo simulator and provides interconnection to external third party software, multiple physics and rendering engines, distributed simulation capabilities and it is ROS compliant. Other than that, [22] extends the Gym interface with ROS compliance and it uses the Gazebo simulator as well. The latest version of this work [13] is compatible with ROS 2 and is extended and applied in more real world oriented examples. All of these works are limited by the low quality graphics provided by the Gazebo simulator, rendering them less useful for DRL algorithms that rely on visual input. Finally, Isaac Gym [9] is a powerful software development toolkit providing photorealistic rendering, parallelization and is packed as a unified framework for DRL and robotics. However, its closed source nature can render it difficult to use, especially on scenarios that deviate from its original use cases. To the best of our knowledge this is the first work which t provide a generic OpenAI Gym interface for Webots, standardizing the way DRL algorithms interface with Webots and provide easy access to a state-of-the-art simulator.

3   Deepbots

Deepbots follows the same agent-environment loop as the OpenAI Gym framework, with the only difference being that the agent, which is responsible for choosing an action, runs on the supervisor and the observations are acquired by the robot. This master-minion protocol is not problem-specific and thus has the advantage of generalization, due to the fact that it can be used in more than one examples. That makes it easier to construct various use cases and utilize them as benchmarks. In this way, the deepbots framework acts as a wrapper, meaning that it wraps up and hides certain operations from the users, so that they are able to focus on the DRL task, rather than handling all the technical simulator-specific details. At the same time, deepbots also enriches the training pipeline with live monitoring features, which helps researchers get early observations about the fundamental parts of the training process. All these features contribute into providing a powerful DRL-oriented abstraction over Webots, allowing researchers to quickly model different use cases and simulation environments, as well as employ them to develop sophisticated DRL algorithms.

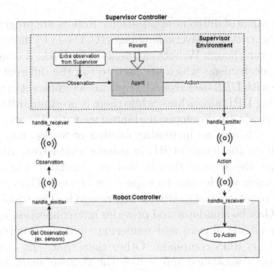

Fig. 2. Deepbots supervisor-controller communication

Before describing deepbots in detail, it is useful to briefly review the way Webots handles various simulation tasks. Webots represents scenes with a tree structure in which the root node is the world and its children nodes are the different items in the world. Consequently, a robot should be a node under the root node which contains a controller. Controllers are scripts responsible for the node's functionality. A robot for example has a controller in order to read values from sensors and act accordingly. For RL purposes, it is necessary to include a special supervisor node which has full knowledge of the world and it can get information for and modify every other node. For example, the supervisor can be used to move items in the world or measure distances between a robot and a target.

With respect to the aforementioned logic, deepbots gives the ability to easily construct DRL tasks with minimal effort. The basic structure of deepbots communication scheme is depicted in Fig. 2, where the supervisor controller is the script of the supervisor and the robot controller is the script of the robot. The communication between supervisor and robot is achieved via emitters/receivers, which broadcast and receive messages respectively. Without loss of generality, the supervisor is a node without mass or any physical properties in the simulation. For example, in a real case scenario, a supervisor could be a laptop which transmits actions to the robot, but without interacting with the actual scene. Furthermore, the emitter and the receiver could be any possible device, either cable or wireless, properly set up for this task.

Deepbots works as follows: first of all the simulator has to be reset in the initial state. On the one hand, the robot collects the first set of observations and by using its emitter sends the information to the supervisor node. On the other hand, the supervisor receives the observations with its receiver component and in turn passes them to the agent. In this step, if needed, the supervisor can augment the observation with extra information, e.g., Euclidean distances with respect to some ground truth objects, which are unavailable to the robot and its sensors. Except for the observation, the supervisor can pass the previous action, reward and/or any other state information to the agent. It should be mentioned that deepbots is not bound to any DL framework and the agent can be written in any Python-based DL framework, e.g., PyTorch, TensorFlow, Numpy, etc. After that, the agent produces an action which is going to be transmitted to the robot via the emitter. Finally, the robot has to perform the action which was received, with its actuators. The aforementioned procedure is performed iteratively until the robot achieves its objective or for a certain number of epochs/episodes, or whatever condition is needed by the use case.

```
1  class FindTargetSupervisor(SupervisorEmitterReceiver):
2      def get_observation(self):...
3      def get_reward(self, action):...
4      def is_done(self):...
5      def reset(self):...
6      def step(self, action):...
7
8      env = FindTargetSupervisor()
9      env = TensorboardLogger(env)
10     agent = DDPG(...)
11     for i in range(EPOCHS):
12         done = False
13         score = 0
14         obs = env.reset()
15         while not done:
16             act = agent.choose_action(obs)
17             obs, reward, done, info = env.step(act)
18             agent.remember(obs, action, reward, done)
19             agent.learn()
20             score += reward
```

Code Example 1.1: Supervisor controller code example

In order to implement an agent, the user has to implement two scripts at each side of the communication channel and the framework handles the details. On the supervisor side, the user has to create a Gym environment with the well known abstract methods and train/evaluate the DRL model, as shown in Code Example 1.1. While on the other side, a simple script has to be written for reading values from sensors and translating messages to the actual values needed by the actuators. A typical script for this task is shown in Code Example 1.2. The deepbots framework runs all the essential operations needed by the simulation, executes the step function and handles the communication between the supervisor and the robot controller in the background. In addition, by following the framework workflow, cleaner code is achieved, while the agent logic is separated from the actuator manipulation and it is closer to the physical cases. Furthermore, the framework logic is similar to ROS and can be integrated with it with minimal effort.

```
1  class FindTargetRobot(RobotEmitterReceiver):
2    def create_message(self):
3      message = []
4        for rangefinder in self.rangefinders:
5          message.append(rangefinder.value())
6      return message
7    def use_message_data(self, message):
8      gas = float(message[1])
9      wheel = float(message[0])
10     ...
11 robot_controller = FindTargetRobot()
12 robot_controller.run()
```

Code Example 1.2: Robot controller code example

As the deepbots framework mostly aims to be a user-friendly tool for educational and research purposes, it has different levels of abstraction. An overview of the abstraction level class diagram of deepbots is provided in Fig. 3. For example, users can choose if they would use JSON emitters and receivers or if they want to go on with an implementation from scratch. At the top of the abstraction hierarchy is the *SupervisorEnv*, which is the OpenAI Gym abstract class. Below that level is the actual implementation which resolves the communication between the supervisor and the robot. Similarly, the robot has also different levels of abstraction. A user can choose among certain types of message formats to transmit actions and observations. Extra features can be added to the framework as decorator classes by implementing the OpenAI Gym interface, as demonstrated in line 9 of Code Example 1.1. This design pattern could be used to stack different controls, monitoring and other functionalities.

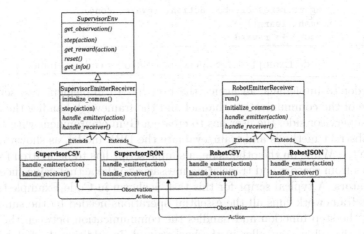

Fig. 3. Abstraction level class diagram

4   Example Environments

Deepbots contains a collection of ready-to-use environments, which showcase uses of the framework in toy or complicated examples. On the one hand, the community can contribute new environments and use cases to enrich the existing collection. On the other hand, this collection can be used by researchers to benchmark RL algorithms in Webots. Three environments of varying complexity are presented in this Section.

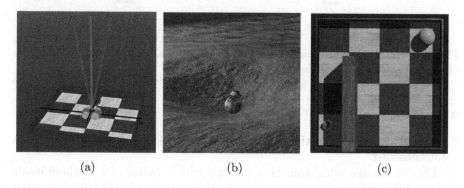

(a)   (b)   (c)

Fig. 4. (a) CartPole: the x axis is the cart motion axis, the y axis is the pole rotation axis, (b) PitEscape: the robot inside the pit, (c) FindBall: the robot searching for the yellow ball (Color figure online)

4.1   CartPole

The CartPole example is based on the problem described in [2] and adapted to Webots. In the world exists an arena, and a small four wheeled cart that has a long pole connected to it by a free hinge, as shown in Fig. 4. The hinge contains a sensor to measure the angle the pole has off vertical. The pole acts as an inverted pendulum and the goal is to keep it vertical by moving the cart forward and backward. This task is tackled with the discrete version of the Proximal Policy Optimization (PPO) RL algorithm [19].

The observation contains the cart position and velocity on the x axis, the pole angle off vertical and the pole velocity at its tip. The action space is discrete containing two possible actions for each time step, move forward or move backward. For every step taken, the agent is rewarded with +1 including the termination step. Each episode is terminated after a maximum 200 steps or earlier if the pole has fallen ±15° off vertical or the cart has moved more than ±0.39 m on the x axis. To consider the task solved, the agent has to achieve an average score of over 195.0 in 100 consecutive episodes.

The learning curve using the PPO algorithm, as well as the average action probability over the training process are depicted in Fig. 5. The actor and critic consist of small two-layered neural networks with 10 ReLU neurons on each layer and the agent was able to solve the problem after running for a simulated time of about 2.5 h.

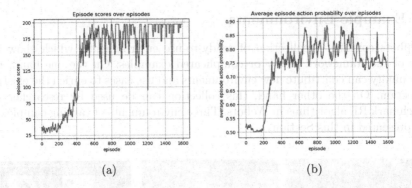

(a)   (b)

Fig. 5. CartPole: (a) reward accumulated for each episode, and (b) average probability of selected actions per episode

4.2 Pit Escape

The Pit Escape example that can be seen in Fig. 4 is a problem taken from *robotbenchmark*[1].

The Pit Escape world comprises of a pit with a radius of 2.7, where inside it lies a BB-8 robot with a spherical body that acts as a wheel [16]. The robot contains a pitch and a yaw motor and can move by rolling forward or backward (pitch) or by turning left and right (yaw). The task is to escape from the pit, but its motors are not powerful enough to escape by just moving forward, so it needs to move in some other way. This task is also tackled with the discrete version of the PPO RL algorithm [19].

This problem is very similar to the Mountain Car one [18], but in three dimensions and has more complex observation and action spaces. The robot contains a gyroscope and an accelerometer which provide the observation. Thus, the observation contains the robot orientation and acceleration in the x, y, z axes, i.e., a total of 6 values. The action space is discrete containing 4 possible actions for each time step. With each action the robot can set its motor speeds to their maximum or minimum values. Each episode lasts 60 s and the reward function is based on a metric M:

$$M = \begin{cases} 0.5\frac{d}{R} & d < R \\ 0.5 + 0.5\frac{T-t}{T} & d > R \end{cases}, \tag{1}$$

where d is the maximum distance achieved from the center of the pit until now in the episode, R is the radius of the pit, T is the maximum time allowed per episode (60 s), and t is the time until now in the episode. M only changes when a higher distance from the center is achieved during the episode. For each time step, based on the change between the previous step and current step metrics, the reward R_i for step i is calculated as $R_i = M_{old} - M$, where M_{old} is the previous step metric and M is the current step metric. An episode terminates

---

[1] Robotbenchmark, https://robotbenchmark.net.

after 60 s or if the robot has escaped the pit, which is calculated by the distance between the robot and the pit center.

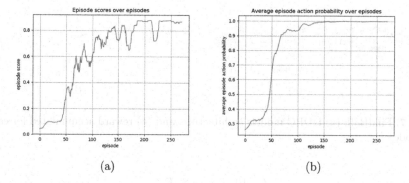

Fig. 6. Pit Escape: (a) reward accumulated for each episode, and (b) average probability of selected actions per episode

Two-layered networks with 60 ReLU neurons on each layer were used for the actor and critic models, while the learning curves are provided in Fig. 6. The agent achieved an average episode score of over 0.8 after training for a simulated time of about 3 h.

4.3   Find the Ball and Avoid Obstacles

The last example is a typical find target and avoid obstacles task with a simple world configuration. For this task the E-puck robot is used [7], which is a compact mobile robot developed by GCtronic and EPFL and is included in Webots. The world configuration contains an obstacle and a target ball. Different world configurations with incremental difficulty have been used in the training sessions for better generalization. It has been observed that the convergence of training algorithms can be improved by incrementing the difficulty of the problems [10]. The E-puck robot uses 8 IR proximity distance sensors and it has two motors for moving. The agent, apart from the distance sensor values, also receives the Euclidean distance and angle from the target. Consequently, the observation the agent gets is an one-dimensional vector with 10 values. On the other hand, the actuators are motors, which means that the outputs of the agent are two values controlling the forward/backward movement and left/right turning respectively (referred to as gas and wheel).

In order to deal with the continuous action space problem, the Deep Deterministic Policy Gradient (DDPG) algorithm was used to tackle this task [12]. The architecture of the models is described in Fig. 7. The reward function used for training the agent is calculated as:

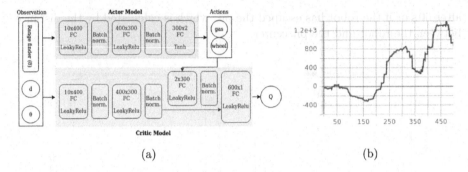

(a)   (b)

Fig. 7. FindBall: (a) DDPG models architecture, and (b) reward accumulated for each episode

$$R = \begin{cases} -10 & s > T_{steps} \\ +10 & d < T_{distance} \\ -1 & crashed \\ \frac{1}{d} - \frac{T_{steps}}{s} & otherwise \end{cases}, \qquad (2)$$

where s is the current step, T_{steps} the maximum allowed steps, d the current distance from target and $T_{distance}$ the minimum distance between the robot and the target which is considered as reaching the goal. This reward function takes into account both the distance from the target and the number of steps elapsed, while when the robot crashes on an obstacle or does not find the target after certain steps, it provides a negative reward and the episode is terminated.

The agent has been trained for 500 episodes and the accumulated reward is presented in Fig. 7. The training session lasted for about 1 h of wall clock time and about 3 h of simulated time. Although the agent solved the problem, it fails to generalize in more complicated scenes, highlighting the challenging nature of this baseline, that can be used for benchmarking future DRL algorithms.

5   Conclusions

Even though there have been attempts to formalize the use of RL in robotic simulators, none of them targets the state-of-the-art simulator Webots. The deepbots framework comes to fill that gap for anyone who wants to apply RL and DRL in a high fidelity simulator. Deepbots provides a standardized way to apply RL on Webots, by focusing only on parts that are important for the task at hand. Deepbots can fit a high variety of use cases, both research and educational, and can be extended by the community due to its open-source nature. Together with Webots, it provides a test bed for algorithm research and task solving with RL, as well as a practical platform for students to experiment with and learn about RL and robotics.

Acknowledgments. This project has received funding from the European Union's Horizon 2020 research and innovation programme under grant agreement No 871449 (OpenDR). This publication reflects the authors' views only. The European Commission is not responsible for any use that may be made of the information it contains.

References

1. Actin. https://www.energid.com/actin. Industrial-Grade Software for Advanced Robotic Solutions
2. Barto, A., Sutton, R., Anderson, C.: Neuron like elements that can solve difficult learning control problems. IEEE Trans. Syst. Man Cybern. **13**, 834–846 (1970)
3. Brockman, G., et al.: OpenAI Gym (2016)
4. Diankov, R.: Automated construction of robotic manipulation programs. Ph.D. thesis, Carnegie Mellon University, Robotics Institute, August 2010. http://www.programmingvision.com/rosen_diankov_thesis.pdf
5. Dulac-Arnold, G., Mankowitz, D.J., Hester, T.: Challenges of real-world reinforcement learning. CoRR abs/1904.12901 (2019). http://arxiv.org/abs/1904.12901
6. Ferigo, D., Traversaro, S., Metta, G., Pucci, D.: Gym-ignition: reproducible robotic simulations for reinforcement learning (2019)
7. Gonçalves, P., et al.: The e-puck, a robot designed for education in engineering. In: Proceedings of the Conference on Autonomous Robot Systems and Competitions, vol. 1, January 2009
8. Goodfellow, I.G., Bengio, Y., Courville, A.C.: Deep learning. Nature **521**, 436–444 (2015)
9. Gym, I.: https://www.nvidia.com/en-in/deep-learning-ai/industries/robotics
10. Hacohen, G., Weinshall, D.: On the power of curriculum learning in training deep networks (2019)
11. Koenig, N., Howard, A.: Design and use paradigms for gazebo, an open-source multi-robot simulator. In: Proceedings of the 2004 IEEE/RSJ International Conference on Intelligent Robots and Systems, vol. 3, pp. 2149–2154 (2004)
12. Lillicrap, T.P., et al.: Continuous control with deep reinforcement learning (2015)
13. Lopez, N.G., et al.: Gym-gazebo2, a toolkit for reinforcement learning using ROS 2 and Gazebo (2019)
14. Metta, G., Sandini, G., Vernon, D., Natale, L., Nori, F.: The iCub humanoid robot: an open platform for research in embodied cognition. In: Performance Metrics for Intelligent Systems (PerMIS) Workshop, January 2008
15. Michel, O.: Webots: professional mobile robot simulation. J. Adv. Robot. Syst. **1**(1), 39–42 (2004)
16. Michel, O.: WebotsTM: professional mobile robot simulation. Int. J. Adv. Robot. Syst. **1**, 40–43 (2004)
17. Mnih, V., et al.: Playing Atari with deep reinforcement learning (2013)
18. Moore, A.: Efficient memory-based learning for robot control, June 2002
19. Schulman, J., Wolski, F., Dhariwal, P., Radford, A., Klimov, O.: Proximal policy optimization algorithms (2017)
20. Steckelmacher, D., Plisnier, H., Roijers, D.M., Nowé, A.: Sample-efficient model-free reinforcement learning with off-policy critics (2019)
21. Todorov, E., Erez, T., Tassa, Y.: MuJoCo: a physics engine for model-based control. pp. 5026–5033, October 2012
22. Zamora, I., Lopez, N.G., Vilches, V.M., Cordero, A.H.: Extending the OpenAI Gym for robotics: a toolkit for reinforcement learning using ROS and Gazebo (2016)

Innovative Deep Neural Network Fusion for Pairwise Translation Evaluation

Despoina Mouratidis[1]($\boxtimes$) (ID), Katia Lida Kermanidis[1] (ID), and Vilelmini Sosoni[2] (ID)

[1] Department of Informatics, Ionian University, Tsirigoti Square 7, 49100 Corfu, Greece
{c12mour,kerman}@ionio.gr
[2] Department of Foreign Languages, Translation and Interpreting, Ionian University, Tsirigoti Square 7, 49100 Corfu, Greece
sosoni@ionio.gr

Abstract. A language independent deep learning (DL) architecture for machine translation (MT) evaluation is presented. This DL architecture aims at the best choice between two MT (*S1, S2*) outputs, based on the reference translation (*Sr*) and the annotation score. The outputs were generated from a statistical machine translation (SMT) system and a neural machine translation (NMT) system. The model applied in two language pairs: English - Greek (EN-EL) and English - Italian (EN-IT). In this paper, a variety of experiments with different parameter configurations is presented. Moreover, linguistic features, embeddings representation and natural language processing (NLP) metrics (BLEU, METEOR, TER, WER) were tested. The best score was achieved when the proposed model used source segments (*SSE*) information and the NLP metrics set. Classification accuracy has increased up to 5% (compared to previous related work) and reached quite satisfactory results for the Kendall τ score.

Keywords: Machine learning · Machine translation evaluation · Deep learning · Neural network architecture · Pairwise classification

1 Introduction

Deep neural networks are demonstrating a large impact on NLP. NMT [2, 14, 26, 28], in particular, has gained increasing popularity since it has shown remarkable results in several tasks and its effective approach has had a strong influence on other related NLP tasks, such as dialogue generation [8].

The evaluation of MT systems is a vital field of research, both for determining the effectiveness of existing MT systems (evaluation of the classification performance) and for guiding the MT systems modeling. Progress in the field of MT relies on assessing the quality of a new system through systematic evaluation, such that the new system can be shown to perform better than pre-existing systems. The difficulty arises in the definition of a better system. When assessing the quality of a translation, there is no single correct answer; rather, there may be any number of possible correct translations. In addition,

© IFIP International Federation for Information Processing 2020
Published by Springer Nature Switzerland AG 2020
I. Maglogiannis et al. (Eds.): AIAI 2020, IFIP AICT 584, pp. 76–87, 2020.
https://doi.org/10.1007/978-3-030-49186-4_7

when two translations are only partially correct -but in different ways- it is difficult to distinguish quality.

Many methods for MT evaluation have been employed. There are metrics that focus on the MT output evaluation, such as BLEU [18], METEOR [4], TER [24] and WER [25]. BLEU score is maybe the most famous and widely-used metric in MT evaluation. The closer an MT output is to the professional translation, the higher the BLEU score is. The BLEU score suffers from several shortcomings i.e. it doesn't handle morphologically rich languages well and it doesn't map well to human judgements. Several other metrics, that address these issues, are used, such as METEOR. The METEOR score has a good correlation with human judgement at the segment level. It is based on the alignment between the MT outputs and the professional translation. Alignments are based on synonym and paraphrase matches between words and phrases. The translation error rate (TER) and word error rate (WER) are other commonly-used metrics. They are based on the matching of the MT outputs with the professional translation. They measure the minimum number of edits needed to change the original output translation into the professional translation. Other metrics focus on performance evaluation. In some studies [15, 17], parallel corpora are used and showed that certain string-based features, e.g. the length of the segments, and similarity-based features e.g. the ratio of common suffixes shared between the MT outputs and the reference, could improve the MT system performance. They considered the task as a classification problem and they used Random Forest (RF) as classifier.

NMT can potentially perform end-to-end translation, though many NMT systems are still relying on language-dependent pre- and post-processors, which have been used in traditional SMT systems. Moses [11], a toolkit for SMT, implements a reasonably useful pre- and post-processor. A language dependent processing also makes it hard to train multilingual NMT models.

It is important for the NLP community to develop a simple, efficient and language independent framework for automatic MT evaluation. A few studies have been reported using learning frameworks. Duh [5] uses a framework for ranking translations in parallel settings, given information of translation outputs and a reference translation. This study showed that ranking achieves higher correlation to human judgments when the framework makes use of a ranking specific feature set and of BLEU score information. They have tested the framework performance using Support Vector Machine (SVM). Another important work is presented by [7] who used syntactic and semantic information about the reference and the machine-generated translation as well, by using pre-trained embeddings and the BLEU translations scores. They used a feedforward neural network (NN) to decide which of the MT outputs is better. A learning scheme to classify machine-generated translations using information from numerous linguistic features and hand-crafted word embeddings from two MT outputs and one reference translation is presented from [16]. They used a convolutional NN to choose the right translation among two provided.

In this paper, we introduce a learning schema, for evaluating MT, similar to that of a preliminary study [16], but we extend it to a new level, both in terms of number of feature and their representation and learning framework as well.

Compared to that study, the present approach includes the following novelties:

- the utilization of a deeper NN architecture. More hidden layers and different types were tested (Dense and LSTM layers).
- the inclusion of an NLP metric set (BLEU score, METEOR score, TER, WER).
- the use of the linguistic information from the *SSE* in EN. 18 string-based features were calculated and used as an extra input to the DL architecture.
- the accuracy exploration of different inputs to the hidden layers (the NLP set and the string-based features).

To the best of the authors' knowledge, this is the first time that information of the *SSE* combined with handcrafted features, embeddings and a set of NLP metrics are used from a DL architecture for a classification task.

2 Materials and Methods

The current section presents the corpora, the features and NLP set as well as the DL architecture used in the experiments.

2.1 Dataset

The dataset used in the experiments consists of parallel corpora in the language pairs EN-EL and EN-IT. The dataset is part of the test sets developed in the TraMOOC project [12]. They are educational corpora from video lectures and they contain mathematical expressions, URLs and many special characters, such as /, @, #. The corpora are described in detail by [15, 17]. The EN-EL corpora consists of 2686 segments and the EN-IT consist of 2745 segments. Two MT outputs were used - one generated by the Moses SMT toolkit [11] and the other generated by the NMT Nematus toolkit [22]. Both models trained on in- and out-of-domain data. In- and out-of-domain data included widely known corpora e.g. TED, OPUS. In order to improve the classification, a professional translation is provided for every segment. More details on the training datasets can be found in [27].

2.2 The Feature Set Used

The feature set used is based on linguistics features divided in three categories: i) string similarity features, such as ratios between words of *S1*, *S2* and *Sr*, word distances (e.g. Dice distance [20]), percentage of segments similarity, ii) features finding the percentage of the noise in the data set (e.g. repeated words) and iii) features using length factor (LF) [21]. More details on the feature set used can be found in [17]. In this work, in order to check if the information from *SSE* will help the accuracy, additional features from the *SSE* in the EN language are used. Based on the other features, it is observed that features containing ratios are more effective to the classifier. These features are: 1) the words and character length of the *SSE*, 2) the ratio between these lengths in the *SSE* and the two MT outputs, 3) the longest word length, 4) the ratio between longest words from *SSE* and the two MT outputs and *Sr* translation.

2.3 Word Embeddings

The use of word embeddings helped us to model the relations between the two translations and the reference. In these experiments, hand-crafted embeddings were used,
for the two MT outputs and the reference translation as well for both language pairs.
The encoding function used is the one-hot function. The size, in number of nodes, of
the embedding layer is 64 for both languages. The input dimensions of the embedding
layers are in agreement with the vocabulary of each language (taking into account the
most frequent words): 400 for the EN-EL language pair and 200 for the EN-IT language
pair. The embedding layer used is the one provided by Keras [10] with TensorFlow as
backend [1].

2.4 The NLP Metrics Used

The NLP set used in these experiments contains the BLEU score, METEOR, TER and
WER. To calculate the BLEU score, an implementation of the BLEU score from the
Python Natural Language Toolkit library [13] is used. For the calculation of the other
three metrics, the code from GitHub [6] is used. All metrics were calculated for ($S1$, $S2$),
($S1$, Sr), ($S2$, Sr).

2.5 The DL Schema

This study approaches the MT evaluation problem as a classification task. In particular,
two volunteer linguists-annotators chose the better MT output. The linguists annotate
the corpora as follows: $Y = 0$ if $S1$ is better than $S2$, and $Y = 1$ if $S2$ is better than $S1$
for both language pairs. Where Y is the output, i.e. the label of the classification class.
This information is used as the 'ground truth'. As an input to the learning schema, the
vectors ($S1$, $S2$, Sr) were used, in a parallel setting. The embedding layer (as described
in Sect. 2.3) is applied and the respective embeddings $EmbS1$, $EmbS2$ and $EmbSr$ were
created. The embeddings $EmbS1$, $EmbS2$ and $EmbSr$ were contracted in a pairwise
setting, and the vectors ($EmbS1$, $EmbS2$), ($EmbS1$, $EmbSr$) and ($EmbS2$, $EmbSr$) were
created. These vectors are the input to the hidden layers $h12$, $h1r$, $h2r$ respectively. Using
hidden layers $h1r$ and $h2r$, the similarity between the two MT outputs and the professional
translation (Sr) is explored. It is important to investigate the similarity between $S1$ and
$S2$, so an extra hidden layer $h12$ is added. Interestingly, it is often observed that the MT
outputs were more similar to each other than to the Sr. Every hidden layer $h12$, $h1r$, $h2r$,
got as an extra input 2D matrixes $H_{12}[i, j]$, $H_{1r}[i, j]$, $H_{2r}[i, j]$, where i is the number of
segments and j is the number of features. These matrices contain information about (i)
the NLP set for $S1$-$S2$, $S1$-Sr, $S2$-Sr (as described in Sect. 2.4) or (ii) information about
linguistic features of the SSE, i.e. n-grams, or (iii) the combination of the previous two
options. The outputs of the hidden layers $h12$, $h1r$, $h2r$ are grouped and became the input
to the last layer of the NN model. An extra 2D $A[i, j]$ matrix with hand-crafted features
(string-based) (as described in Sect. 2.2) was added to this last layer.

The model of the DL architecture is shown in Fig. 1.

A suitable function to describe the input-output relationship in the training data
should be selected. The output label is modeled as a random variable in order to minimize

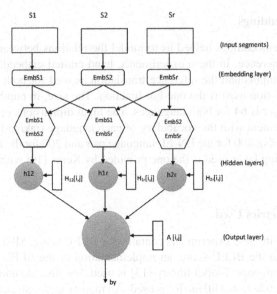

Fig. 1. Proposed model architecture

the discrepancy between the predicted and the true labels – maximum likelihood estimation. The binary classification problem is modeled as a Bernoulli distribution (Eq. 1)

$$Y \sim \text{Bernoulli}(Y/b_y) \qquad (1)$$

Where b_y is the sigmoid function $\sigma(w^T x + b)$, w^T and b are network's parameters.

Finally, the MaxAbsScaler [19] is used, as a preprocessing method for *EmbS1*, *EmbS2*, *EmbSr* and matrices $H_{12}[i, j]$, $H_{1r}[i, j]$, $H_{2r}[i, j]$, $A[i, j]$ as well. Every feature is scaled by its maximum absolute value.

3   Experimental Setup and Results

This section describes the details about experiments and its results.

3.1   Network Parameters

After experimentation, in order to test the proposed DL architecture, the model architecture for the experiments is defined as follows (Table 1).

3.2   Evaluation Scores

There are many machine learning evaluation metrics. In this study, commonly used metrics in classification (precision, recall and F-score) were used for the model performance evaluation. The first score (precision) shows the number of the correctly predictive values, the second score (recall) shows the percentage of total results correctly classified

Table 1. Model parameters

	Proposed NN	+NLP	+SSE	+NLP+SSE
Number of LSTM layers/Hidden units	2/100	2/400	2/800	2/400
Dropout of LSTM layers	0.2	0.7	0.7	0.7
Size of dense layers/Hidden units	3/50	3/50, 1/400	3/50, 1/800	3/50, 1/400
Dropout of dense 4 layer	0.2	0.7	0.7	0.7
Activation function of dense layers	Relu	Relu, linear	Relu, linear	Relu, linear
Output layer	Activation sigmoid			
Learning rate	0.01	0.01	0.01	0.05
Activation function of dense layers	Softmax			
Loss function	Binary cross entropy			
Optimizer	Adam			
Batch size	256	128	64	64
Epochs	10	10	6	20

by the model. However, because of the unbalanced precision and recall, F-score (F1), which is a harmonic mean of precision and recall, is used. It is important to analyze the relationship between the MT outputs and the human translation, using a statistic metric - Kendall τ [9]. It is a non-parametric test used to measure the ordinal association between the two MT outputs. Kendall τ is calculated for every language pair and the macro average across all language pairs.

3.3 Results

The main results of the experiments are shown in Table 2. Different experiments were tested in the same DL architecture - using different information. The NLP set gave 67% accuracy for EN-EL and 60% for EN-IT. Subsequently, the goal was to verify if the *SSE* information can improve the model accuracy. Indeed, an increase of 2% of the classification accuracy for EN-EL and EN-IT is observed. Better accuracy results are reported when the proposed NN model uses both the information from the NLP set and *SSE* (72% accuracy for EN-EL/70% for EN-IT). It's quite interesting that when the proposed NN model is used, without using any extra information in the hidden layers, it correctly classifies all the instances for the NMT class. Nevertheless, this model cannot be considered as the best, because the number of the correctly classified instances for the SMT class was low. The 2D matrixes $H12[i, j]$, $H1r[i, j]$, $H2r[i, j]$ utilization in every hidden layer $h12$, $h1r$, $h2r$ gave balance between the correct instances.

It is important to have balance of accuracy performance for both classes, so the F1 score is used. In order to make a direct comparison with other models [3, 23], additional experiments were run, using, for some of them, the WEKA framework as backend [22] for the SVM and RF classifiers. It is observed that the proposed model achieves a better F1 score 4% compared with the RF, and 5% with SVM (Fig. 2).

Table 2. Accuracy percentage for SMT and NMT for both languages pairs.

Model	Precision		Recall		Precision		Recall	
Language pair	EN-EL				EN-IT			
	SMT	NMT	SMT	NMT	SMT	NMT	SMT	NMT
Proposed NN + NLP set	58%	69%	40%	83%	50%	65%	40%	80%
Proposed NN + SSE	69%	74%	44%	90%	55%	68%	42%	84%
Proposed NN + NLP set + SSE	68%	75%	50%	92%	62%	70%	44%	87%

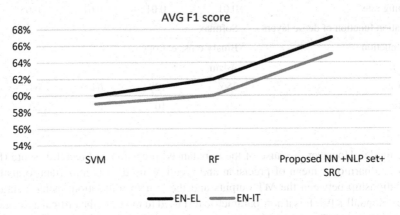

Fig. 2. Average F1 comparison between the proposed model and other works.

Table 3 shows the Kendall τ results for different models. Firstly, Kendall τ is presented for four commonly used metrics in MT evaluation (NLP set), comparing the MT outputs $S1$, $S2$ with the reference Sr. These metrics achieved Kendall τ between 14–20. However, when they were used as extra input to the hidden layers, they led to significant improvements. In Table 2, Kendall τ values are presented for the model using different configuration setups. The NN itself achieves lower τ value compared to the other NN architectures, something which should not be surprising because this architecture does not use any further linguistic information. The NLP set utilization in the NN gets Kendall τ average (AVG) for both languages 27 points. This is because NLP metrics contain significant linguistic information about the languages (i.e. similarity scores, length). An increase up to 2.5 points is observed using information about the SSE (in English). Moreover, the Kendall τ reaches its highest value when both the NLP set and SSE information were applied (36 for EL/32 for IT).

Table 3. Kendall τ for every language pair and their average.

System	EL	IT	AVG
NLP metrics set			
BLEU	17	14	15.5
METEOR	20	18	19
WER	18	16	17
TER	19	17	18
DL architecture			
Proposed NN + NLP set	29	25	27
Proposed NN +SSE	31	28	29.5
Proposed NN + NLP set + NLP set + SSE	36	32	34

3.4 Linguistic Analysis

Linguistic analysis helps us to understand better the reasons why the MT output that belongs to NMT class yields higher accuracy and Kendall τ scores in both languages pairs. In Table 4, two cases are presented in the EN-EL language pair that the model didn't classify correctly.

ID 1:

- In this segment, *S2* made two serious mistakes. In the literal sense, the compound word *bandwagon* is a *wagon used for carrying a band in a parade or procession*. As a metaphor, the word *bandwagon* is used for *an activity, cause, that is currently fashionable or popular and attracting increasing support*. *S2* "didn't know" the metaphorical meaning of the word, so it has erroneously translated only the second part of the compound word in question: *wagon as* άμαξα (carriage, coach). Moreover, it is surprising that *S2* didn't even translate the first part of that compound word (*band*).
- *S2* has the phrase *gut feeling*. Gut feeling is an idiom, meaning *an instinct or intuition, an immediate or basic feeling or reaction without a logical rationale*. *S2* has literally translated the phrase: το ένστικτο του εντέρου (!) (the instinct of the gut). Even though in English there is also the idiom *gut instinct*, as a synonym of *gut feeling*, in Greek the literal translation of *gut instinct* is non-sensical.
- Finally, *S2* also made a slight mistake. It erroneously translated the adverb phrase *by habit* (habitually) literally: από τη συνήθεια (from the habit).
- *S1* has erroneously translated the above adverbial phrase *by bandwagon* as με ρεύμα, being unclear as to the precise meaning of the word ρεύμα, as in Greek this is a polysemous term that may refer to: electricity, drift, current, stream. With the preposition με, the Greek version is closer to the first meaning: with electricity (!), but this is nonsensical.

Table 4. Examples of EN-EL segments.

ID	SSE	S1	S2	Sr
1	Decisions are often taken by habit, by bandwagon (everybody's doing it, so it must be right), by gut feeling	Οι αποφάσεις λαμβάνονται συχνά από συνήθεια, με ρεύμα (όλοι το κάνουν, οπότε πρέπει να είναι σωστό), από ένστικτο	Οι αποφάσεις συχνά λαμβάνονται από τη συνήθεια, με την άμαξα (όλοι το κάνουν, άρα πρέπει να είναι σωστό), με το ένστικτο του εντέρου	Οι αποφάσεις παίρνονται συνήθως λόγω συνήθειας, λόγω μαζικής τάσης (όλοι το κάνουν, άρα πρέπει να είναι σωστό), λόγω καλού προαισθήματος
2	According to Robert Pratten, what is the difference between franchise transmedia and portmanteau transmedia?	Σύμφωνα με τον Robert Pratten, ποια είναι η διαφορά μεταξύ transmedia franchise και σύμμειξη transmedia	Σύμφωνα με τον Ρόμπερτ Πράτεν, ποια είναι η διαφορά μεταξύ των τρανζίστορ και των τρανζίσον	Σύμφωνα με τον Robert Pratten, ποια είναι η διαφορά μεταξύ μεθοδολογίας franchise transmedia και μεθοδολογίας portmanteau transmedia

ID2:

- *S1* has not localized the proper noun *Robert Pratten* and rightly so, as this is the most common choice.
- *S1* did not at all translate the first of the two phrases: *francise transmedia* as well as the second word of the second phrase: *portmanteau transmedia*. *S1* has only translated the first word of this phrase: *portmanteau*, without, nevertheless, adopting the very common sense of the word: *bag, luggage, valise*, but a special and relatively rare one: σύμμειξη (*compounding, blending*). The professional linguist did not at all translate these phrases.
- On the contrary, *S2* translated the same phrases in a completely erroneous way: τρανζίστορ (*transistor*) and τρανζίσον (no meaning in Greek) respectively. *S2* translated these phrases incompletely and erroneously, obviously "misled" by the prefix: *–trans* of *transmedia*.
- Neither *S1* nor *S2* identified that *franchise transmedia* and *portmanteau transmedia* are methodologies (methods, techniques, approaches), as professional linguist (Reference) did.

4   Conclusion and Future Work

In this study, it is presented a DL architecture for classifying the best MT output between two options provided (one from an SMT model and the other from an NMT model), given a reference translation and an annotation schema, as well. It is worth mentioning that the translation was from EN to EL, and EN to IT which increased the task complexity, since the Greek and Italian languages are both morphologically rich languages. Well known NLP metrics were calculated and became extra inputs to the NN. Also, linguistics features from the *SSE* were used. The model's accuracy performance was tested in configurations. When the NN combines embeddings, the NLP set (BLEU, METEOR, TER, WER) and *SSE* information (i.e. some ratios) achieved better accuracy results (increase up to 5%) and a higher Kendall τ score (increase up to 4 points) compared to related work. A linguistic analysis is also provided in order to explain linguistically the above results.

In future work, it is important to study other aspects which are likely to improve the DL architecture accuracy, such as a) a different NN configuration (e.g. different kinds of NN layers, batch normalization, learning rate), b) a feature selection method to reject the features that aren't effective for the model and c) a feature importance method to apply the proper feature weights during the NN training. In addition, it worth exploring the reasons for which the proposed model presents low accuracy values in the EN-IT pair, even though it is language independent. Finally, the model will be tested with another dataset, including in- and out-of-domain data.

Acknowledgments. This project has received funding from the GSRT for the European Union's Horizon 2020 research and innovation program under grant agreement No 644333.

References

1. Abadi, M., et al.: Tensorflow: a system for large-scale machine learning. In: 12th {USENIX} Symposium on Operating Systems Design and Implementation ({OSDI} 2016), USA, pp. 265–283. USENIX Association (2016)
2. Bahdanau, D., Cho, K., Bengio, Y.: Neural machine translation by jointly learning to align and translate. In: Proceedings of 3th International Conference on Learning Representations, San Diego, pp. 1–15. ICLR (2015)
3. Barrón-Cedeño, A., Màrquez Villodre, L., Henríquez Quintana, C.A., Formiga Fanals, L., Romero Merino, E., May, J.: Identifying useful human correction feedback from an on-line machine translation service. In: Proceedings of 23rd International Joint Conference on Artificial Intelligence, Beijing, pp. 2057–2063. AAAI Press (2013)
4. Denkowski, M., Lavie, A.: Meteor universal: language specific translation evaluation for any target language. In: Proceedings of the 9th Workshop on Statistical Machine Translation, Baltimore, Maryland, USA, pp. 376–380. ACL (2014)
5. Duh, K.: Ranking vs. regression in machine translation evaluation. In: Proceedings of the 3rd Workshop on Statistical Machine Translation, Columbus, Ohio, pp. 191–194. ACL (2008)
6. GitHub. https://github.com/gcunhase/NLPMetrics. Accessed 20 Feb 2020
7. Guzmán, F., Joty, S., Màrquez, L., Nakov, P.: Pairwise neural machine translation evaluation. arXiv preprint arXiv:1912.03135. In: Proceedings of the 53rd Annual Meeting of the Association for Computational Linguistics and the 7th International Joint Conference on Natural Language Processing, Beijing, China, pp. 805–814. ACL (2015)

8. Jaitly, N., Sussillo, D., Le, Q.V., Vinyals, O., Sutskever, I., Bengio, S.: A neural transducer. Cornell University Library. arXiv preprint arXiv:1511.04868 (2015)

9. Kendall, M.: A new measure of rank correlation. Biometrika **30**(1/2), 81–93 (1938)

10. Keras: Deep learning library for theano and tensorflow. https://keras.io/k7.8. Accessed 20 Feb 2020

11. Koehn, P., et al.: Moses: open source toolkit for statistical machine translation. In: Proceedings of the 45th Annual Meeting of the ACL on Interactive Poster and Demonstration Sessions, Prague, pp. 177–180. ACL (2007)

12. Kordoni, V., et al.: TraMOOC (translation for massive open online courses): providing reliable MT for MOOCs. In: Proceedings of the 19th Annual Conference of the European Association for Machine Translation (EAMT), Riga, pp. 376–400. European Association for Machine Translation (EAMT) (2016)

13. Loper, E., Bird, S.: NLTK: the natural language toolkit. In: Proceedings of the ACL 2002 Workshop on Effective Tools and Methodologies for Teaching Natural Language Processing and Computational Linguistics, USA, pp. 63–70. ACL (2002)

14. Luong, M.T., Pham, H., Manning, C.D.: Effective approaches to attention-based neural machine translation. In: Proceedings of the 2015 Conference on Empirical Methods in Natural Language Processing, Lisbon, Portugal, pp. 1412–1421. ACL (2015)

15. Mouratidis, D., Kermanidis, K.L.: Automatic selection of parallel data for machine translation. In: Iliadis, L., Maglogiannis, I., Plagianakos, V. (eds.) AIAI 2018. IAICT, vol. 520, pp. 146–156. Springer, Cham (2018). https://doi.org/10.1007/978-3-319-92016-0_14

16. Mouratidis, D., Kermanidis, K.L.: Comparing a hand-crafted to an automatically generated feature set for deep learning: pairwise translation evaluation. In: 2nd Workshop on Human-Informed Translation and Interpreting Technology, Varna, Bulgaria, pp. 66–74. HiT-IT (2019)

17. Mouratidis, D., Kermanidis, K.L.: Ensemble and deep learning for language-independent automatic selection of parallel data. Algorithms **12**(1), 12–26 (2019)

18. Papineni, K., Roukos, S., Ward, T., Zhu, W.J.: BLEU: a method for automatic evaluation of machine translation. In: Proceedings of the 40th Annual Meeting on Association for Computational Linguistics, Philadelphia, pp. 311–318. Association for Computational Linguistics (2002)

19. Pedregosa, F., et al.: Scikit-learn: machine learning in python. J. Mach. Learn. Res. **12**, 2825–2830 (2011)

20. Peris, Á., Cebrián, L., Casacuberta, F.: Online learning for neural machine translation post-editing. Cornell University Library. arXiv preprint 1, pp. 1–12. arXiv:1706.03196 (2017)

21. Pouliquen, B., Steinberger, R., Ignat, C.: Automatic identification of document translations in large multilingual document collections. In: Proceedings of the International Conference Recent Advances in Natural Language Processing (RANLP), Borovets, pp. 401–408. Recent Advances in Natural Language Processing (RANLP) (2003)

22. Sennrich, R., et al.: Nematus: a toolkit for neural machine translation. In: Proceedings of the EACL 2017 Software Demonstrations, Valencia, pp. 65–68. ACL (2017)

23. Singhal, S., Jena, M.: A study on WEKA tool for data preprocessing, classification and clustering. Int. J. Innov. Technol. Explor. Eng. (IJITEE) **2**(6), 250–253 (2013)

24. Snover, M., Dorr, B., Schwartz, R., Micciulla, L., Makhoul, J.: A study of translation edit rate with targeted human annotation. In: Proceedings of the 7th Conference of the Association for Machine Translation in the Americas, Cambridge, pp. 223–231. The Association for Machine Translation in the Americas (2006)

25. Su, K.Y., Wu, M.W., Chang, J.S.: A new quantitative quality measure for machine translation systems. In: Proceedings of the 14th Conference on Computational Linguistics, Nantes, France, vol. 2, pp. 433–439. Association for Computational Linguistics (1992)

26. Vaswani, A., et al.: Attention is all you need. In: 31st Conference on Neural Information Processing Systems, Long Beach, CA, USA, pp. 5998–6008. NIPS (2017)

27. Sosoni, V., et al.: Translation crowdsourcing: creating a multilingual corpus of online educational content. In: Proceedings of the 11th International Conference on Language Resources and Evaluation, Japan, pp. 479–483. European Language Resources Association (2018)
28. Wu, Y., et al.: Google's neural machine translation system: bridging the gap between human and machine translation. arXiv preprint arXiv:1609.08144 (2016)

Introducing an Edge-Native Deep Learning Platform for Exergames

Antonis Pardos[1]🆔, Andreas Menychtas[1,2](✉)🆔, and Ilias Maglogiannis[2]🆔

[1] BioAssist SA, 1 G. Mpakou Street, 11524 Athens, Greece
{antonispardos,amenychtas}@bioassist.gr
[2] Department of Digital Systems, University of Pireas,
80 M. Karaoli & A. Dimitriou Street, 18534 Pireas, Greece
imaglo@unipi.gr

Abstract. The recent advancements in the areas of computer vision and deep learning with the development of convolutional neural networks and the profusion of highly accurate general purpose pre-trained models, create new opportunities for the interaction of humans with systems and facilitate the development of advanced features for all types of platforms and applications. Research, consumer and industrial applications increasingly integrate deep learning frameworks into their operational flow, and as a result of the availability of high performance hardware (Computer Boards, GPUs, TPUs) also for individual consumers and home use, this functionality has been moved closer to the end-users, at the edge of the network. In this work, we exploit the aforementioned approaches and tools for the development of an edge-native platform for exergames, which includes innovative gameplay and features for the users. A prototype game was created using the platform that was deployed in the real-world scenario of a rehabilitation center. The proposed approach provides advanced user experience based on the automated, real-time pose and gesture detection, and in parallel maintains low-cost to enable wide adoption in multiple applications across domains and usage scenarios.

Keywords: Deep learning · Edge computing · Exergames · Computer vision · Convolutional neural networks · Rehabilitation

1 Introduction

The computer games landscape is very rich nowadays and is continuously expanding with new approaches which are based on innovative technologies for human-computer interaction, provide advanced game-play, are available on multiple platforms and devices, and target different user groups. Therefore, the users can find in the market a variety of games, from the traditional video games, in which a player is seated in front of the screen and controls the game with a controller such as keyboard, mouse or gamepads, to the virtual and augmented reality games which require special equipment. In this work, we focus on the *Serious Games* [16] category and particularly the exergames, that has great scientific

© IFIP International Federation for Information Processing 2020
Published by Springer Nature Switzerland AG 2020
I. Maglogiannis et al. (Eds.): AIAI 2020, IFIP AICT 584, pp. 88–98, 2020.
https://doi.org/10.1007/978-3-030-49186-4_8

value and impact, and serves purposes beyond the game itself and the satisfaction of the player [15]. Active Video Games (AVG) or exergames are based on the technology that monitors body movement or reactions and have been credited with upgrading the game stereotype as a sedentary activity, promoting a more active lifestyle. Golstein et al. [6] showed that there was a significant improvement in the reaction time of 69–90 year old when they were playing video games for 5 h each week, for 5 consecutive weeks.

There several works in the literature highlighting the positive effects of exergames, both generally, regarding the increase of the physical activity and improvement of well-being, and also for specific use case scenarios, such deployments in schools, elder houses, rehabilitation centers, and more, where the benefits in the certain context are better measurable with the direct involvement of scientific personnel and experts [8,13]. In technical level, these exergames are usually based on the use of particular platforms such as Nintendo Wii, and require the use of peripherals that the players hold or wear and act as game controls. The controls (boards, sticks, patches) capture the characteristics of the user's movement and interpret it to events or gestures through which the game is controlled.

The recent technological advancements in the areas of computer vision and deep learning enable the real-time analysis of images and streaming video to solve several classification problems, from object detection and face recognition, to emotion analysis and pose detection, which in turn are applied to real-world scenarios for providing advanced interactivity with systems and creating rich user experiences. Particularly, the detection body pose or face landmarks is applied in several entertainment, medical and business use cases exploiting the increased computational capabilities of the available hardware nowadays and the advanced features of the computer vision and deep learning frameworks and models. In this work we propose an exergame platform which runs on low-cost commodity hardware and utilizes deep learning pose detection based on Convolutional Neural Networks - CNN for human-game interaction. Nowadays, the exegames are implemented using expensive, special purpose systems, which are designed and implemented specifically for the a particular game. The proposed solution, is a modular and extensible platform that is based on commodity hardware and uses generic purpose software and tools, not only allowing for the inexpensive implementation of several different games that exploit computer vision and deep learning on top of the platform, but also enabling its use and the adoption of the overall concept, in other fields and use cases where human-computer interaction is required.

The rest of the document is structured as follows. Section 2 analyses the related works and the technological and scientific baseline for the proposed platform. The platform of the overall system is presented in Sect. 3 along with the hardware elements, the software frameworks and the techniques that were deployed. Section 4 describes the use of the system in a real scenario as well as initial results of its operation. Finally, Sect. 5 concludes the manuscript and presents future extensions and improvements.

2   Related Work and Background Information

Studies have shown that exergames can offer significant benefits to the mental and physical health of people who play especially for the elderly, children, adolescents and people with disabilities [2,3,14], such as the significant decrease in BMI (Body Mass Index) in some cases. Games that enhance physical activity have been developed in the past. Some examples are: "Dance Dance Revolution" [17], introduced in Japan in 1998. Players standing on a "dance platform" or stage and hit colored arrows laid out in a cross with their feet to musical and visual cues. Players are judged by how well they time their dance to the patterns presented to them and are allowed to choose more music to play to if they receive a passing score. "Active Life Outdoor Challenge" [5], a video game for the Wii platform, where players use a mat in conjunction with the Wii remote in order to complete a variety of mini games. "The Think & Learn Smart Cycle" [1] is a stationary bike that hooks up to a tablet via Bluetooth for preschoolers to play different learning games. There are games about Letters & Phonics, Spelling & Vocabulary, and Reading & Rhyming. In addition to learning, this game also keeps them active because the faster they pedal, the faster the on-screen action. "Wii Fit" [5], It is an exercise game with several activities using the Wii Balance Board peripheral, a device that tracks the user's center of balance.

In order to be effective, a game of this type must always take into account the user himself. Children, adolescents, elderly or people with disabilities have different needs, which if properly defined will greatly contribute to the success of the game. Another key component to the success of these games after identifying users' needs, is that the game should entertain and stimulate the user's interest, while convincing him for more exercise. A game that captures the attention of the user is often the impulse for a longer training period, examples like a fixed bike on which a user controls the flow of a game while cycling, resulting in more power being consumed by another user who did the same without controlling a game. Graphics are also an important factor in the success of the game. Seniors need games with simple graphics, easy information and more time available for understanding processes [3]. On the other side are the younger users who are expecting games with rich graphics and speed. These games may be adapted in accordance with user's physical condition improvement, or user needs as mentioned above. Through this process significant benefits have been observed in restoring the development of children and young people (5–25 years old) with mobility problem [11].

Studies have shown that after a few weeks the frequency the user plays is decreased and as a result there is no significant improvement in the user's physical condition [14]. This is why it is prominent to be the right evaluation of the target group in order the main strategy and design of the game to be drawn. Another crucial factor that limits the scope of these games is the hardware that must be applied. Examples of the exergames we saw above make use of special peripherals and this makes the game more complicated in terms of both architecture and cost. However, there should be a balance in the length and frequency the user is involved with the game, and extra care must be taken to avoid overuse as there is possibility of injury [14]. To avoid such situations, clear instructions should be given to the user, as well as a game play program. A key factor in using these games is their design and their strategic goal. For example, some games require the user to carry on-screen objects with their hands, while others, such as the "Fish'n Steps" project encourage the community to walk around in order to grow a virtual aquarium fish. In all cases the player is evaluated according to the strategy of the game and the result he wants to achieve. For example he can be scored according to the accuracy, speed, or the "Quantity" of movement.

3   System Architecture and Implementation

The objective of this work is to deliver a stand-alone exergame solution, exploiting computer vision and deep learning techniques for the human - game interaction. The proposed system combines state-of-the-art hardware and software technologies and tools in order to deliver advanced exergames features and performance, without compromising the requirements for cost and user experience. An overview of the system is depicted in Fig. 1.

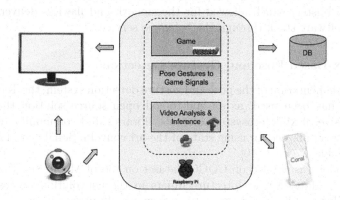

Fig. 1. Overview of the system.

The main components of the system include a) the core game implementation, b) the video analysis and inference mechanism, and c) a custom mapper for interpreting the users' pose and gestures to events and signals for controlling the game. In addition, a database for storing the game results and managing the users has been integrated into the system. The game implementation is based on the popular Python library PyGame while the video analysis and inference on OpenCV and Tensorflow framework.

One of the main challenges while designing the system was to address the contradicting requirements considering on one hand the low-cost and one the other hand the nature of the exergame which required satisfying game speed, graphics and control. The most common computer board that could be used in order to meet the requirements for low-cost, since custom hardware was not an option, is Raspberry Pi Model 3 [12]. There were two technical difficulties following with this approach though: a) to perform the video analysis and deep learning inference operations in such hardware, like in any edge computing architecture and b) the remote management and maintenance of the edge device. The first one was addressed by using the Coral USB [4] Tensor Processing Unit (TPU), a special circuit designed to achieve high computational speeds in neural network applications which is compliant with the Tensorflow tools and the deep learning models applied. The TPU is designed for the performance phase, when systems with compiled models are presented with real-world data and are expected to behave appropriately using a version of TensorFlow called TensorFlow Lite. In order to facilitate remote management and maintenance, the deployment and operation was based on the Balena.io remote management solution which allows for dockerization of the software elements and their instantiation through well-defined DevOps processes [9].

Additional hardware peripherals were integrated for the interaction with the users; a Raspberry Pi Camera Module v2 is used for capturing images as the input sensor for controlling the game as well as a TV monitor connected via a HDMI. On software level, the proposed solution was based on Linux Debian 10 (codename: buster) which allowed for the smooth and flawless delivery of the complete software stack through the balena.io services.

3.1   CNN-Based Pose and Gesture Detection

For the implementation of the pose and gesture detection system, the TensorFlow framework has been used, as an end-to-end open source solution that has a comprehensive, flexible ecosystem of tools, libraries and community resources. This framework allowed for using state-of-the-art convoluntional neural networks and models like Posenet [10].

The model was trained on COCO dataset on top of MobileNet V1 network architecture. MobileNet architecture differs in the convolution process. Standard convolution filters and computes a new set of outputs in one step which costs in speed and time. In MobileNet, this is a two step procedure which is factorized in a depthwise convolution for filtering and a 1×1 convolution for combining. This factorization reduce model size and computation drastically [7]. Posenet provides an average keypoint precision of 0.665 to 0.687 depending on single or multi-scale inference. For person detection and pose estimation, Posenet adopts the bottom-up approach by localizing identity-free semantic entities (landmarks) and grouping them into person instances. The model learns to predict the relative displacement between each pair of keypoints starting from the most confident detection of a distinguished landmark such as nose. Totally seventeen (17) heatmaps have been developed, each one corresponding to a keypoint along with offset vectors. In the proposed approach, this procedure runs on the TPU module that is connected on Raspberry pi. Each keypoint is related to a heatmap which can be decoded and give the highest confidence areas in the image, corresponding to the keypoint. The offset vector is a 3D tensor with size

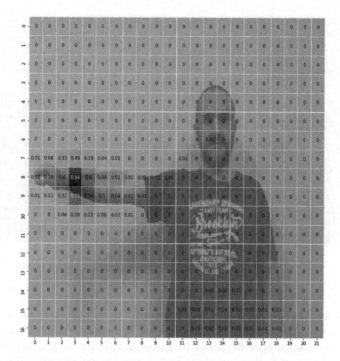

Fig. 2. Example of heatmap #10 that represents keypoint 'Right Wrist'.

Width × Height × 34 (two coordinates for every keypoint) and used to predict the exact location of each keypoint in the image. Models varying in the number of operations per layer have been obtained by the Posenet. The larger the size of ops in layers the more accurate the model at the cost of speed. The keypoints we get from Posenet are: (1) Nose, (2) Left Eye, (3) Right Eye, (4) Left Ear, (5) Right Ear, (6) Left Shoulder, (7) Right Shoulder, (8) Left Elbow, (9) Right Elbow, (10) Left Wrist, (11) Right Wrist, (12) Left Hip, (13) Right Hip, (14) Left Knee, (15) Right Knee, (16) Left Ankle, (17) Right Ankle. An example of Right Wrist landmark detection heatmap is shown in Fig. 2.

3.2 Exergame Engine Implementation

As already mentioned the exergame engine is based on PyGame, which is a cross platform set of modules that is used for writing video games in Python and includes computer graphics and sound libraries. The various software components were designed in order to connect neural network landmarks output with the control of the game. The deep learning framework provides for each analysis a dictionary in the following format: $\{keypoint_1 : (x, y, score), ..., keypoint_{17} : (x, y, score)\}$, where x, y are the pixel coordinates of input image. Following the analysis, set of rules are applied to the result to inference the user's gesture from landmarks. Every gesture is an event associated with a signal which in turn

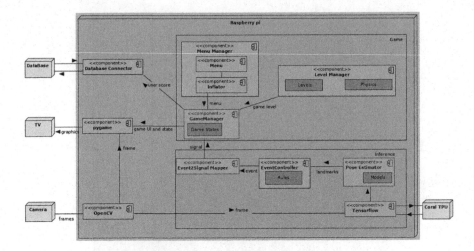

Fig. 3. Exergame platform component diagram

controls the game's functions and menu options. With rules we can identify the pose or gesture of the user and with signals we can control the game flow. More specifically, we assign specific poses to game signals for each control function like *stop, pause, next button, previous button* and we set the rules for recognizing that poses. When a pose is estimated a related signal broadcasts and the function that is related with that signal is executed. Signals control the game logic through the *Game Manager* component. This is the key component that listens to the signals produced after the gesture recognition process as is shown in the component diagram illustrated in Fig. 3. What happens in practise is that a gesture is recognized and the event that is associated with this gesture is passed through *Event2SignalMapper* and a signal is generated. *Game Manager* receives the signal and updates Menu's selected button through the *Inflator* update. The *Inflator* updates the container and displays it on the screen surface.

4   The Game in Practice

In order to assess the capabilities of the game engine and the overall system a prototype game has been developed. The story of the game is as follows: The player tries to pop bubbles with one part of his body, which in this case was the nose for two main reasons: a) it is part of the body that achieves a high score of confidence and b) it forces the player to move his whole body to pop the bubbles. The deep learning framework returns the nose coordinates and a gun-sight icon is drawn on the screen to these coordinates.

In Fig. 4 two sample levels of the game are presented, each one of which has a time limit of two minutes. In the first level the player has to move the gun-sight in a bubble and hold it there for 5 s.

Fig. 4. (a) The user interface of the prototype game.

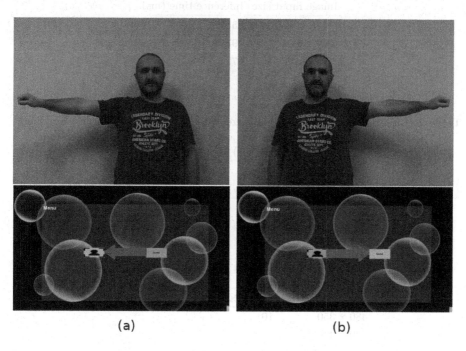

Fig. 5. Navigation on menu buttons: (a) Previous, (b) Next.

Following this approach, the player experiences how the gun-sight can be moved through the game and how the game perceives the player's movements. At the next level the player has to pop a number bubbles which are appearing gradually. Every effort of the player and the score of each level is kept in a database. The game also includes a main menu which is also controlled in as similar way. More specifically, the user can navigate right or left as presented in Fig. 5 and with other similar gestures, the *OK* and *Cancel* events are triggered.

As part of the system evaluation, extensive measurements have been performed on different aspects of the proposed implementation. During the game different models have been tested in respect to input size Table 1, taking into consideration the input sizes that are accepted in each model, which considerable affects the inference time of the deep learning framework. In Table 2, we measure the total cycle time (in milliseconds) in respect to frame sampling in a model with image input size 480 × 350. Finally, in Table 3 we depicts the frame rate of the game with 200 ms frame sampling in respect to different video input sizes.

Table 1. Inference time for different inputs with 200 ms frame sampling.

Image input size	Inference time (ms)
480 × 350	51
640 × 480	97
1200 × 720	272

Table 2. Total processing time for input size 480 × 350 for different frame samplings.

Frame sampling to model (ms)	Total processing time (ms)
300	80–85
200	90–100
150	120

Table 3. Frame rate for different inputs and sampling rate 5 Hz.

Image input size	Achieved game frame rate (fps)
480 × 350	10
640 × 480	5
1200 × 720	2

5   Conclusions and Future Work

The research presented in this work focused on the implementation of an innovative exergame engine exploiting state-of-the-art software technologies and the use of low-cost, commodity hardware. This was achieved by making use of methodologies and frameworks which are based on edge computing and computer vision using deep learning and convolutional neural networks. Concerning the deployment on user's environment, a Raspberry Pi device was used, bundled with a USB tensor accelerator enabling the execution of the deep learning models. The experimentation with a prototype in a real scenario presented significant results, both for the user's perception and for the system performance. The future plans include the use of more effective game control approaches based on better models for pose detection and video analysis techniques, and the support of multi-user gameplay. In addition, given the fact that the proposed exergame engine belongs to the family of serious games and targets specific user groups and use cases, more aspects of the users' physical and mental condition will be taken into consideration such as their biosignal measurements and their emotions during the game. This is expected to increase the effectiveness of the games allowing for better analysis of their health status and providing at the same time more incentives through the advanced gameplay and the enhanced game experience.

Acknowledgement. This research has been co-financed by the European Union and Greek national funds through the Operational Program Competitiveness, Entrepreneurship and Innovation, under the call RESEARCH – CREATE – INNOVATE (project code: SISEI Smart Infotainment System with Emotional Intelligence T1EDK-01046).

References

1. Al-Hrathi, R., Karime, A., Al-Osman, H., El Saddik, A.: Exerlearn bike: an exergaming system for children's educational and physical well-being. In: 2012 IEEE International Conference on Multimedia and Expo Workshops, pp. 489–494. IEEE (2012)
2. Benzing, V., Schmidt, M.: Exergaming for children and adolescents: strengths, weaknesses, opportunities and threats. J. Clin. Med. **7**(11), 422 (2018)
3. Brox, E., Fernandez-Luque, L., Tøllefsen, T.: Healthy gaming-video game design to promote health. Appl. Clin. Inform. **2**(2), 128–142 (2011)
4. Cass, S.: Taking AI to the edge: Google's TPU now comes in a maker-friendly package. IEEE Spectr. **56**(5), 16–17 (2019)
5. Deutsch, J.E., et al.: Nintendo wii sports and wii fit game analysis, validation, and application to stroke rehabilitation. Top. Stroke Rehabil. **18**(6), 701–719 (2011)
6. Goldstein, J., Cajko, L., Oosterbroek, M., Michielsen, M., Van Houten, O., Salverda, F.: Video games and the elderly. Soc. Behav. Pers. Int. J. **25**(4), 345–352 (1997)
7. Howard, A.G., et al.: MobileNets: efficient convolutional neural networks for mobile vision applications. arXiv preprint arXiv:1704.04861 (2017)

8. Matallaoui, A., Koivisto, J., Hamari, J., Zarnekow, R.: How effective is "exergamification"? A systematic review on the effectiveness of gamification features in exergames. In: Proceedings of the 50th Hawaii International Conference on System Sciences (2017)

9. Menychtas, A., Doukas, C., Tsanakas, P., Maglogiannis, I.: A versatile architecture for building IoT quantified-self applications. In: 2017 IEEE 30th International Symposium on Computer-Based Medical Systems (CBMS), pp. 500–505. IEEE (2017)

10. Papandreou, G., Zhu, T., Chen, L.C., Gidaris, S., Tompson, J., Murphy, K.: PersonLab: person pose estimation and instance segmentation with a bottom-up, part-based, geometric embedding model. In: Proceedings of the European Conference on Computer Vision (ECCV), pp. 269–286 (2018)

11. Pope, Z., Zeng, N., Gao, Z.: The effects of active video games on patients' rehabilitative outcomes: a meta-analysis. Prev. Med. **95**, 38–46 (2017)

12. Senthilkumar, G., Gopalakrishnan, K., Kumar, V.S.: Embedded image capturing system using Raspberry Pi system. Int. J. Emerg. Trends Technol. Comput. Sci. **3**(2), 213–215 (2014)

13. Skjæret, N., Nawaz, A., Morat, T., Schoene, D., Helbostad, J.L., Vereijken, B.: Exercise and rehabilitation delivered through exergames in older adults: an integrative review of technologies, safety and efficacy. Int. J. Med. Inform. **85**(1), 1–16 (2016)

14. Street, T.D., Lacey, S.J., Langdon, R.R.: Gaming your way to health: a systematic review of exergaming programs to increase health and exercise behaviors in adults. Games Health J. **6**(3), 136–146 (2017)

15. Styliadis, C., Konstantinidis, E., Billis, A., Bamidis, P.: Employing affection in elderly healthcare serious games interventions. In: Proceedings of the 7th International Conference on PErvasive Technologies Related to Assistive Environments, pp. 1–4 (2014)

16. Susi, T., Johannesson, M., Backlund, P.: Serious games: an overview (2007)

17. Trout, J., Zamora, K.: Using dance dance revolution in physical education. Teach. Elementary Phys. Educ. **16**(5), 22–25 (2005)

Investigating the Problem of Cryptocurrency Price Prediction: A Deep Learning Approach

Emmanuel Pintelas[1], Ioannis E. Livieris[1,2(✉)], Stavros Stavroyiannis[3], Theodore Kotsilieris[2], and Panagiotis Pintelas[1]

[1] Department of Mathematics, University of Patras, Patras, Greece
ece6835@upnet.gr, livieris@upatras.gr, ppintelas@gmail.com
[2] Department of Business Administration, University of the Peloponnese, Antikalamos, Greece
t.kotsilieris@teipel.gr
[3] Department of Accounting and Finance, University of the Peloponnese, Antikalamos, Greece
computmath@gmail.com

Abstract. In last decade, cryptocurrency has emerged in financial area as a key factor in businesses and financial market opportunities. Accurate predictions can assist cryptocurrency investors towards right investing decisions and lead to potential increased profits. Additionally, they can also support policy makers and financial researchers in studying cryptocurrency markets behavior. Nevertheless, cryptocurrency price prediction is considered a very challenging task, due to its chaotic and very complex nature. In this study we evaluate some of the most successful and widely used deep learning algorithms forecasting cryptocurrency prices. The results obtained, provide significant evidence that deep learning models are not able to solve this problem efficiently and effectively. Conducting detailed experimentation and results analysis, we conclude that it is essential to invent and incorporate new techniques, strategies and alternative approaches such as: more sophisticated prediction algorithms, advanced ensemble methods, feature engineering techniques and other validation metrics.

Keywords: Deep learning · CNN · LSTM · BiLSTM · Cryptocurrency price prediction · Time series

1 Introduction

Cryptocurrency is a new type of digital currency which utilizes blockchain technology and cryptographic functions to gain transparency, decentralization and immutability [12]. Bitcoin (BTC) is considered the first and the most popular cryptocurrency, which was invented by an anonymous group or person in 2009. Since then, 4000 alternative cryptocurrencies like Etherium (ETH) and Ripple (XRP) were created proving that the cryptocurrency market has emerged in

© IFIP International Federation for Information Processing 2020
Published by Springer Nature Switzerland AG 2020
I. Maglogiannis et al. (Eds.): AIAI 2020, IFIP AICT 584, pp. 99–110, 2020.
https://doi.org/10.1007/978-3-030-49186-4_9

financial area. BTC, ETH and XRP are the most popular cryptocurrencies, since they almost hold the 79.5% of the global cryptocurrency market capitalization.

Cryptocurrency price prediction can provide a lending hand to cryptocurrency investors for making proper investment decisions in order to acquire higher profits while it can also support policy decision-making and financial researchers for studying cryptocurrency markets behavior. Cryptocurrency price prediction can be considered as a common type of time series problems, like the stock price prediction. Traditional time series methods such as the well-known AutoRegressive Integrated Moving Average (ARIMA) model, have been applied for cryptocurrencies price and movement prediction [13]. However, these models are not able to capture non-linear patterns of very complicated prediction problems in contrast to Deep Learning algorithms which achieve greater performance on forecasting time series problems [17].

Deep Learning (DL) refers to powerful machine learning algorithms which specialize in solving nonlinear and complex problems exploiting most of the times big amounts of data in order to become efficient predictor models. The accurate cryptocurrency price prediction is by nature a significantly challenging and complex problem since its values have very big fluctuations over time following an almost chaotic and unpredictable behavior. Therefore, deep learning techniques may constitute the proper methodology to solve this problem.

Recent research efforts have adopted deep learning techniques for predicting cryptocurrency price. Ji et al. [8] conducted a comparison of state-of-the-art deep neural networks such as Long Short-Term Memory (LSTM), Deep Neural Networks (DNNs), deep residual network and their combinations for predicting Bitcoin price. Their results demonstrated slightly better accuracy of LSTM compared to other models for regression problem while DNNs outperformed all models on price movement prediction. Shintate and Pichl [16] developed a trend prediction classification framework for predicting non-stationary cryptocurrency time series utilizing deep learning. Their results revealed that their proposed model outperformed LSTM baseline model while the profitability analysis showed that simple buy-and-hold strategy was superior to their model and thus it cannot yet be used for algorithmic trading. Their results showed that LSTM was superior to the generalized regression neural architecture concluding that deep learning is a very efficient method in predicting the inherent chaotic dynamics of cryptocurrency prices. Amjad and Shah [3] used live streaming Bitcoin data for predicting price changes (increase, decrease or no-change), building a model based on the most confident predictions, in order to perform profitable trades. The classification algorithms which they used were Random Forest, Logistic Regression and Linear Discriminant Analysis. Their results seem to be very impressive since they achieved a high prediction accuracy (>60–70%) and about $5.33x$ average return on investments on a test set.

In this work, we evaluate the performance of advanced deep learning algorithms for predicting the price and movement of the three most popular cryptocurrencies (BTC, ETH and XRP). The main contribution of this research lies in investigating three major questions: i) *Can deep learning efficiently predict cryptocurrency prices?* ii) *Are cryptocurrency prices a random walk process?* iii) *Is there a proper validation method of cryptocurrency price prediction models?*

Furthermore, it also lies in the recommendation for new algorithms and alternative approaches for the cryptocurrency prediction problem.

The remainder of this research is organized as follows: Sect. 2 performs a brief introduction to the advanced deep learning models utilized in our experiments. Section 3 presents our research methodology and experimental results. Section 4 discusses and answers the three research questions, while Sect. 5 presents our suggestions on possible alternative solutions for the cryptocurrency prediction problem. Finally, Sect. 6 presents our concluding remarks.

2   Brief Description of Advanced Deep Learning Models

Deep learning algorithms constitute one of the most powerful machine learning algorithms categories which have been successfully applied on a multitude of commercial applications. Long Short Term Memory and Convolutional Neural Networks are probably the most popular, successful and widely used deep learning techniques.

Long Short-Term Memory (LSTM) [6] constitute a special type of deep neural networks, which are able to learn long-term dependencies by utilizing feedback connections in order to "*remember*" past network cell states. These networks have become very popular since they have been successfully applied on a wide range of applications and have shown remarkable performance on time series forecasting [5]. More specifically, LSTM networks are composed by a memory cell, an input, output and forget gate. The input gate controls the new stored information into the memory cell, while the forget gate controls the information which must be vanished. Finally, the output gate controls the final output information value which is given after a delay into the forget, input gate utilizing a feedback connection loop. In this way, LSTM is able to create a controlled information flow filtering unnecessary information and thus achieving to learn long term dependencies.

Bidirectional Long Short-Term Memory (BiLSTM) [15] are a special type of recurrent neural networks which connect two LSTM layers of converse directions to the same output, in order to remember past and future network cell states. The principle idea is that each training sequence is presented forwards and backwards into two separate LSTM layers aiming in accessing both past and future contexts for a given time. More specifically, the first hidden layer possesses recurrent connections from the past time steps; while in the second one, the recurrent connections are reversed, transferring activation backwards along the sequence.

Convolutional Neural Networks (CNN) [2] constitute another type of deep neural networks which utilize convolution and pooling layers in order to filter the raw input data and extract valuable features, which will feed a fully connected layer in order to produce the final output. More specifically, they apply convolution operations in the input data and in order to produce new more useful features. The convolutional layers are usually followed by a pooling layer which extracts values from the convolved features producing a lower dimension instance. In fact, a pooling layer produces new features which can be considered as summarized versions of the convolved features produced by the convolutional layer. This implies that pooling operations can significantly assist the network to be more robust since small changes of the inputs, which are usually detected by the convolutional layers, will become approximately invariant.

3   Experimental Methodology

In this work, we evaluate the performance of advanced DL models for predicting the price of BTC, ETH and XRP. The evaluated DL models are constituted by CNN, LSTM, BiLSTM and dense layers. Table 1 depicts our DL models for the best identified topologies. We have to mention that exhaustive and thorough experiments were performed in order to identify the DL topologies which incur the best performance results.

Table 1. Best identified topologies for our deep learning models

Model	Description
$LSTM_1$	LSTM layer with 50 units
$LSTM_2$	Two LSTM layers with 30 and 15 units, respectively
$BiLSTM_1$	BiLSTM layer with 60 units
$BiLSTM_2$	Two BiLSTM layers with 40 and 20 units, respectively
$CNN\text{-}LSTM_1$	Convolutional layer with 64 of filters of size (2,)
	Convolutional layer with 128 of filters of size (2,)
	Max pooling layer with size (2,)
	LSTM layer with 100 units
$CNN\text{-}LSTM_2$	Convolutional layer with 64 of filters of size (2,)
	Convolutional layer with 128 of filters of size (2,)
	Max pooling layer with size (2,)
	LSTM layer with 70 units
	Dense layer with 16 neurons
$CNN\text{-}BiLSTM_1$	Convolutional layer with 64 of filters of size (2,)
	Convolutional layer with 128 of filters of size (2,)
	Max pooling layer with size (2,)
	LSTM layer with 100 units
$CNN\text{-}BiLSTM_2$	Convolutional layer with 64 of filters of size (2,)
	Convolutional layer with 128 of filters of size (2,)
	Max pooling layer with size (2,)
	BiLSTM layer with 70 units
	Dense layer with 16 neurons

We recall that the basic idea of utilizing LSTM and BiLSTM on cryptocurrency price prediction problems, is that they might be able to capture useful long or short sequence pattern dependencies, due to their special architecture design, assisting on prediction performance, while the convolutional layers of a CNN model might filter out the noise of the raw input data and extract valuable features producing a less complicated dataset which would be more useful for the final prediction model [9]. Therefore, we expect that a noticeable performance increase will be achieved by the incorporation of these advanced models comparing to classic machine learning algorithms.

Additionally, the performance of the DL models was compared against that of traditional state-of-the-art ML models: Support Vector Regressor (SVR) [4], 3-Nearest Neighbors (3NN) [1] and Decision Tree Regressor (DTR) [10]. The implementation code was written in Python 3.4 while for all deep learning models we utilized Keras library and Theano as back-end while Scikit-learn library was used for the machine learning models.

For evaluating the regression performance of forecasting models the most common validation metrics are Mean Absolute Error (MAE) and Root Mean Squared Error (RMSE). Since the cryptocurrency price prediction problem can be considered a regression problem, in our experiments we utilized these two evaluation metrics. Nevertheless, we included only the RMSE score in this study, since the MAE score had almost the same behavior with RMSE. Moreover, by comparing the predicted prices of our models, with the real ones, we managed to compute the classification accuracy of price movement direction prediction (if the price will increase or decrease). Therefore, we utilized two additional performance metrics: Accuracy (Acc) and F_1-score (F_1).

3.1 Dataset

For the purpose of this research, we utilized data from Jan-2018 to Aug-2019, concerning the hourly prices in USD and were divided into training set consisting of data from Jan-2018 to Feb-2019 (10176 values) and testing set from Mar-2019 to Aug-2019 (4416 values). This data were taken from www.kraken.com website, which is a trading platform for cryptocurrency exchanges. Also, we utilized four forecasting horizons F (number of past prices taken into consideration), i.e., 4, 9, 12 and 16 h, while in this study we present only the 4 and 9 h horizon results, since for larger horizon values it was identified a decrease in performance. An extended report which includes all experimental results can be found in [14].

3.2 Experimental Results

Tables 2 and 3 present the experimental results of our DL models ML models. CNN-LSTM and CNN-BiLSTM models exhibited the best overall performance among all prediction models. In particular, the CNN-LSTM exhibited the highest RMSE performance for all datasets for every forecasting horizon comparing to other DL models, while the CNN-BiLSTM exhibited the best Acc and F_1 score in most cases. Nevertheless, the performance variations for all DL models seem to be minimal. The 3NN model reported the highest RMSE performance for forecasting horizon 4 among all the ML models on BTC and ETH datasets, while the DTR exhibited the highest RMSE on XRP. For forecasting horizon 9 the DTR outperformed all ML models for all dataset, regarding to RMSE score. Furthermore, the 3NN model managed to achieve the best overall performance in Acc score almost in all cases. In summary, advanced DL models seem to slightly outperform ML models while they did not manage to achieve a noticeable performance increase comparing to our ML models.

Table 2. Performance of DL and ML forecasting models for $F = 4$

Model	BTC			ETH			XRP		
	RMSE	Acc	F_1	RMSE	Acc	F_1	RMSE	Acc	F_1
SVR	0.0209	50.80%	0.484	0.0163	49.14%	0.482	0.0370	47.67%	0.644
3NN	0.0161	49.51%	0.469	0.0149	51.56%	0.481	0.0115	52.33%	0.492
DTR	0.0170	49.22%	0.465	0.0199	50.11%	0.467	0.0157	50.14%	0.504
LSTM$_1$	0.0138	52.86%	0.524	0.0130	53.59%	0.531	0.0106	53.40%	0.509
LSTM$_2$	0.0144	53.63%	0.522	0.0165	52.68%	0.501	0.0116	53.13%	0.514
BiLSTM$_1$	0.0142	52.05%	0.508	0.0130	53.89%	0.538	0.0114	53.21%	0.519
BiLSTM$_2$	0.0132	52.56%	0.519	0.0134	53.30%	0.502	0.0118	52.75%	0.518
CNN-LSTM$_1$	0.0109	53.21%	0.530	0.0106	53.77%	0.533	0.0097	52.42%	0.472
CNN-LSTM$_2$	0.0099	52.50%	0.502	0.0106	54.20%	0.524	0.0097	53.03%	0.436
CNN-BiLSTM$_1$	0.0107	54.51%	0.544	0.0121	53.91%	0.524	0.0119	53.61%	0.534
CNN-BiLSTM$_2$	0.0101	55.43%	0.548	0.0115	54.18%	0.541	0.0117	53.46%	0.506

Table 3. Performance of DL and ML forecasting models for $F = 9$

Model	BTC			ETH			XRP		
	RMSE	Acc	F_1	RMSE	Acc	F_1	RMSE	Acc	F_1
SVR	0.0192	52.57%	0.546	0.0146	49.30%	0.501	0.0292	47.92%	0.457
3NN	0.0197	51.57%	0.484	0.0195	51.50%	0.504	0.0132	54.35%	0.485
DTR	0.0179	49.54%	0.459	0.0228	49.64%	0.465	0.0161	49.89%	0.504
LSTM$_1$	0.0159	51.99%	0.479	0.0158	53.06%	0.500	0.0117	51.34%	0.413
LSTM$_2$	0.0207	52.45%	0.499	0.0208	52.75%	0.527	0.0115	51.66%	0.456
BiLSTM$_1$	0.0170	52.00%	0.489	0.0165	53.31%	0.512	0.0126	52.73%	0.517
BiLSTM$_2$	0.0168	52.97%	0.530	0.0166	53.80%	0.527	0.0121	55.22%	0.534
CNN-LSTM$_1$	0.0119	53.92%	0.536	0.0130	53.92%	0.530	0.0096	51.06%	0.453
CNN-LSTM$_2$	0.0107	54.20%	0.532	0.0124	54.45%	0.537	0.0100	51.54%	0.493
CNN-BiLSTM$_1$	0.0149	53.44%	0.533	0.0158	53.10%	0.522	0.0148	54.01%	0.540
CNN-BiLSTM$_2$	0.0125	54.89%	0.541	0.0152	53.95%	0.533	0.0157	53.95%	0.532

3.3   Forecasting Reliability Evaluation

In the sequel, we evaluate the forecasting reliability of the proposed prediction models, by performing a test of autocorrelation in the residuals [11]. This test examines the presence of autocorrelation between the residuals (differences between predicted and real values). In case autocorrelation exists, then the prediction model may be inefficient since it did not manage to capture all the possible information which lies into the data. To this end, we perform the autocorrelation test to the residuals in order to evaluate the reliability of CNN-BiLSTM for $F = 4$, CNN-LSTM for $F = 4$ and CNN-BiLSTM for $F = 9$ which presented the best overall performance for BTC, ETH and XRP, respectively.

Figures 1, 2 and 3 present the Auto-Correlation Function (ACF) plot of the selected models for BTC, ETH and XRP, respectively. Notice that the confident limits (blue dashed line) are constructed assuming that the residuals follow a Gaussian probability distribution. Clearly, all present ACF plots reveal that some correlation coefficients were not within the confidence limits (dashed lines), violating the assumption of no auto-correlation in the errors. More specifically, the interpretation of Figs. 1 and 2 present that there are significant spikes at lags 1 and 2 while the interpretation of Fig. 3 show that there exist small spikes at lags 1, 2, 6, 7 and 10. Therefore, the presence of correlation indicates that the advanced DL models are unreliable for cryptocurrency price predictors since there exists some significant information left over which should be taken into account for obtaining better predictions.

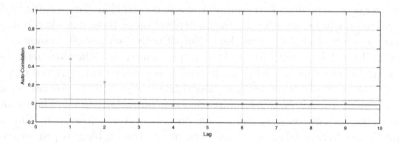

Fig. 1. ACF plots on the residuals for BTC using CNN-BiLSTM for $F = 4$ (Color figure online)

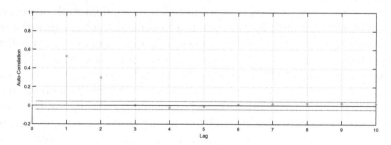

Fig. 2. ACF plots on the residuals for ETH dataset using CNN-LSTM for $F = 4$ (Color figure online)

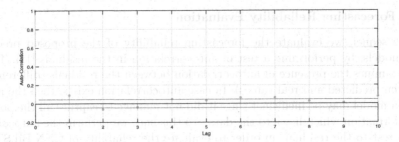

Fig. 3. ACF plots on the residuals for XRP dataset using CNN- BiLSTM for $F = 9$ (Color figure online)

4   Discussion

Following our experiments, this section is dedicated in providing a thorough and sufficiently detailed discussion of our findings with regard to the predefined three research questions: *Can deep learning algorithms efficiently predict cryptocurrency prices? Are cryptocurrency prices a random walk process? Which is a proper validation method of cryptocurrency price prediction models?*

4.1   Can Deep Learning Efficiently Predict Cryptocurrency Prices?

Deep learning algorithms are considered to be the most powerful and the most effective methods in approximating extremely complex and non-linear classification and regression problems, therefore it was expected that a noticeable performance increase will be achieved by the incorporation of these models comparing to classic machine learning algorithms. Surprisingly, our results demonstrated that the utilized DL algorithms, slightly outperformed the other ML algorithms utilized in our experiments, whereas instead a noticeable performance increase was anticipated. So, it is paramount importance to investigate the reason why that happened. To this end, we summarize two possible reasons: The problem we are trying to solve is a random walk process or very close to it, thus any attempt for prediction might be of poor quality or the problem is just too complicated that even advanced deep learning methods cannot find any pattern that would lead to any reliable prediction. Thus, more sophisticated methodologies, techniques and innovative strategies are needed to be investigated.

When a time series prediction problem follows a random walk process or it is so complicated that most models face it as a random process, then the more efficient method to face it, is the employment of present values as the prediction values for the next state [11]. That is exactly what a persistence model does and maybe what most prediction models really do and possibly that's the reason why ML models used in our experiments achieve almost the same performance score compared to the deep learning models used in our experiments. In contrast, the deep learning models may attempt forecasting based on patterns that were traced and as a result are unable to achieve high performance because either those patterns are false or because there exist no such patterns at all, in the case that the cryptocurrency price prediction problem is a random walk process.

Nevertheless, as mentioned before, the DL models did not manage to achieve a noticeable performance score in our experiments, since their score was almost the same with the ML models. Thus, we conclude that these advanced DL models cannot efficiently predict cryptocurrency prices because the utilized datasets with the specific form which we *"fed"* them to our prediction models, probably follow almost a random walk process and thus not sufficient information lies on them in order to perform accurate and reliable future predictions.

4.2 Are Cryptocurrency Prices a Random Walk Process?

Towards the construction of a model which performs reliable and accurate predictions, firstly, we have to identify if the cryptocurrency price prediction problem is a random walk process. In a recent study, Stavroyiannis et al. [18] proved that Bitcoin prices follow a random walk process since their experiments revealed the presence of unit roots, for several time intervals from 1-min to 180-min, and thus reliable profitable trading opportunities may not be possible in Bitcoin markets. However, since this problem is highly affected by time evolution and external changes, these results maybe temporary and reverse in future.

However, there are numerous technical strategies that the majority of the professional traders utilize in order to make trading decisions in stock market and cryptocurrency investments. Most of them seem to be heuristic and empirical strategies which are based on various technical indicators and patterns such as the *"Engulfing Pattern"* and the *"Evening Star"*. A recent study utilized those technical indicators and trading patterns strategies in order to predict stock market and cryptocurrency prices [7]. Their results provide evidence that technical analysis strategies have strong predictive power and thus can be useful in cryptocurrencies markets like Bitcoin.

Therefore, we conclude that the cryptocurrency prices in general are not totally a random walk process but they may be close to it, which means that probably exist some actual patterns on historic data that could assist on forecasting attempts. In other words, we liken this problem as a *"huge sea of random walk points where small hidden islands (patterns) may exist in"*. As a result, more research is required for the discovery of alternative, innovative and more sophisticated methods such as the incorporation of new feature engineering strategies and the creation of new algorithmic and ensemble methods.

4.3 What is a Proper Validation Method of Cryptocurrency Price Prediction Models?

As mentioned above, the most common validation metrics for measuring the performance of most regression algorithms are MAE and RMSE. However, finding a proper validation metric for cryptocurrency price prediction models can be a very complicated and tricky task and cannot be considered an easy and straightforward process. The MAE and RMSE may constitute an incomplete way for validating cryptocurrency price prediction problems since a prediction model may have excellent MAE and RMSE performance but cannot properly predict the cryptocurrency price direction move (classification problem). A cryptocurrency trader or investor may be more interested in the future price direction movement rather than knowing the exact future cryptocurrency price. Profitability analysis for algorithmic trading strategies reveal that classification prediction models were more effective than regression models [8].

Even if we utilize a third evaluation metric which will measure the performance accuracy of cryptocurrency price direction movement, that may still constitute an incomplete method for validating cryptocurrency prediction algorithms. Consider the following example: Suppose we wish to validate 2 cryptocurrency prediction models utilizing a test set of 100 questions, e.g. what is the future price direction movement on the next 100 time steps? The first model answers (predicts) all questions while it answers correctly 52 questions achieving an accuracy score of 52%. The second model answers only 5 from 100 questions but it cannot answer the other 95 questions, while these 5 answers are correct. So, the second model achieves a score of 5%. Thus, an important question is raised, *"which is the best model"*? A cryptocurrency trader or investor will probably choose the second model since it acts in a more reliable way and it would be more valuable for him to possess a model which performs accurate predictions on random times (specified by the model), rather than possessing a model which performs unreliable predictions on every moment (specified by the user).

Therefore, we conclude that finding a proper validation metric for cryptocurrency price prediction models is a very challenging task and thus alternative and new methods for evaluating cryptocurrency prediction models are essential.

5   Revisiting the Problem

One of the most significant steps in order to solve any problem, especially the really hard and challenging ones, lies in finding a proper strategy approach and securing the complete understanding of the problem we try to solve. A proper strategy approach should answer questions such as: should we have to predict prices, price movement direction, price trends, price spikes and so on. Next, should we apply data preprocessing and feature engineering strategies (e.g. which features should we use in order to efficiently train a prediction model?) Also, what is the best prediction model to apply (e.g. DNNs, other sophisticated prediction models, ensemble models and so on) and finally, which is a proper method to validate this model? All these issues, considered as discrete steps in the process, should be taken into serious consideration since each one of them can significantly contribute to any prediction attempt in order to efficiently approximate the problem.

These steps are not a straightforward process, since we should always have to consider its chaotic and extremely complicated nature with respect to its practical contribution after a possible solution. For example, it may be an easier task to solve and possibly more beneficial for the investment and trading world to predict if the price will just increase or decrease (classification problem for price direction movement prediction) rather than predicting the exact value of cryptocurrency price. Some strategy approaches examples are presented in brief below.

Instead of adopting a specific time interval, one could utilize various time intervals of higher and lower frequency historic datasets for predicting the prices on a specific future interval in order to utilize and exploit in a more efficient way all possible information that a historic dataset may contain. Another approach could be instead of predicting the price or the movement direction on one discrete future time value, to predict the average and movement direction price or peak price inside a future time window frame (this approach would be more similar to a trend prediction problem).

Pattern identification and recognition could be another approach. This approach would be more similar to a pattern detection framework in which the model would detect specific pattern areas in order to perform a prediction. More specifically, if we are able to identify the feature characteristics of possible useful patterns that a prediction model found, then we could filter out useless sequence inputs which have no predictive information and then utilize only the useful sequence inputs which will possibly assist on reliable and accurate predictions. In this case the prediction model will perform prediction operations only when the input sequence falls into the same category with the chosen patterns. This framework would be more similar to the way that a professional trader often acts, who performs investment decisions based on his/her personal chosen patterns and indicators recipes on technical analysis of historic price charts.

Finally, another approach could be the investigation of heuristic patterns and other financial indicators which professional traders and bankers utilize in their trading and financial technical analysis. It is essential to identify how these methods actually assist predictions and investment decisions in a more mathematic way (if they actually work) and maybe incorporate these techniques in a machine learning framework for developing co-operative prediction models. That could be an effective cryptocurrency prediction framework.

6   Conclusions

In this work, we evaluated advanced DL models for predicting cryptocurrency prices and also investigated three research question concerning this problem in a review and discussion approach. Our results revealed that the presented models are inefficient and unreliable cryptocurrency price predictors, probably due to the fact that this problem is a very complicated one, that even advanced deep learning techniques such as LSTM and CNNs are not able to solve efficiently. Also, based on our experimental results and investigation regarding to our research questions about cryptocurrency price problem, we conclude that cryptocurrency prices follow almost a random walk process while few hidden patterns may probably exist in, where an intelligent framework has to identify them in order for a prediction model to make accurate and reliable forecasts. Therefore, new sophisticated algorithmic methods, alternative approaches, new validation metrics should be explored.

Finally, since cryptocurrency datasets follow typical time-series patterns, one may logically conclude that the research questions posed in this work and our concluding remarks and proposals apply to all application domains in which the datasets demonstrate time-series behavior.

References

1. Aha, D.W.: Lazy Learning. Springer, Heidelberg (2013)
2. Aloysius, N., Geetha, M.: A review on deep convolutional neural networks. In: 2017 International Conference on Communication and Signal Processing (ICCSP), pp. 588–592. IEEE (2017)
3. Amjad, M., Shah, D.: Trading Bitcoin and online time series prediction. In: NIPS 2016 Time Series Workshop, pp. 1–15 (2017)
4. Deng, N., Tian, Y., Zhang, C.: Support Vector Machines: Optimization Based Theory, Algorithms, and Extensions. Chapman and Hall/CRC, Boca Raton (2012)
5. Ismail Fawaz, H., Forestier, G., Weber, J., Idoumghar, L., Muller, P.-A.: Deep learning for time series classification: a review. Data Min. Knowl. Discov. **33**(4), 917–963 (2019). https://doi.org/10.1007/s10618-019-00619-1
6. Graves, A.: Long short-term memory. In: Graves, A. (ed.) Supervised Sequence Labelling with Recurrent Neural Networks. SCI, vol. 385, pp. 37–45. Springer, Heidelberg (2012). https://doi.org/10.1007/978-3-642-24797-2_4
7. Huang, J.Z., Huang, W., Ni, J.: Predicting Bitcoin returns using high-dimensional technical indicators. J. Finan. Data Sci. **5**(3), 140–155 (2019)
8. Ji, S., Kim, J., Im, H.: A comparative study of Bitcoin price prediction using deep learning. Mathematics **7**(10), 898 (2019)
9. Livieris, I.E., Pintelas, E., Pintelas, P.: A CNN-LSTM model for gold price time series forecasting. Neural Comput. Appl. (2020)
10. Loh, W.Y.: Classification and regression tree methods. Wiley StatsRef: Statistics Reference Online (2014)
11. Montgomery, D.C., Jennings, C.L., Kulahci, M.: Introduction to Time Series Analysis and Forecasting. Wiley, Hoboken (2015)
12. Narayanan, A., Bonneau, J., Felten, E., Miller, A., Goldfeder, S.: Bitcoin and Cryptocurrency Technologies: A Comprehensive Introduction. Princeton University Press, Princeton (2016)
13. Olvera-Juarez, D., Huerta-Manzanilla, E.: Forecasting Bitcoin pricing with hybrid models: a review of the literature. Int. J. Adv. Eng. Res. Sci. **6**(9), 161–164 (2019)
14. Pintelas, E., Livieris, I.E., Stavroyiannis, S., Kotsilieris, T., Pintelas, P.: Fundamental research questions and proposals on predicting cryptocurrency prices using DNNs. Technical report, TR20-01, University of Patras (2020)
15. Schuster, M., Paliwal, K.K.: Bidirectional recurrent neural networks. IEEE Trans. Signal Process. **45**(11), 2673–2681 (1997)
16. Shintate, T., Pichl, L.: Trend prediction classification for high frequency Bitcoin time series with deep learning. J. Risk Finan. Manag. **12**(1), 17 (2019)
17. Siami-Namini, S., Tavakoli, N., Namin, A.S.: A comparison of ARIMA and LSTM in forecasting time series. In: 2018 17th IEEE International Conference on Machine Learning and Applications (ICMLA), pp. 1394–1401. IEEE (2018)
18. Stavroyiannis, S., Babalos, V., Bekiros, S., Lahmiri, S., Uddin, G.S.: The high frequency multifractal properties of Bitcoin. Phys. A: Stat. Mech. Appl. **520**, 62–71 (2019)

Regularized Evolution for Macro Neural Architecture Search

George Kyriakides$^{(\boxtimes)}$ (iD) and Konstantinos Margaritis (iD)

University of Macedonia, 55236 Thessaloniki, Greece
{ge.kyriakides,kmarg}@uom.edu.gr

Abstract. Neural Architecture Search is becoming an increasingly popular research field and method to design deep learning architectures. Most research focuses on searching for small blocks of deep learning operations, or micro-search. This method yields satisfactory results but demands prior knowledge of the macro architecture's structure. Generally, methods that do not utilize macro structure knowledge perform worse but are able to be applied to datasets of completely new domains. In this paper, we propose a macro NAS methodology which utilizes concepts of Regularized Evolution and Macro Neural Architecture Search (DeepNEAT), and apply it to the Fashion-MNIST dataset. By utilizing our method, we are able to produce networks that outperform other macro NAS methods on the dataset, when the same post-search inference methods are used. Furthermore, we are able to achieve 94.46% test accuracy, while requiring considerably less epochs to fully train our network.

Keywords: NAS · Neural Architecture Search · Fashion MNIST

1 Introduction

As the amount of available raw data has increased over the years, in conjunction with the steady increase in available processing power, machine learning has also increased in popularity as a modelling and predictive tool. This has spurred new research in the field, which in turn has enabled the implementation and automation of tasks previously considered exceptionally hard for a computer. Especially in the past decade, neural networks in the form of deep learning have been an increasingly popular tool for such industrial and research applications [2, 4, 6, 20].

Deep learning demands specific neural architectures in order to be successfully and effectively implemented. Traditionally, these architectures were hand-crafted by human experts, a laborious and time-consuming process. More recently, there have been significant efforts to automate the process of designing them. Thus, a new research field has emerged from these efforts, known as Neural Architecture Search (NAS). Many state-of-the-art architectures have been produced as the research in the field has progressed [11, 13, 16, 17, 22]. The process of designing the architectures requires significantly more computational resources than training a single architecture, but can be more resource-efficient in the long run, especially when low-power hardware is targeted [19].

© IFIP International Federation for Information Processing 2020
Published by Springer Nature Switzerland AG 2020
I. Maglogiannis et al. (Eds.): AIAI 2020, IFIP AICT 584, pp. 111–122, 2020.
https://doi.org/10.1007/978-3-030-49186-4_10

Although most methodologies are able to produce state-of-the-art architectures, they rely on a pre-determined architectural macro structure (outer skeleton). Instead of designing the entire network, a small network block (called cell) is designed, and several cells are stacked in order to produce the final network, as dictated by the skeleton. This can essentially be called a micro-search, as the micro-structure of the network's cells is optimized. Although this philosophy is both effective and efficient, it requires prior knowledge of an appropriate skeleton. Thus, it is only applicable in areas where such prior knowledge has been obtained, usually through human experimentation. This greatly reduces the applicability of such methods to domains that have already been studied to a certain degree.

In this paper, we explore an algorithm created by combination of Regularized Evolution from [17] and genetic encoding from DeepNEAT [13], in order to generate architectures for the Fashion-MNIST dataset [21], by conducting a macro-architecture search. We aim to evaluate the proposed method's ability to generate good architectures without any prior knowledge about the application's domain, as well as compare it to other macro-architecture search methodologies. We first briefly present the two algorithms from which we borrow elements from, as well as the dataset and various NAS algorithms tested on it. Following, we explain our methodology and our experimental results. Finally, we present our research's limitations and discuss our findings.

2   Related Work

As previously mentioned, for our method we utilize a combination of Regularized Evolution and DeepNEAT, in order to formulate a macro NAS methodology. As such, in this section we present the basics of the two algorithms, as well as the Fashion-MNIST dataset.

2.1   Regularized Evolution

In the corresponding paper [17], the authors propose the usage of an evolutionary algorithm in order to generate architectures of increasingly better performance. Instead of utilizing tournament selection when updating the populations, the oldest solution in the population is discarded. Following, by randomly sampling N individuals, the best is selected to generate an offspring. The offspring is added to the population, while the oldest individual is discarded. This forces the algorithm to retain architectures that consistently perform well. An architecture which was trained to a high test accuracy by luck will not be able to survive for very long. The authors were able to produce better architectures than the state-of-the-art for the CIFAR10 dataset in the NASNet Search Space [23]. The search space restricts the algorithm's ability to produce arbitrary architectures. Instead, a pre-defined skeleton of repeated cells is used, while the algorithm designs the individual cells.

2.2 DeepNEAT

Even before the advent of deep learning, efforts where made in order to optimize both the design, as well as the weights of neural networks. One such example is the NEAT algorithm [18], which utilized generational evolution in order to evolve simple neural networks. The method was more recently extended in order to design deep neural architectures, in the forms of DeepNEAT and CoDeep-NEAT [13]. DeepNEAT is able to design macro-architectures, by encoding each layer's parameters, as well as connections between layers to different genes. In order to achieve crossover, each gene is marked with an indicator, which lets homologous genes align. In the paper, the researchers are able to utilize CoDeep-NEAT which employs co-evolution of cells and skeletons, in order to generate image recognition and captioning architectures with success.

2.3 Fashion-MNIST

Fashion-MNIST [21] is a benchmarking dataset for image recognition tasks. The dataset contains a number of labeled fashion items. Each image is a 28 × 28 pixel grayscale image of various clothing items. In total there are 60,000 items in the training set and 10,000 in the test set. There are a total of 10 different classes in the dataset. An example of the dataset can be seen in Fig. 1. It has been proposed as a direct drop-in replacement for the popular MNIST dataset [10].

A number of NAS methodologies have been able to produce state-of-the-art networks for this specific dataset. The latest NAS paper which was able to produce a new state-of-the-art is [14]. By using the experts advice framework, it is able to achieve a 96.36% test set accuracy, higher than any other NAS paper. Other methodologies are presented and compared in [15], which are able to produce very high-performing architectures. Nonetheless, these are all micro-search methods. Macro-search, although not as successful, is able to produce sufficient architectures, taking into account that the search space usually becomes unbound. One such methodology is described in [1], which combines Genetic Algorithms and Dynamic Structured Grammatical Evolution, in order to evolve the architectures. The algorithm is able to produce high-performing architectures, although some post-search training and inference techniques are utilized, in order to boost the network's performance. Another study employs the Ant Colony Optimization algorithm in order to design neural architectures [3] with marginally less success, although the concept of weight re-usability that the algorithm enables is very interesting. Finally, in [7], the authors propose Bayesian Optimization as a means to guide through the search space, while using network morphism first proposed in [5].

3 Methodology

In this paper, we utilize concepts from DeepNEAT and Regularized Evolution in order to generate convolutional architectures for the Fashion-MNIST dataset.

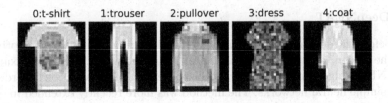

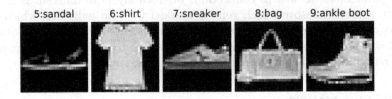

Fig. 1. Sampled Fashion-MNIST images, one for each class.

In order to do so, we first implement the genes that will encode the information regarding the connections, as well as the nodes of each architecture. This is inspired by the work done in [13]. Following, we define the rules of evolution, taken from [17]. Finally, we explain how a genome (a collection of genes) is translated to a functioning neural network and evaluated.

3.1 Architecture Representation

In order to represent a neural architecture, we utilize layer and connection genes. These define the network's topology, as well as each layer's functionality and parameters.

Layer Genes. Each layer gene dictates the layer implemented by each node. Originally, DeepNEAT proposes the use of a Convolution, followed by a Dropout and possibly a MaxPool layer. In this configuration, the gene contains information about the kernel size, number of filters, dropout probability, as well as the existence of the MaxPool layer. As this can greatly increase the search space size (in the original paper, there were 896 possible combinations of kernel size, number of filters, and MaxPooling), we utilize only 12 discrete layer choices. First, we define a fixed number of channels for all architectures, as in [17]. Each node can be either a convolution or a pooling layer, with kernels of sizes 2×2, 3×3, and 5×5. Furthermore, each pooling layer can be either a MaxPooling or an AveragePooling layer. Finally, each convolution layer can either have filters equal to the initially defined number of channels, or half of that. The available choices are depicted in Table 1. Thus, we are able to reduce the number of choices to 12 for each layer gene. As we only employ the mutation operator, we mutate a gene by selecting a different layer setup from the 12 available.

Table 1. Available layer choices.

Layer selection	Layer type	Kernel size	Operation
CONV_2.H	Convolution	2 × 2	Filters Number = Half Channels
CONV_3.H		3 × 3	
CONV_5.H		5 × 5	
CONV_2.N	Convolution	2 × 2	Filters Number = All Channels
CONV_3.N		3 × 3	
CONV_5.N		5 × 5	
POOL_2.A	Pooling	2 × 2	Average
POOL_3.A		3 × 3	
POOL_5.A		5 × 5	
POOL_2.M	Pooling	2 × 2	Max
POOL_3.M		3 × 3	
POOL_5.M		5 × 5	

Connection Genes. Connection genes dictate the network's topology. Each gene contains three values; the origin node, the destination node, as well as a boolean flag, indicating whether the connection is active. The mutation operator reverses the connection's status. Namely, it disables the connection if it is active and vice versa. Although in [17] the authors allow a specific connection to change either its origin or destination, we choose not to. Our choice does not hinder the development of non-sequential architectures i.e. having two or more layers with the same inputs, or a layer with multiple inputs. This will be further analyzed in the following subsection.

3.2 Evolving Architectures

Following the basics of [17], we first initialize a population of P individuals. Each individual consists of 4 random layers. Note that this does not pose a lower limit to the number of layers in the population, as connections can be disabled and thus layers can be deactivated, by disabling its incoming connections. In order to evolve our architectures, we utilize the mutation operator, with a probability of p_{mutate}. Furthermore, we define a probability p_{add} to add a new node to the network, as well as a probability $p_{identity}$ to leave the network intact. At each evolution cycle (offspring creation and replacement of the oldest individual), N individuals are sampled from the population. The best out of N is selected and its offspring is generated through mutation. After evaluating its fitness, the offspring is inserted into the population and the oldest individual is discarded. We utilize the generated network's accuracy on the test set as an individual's fitness.

Adding Nodes. In order to insert a new node into the network, a random connection is selected. The new node is placed between the connection's origin and destination nodes, while the connection is disabled. As a layer gene's mutation allows the re-activation of the connection, the original destination node can have multiple inputs. Likewise, the original origin node can have multiple outputs. This can be seen in Fig. 2. Each node is assigned a unique number which indicates when the node was added. Node −2 is the network's input, while 99 is the network's output. Here it is evident how a single node may have multiple inputs and outputs. Node −2 has developed multiple outputs, while node 99 has developed multiple inputs.

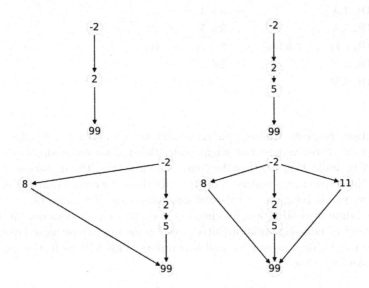

Fig. 2. Three node additions to the same genome.

3.3   Implementing the Networks

As our genomes describe the convolutional section of the network, we define a small outer skeleton, in order to be able to utilize the architecture for classification. We add three fully connected layers (FC) after the output node, with decreasing number of filters. The first FC layer (FC1) has number of filters equal to a quarter of the input features. Layer FC2 has half of the filters FC1 has. Finally, FC3 has 10 filters, equal to the number of classes in the dataset.

When a gene represents a convolution layer, the ReLU activation function is used after the layer, clipped to a maximum value of 6 (ReLU6). Following, a Batch Normalization layer is employed, along with a Dropout2d layer, with 0.1 dropout probability. In case a node has multiple inputs, all of them are scaled to the same size and summed pixel-wise. In case there are inputs with different numbers of channels, a 1×1 convolution is utilized in order to scale the number of channels, before summing the inputs.

4 Experiments

4.1 Experimental Setup

For our experiments, we employed a population size of $P = 100$ genomes. A total of 400 cycles were utilized, in order to evolve the networks. At each cycle, $N = 25$ individuals were selected as parents. Each network was trained for 10 epochs on the train set and then evaluated on the test set. In order to train the networks we utilized the Adam optimizer [8]. No data augmentation was performed on the dataset during the search phase. We chose 64 as the standard number of channels, meaning that CONV_2.H, CONV_3.H, and CONV_5.H will each contain 32 filters. The probability to add a new node to a network was set to $p_{add} = 0.25$, while the probability to leave a network intact was set to $p_{identity} = 0.05$. Thus, the probability to mutate either a node or a connection was $p_{mutate} = 0.7$.

In order to establish a baseline for our method, we also conducted a random search. For this purpose, we repeated the experiment, but selected a single random individual from the population to generate the offspring each time, instead of 25. Thus, the algorithm reduced to a random search. Both experiments were conducted by utilizing 4 NVIDIA Tesla V100 GPU cards. The experiments were implemented using the NORD framework [9]. All code is available on github[1], along with the PyTorch state dictionaries of the fully trained networks.

4.2 Results

Evolutionary search (ES) seems to outperform random search (RS) significantly. It is able to find better models, as well as do so more consistently. This can be seen in Fig. 3, where the best-performing architecture found so far is depicted for both ES and RS. ES is able to continuously improve its best-performing architecture, while RS struggles to find better architectures (left). Furthermore, ES's empirical cumulative distribution function (right) exerts first-order stochastic dominance over ES. This means that it is able to find solutions as good as random search or better, with higher probability. ES was able to find architectures with up to 93.62% test accuracy, which consisted of two CONV_2.N layers, followed by a CONV_3.N, a CONV_5.N, a POOL_5.A, and a CONV_3.H layers, all of them sequentially connected. Random Search was able to find architectures with up to 91.91% test accuracy, although these architectures consisted of a CONV_3.H, CONV_5.H, a CONV_3.H, and a POOL_5.M layer, again connected sequentially.

Most of the best architectures found by ES consisted of sequentially connected convolutions with full channel count, followed by a pooling layer and a reduced-channel convolution. Some architectures developed fanning inputs and outputs, with satisfactory performance. The most prominent features were two sequential CONV_2.N layers, followed by a CONV_3.N, a CONV_5.N, and a

---

[1] https://github.com/GeorgeKyriakides/nord/tree/AIAI2020/nord.

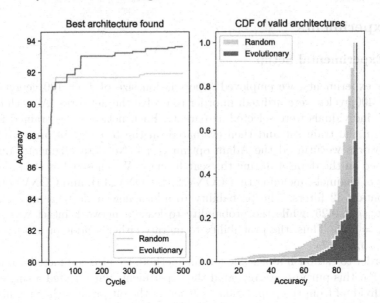

Fig. 3. Best architecture's performance (left) and performance empirical cumulative distributions (right).

POOL_5.A layer. In Fig. 4 each evolved genome is plotted, with alpha (opacity factor) equal to $\frac{1}{400}$ (as we have a total of 400 evolved architectures). Thus, more dominant patterns can be visually distinguished. Here, OUTPUT layers denote the output of the convolutional part of the network, i.e. the input to the fully connected layers. On average, valid evolved architectures contained 5.5 nodes, with a standard deviation of 1.1. Invalid architectures include those that did not contain at least a single path from the input to the output layer, or the intermediate operations were not able to be computed (e.x. a kernel size greater than the input tensor size). Here, it is easy to distinguish the pattern of 4 convolutions of increasing kernel size, as well as the existence of branching layer outputs. Another feature that seems to appear frequently is a convolution layer with kernel size 3, which branches out of the main network's structure and is connected directly to the output layer. Finally, it is interesting to note that the dominant pattern seen in Fig. 4 evolved from many different genomes by mutation, and was then established due to its performance.

We chose the architecture with the best average performance, out of the architectures sampled at least 5 times, in order to further train and evaluate it. We select the average best, as we would like to avoid choosing a network trained to a high accuracy due to luck (initialization weights, sequence of non-deterministic operations). This is also in line with the work done in [12], as similar levels of variance are recorded (0.2% on average). The chosen network closely follows the dominant architecture (first 5 sequential layers in 4), followed by two CONV_3.H layers branching out after the pooling operation. By utilizing data augmentation (horizontal flipping and random erasing) on the train set,

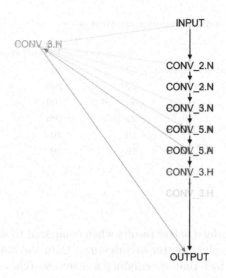

Fig. 4. Overlay of all genomes' architectures.

setting the number of channels to 256 (3.1 million parameters), as well as training the architecture for 20 epochs, we were able to achieve 94.46% test accuracy with this architecture. With a more modest number of 128 channels (791,338 parameters), we were able to achieve a performance of 94.26% accuracy. We call the architecture REMNet (Regularized Evolution for Macro-NAS Net) for ease of reference.

4.3   Discussion

When compared to networks found by similar macro NAS methodologies, applied to the same dataset, the network found by our method seems to perform marginally better. There are post-search training and inference techniques that produce even better-performing networks, such as test-set data augmentation and ensembles of neural networks, used in [1]. By utilizing those techniques, the authors are able to boost their network's performance to 94.7%, with test set augmentation and to 95.26% with ensembles. As we do not utilize any of these, we do not think it is fair to compare our results to those obtained by using them.

   A summary of results obtained by similar methods is depicted in Table 2. One interesting detail is that our methodology needs less epochs to fully train the final architecture, compared to other methods. This can be attributed to two facts. First, we restrict the search phase training epochs to only 10. This seems to pose a regulatory effect on the network's structure, as networks that need a large number of epochs to train have poor performance when evaluated. This has also been observed in [13]. Furthermore, we utilize optimizers with adaptive learning rate. Thus, we do not need to employ complex learning rate policies, as in [1].

Table 2. Performance of similar algorithms

Method	Final accuracy	Search epochs	Training epochs
DENSER [1]	*94.23%* (**94.70%**, test set augment.)	**10**	400
Auto-Keras [7]	93.28%	200	200
DeepSwarm [3]	93.56%	50	100
NASH [5][a]	91.95%	20–105	205
REMNet-256	**94.46%**	**10**	**20**
REMNet-128	*94.26%*	**10**	**20**

[a] As implemented in [7]

4.4 Limitations

Although our method is able to produce better results when compared to similar methodologies, it is not able to produce better architectures than the state-of-the-art. This is partly due to the fact that we conduct a macro-search, and as such explore a greater search space. Furthermore, we do not implement any other methodology to directly compare ours, and as such comparisons may be unfair either to our advantage or disadvantage.

5 Conclusions and Future Work

As Deep Learning is becoming more popular and widely adopted, Neural Architecture Search is also increasingly studied. Many methods are able to produce state-of-the-art networks, by utilizing previous knowledge about the network's macro-structure, and thus conducting NAS on the micro (cell) level.

In this paper we have proposed a methodology for macro NAS, by combining elements of regularized evolution utilized for micro NAS [17], and DeepNEAT [13], utilized for macro-search. Compared to [17] we were able to conduct search not only on a cell-level, but also on a macro-architecture level. Compared to [13], we had less parameters to optimize and less operations to define, as we did not utilize crossover and we restricted the parameter space to 12 discrete values.

By applying our method on the Fashion-MNIST dataset, we were able to out-perform other macro NAS methodologies, when the post-search evaluation methods were similar. We designed networks capable of up to 94.46% accuracy on the test set, while requiring relatively few epochs to train. As such, we hope to continue our experimentation, in order to expand it to other datasets and network types, as well as study the generated architectures further.

Acknowledgements. This work was supported by computational time granted from the Greek Research & Technology Network (GRNET) in the National HPC facility - ARIS - under project ID DNAD. Furthermore, this research is funded by the University of Macedonia Research Committee as part of the "Principal Research 2019" funding program.

This paper's first author was supported by the Hellenic Foundation for Research and Innovation (HFRI) under the HFRI PhD Fellowship grant (Award Number: 20540).

References

1. Assunçao, F., Lourenço, N., Machado, P., Ribeiro, B.: Denser: deep evolutionary network structured representation. Genet. Program Evolvable Mach. **20**(1), 5–35 (2019)
2. Baccouche, M., Mamalet, F., Wolf, C., Garcia, C., Baskurt, A.: Sequential deep learning for human action recognition. In: Salah, A.A., Lepri, B. (eds.) HBU 2011. LNCS, vol. 7065, pp. 29–39. Springer, Heidelberg (2011). https://doi.org/10.1007/978-3-642-25446-8_4
3. Byla, E., Pang, W.: DeepSwarm: optimising convolutional neural networks using swarm intelligence. In: Ju, Z., Yang, L., Yang, C., Gegov, A., Zhou, D. (eds.) UKCI 2019. AISC, vol. 1043, pp. 119–130. Springer, Cham (2020). https://doi.org/10.1007/978-3-030-29933-0_10
4. Cheng, Z., Yang, Q., Sheng, B.: Deep colorization. In: Proceedings of the IEEE International Conference on Computer Vision, pp. 415–423 (2015)
5. Elsken, T., Metzen, J.H., Hutter, F.: Simple and efficient architecture search for convolutional neural networks. arXiv preprint arXiv:1711.04528 (2017)
6. He, K., Wang, Y., Hopcroft, J.: A powerful generative model using random weights for the deep image representation. In: Advances in Neural Information Processing Systems. pp. 631–639 (2016)
7. Jin, H., Song, Q., Hu, X.: Auto-keras: an efficient neural architecture search system. In: Proceedings of the 25th ACM SIGKDD International Conference on Knowledge Discovery & Data Mining, pp. 1946–1956 (2019)
8. Kingma, D.P., Ba, J.: Adam: a method for stochastic optimization. arXiv preprint arXiv:1412.6980 (2014)
9. Kyriakides, G., Margaritis, K.G.: Towards automated neural design: an open source, distributed neural architecture research framework. In: Proceedings of the 22nd Pan-Hellenic Conference on Informatics, pp. 113–116 (2018)
10. LeCun, Y., Bottou, L., Bengio, Y., Haffner, P.: Gradient-based learning applied to document recognition. Proc. IEEE **86**(11), 2278–2324 (1998)
11. Liu, C., et al.: Progressive neural architecture search. In: Proceedings of the European Conference on Computer Vision, ECCV, pp. 19–34 (2018)
12. Liu, H., Simonyan, K., Vinyals, O., Fernando, C., Kavukcuoglu, K.: Hierarchical representations for efficient architecture search. arXiv preprint arXiv:1711.00436 (2017)
13. Miikkulainen, R., et al.: Evolving deep neural networks. In: Artificial Intelligence in the Age of Neural Networks and Brain Computing, pp. 293–312. Elsevier (2019)
14. Nayman, N., Noy, A., Ridnik, T., Friedman, I., Jin, R., Zelnik, L.: Xnas: neural architecture search with expert advice. In: Advances in Neural Information Processing Systems, pp. 1975–1985 (2019)
15. Noy, A., et al.: Asap: Architecture search, anneal and prune. arXiv preprint arXiv:1904.04123 (2019)
16. Pham, H., Guan, M.Y., Zoph, B., Le, Q.V., Dean, J.: Efficient neural architecture search via parameter sharing. arXiv preprint arXiv:1802.03268 (2018)
17. Real, E., Aggarwal, A., Huang, Y., Le, Q.V.: Regularized evolution for image classifier architecture search. In: Proceedings of the AAAI Conference on Artificial Intelligence, vol. 33, pp. 4780–4789 (2019)
18. Stanley, K.O., Miikkulainen, R.: Evolving neural networks through augmenting topologies. Evol. Comput. **10**(2), 99–127 (2002)

19. Wu, B., et al.: Fbnet: hardware-aware efficient convnet design via differentiable neural architecture search. In: Proceedings of the IEEE Conference on Computer Vision and Pattern Recognition, pp. 10734–10742 (2019)
20. Wu, B., Iandola, F., Jin, P.H., Keutzer, K.: Squeezedet: unified, small, low power fully convolutional neural networks for real-time object detection for autonomous driving. In: Proceedings of the IEEE Conference on Computer Vision and Pattern Recognition Workshops, pp. 129–137 (2017)
21. Xiao, H., Rasul, K., Vollgraf, R.: Fashion-MNIST: a novel image dataset for benchmarking machine learning algorithms. arXiv preprint arXiv:1708.07747 (2017)
22. Zoph, B., Le, Q.V.: Neural architecture search with reinforcement learning. arXiv preprint arXiv:1611.01578 (2016)
23. Zoph, B., Vasudevan, V., Shlens, J., Le, Q.V.: Learning transferable architectures for scalable image recognition. In: Proceedings of the IEEE conference on computer vision and pattern recognition, pp. 8697–8710 (2018)

Using Multimodal Contextual Process Information for the Supervised Detection of Connector Lock Events

David Bricher[✉] and Andreas Müller

Institute of Robotics, Johannes Kepler University, Linz, Austria
`david.bricher@bmw.com, a.mueller@jku.at`

Abstract. The field of sound event detection is a growing sector which has mainly focused on the identification of sound classes from daily life situations. In most cases these sound detection models are trained on publicly available sound databases, up to now, however, they do not include acoustic data from manufacturing environments. Within manufacturing industries, acoustic data can be exploited in order to evaluate the correct execution of assembling processes. As an example, in this paper the correct plugging of connectors is analyzed on the basis of multimodal contextual process information. The latter are the connector's acoustic properties and visual information recorded in form of video files while executing connector locking processes.

For the first time optical microphones are used for the acquisition and analysis of connector sound data in order to differentiate connector locking sounds from each other respectively from background noise and sound events with similar acoustic properties. Therefore, different types of feature representations as well as neural network architectures are investigated for this specific task.

The results from the proposed analysis show, that multimodal approaches clearly outperform unimodal neural network architectures for the task of connector locking validation by reaching maximal accuracy levels close to 85%. Since in many cases there are no additional validation methods applied for the detection of correctly locked connectors in manufacturing industries, it is concluded that the proposed connector lock event detection framework is a significant improvement for the qualitative validation of plugging operations.

Keywords: Connector lock detection · Manufacturing sound events · Sound event detection · Applied machine learning · Neural networks · Optical microphone · Deep learning

1 Introduction

Although the degree of automation in manufacturing industries is constantly increasing, the correct plugging and locking of cable harness connectors is still a challenging task for machines [5, 11] and is therefore mainly carried out manually.

© IFIP International Federation for Information Processing 2020
Published by Springer Nature Switzerland AG 2020
I. Maglogiannis et al. (Eds.): AIAI 2020, IFIP AICT 584, pp. 123–134, 2020.
https://doi.org/10.1007/978-3-030-49186-4_11

As the human failure rate for manual working tasks typically exceeds the failure rate of automated processes, additional validation methods are introduced, such that possible errors can be detected before leaving the manufacturing line.

In many cases the validation of correctly assembled parts can be achieved with visual inspection, but the investigation of correctly assembled connector cables solely based on image data is rather inefficient due to multiple reasons, e.g. occlusion by other assembling parts, variation of connector positions or drill of connector cables. Moreover, the validation of plugging processes is often not carried out at all.

Thus, it is of great interest to find multimodal contextual process information which can be used for the validation of correctly executed connector plugging processes. The most common errors occurring are those where the connectors have not been locked properly. A potential approach to assess, whether a locking has been correctly performed, is to analyze the inherent acoustic properties of the plugging event. Consequently, the presence and correct classification of locking sound events can provide information on the qualitative execution of plugging processes.

The field of sound event detection is a challenging sector, whose main target is to mimic the human capability of distinguishing different acoustic events and correctly classifying them. In the last decade multiple approaches (e.g. Gaussian mixture models (GMM) [2,16], hidden Markov models (HMM) [7,14], support vector machines (SVM) [17], random forests [3] or different deep neural network (NN) architectures [4,9,10,12]) have been applied to this field but they have been mostly evaluated on publicly available benchmark datasets, which are composed of audio scenes from everyday life (e.g. TUT sound event databases [15] or DCASE databases [13]). Up to now, there are no datasets which specialise on sound events from manufacturing environments. Especially, the task of connector locking detection has been hitherto comparatively little explored [1].

For this reason, this paper investigates the performance of different neural network architectures in order to distinguish different connector locking events from each other respectively from other "fake" events with similar acoustic properties in a manufacturing environment. Due to the short sound duration of connector plugging events, for the first time, an optical membrane-free sensor with high sampling rates is used for the acoustic data acquisition. In order to increase the robustness of the connector locking assessment, multimodal sensor data are extracted in order to improve the classification accuracy. In particular, in addition to sound data, video data obtained by capturing the workflow during the execution of manufacturing working tasks is used. In this paper, a neural network based framework for assessment of connector plugging is presented. The processing of sound and video data is discussed in detail and the performance of all different network architectures is analyzed.

a)   b)

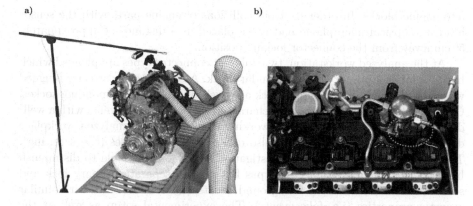

Fig. 1. (a) Experimental setup for multimodal data acquisition. (b) The analyzed connector types 1 and 2 are highlighted with a green and a red circle. (Color figure online)

2 Optical Microphones

The working principle of the used membrane-free optical microphone exploits the properties of a rigid Fabry-Pérot interferometer [8]. Laser light is transmitted through an optical fibre to a semi-reflective two-mirror system within the sensor head. Sound events from the external environment can penetrate the etalon through a small aperture window. Subsequently, the changing pressure of the sound wave causes fluctuations of the refractive index within the laser propagating medium, i.e. the change of the refractive index is leading to slight wavelength shifts of the transmitted and reflected laser light in the Fabry-Pérot interferometer. In order to draw conclusion on the occurring sound event, the outgoing laser light from the etalon interferes with the incoming laser beam which is leading to detectable laser intensity fluctuations. These intensity fluctuations are transformed to analogue voltage signals which can be used for the examination of the acoustic signal.

In contrast to state-of-the-art membrane microphones, optical microphones offer linear frequency responses from 10 Hz up to 1 MHz and allow sampling rates of up to 4 MHz. Thus, it is possible to resolve sound events with time durations below 1 ms (e.g. connector locking events) very accurately.

3 Sound Data Acquisition and Representation

3.1 Data Generation

Since there are no publicly available datasets for the task of connector lock detection, it is mandatory to acquire a sufficiently large set of locking sound data in order to train a supervised machine learning model. To this end, the optical microphone has been installed at the final assembly line of an engine manufacturing plant. At the considered workstation, electric connectors must be plugged into engine blocks. In order to avoid collisions of engine parts with the sensor head, the optical microphone had to be placed at a distance of approximately 50 cm away from the connector locking position.

At the analyzed workstation, two different connector types are plugged which both comprise a primary and a secondary lock. A connector is correctly plugged when the primary and secondary lock are pushed into the corresponding socket. Consequently, a correct plugging is characterized by two click events with a well-defined temporal separation. The working contents of the analyzed workplace do not only include plugging but also other working processes (e.g. screwing). Thus, it is a main aim of this investigation to not only be able to distinguish between two different connector types but also to separate primary lock and secondary lock events from background events and sound events with similar acoustic properties (i.e. fake events). The experimental setup as well as the analyzed connector types are depicted in Fig. 1.

The start and stop of the sound measurements have been triggered with digital outputs whenever a new engine is conveyed to the workstation. In order to assign the acoustic signal to the corresponding sound source, video files of the working tasks have been captured simultaneously for each acoustic sound sample taken.

3.2   Data Annotation

In order to train a supervised machine learning model, the generated sound data have to be annotated. Thus, it is mandatory to determine at which point in time the locking events occur. For this reason, those local maxima from the analogue signal need to be determined, whose signal-to-noise ratio exceeds a predefined threshold level A_{thres} and whose minimal temporal separation lies above a threshold t_{sep}.

Within the proposed analysis the datasets have been labeled by hand, i.e. the determined maxima at the given time instances are classified according to the executed work step recorded on the corresponding video files. Within the proposed analysis the following classes have been considered: background (BG), primary lock connector 1 (CP1), secondary lock connector 1 (CS1), primary lock connector 2 (CP2), secondary lock connector 2 (CS2), fake event (FE). The background events are generated from recorded data before the locking process of connector 1 has been initialized. Those maxima which could not be assigned to a connector locking event are classified as fake events. Subsequently, in sum 1,223 data samples have been generated, of which 988 samples have been used for training, while the remaining 235 data samples are used for testing. The distribution of data classes is given in Table 1.

Table 1. Distribution of data classes

Classes	Distribution [%]
Background (BG)	26
Primary lock connector 1 (CP1)	15
Secondary lock connector 1 (CS1)	16
Primary lock connector 2 (CP2)	15
Secondary lock connector 2 (CS2)	10
Fake event (FE)	18

3.3 Feature Generation

Before training the model, the feature representation that is best suited for the classification task should be generated from the input data. The information about a sound event class is typically not stored in a single time frame but over a consecutive temporal context, i.e. the feature representation comprises a temporal sequence of feature vectors. The analyzed connector locking sound events typically last for approximately 1 ms. In order to avoid a cropping of the locking sound signal, a maximal feature time duration of 3 ms is chosen (1 ms before and 2 ms after the sound peak maximum). By applying a sample rate of 4 MHz a time window of 3 ms corresponds to a feature length n_F of 12,000 for the extracted analogue signal. In order to determine the optimal feature length, n_F is treated as a hyperparameter and optimized by means of a grid search. Thus, n_F highly depends on the feature space chosen, i.e. in frequency space the feature length is set together by the number of frequency contributions times the analyzed time steps.

Since the use of optical microphone data for audio event detection is so far an unexplored discipline, three different types of input features from time and frequency domain are investigated in terms of their locking event detection performance:

a) *Analogue time signal*: Due to the high sampling rate of the optical microphone, the time signal can be resolved very accurately. The feature representation of the connector locking event is described by 12,000 input features.
b) *Log-STFT signal*: From the logarithmic frequency spectrum of the Short-time Fourier Transform (STFT) follows, that primary and secondary locking events show frequency contributions up to 100 kHz, which correspond to the first 40 amplitude contributions from the log-STFT spectrum. In total, this gives a log-STFT feature length $n_F = 6,000$.
c) *MFCC features*: In the field of sound event detection the frequency information from the STFT is further processed in order to find an optimal set of sound event features. In many cases the spectogram is transformed to the mel scale, which shows higher frequency resolutions in lower frequency domains. By applying Discrete Cosine Transforms over the mel spectogram, Mel Frequency Cepstral Coefficients (MFCC) can be generated which are often used

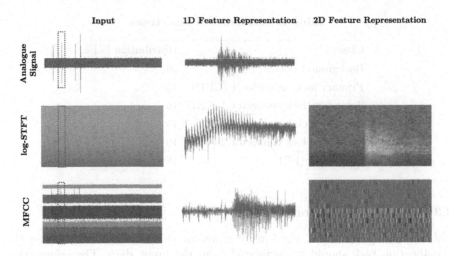

Fig. 2. Comparison of input data and feature representations of a locking event generated from analogue signal, log-STFT signal and MFCC features. The locking events are highlighted in the input data with a red dashed box. (Color figure online)

for the spectral representation of acoustic signals [6]. Within this paper the first 13 MFCCs are considered which correspond to an overall MFCC feature length of 1,950.

All of the investigated features are normalized to a range $[-1,1]$. In contrast to the other feature representations analyzed, the MFCC features are standardized with mean zero and a standard deviation of one before getting normalized. The feature representations generated on the basis of their input signals are exemplified in Fig. 2.

4   Data Processing

Since manufacturing environments are well-known for machinery noise, it is questionable whether the correct locking of connector cables can be determined solely from the acoustic properties of clicking events. Thus, in the proposed work the performance of unimodal as well as multimodal neural network architectures are both investigated for the specific task of detecting connector locking events. The different types of neural network architectures analyzed are introduced in the following.

4.1   Unimodal Neural Network Architectures

Neural network architectures tend to outperform other approaches for sound event detection, and the following four different architecture types are considered in the proposed work.

a) *Neural Network*: The chosen feed-forward neural network architecture (NN) is composed of multiple hidden layers followed by the output layer used for classification.

b) *1D-Convolutional Neural Network*: One-dimensional convolutional neural networks (CNN) are investigated in order to capture temporal correlation effects of the input features. The CNN architecture is set up by multiple convolutional hidden layers followed by pooling layers.

c) *2D-Convolutional Neural Network*: Two-dimensional CNN architectures are the state-of-the-art approach for image data classification, and the log-STFT spectrum as well as the MFCC features are not only analyzed with a one-dimensional but also with a two-dimensional feature representation. Consecutive 2D convolutional layers are applied in connection with max pooling layers, which are fed into fully connected layers, before being transferred to the final classification layer.

d) *Combined Model (NN+CNN)*: In order to make use of the acoustic signal information in time and frequency domain, a combined model is introduced, which makes use of a NN for analogue signal features, while the frequency features are fed into a one-dimensional CNN branch. Both networks are merged into two joint fully connected layers, which are followed by the output layer for classification.

Table 2. Hyperparameter choices for neural network architecture optimization

Hyperparameter	Range
Feature length n_F fraction	1/10, 1/6, 1/2, 1
Number of layers	1, 2, 4, 6
Number of neurons per layer and input feature length	1, 2, 4, 6
Learning rate	0.01, 0.001, 0.0001

All of the proposed architectures are analyzed for different sets of hyperparameters. A grid search is carried out for all network types in order to find the optimal choice of hyperparameters. The range of used hyperparameters for the grid search are given in Table 2.

4.2 Multimodal Neural Network Architectures

In order to further improve the validation performance of the described neural network architectures for connector locking event classification, it is beneficial to extend the amount of gathered information from the plugging processes. Further, using multiple input sources is beneficial in order to better distinguish and characterize similar manufacturing working steps. These different feature representations can all be fed individually into separate branches of a multimodal

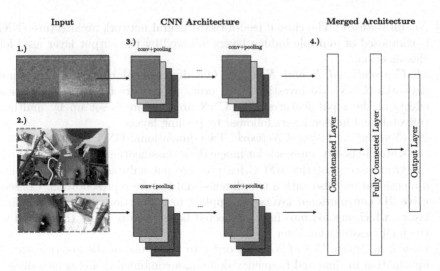

Fig. 3. Multimodal neural network architecture using log-STFT features as input.

neural network architecture. As an example the processing steps of a multi-modal architecture are shown for log-STFT features in Fig. 3 and explained in the following.

1.) For the proposed task the multimodal neural network processes the sound features in the first branch of the network - the audio branch.
2.) The second branch - the visual branch - processes the recorded video files during process execution. As the plugging processes are carried out at the same location for all engine types, small image patches ($100 \times 100 \times 3$) centered at the two connector locking positions are cropped from the video files and merged into one image patch ($100 \times 200 \times 3$). These images are chosen from the video file in accordance with the instance in time when an acoustic peak is determined.
3.) The visual branch consists of a 2D-CNN architecture which is composed of several blocks of convolutional and pooling layers that are followed by fully connected layers.
4.) The last fully connected layer of the visual branch is merged with the last fully connected layer of the audio branch. The concatenated layer is again fed into fully connector layers followed by the output layer which is then used for classification.

5   Experimental Results

The trained models for connector lock event detection are validated on the basis of the following two evaluation metrics: *Accuracy* and the F_1-*score*. The accuracy is calculated as

$$Acc = \frac{TP + TN}{TP + FP + TN + FN} \tag{1}$$

with TP, TN, FP, FN corresponding to the number of true positive, true negative, false positive and false negative predictions. The F_1-score is determined as

$$F_1 = 2 \cdot \frac{PRE \cdot REC}{PRE + REC} \tag{2}$$

with PRE being the classification precision and REC the classification recall.

The overall results of the unimodal (sound only) and multimodal (sound and visual) neural network architectures are given in Table 3. With only reaching maximal accuracy values close to 75% (for the log-STFT with 2D-CNN architecture), one can deduce, that the sole use of acoustic information processed by the analyzed neural network architectures is not robust enough to classify connector locking events in a manufacturing environment. From the results follow that log-STFT and MFCC feature representations are clearly preferable over the sampled analogue signal for the task of connector locking classification. With regard to the investigated neural network architectures, 2D-CNNs exceed the performance of all other architectures.

Table 3. Accuracy and F_1 results obtained for connector locking detection using the proposed unimodal and multimodal neural network architectures.

Method	Unimodal architectures						Multimodal architectures			
	Analogue Signal		log-STFT		MFCC		log-STFT		MFCC	
	Acc	F1	Acc	F1	Acc	F1	Acc	F1	Acc	F1
NN	0.54	0.48	0.64	0.55	0.66	0.62	0.63	0.61	0.72	0.68
CNN (1D)	0.62	0.57	0.66	0.61	0.72	0.69	0.74	0.71	0.84	0.82
CNN (2D)	-	-	0.75	0.71	0.69	0.62	0.84	0.81	0.84	0.82
NN + CNN	-	-	0.66	0.61	0.70	0.64	-	-	-	-

Compared to the unimodal results, the investigated multimodal approaches outperform all investigated unimodal approaches with maximal accuracy levels close to 85% (for the log-STFT/MFCC and 2D-CNN architecture). Thus, exploiting multiple process information can help to describe complex tasks like the correct locking of connector cables. A more detailed evaluation of the results (illustrated by the confusion matrix of Fig. 4) shows, that the occurring error can be mainly attributed to mispredictions of primary and secondary locking events for both connector types. Apart from these false predictions, there are still a few cases where fake events get predicted as locking events. This scenario is definitely more problematic than connector lock mispredictions, because in the worst case the plugging process would be classified as correctly executed,

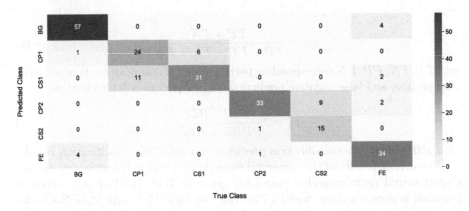

Fig. 4. Confusion matrix for the evaluation of the 235 test data samples using the best performing log-STFT multimodal network architecture.

although the connector might not even be plugged at all. In practice this case would only occur, when two consecutive fake events are wrongly classified as primary and secondary lock events of a specific connector type. Nevertheless, since most connector plugging processes are not validated at all at the assembly lines, the proposed connector lock detection framework can lead to a significant improvement of the quality validation in manufacturing. By integrating the framework in series production, additional data can be collected and the model can be further optimized.

6   Conclusion

In this paper, the task of connector locking detection has been investigated by making use of optical microphones. Different unimodal and multimodal neural network architectures have been trained in order to distinguish primary and secondary locking events of two different connector types from events with similar sound characteristics respectively from each other and hence to assess whether connectors were correctly plugged. The obtained results indicate that multimodal neural network architectures making use of acoustic and visual process information clearly outperform unimodal approaches that only take into account sound features and achieve connector locking classification accuracy scores close to 85%. Since currently there is no check for correct connector plugging, the proposed framework can help directly to increase the quality of process execution in manufacturing.

It would be of great interest to analyze, if additional contextual process information from sensor data (e.g. the force or pressure measured at the thumb during the plugging process) could help to eliminate the occurring mispredictions and would thereby allow a more robust use in manufacturing. Furthermore, the used laser microphone had to be positioned at a comparatively high separation distance which is accompanied by an attenuation of high-frequency components and thus partly annihilate the advantage of the investigated optical sensor. Instead, one could install small membrane microphones in the glove of an employee and thereby extract sound events better which only occur in the close vicinity of the microphone. Thus, the acquired sound data could potentially lead to a more robust detection of connector lock events in combination with the proposed multimodal framework.

References

1. Aoyagi, M., Ueno, T., Okuda, M.: Automatic detection system for complete connection of a waterproof soft-shell electronic connector with a sliding locking device. IEEE Sens. J. **9**(3), 285–292 (2009). https://doi.org/10.1109/JSEN.2008.2012225
2. Atrey, P.K., Maddage, N.C., Kankanhalli, M.S.: Audio based event detection for multimedia surveillance. In: 2006 IEEE International Conference on Acoustics Speech and Signal Processing Proceedings, vol. 5, pp. 813–816, May 2006. https://doi.org/10.1109/ICASSP.2006.1661400
3. Barchiesi, D., Giannoulis, D., Stowell, D., Plumbley, M.D.: Acoustic scene classification: classifying environments from the sounds they produce. IEEE Signal Process. Mag. **32**(3), 16–34 (2015). https://doi.org/10.1109/MSP.2014.2326181
4. Cakir, E., Heittola, T., Huttunen, H., Virtanen, T.: Polyphonic sound event detection using multi label deep neural networks. In: 2015 International Joint Conference on Neural Networks, IJCNN, pp. 1–7, July 2015. https://doi.org/10.1109/IJCNN.2015.7280624
5. Cho, H., Kim, Y., Kim, B., Song, J.: A strategy for connector assembly using impedance control for industrial robots. In: 2012 12th International Conference on Control, Automation and Systems, pp. 1433–1435, October 2012
6. Davis, S., Mermelstein, P.: Comparison of parametric representations for monosyllabic word recognition in continuously spoken sentences. IEEE Trans. Acoust. Speech Signal Process. **28**(4), 357–366 (1980). https://doi.org/10.1109/TASSP.1980.1163420
7. Eronen, A.J., et al.: Audio-based context recognition. IEEE Trans. Audio Speech Lang. Process. **14**(1), 321–329 (2006). https://doi.org/10.1109/TSA.2005.854103
8. Fischer, B.: Optical microphone hears ultrasound. Nat. Photonics **10**, 356–358 (2016). https://doi.org/10.1038/nphoton.2016.95
9. Hayashi, T., Watanabe, S., Toda, T., Hori, T., Le Roux, J., Takeda, K.: Duration-controlled LSTM for polyphonic sound event detection. IEEE/ACM Trans. Audio Speech Lang. Process. **25**(11), 2059–2070 (2017). https://doi.org/10.1109/TASLP.2017.2740002
10. Hershey, S., et al.: CNN architectures for large-scale audio classification. In: 2017 IEEE International Conference on Acoustics, Speech and Signal Processing, ICASSP, pp. 131–135, March 2017. https://doi.org/10.1109/ICASSP.2017.7952132
11. Jorg, S., Langwald, J., Stelter, J., Hirzinger, G., Natale, C.: Flexible robot-assembly using a multi-sensory approach. In: Proceedings 2000 ICRA. Millennium Conference. IEEE International Conference on Robotics and Automation Symposia Proceedings, vol. 4, pp. 3687–3694, April 2000. https://doi.org/10.1109/ROBOT.2000.845306. (Cat. No.00CH37065)

12. Marchi, E., Vesperini, F., Eyben, F., Squartini, S., Schuller, B.: A novel approach for automatic acoustic novelty detection using a denoising autoencoder with bidirectional LSTM neural networks. In: 2015 IEEE International Conference on Acoustics, Speech and Signal Processing, ICASSP, pp. 1996–2000, April 2015. https://doi.org/10.1109/ICASSP.2015.7178320

13. Mesaros, A., et al.: DCASE 2017 challenge setup: tasks, datasets and baseline system. In: DCASE 2017 - Workshop on Detection and Classification of Acoustic Scenes and Events. Munich, Germany, November 2017. https://hal.inria.fr/hal-01627981

14. Mesaros, A., Heittola, T., Eronen, A., Virtanen, T.: Acoustic event detection in real life recordings. In: 2010 18th European Signal Processing Conference, pp. 1267–1271, August 2010

15. Mesaros, A., Heittola, T., Virtanen, T.: TUT database for acoustic scene classification and sound event detection. In: 2016 24th European Signal Processing Conference, EUSIPCO, pp. 1128–1132, August 2016. https://doi.org/10.1109/EUSIPCO.2016.7760424

16. Oldoni, D., De Coensel, B., Rademaker, M., De Baets, B., Botteldooren, D.: Context-dependent environmental sound monitoring using SOM coupled with LEGION. In: The 2010 International Joint Conference on Neural Networks, IJCNN, pp. 1–8, July 2010. https://doi.org/10.1109/IJCNN.2010.5596977

17. Rakotomamonjy, A., Gasso, G.: Histogram of gradients of time-frequency representations for audio scene classification. IEEE/ACM Trans. Audio Speech Lang. Process. **23**(1), 142–153 (2015). https://doi.org/10.1109/TASLP.2014.2375575

A Machine Learning Model to Detect Speech and Reading Pathologies

Fabio Fassetti[1]([✉]) and Ilaria Fassetti[2]

[1] DIMES, University of Calabria, Rende, Italy
f.fassetti@dimes.unical.it
[2] Therapeia Rehabilitation Center, Rende, Italy
ilaria.fassetti@gmail.com

Abstract. This work addresses the problem of helping speech therapists in interpreting results of tachistoscopes. These are instruments widely employed to diagnose speech and reading disorders. Roughly speaking, they work as follows. During a session, some strings of letters, which may or not correspond to existing words, are displayed to the patient for an amount of time set by the therapist. Next, the patient is asked for typing the read string. From the machine learning point of view, this raise an interesting problem of analyzing the sets of input and output words to evaluate the presence of a pathology.

Keywords: Tachistoscope · Deep neural networks · Dyslexia

1 Introduction

This work aims at helping in analyzing results of the tachistoscope [5], a widely used diagnostic tool by speech therapists for reading or writing disorders.

Tachistoscopes [1] are employed to diagnose reading disorders related to many kinds of dyslexia and disgraphia and, also, to increase recognition speed and to increase reading comprehension.

In more details, tachistoscopes are devices that display a word for several seconds and ask the patient to write down the word. Many parameters can be set during the therapy session, like the duration of displayed word, the length of the word, the structural easiness of the word, how much the word is common and others [7]. Among them a relevant role is played by the possibility for the therapist of choosing existent or non existing words. Indeed, the presence of non existing words avoids the help coming from the semantic interpretation [4].

Hence, a tachistoscope session provides the set of configuration parameters, the set of displayed words and the set of words as typed by the patient. The set of produced words obviously depends on the pathologies affecting the patient

© IFIP International Federation for Information Processing 2020
Published by Springer Nature Switzerland AG 2020
I. Maglogiannis et al. (Eds.): AIAI 2020, IFIP AICT 584, pp. 135–142, 2020.
https://doi.org/10.1007/978-3-030-49186-4_12

but, in many cases, more pathologies can simultaneously be present and this, together with the many words to be analyzed, make diagnosis a tedious and very hard task [2,3,6].

This work faces the challenging problem of recognizing pathologies involved and helping experts by suggesting which pathologies and to what extent are likely to affect the patient.

To the best of our knowledge this is the first work attempting to tackle this problem. Authors of [5] address the question of mining tachistoscope output but from a totally different point of view. Indeed, there, the focus is on individuating patterns characterizing errors and not on pathologies.

However, the proposed framework is more widely applicable and, then, this work aims at providing general contributions to the field.

The paper is organized as follows. Section 2 presents preliminary notions, Sect. 3 details the technique proposed to face the problem at hand, Sect. 4 reports experiments conducted to show the effectiveness of the approach and, finally, Sect. 5 depicts conclusions.

2 Preliminaries

Let Σ be the considered alphabet composed, for English, by 26 letters plus stressed vowels and, then, by 32 elements. A word w is an element of Σ^*. Let $\mathcal{W}$ be a set of words, for each pathology p, let $\mathcal{F}_p$ be a transfer function that converts a word $w \in \mathcal{W}$ in a new word w_p due to the pathology p.

Tachistoscope. The tachistoscope is an instrument largely employed for detecting many reading and writing disorders. In a session, the user is provided with a sequence of trials each having a word associated with it. The trial consists in two phases: *visualization phase* and *guessing phase*. During the visualization phase a word is shown for some milliseconds (typically ranging from 15 to 1,500). To this phase, the guessing phase follows. During this phase, the word disappears and the user has to guess the word by typing it. After that the word disappears, it could be substituted by a set of '#' to increase the difficulty since the user loses the visive memory. In such a case, the word is said "masked".

The therapist, other that the visualization time, can impose several settings about the word, in particular, (*i*) the *frequency* of the word, which represents how much the word is of current use; (*ii*) the *length* of the word, which represents how much long is the word; (*iii*) the *easiness* of the word, which represents how much difficult is reading and writing the word (for example, each consonant is followed by a vowel); (*iv*) the *existence* of the word, which represents the existence of the word in the dictionary. As for the existence of the word, the tachistoscope is able to show both existing words and non-existing words which are random sequences of letters with the constraints that (*i*) each syllable has to appear in at least one existing word and (*ii*) each pair of adjacent letters has to appear in at least one existing word. Such constraints aim at generating readable sequences of letters.

3 Technique

In this section, the main problem addressed in this work is formally presented.

Definition 1 (Problem). *Let $\mathcal{W}^{in}$ be a set of input instances and $\mathcal{W}^{out}$ the set of output instances, provide the likelihood λ that the patient producing $\mathcal{W}^{out}$ is affected by pathology p.*

In this scenario, the available examples provide as information that w^{in} is transformed in w_i^{out} when individuals affected by the pathology p_i elaborate w^{in}.

For a given pathology p, experts provide sets $\mathcal{W}_p^{in}$ and $\mathcal{W}_p^{out}$, which are, respectively, the sets of correct and erroneous words associated with individuals affected by pathology p, where for each word $w \in \mathcal{W}_p^{in}$ there is an associated word in $\mathcal{W}_p^{out}$.

For a given patient, tachistoscope sessions provide two sets of words $\mathcal{W}^{in}$ and $\mathcal{W}^{out}$, where the former one is composed by the correct words submitted to the patient, the latter one is composed by the words, possibly erroneous, typed by the patient. Such sets represent the input of the proposed technique.

The aim of the technique is to learn the transfer function $\mathcal{F}_p$ by exploiting sets $\mathcal{W}_p^{in}$ and $\mathcal{W}_p^{out}$, so that the system is able to reconstruct how a patient would have typed a word $w \in \mathcal{W}^{in}$ if affected by p. Hence, by comparing the word of $\mathcal{W}^{out}$ associated with w and the word $\mathcal{F}_p(w)$, the likelihood for the patient to be affected by p can be computed. Thus, the technique has the following steps:

1. encode words in a numeric feature space;
2. learn function $\mathcal{F}_p$;
3. compute the likelihood of pathology involvement.

Each step is detailed next.

3.1 Word Encoding

The considered features are associated with the occurrences in the word of letters in Σ and pairs of letters in $\Sigma \times \Sigma$. Also, there are six features, said *contextual*: *time, masking, existence, length, frequency* and *easiness*, related to the characteristics of the session during with the word is submitted to the patient. Thus, the numeric feature space $\mathcal{S}$ consists in $32 + 32 \cdot 32 + 6$ components where the i-th component $\mathcal{S}_i$ of $\mathcal{S}$, with $i \in [0, 32)$, is associated with a letter in Σ, the i-th component $\mathcal{S}_i$ of $\mathcal{S}$, with $i \in [32, 32 + 32 \cdot 32)$, is associated with a pair of letters in $\Sigma \times \Sigma$ and the i-th component $\mathcal{S}_i$ of $\mathcal{S}$, with $i \in [32 + 32 \cdot 32, 32 + 32 \cdot 32 + 6)$, is associated with a contextual feature.

Given a word $w \in \Sigma^*$, the encoding of w in a vector $v \in \mathcal{S}$ is such that the i-th component $v[i]$ of v is the number of occurrences of the letter or the pair of letters associated with $\mathcal{S}_i$ in w, for any $i \in [0, 32 + 32 \cdot 32)$, while, as for the contextual features, the value v assumes on these depends on the settings of the session as described in Sect. 2.

In the following, since there is a one to one correspondence between letters, pairs of letters and components, for the sake of readability, the notation $v[s]$, with $s \in \Sigma$ or $s \in \Sigma \times \Sigma$, is employed instead of $v[i]$, with i such that $\mathcal{S}_i$ is associated with s.

Hence, for example, $w =$ "*paper*" is encoded in a vector v such that $v['P'] = 2$, $v['A'] = 1$, $v['E'] = 1$, $v['R'] = 1$, $v['PA'] = 1$, $v['AP'] = 1$, $v['PE'] = 1$, $v['ER'] = 1$, and 0 elsewhere. Note that, as a preliminary steps, all words are uppercased.

Analogously, the notation $v['time']$, $v['masking']$, $v['existence']$, $v['length']$, $v['frequency']$ and $v['easiness']$ is employed to indicate the value the vector assumes on the components associated with contextual features.

3.2 Learning Model

Let $n = |\mathcal{S}|$ be the number of considered features. After some trials, the architecture of the neural network for this phase is designed as an autoencoder with five dense layers configured as follows:

layer 1:   kind: *Dense*,   neurons: n,   activation: '*relu*';
layer 2:   kind: *Dense*,   neurons: $2n$,   activation: '*relu*';
layer 3:   kind: *Dense*,   neurons: n,   activation: '*relu*'.

Differently from classical auto encoders, latent space is larger than input and output ones this is due to the fact that here it is not relevant to highlight characterizing features shared by the input instances, since this could lead to exclude transformed features. The subsequent experimental section motivates this architecture, through an analysis devoted to model selection.

3.3 Likelihood Computation

The trained neural network provides how an input word would be transformed when typed by patients affected by a given pathology. Thus, for a given pathology p, the input word w, the word w' typed by patient and the word w^* returned by the model for the pathology p are available. The likelihood that the patient is affected by pathology p is given by the following formula, where w is employed as reference word, v_w, $v_{w'}$ and v_{w^*} are the encodings of w, w' and w^*.

$$\eta(w, w', w^*) = \sum_{i:v_w[i] \neq v_{w'}[i] \wedge v_w[i] \neq v_{w^*}[i]} \frac{\left| v_{w'}[i] - v_{w^*}[i] \right|}{\max\{v_{w'}[i], v_{w^*}[i]\}}. \tag{1}$$

Equation 1 is designed to measure the differences between w' and w^* by skipping letters where both w' and w^* agree with w, moreover it is able to highlight that both w' and w^* differ from w and that they agree on the presence of a certain letter, namely each letter in w' is in w^* and vice versa.

For example, let

$$w = paperapa,$$
$$w^i = qaqeraqa,$$
$$w^{ii} = qaeraqa,$$
$$w^{iii} = lalerala,$$
$$w^{iv} = pqperqpq,$$

where w is the reference word. The distances are the following:

- $\eta(w, w^i, w^{ii}) = 2.8$, since both w^i and w^{ii} differ from w for the same letter p, both of them substitute it with q;
- $\eta(w, w^i, w^{iii}) = 8$, since both w^i and w^{iii} differ from w for the same letter p, even if different substitutions are applied;
- $\eta(w, w^i, w^{iv}) = 10$, since both w^i and w^{iv} differ from w for a different letter, even if the same substituting letter appears.

4 Experiments

The conducted experiments are for Italian language since the rehabilitation center involved in this study provide input data in this language. However, the work has no dependencies with the language. The whole framework is written in *python* and uses *tensorflow* as underlying engine.

Figure 1 reports results for four different pathologies and for the following configurations:

Configuration 1

layer 1:	kind: *Dense*,	neurons: n,	activation: *'relu'*;
layer 2:	kind: *Dense*,	neurons: $n/2$,	activation: *'relu'*;
layer 3:	kind: *Dense*,	neurons: $n/4$,	activation: *'relu'*;
layer 4:	kind: *Dense*,	neurons: $n/2$,	activation: *'relu'*;
layer 5:	kind: *Dense*,	neurons: n,	activation: *'relu'*.

Configuration 2

layer 1:	kind: *Dense*,	neurons: n,	activation: *'relu'*;
layer 2:	kind: *Dense*,	neurons: $n/2$,	activation: *'relu'*;
layer 3:	kind: *Dense*,	neurons: n,	activation: *'relu'*.

Configuration 3

| layer 1: | kind: *Dense*, | neurons: n, | activation: *'relu'*; |

Configuration 4

layer 1:	kind: *Dense*,	neurons: n,	activation: *'relu'*;
layer 2:	kind: *Dense*,	neurons: $2n$,	activation: *'relu'*;
layer 3:	kind: *Dense*,	neurons: n,	activation: *'relu'*.

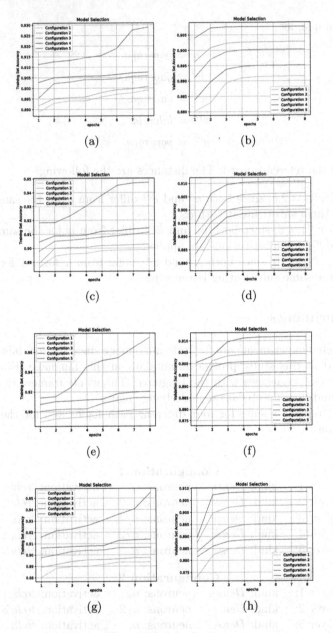

Fig. 1. Model selection

Configuration 5

layer 1:   kind: *Dense*,   neurons: n,   activation: *'relu'*;
layer 2:   kind: *Dense*,   neurons: $2n$,   activation: *'relu'*;
layer 3:   kind: *Dense*,   neurons: $4n$,   activation: *'relu'*;
layer 4:   kind: *Dense*,   neurons: $2n$,   activation: *'relu'*;
layer 5:   kind: *Dense*,   neurons: n,   activation: *'relu'*.

Plots on the left report the accuracies for the considered configurations. It is worth to note that, differently from classical auto encoders, in this case better results are achieved if the latent space is larger than input and output ones. This is due to the fact that it particularly relevant here to highlight all the hidden features and not just the more significant ones as in classical scenarios, where most features are shared and than can be considered as noise. However, too layers lead to obtain an over fitting network, how the right plot of Fig. 1 show. Thus, even if in the training set configuration 5 achieves better results, in the validation set, reported in figure, better results are achieved by configuration 4.

The second group of experiments concerns the ROC analysis to show the capability of the framework in detecting pathologies.

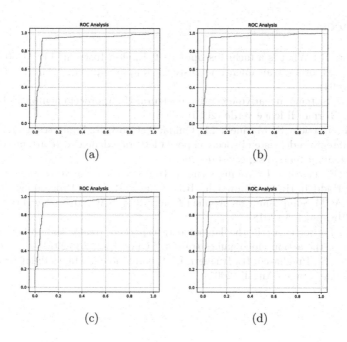

Fig. 2. Accuracy

In particular, for each pathology p, consider the set W of input words, the set W' of words typed by a patient affected by p, the set W_p of words as transformed by the learned model for pathology p, and a set W^* of words generated by applying random modification to words in W. The aim is to evaluate the capability of the method in distinguish words in W' from word in W^*. Thus, for any input word $w \in W$, consider the associated output word $w' \in W'$, the associated word $w_p \in W_p$ transformed by the proposed model for pathology p and the associated word $w^* in W^*$ and consider the values of $\eta(w, w', w_p)$ and $\eta(w, w', w^*)$. By exploiting these values, the ROC analysis can be conducted and Fig. 2 reports results.

5   Conclusions

This work addresses the problem of mining knowledge from the output of the tachistoscope, a widely employed tool of speech therapists to diagnose disorders often related to dyslexia and dysgraphia. The topic is to recognize which pathology is likely to be involved in analyzing patient session output. The idea is to compare typed words with those that the model of a certain pathology produces, in order to evaluate the likelihood that the patient is affected by that pathology. Preliminary experiments show that the approach is promising.

References

1. Benschop, R.: What is a tachistoscope? Hist. Explor. Instrum. **11**, 23–50 (1998)
2. Benso, F.: Teoria e trattamenti nei disturbi di apprendimento. Tirrenia (Pisa) Del Cerro (2004)
3. Benso, F.: Sistema attentivo-esecutivo e lettura. Un approccio neuropsicologico alla dislessia. Torino: Il leone verde (2010)
4. Benso, F., Berriolo, S., Marinelli, M., Guida, P., Conti, G., Francescangeli, E.: Stimolazione integrata dei sistemi specifi ci per la lettura e delle risorse attentive dedicate e del sistema attentivo supervisore (2008)
5. Fassetti, F., Fassetti, I.: Mining string patterns for individuating reading pathologies. In: Haddad, H.M., Wainwright, R.L., Chbeir, R. (eds.) Proceedings of the 33rd Annual ACM Symposium on Applied Computing, SAC 2018, Pau, France, 09–13 April 2018, pp. 1–5 (2018)
6. Gori, S., Facoetti, A.: Is the language transparency really that relevant for the outcome of the action video games training? Curr. Biol. **23** (2013)
7. Mafioletti, S., Pregliasco, R., Ruggeri, L.: Il bambino e le abilità di lettura. Il ruolo della visione, Franco Angeli (2005)

Forecasting Hazard Level of Air Pollutants Using LSTM's

Saba Gul[iD] and Gul Muhammad Khan[✉][iD]

National Center of AI, University of Engineering and Technology, Peshawar, Pakistan
{sabagul,gk502}@uetpeshawar.edu.pk

Abstract. The South Asian countries have the most polluted cities in the world which has caused quite a concern in the recent years due to the detrimental effect it had on economy and on health of humans and crops. PM 2.5 in particular has been linked to cardiovascular diseases, pulmonary diseases, increased risk of lung cancer and acute respiratory infections. Higher concentration of surface ozone has been observed to have negatively impacted agricultural yield of crops. Due to its deleterious impact on human health and agriculture, air pollution cannot be brushed off as a trivial matter and measures must be taken to address the problem. Deterministic models have been actively used; but they fall short due to their complexity and inability to accurately model the problem. Deep learning models have however shown potential when it comes to modeling time series data. This article explores the use of recurrent neural networks as a framework for predicting the hazard levels in Lahore, Pakistan with 95.0% accuracy and Beijing, China with 98.95% using the time series data of air pollutants and meteorological parameters. Forecasting air quality index (AQI) and Hazard levels would help the government take appropriate steps to enact policies to reduce the pollutants and keep the citizens informed about the statistics.

Keywords: Air pollution · AQI · Forecasting · LSTM's

1 Introduction

Air pollution has been brushed off for quite some time as a trivial subject but the current research suggest the relentless damage it can cause humans and crop yield. In particular the cities of South Asian countries such as China and India have made to the list of most polluted cities and the cities of Pakistan are joining the list due to increase in levels of particulate matter and toxic fumes from the industries [17].

---

*Supported by NCAI, UET-P.

I. Maglogiannis et al. (Eds.): AIAI 2020, IFIP AICT 584, pp. 143–153, 2020.
https://doi.org/10.1007/978-3-030-49186-4_13

Exposure to higher concentration of surface ozone can trigger allergic reactions such as asthma and cause inflammation of air ways due to oxidative stress [11,12]. PM2.5 has been associated with 4 to 8% increase in cardiopulmonary diseases and lung cancer [10]. Air pollution has been linked to cardiovascular diseases in urban communities. Most of the hospitalized patients suffering from diseases like angina, myocardial infarction and heart failure have been put in such a situation, due to the long-term exposure to combustion-derived nanoparticles that incorporate reactive organic and transition metal components [13]. Moreover, studies suggest that high concentration of surface ozone has a detrimental effect on crop yield [14,15]. In recent years, due to increase in awareness of the bleak consequences of air pollutants, forecasting of air pollutants and their impact on human and crops has become an active area of research. Several deterministic and non-deterministic models were explored to model the behavior of pollutants [7,16]. Deep leaning models have had quite some success when it comes to modeling the problem and forecasting air pollutants. The meteorological parameters due to their conducive behaviour in pollutant dissemination and pollutant concentrations were used to forecast Hazard levels. Since the parameters used for modeling are time series, so the recurrent neural networks, Long Short Term Memory (LSTM) networks are employed due to their ability to accurately capture temporal trends [5–7].

The two major contributions of this article are the following:

1. Provide a dataset comprising of Lahore, Pakistan meteorological and pollutants statistics.
2. Employ deep learning model to develop a forecasting and classification system for assessing air quality.

2   Literature Survey

An LSTM model is trained in [1] on sensor data of Aerosol Optical Depth (AOD), meteorology and particulate matter which can provide quite accurate prediction of the concentrations of harmful gases (80% $PM_{2.5}$ variability). The system has been successfully deployed in Beijing, China and has helped in bringing down the pollution in Beijing by 23%.

A supervised regression model is developed based on historical data of air pollution in Sydney [2] which surpassed its contemporary ANN's in terms of accuracy in prediction and has high spatial resolution.

Forecasting air pollution is done through Multi-channel Ensemble framework through supervised extraction and learning which out performs its contemporary state of the art systems [3]. $PM_{2.5}$, PM_{10}, SO_2, CO, NO_x and ozone levels are predicted quite accurately.

In [4] attempts are made to model the complex relation between different parameters and its individual impact on pollutant levels using deep distribution fusion network while the spatial correlation is modelled using deep neural network. The system, deep air out performs ten state of the art baseline models and achieves an average accuracy of 81.1%, 63%, 46% in 1–6 h, 7–48 h, sudden changes when deployed in 300+ cities of China.

Real time air-pollution predication is carried out in Daegu city, Korea [5] by processing the big data received from the air quality sensing modules installed on taxis. The spatial distribution of the pollutant levels is fed to a CNN model. For accurate processing of the temporal data; LSTM is used with a NN in parallel to cater for the meteorological factors effecting pollutant concentration. The testing results in an accuracy of 74% in real time over the data collected over a span of four months.

Spatial-temporal information is used in [6] to predict air quality using a combination of neural networks called ST-DNN which attempts to model the correlations between several meteorological conditions, elevation space and PM levels. LSTM is used to model long term temporal relations i.e. historical time series relation; CNN extracts the relationship between terrain information and pollutant levels while ANN is used with the current data and thereby models high frequency information. When evaluated on Taiwan and Beijing dataset, the network outperformed the baseline and comparative networks under consideration.

In order to enact policies to alleviate the pollution levels, accurate prediction is needed to carry out informed decisions. The temporal data of pollutants along with meteorological data is processed by a recurrent model [7], LSTM to forecast air pollutants since LSTMs have the ability to capture sequential relations. The frame work can predict air pollution 5–10 h in the future quite well but as the future time steps are increased beyond 10 h, we see degradation in performance. Since short term data of 6–10 h is needed to predict future time steps, power consumption can be reduced by turning the sensors on at specific intervals to collect data.

Artificial neural network (ANN) is used to predict PM_{10} concentration at 6 subways in Seoul, Korea [8]. Due to impracticality of monitoring PM_{10} directly at the crowded stations, PM_{10} concentrations are obtained from public data service near subway stations (PM_{10} out). In addition, it is observed that the shape and depth of the platform at the subway stations play an important role in influencing the model performance. The framework was able to predict PM_{10} concentrations at the platforms with an accuracy 67–80% depending upon parameters; inflow of PM_{10} (PM_{10} in), outflow of PM_{10} (PM_{10} out), ventilation operation, shape and depth of platform.

In [9], air pollution is forecasted using spatio-temporal data of city of Tehran, Iran obtained over a span of 10 years. Several machine learning methods such as; regression support vector machine, geographically weighted regression, artificial neural network and auto-regressive nonlinear neural network are evaluated on two datasets, one of which is cleaned via Savitzky-Golay filter while the other dataset was noisy due to missing entries. On both datasets, nonlinear autoregressive exogenous (NARX) neural network displays superior performance with exceptional performance over the former dataset.

3   Prediction Model Framework

In this article, we use a recurrent neural network that is; long short-term memory (LSTM), to capture the temporal trends of pollutant data. LSTM's perform better on sequential data as it takes the historical events into account by taking the output at instant t-1 as input in addition to inputs at t. This characteristic introduces the concept of Memory in neural networks which is of import when it comes to analyzing data of pollutants as it varies temporally.

Equation 1 and 2 describe the working of an RNN; where H is the tanh activation function, W defines the weight matrices between hidden and input layer (W_{xh}), hidden and hidden layer (W_{hh}), hidden and output layer (W_{hy}), x_t the input sequence, h_t the hidden vector of a module at instant t and b the bias to compute output y_t by iterating across these equations from t = 1 to T.

$$h_t = H(W_{xh}h_{xt} + W_{hh}h_{t-1} + b_h) \tag{1}$$

$$y_t = W_{hy}h_t + b_y \tag{2}$$

Though, RNN perform better when the sequences are short but suffer inherently from exploding gradient problem when working with data having long term dependencies. This problem is tackled by LSTM's which due do its gated memory architecture resolves the issue of vanishing and exploding gradients and is able to retain information for an extended period of time. Equation 3, 4, 5, 6 describes the input, forget and output gate and cell activation vectors of LSTM architecture respectively. Where σ is the sigma activation function.

$$i_t = \sigma(W_{xi}x_t + W_{hi}h_{t-1} + b_i) \tag{3}$$

$$f_t = \sigma(W_{xf}x_t + W_{hf}h_{t-1} + b_f) \tag{4}$$

$$o_t = \sigma(W_{xo}x_t + W_{ho}h_{t-1} + b_o) \tag{5}$$

$$c_t = f_t * c_{t-1} + i_t * tanh(W_{xg}x_t + b_g) \tag{6}$$

$$h_t = o_t * tanh(c_t) \tag{7}$$

3.1   Employed Datasets

The architecture was evaluated using two datasets, the modified UCI dataset published by [7] and on a dataset we introduced with parameters recorded in Lahore, Pakistan. The modified UCI dataset has meteorological data of wind speed, direction, air pressure, temperature, dew point, wind speed, cumulative rain hours and cumulative snow hours. Pollutant data of only $PM_{2.5}$ is recorded 25 times throughout the day. The parameters are collected over a span of 7 years with 43,825 samples from 2010 to 2017 across 35 different stations in Beijing, China. We have taken average of the data per day and on the basis of $PM_{2.5}$ concentration, we calculate the AQI value which is determined by the standard formula developed by environment protection agency (EPA), US. Based on the AQI value, a column of hazard level is added to the dataset. The information of date, hour, day, month, and year in the dataset is removed and pre-processed using normalization.

We obtained the time series pollutants data from environmental protection agency (EPA) Punjab, Pakistan for a span of 2 years from 2017 to 2019. The data of air pollutants is received from 6 stations across the city which includes particulate matter (PM_{10}, $PM_{2.5}$), Nitrogen dioxide, Sulphur dioxide and surface ozone. The meteorological parameters play an instrumental role towards pollution dissemination and concentration in a particular region, thus the meteorological department of Pakistan was contacted to obtain the statistics of wind direction, temperature, barometric pressure, humidity, visibility and type of weather. The data of air pollutants and meteorological statistics are combined and pre-processed to form a dataset of 1500 samples for monitoring and predicting the hazard levels in the form of AQI. We have categorized the hazard into six levels according to the pollutants concentration defined by air quality index (AQI) values set by EPA, US as described in Fig. 1.

Air Quality Index Levels of Health Concern	Numerical Value	Meaning
Good	0 to 50	Air quality is considered satisfactory, and air pollution poses little or no risk.
Moderate	51 to 100	Air quality is acceptable; however, for some pollutants there may be a moderate health concern for a very small number of people who are unusually sensitive to air pollution.
Unhealthy for Sensitive Groups	101 to 150	Members of sensitive groups may experience health effects. The general public is not likely to be affected.
Unhealthy	151 to 200	Everyone may begin to experience health effects; members of sensitive groups may experience more serious health effects.
Very Unhealthy	201 to 300	Health alert: everyone may experience more serious health effects.
Hazardous	301 to 500	Health warnings of emergency conditions. The entire population is more likely to be affected.

Fig. 1. Air Quality Index set by environment protection agency, US

3.2 Network Architecture

The frame work comprises of three layers; a single LSTM layer followed by two dense layers with activations of Tanh and softmax respectively.

The network is evaluated on Lahore dataset by using metrics of sparse categorical cross entropy and accuracy. Batch size of 16 is used with adam as an optimizer and the network is trained for 300 epochs with a data split of 70/15/15 for training, validation and testing.

For modified UCI dataset, the network is trained for 300 epochs with a data split of 70/15/15, batch size of 8 and adamax as an optimizer. Python packages of Keras, tensorflow, Scikit-Learn and Pandas are used to model the network. Early stopping techniques are used by observing the loss on the validation data to reduce over-fitting by curtailing the training period.

3.3   Tuning Network Hyper-Parameters

The hyper-parameters of LSTM model is then tuned based on data to configure optimal parameters. Batch size, Numbers of training epochs, optimizer, learning rate and type of activation function are some of the hyper-parameters tuned by employing grid search algorithm (GSA) to improve performance of the model. We started with tuning the number of training iterations and batch size simultaneously. The model was modified based on these optimal hyper-parameters and the grid search algorithm was run again to find an appropriate optimizer. The model was then tuned based on these parameters to find an activation function that boosts the performance of the LSTM model using grid search algorithm.

According to the results of grid search algorithm, for Lahore dataset, the optimal hyper-parameters for the LSTM model are listed in Table 1, 2, 3 and 4. In Table 2, we select tanh as an activation as it gives better performance with all the other parameters tuned.

Table 1. Selection of training iterations and Batch size using GSA on Lahore dataset

Batch size	Epoch number	Accuracy
16	*300*	*0.92832*
8	350	0.92115
16	350	0.91398
8	300	0.91398
32	350	0.91039
32	300	0.91039
64	350	0.89964

Table 2. Optimal activation function selection using GSA on Lahore dataset

Activation function	Accuracy
softsign	0.92832
tanh	*0.92473*
hard sigmoid	0.92115
linear	0.92115
relu	0.91039
softplus	0.91039
sigmoid	0.90323
softmax	0.82079

The results of grid search algorithm for modified UCI dataset are tabulated in Table 5, 6, 7 and 8.

The optimized hyper-parameters highlighted in italics are reconfigured by incorporating early stopping criterion using the validation set which improves the performance of the model employed.

Table 3. Results of optimizer selection using GSA on Lahore dataset

Optimizer	Accuracy
Adam	*0.931899*
Adadelta	0.92473
Nadam	0.91756
Adamax	0.91398
RMSprop	0.91039
Adagrad	0.88889
SGD	0.55197

Table 4. Optimal learning rate selection using GSA on Lahore dataset

Learning rate	Accuracy
0.002	*0.935454*
0.001	0.921169
0.01	0.914026
0.2	0.896169
0.1	0.889026
0.3	0.462922

Table 5. Selection of training iterations and Batch size using GSA on modified UCI dataset

Batch size	Epoch number	Accuracy
8	*300*	*0.98609*
32	300	0.98609
64	500	0.98510
16	300	0.98411
64	300	0.98361
32	500	0.98312
16	500	0.98262
8	500	0.98213
8	100	0.97120

Table 6. Optimal activation function selection using GSA on modified UCI dataset

Activation function	Accuracy
tanh	*0.98709*
hard sigmoid	0.98560
sigmoid	0.98560
linear	0.98312
relu	0.98262
softsign	0.98262
softplus	0.98064
softmax	0.87388

Table 7. Results of optimizer selection using GSA on modified UCI dataset

Optimizer	Accuracy
Adamax	*0.98759*
Adadelta	0.98461
RMSprop	0.97617
Adam	0.97567
Nadam	0.97368
Adagrad	0.95680
SGD	0.94836

Table 8. Optimal learning rate selection using GSA on modified UCI dataset

Learning rate	Accuracy
0.002	*0.972691*
0.001	0.951837
0.01	0.943893
0.1	0.936941
0.2	0.935452
0.3	0.767130

4   Result and Analysis

The hyper-parameters are tuned using grid search algorithm on the training set and on the validation data with respect to the categorical cross entropy error. It is observed that the model performs best with batch size of 8 with training of 300 epochs on the Beijing dataset and batch size of 16 with training of 300 epochs on Lahore dataset. Moreover, adam and adamax are employed as an optimizers for Lahore and Beijing datsets respectively which helps in convergence at a faster pace. Figure 2 shows that model when trained for 300 epochs on the modified UCI dataset attains a maximum validation accuracy of 98.9583% at epoch 288 and an accuracy of 98.95% on the test set. Thus the temporal characteristic of the data is modeled quite accurately using the recurrent network architecture. Figure 3 depicts the prediction model performance on the test set.

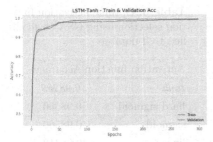

val_loss	val_acc	loss	acc	epoch
0.035727	0.989583	0.022682	0.993049	295
0.035713	0.989583	0.022635	0.993049	296
0.035673	0.989583	0.022582	0.993545	297
0.035627	0.989583	0.022534	0.993545	298
0.035617	0.989583	0.022476	0.993545	299

(a) Training and Validation accuracy

(b) Loss and accuracy results of training/dev set during the final epochs of modified UCI dataset

Fig. 2. Network training results on modified UCI dataset

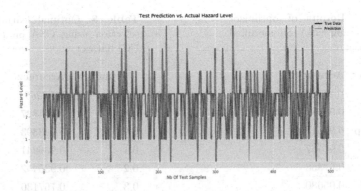

Fig. 3. Actual Vs. Predicted values of employed architecture on modified UCI dataset

The second dataset comprises of parameters recorded over a span of 2 years, thus after tuning the hyper-parameters, we train the network with a batch size of

16 for 300 epochs with early stopping criterion to avoid over-fitting. The results of training are described in Fig. 4 with maximum accuracy achieved at epoch 143 on the validation set. On the second dataset, an accuracy of 95.0% is achieved on the test set as depicted in Fig. 5. The deterioration in performance of the LSTM model for the Lahore dataset is due to the limited time series data required to infer the trends.

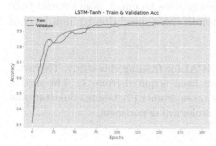

val_loss	val_acc	loss	acc	epoch
0.123860	0.95	0.117020	0.964158	195
0.123876	0.95	0.116676	0.964158	196
0.123878	0.95	0.116334	0.964158	197
0.123885	0.95	0.115992	0.964158	198
0.123915	0.95	0.115654	0.964158	199

(a) Training and Validation accuracy

(b) Loss and accuracy results of training/dev set during the final epochs

Fig. 4. Network training results on Lahore, Pakistan dataset

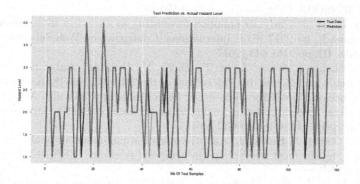

Fig. 5. Actual Vs. Predicted values of employed architecture on Lahore, Pakistan dataset

5 Conclusion

A model for forecasting hazard level has been devised and its performance is evaluated on the meteorological and pollutant data of two of the most polluted cities in the world; Beijing, China and Lahore, Pakistan. It is observed that despite different topography and meteorological information, the proposed network models the complexity of the diverse temporal information quite well. The proposed architecture after employing GSA optimization is able to forecast the hazard levels of the next 24 h with an accuracy of 95.0% on the data recorded in Lahore, Pakistan and 98.95% on Beijing, China dataset due to ability of LSTM's to model temporal data and is thus able to learn the trends of air pollutants. This is an effective measure for the people going out to take necessary precautions and assist the environment protection agencies to enact policies and take steps towards reducing the health and economic risk caused due to high level of pollutants.

Acknowledgment. We would like to thank NCAI for funding this study. The modified UCI data-set employed in our study have been acquired from [7]. The second dataset was created by the data of pollutants taken from EPA lahore, Pakistan and meteorological parameters from Pakistan meteorological department.

References

1. Han, Y., Lam, J.C.K., Li, V.O.K.: A Bayesian LSTM model to evaluate the effects of air pollution control regulations in China. In: 2018 IEEE International Conference on Big Data (Big Data), Seattle, WA, USA, pp. 4465–4468 (2018)
2. Hu, K., Sivaraman, V., Bhrugubanda, H., Kang, S., Rahman, A.: SVR based dense air pollution estimation model using static and wireless sensor network. In: 2016 IEEE SENSORS, Orlando, FL, pp. 1–3 (2016)
3. Zhang, C., et al.: Early air pollution forecasting as a service: an ensemble learning approach. In: 2017 IEEE International Conference on Web Services (ICWS), Honolulu, HI, pp. 636–643 (2017)
4. Yi, X., Zhang, J., Wang, Z., Li, T., Zheng, Y.: Deep distributed fusion network for air quality prediction. In: Proceedings of the 24th ACM SIGKDD International Conference on Knowledge Discovery & Data Mining, pp. 965–973. ACM (2018)
5. Le, D.: Real-time air pollution prediction model based on spatiotemporal big data. arXiv preprint arXiv:1805.00432 (2018)
6. Soh, P., Chang, J., Huang, J.: Adaptive deep learning-based air quality prediction model using the most relevant spatial-temporal relations. IEEE Access **6**, 38186–38199 (2018)
7. Reddy, V., Yedavalli, P., Mohanty, S., Nakhat, U.: Deep air: forecasting air pollution in Beijing, China (2018)
8. Park, S., et al.: Predicting PM10 concentration in Seoul metropolitan subway stations using artificial neural network (ANN). J. Hazard. Mater. **341**, 75–82 (2018). https://doi.org/10.1016/j.jhazmat.2017.07.05010.1016/j.jhazmat.2017.07.050. ISSN 0304-3894
9. Delavar, M., et al.: A novel method for improving air pollution prediction based on machine learning approaches: a case study applied to the capital city of Tehran. ISPRS Int. J. Geo Inf. **8**, 99 (2019). https://doi.org/10.3390/ijgi8020099
10. Pope III, C., et al.: Lung cancer, cardiopulmonary mortality, and long-term exposure to fine particulate air pollution. JAMA **287**(9), 1132–1141 (2002)
11. Kim, K.-H., Jahan, S.A., Kabir, E.: A review on human health perspective of air pollution with respect to allergies and asthma. Environ. Int. **59**, 41–52 (2013)

12. Kelly, F.J.: Oxidative stress: its role in air pollution and adverse health effects. Occup. Environ. Med. **60**(8), 612–616 (2003)
13. Mills, N.L., et al.: Adverse cardiovascular effects of air pollution. Nat. Rev. Cardiol. **6**(1), 36 (2009)
14. Chuwah, C., van Noije, T., van Vuuren, D.P., Stehfest, E., Hazeleger, W.: Global impacts of surface ozone changes on crop yields and land use. Atmos. Environ. **106**, 11–23 (2015)
15. Lin, Y., et al.: Impacts of O3 on premature mortality and crop yield loss across China. Atmos. Environ. **194**, 41–47 (2018)
16. Bai, L., Wang, J., Ma, X., Haiyan, L.: Air pollution forecasts: an overview. Int. J. Environ. Res. Public Health **15**(4), 780 (2018)
17. World air quality report. https://www.iqair.com/world-most-polluted-cities

Fuzzy Algebra/Systems

Hypotheses Tests Using Non-asymptotic Fuzzy Estimators and Fuzzy Critical Values

Nikos Mylonas and Basil Papadopoulos[(✉)]

Department of Civil Engineering, Democritus University of Thrace, Building
A'-Campus Xanthi-Kimmeria, 67100 Xanthi, Greece
{nimylona,papadob}@civil.duth.gr

Abstract. In fuzzy hypothesis testing we use fuzzy test statistics produced by fuzzy estimators and fuzzy critical values. In this paper we use the non-asymptotic fuzzy estimators in fuzzy hypothesis testing. These are triangular shaped fuzzy numbers that generalize the fuzzy estimators based on confidence intervals in such a way that eliminates discontinuities and ensures compact support. Our approach is particularly useful in critical situations, where subtle fuzzy comparisons between almost equal statistical quantities have to be made. In such cases the hypotheses tests that use non-asymptotic fuzzy estimators give better results than the previous approaches, since they give us the possibility of partial rejection or not of H_0.

Keywords: Fuzzy hypothesis tests · Non-asymptotic fuzzy estimators

1 Introduction

The use of fuzzy hypothesis tests is necessary: 1) in cases of samples of crisp data for which the value of the test statistic is very close to critical value, the crisp test is unstable, since small changes in few observations drive from rejection to no rejection or vice-versa, 2) in cases in which the available observations are fuzzy.

In a fuzzy test of a null hypothesis of the form $H_0 : U = U_0$ with alternative $H_1 : U \neq U_0$ for a parameter u we use a fuzzy test statistic, which is constructed using a fuzzy estimator. Since the test statistic is fuzzy, the critical region will be determined by fuzzy critical values CV_i, $i = 1, 2$. The $a-$cuts of CV_i are found in each case as described in [5] and presented in Sects. 2 and 3. So, H_0 is rejected or not rejected in a certain significance level with the help of a fuzzy statistic U which is constructed using a fuzzy estimator and fuzzy inequalities between U and the fuzzy critical values, like $U < CV$, $U > CV$ or $U \approx CV$.

© IFIP International Federation for Information Processing 2020
Published by Springer Nature Switzerland AG 2020
I. Maglogiannis et al. (Eds.): AIAI 2020, IFIP AICT 584, pp. 157–166, 2020.
https://doi.org/10.1007/978-3-030-49186-4_14

In our new approach, from the various types of existing fuzzy estimators we use the non-asymptotic fuzzy estimators [11], which are more convenient since they are triangular shaped fuzzy numbers with compact support without discontinuities.

We apply this concept in Sects. 2 and 3 testing hypotheses for: 1) the mean and 2) the variance of a normal distribution and compare results with these of the respective tests with fuzzy statistics constructed with the estimators of Buckley [5].

1.1   Ordering Fuzzy Numbers

The fuzzy hypotheses tests are based on ordering fuzzy numbers, for which we will use one of the several procedures used [7], according to which the degree $v(A \leq B)$ of the inequality $A \leq B$ that counts the degree to which the fuzzy number A is less or equal than the fuzzy number B is

$$v(A \leq B) = \max \{\min(A(x), B(y)) \quad x \leq y\} \tag{1}$$

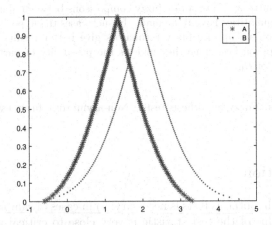

Fig. 1. Ordering triangular shaped fuzzy numbers

If $v(A \leq B) = 1$ we define the truth-value (degree of confidence) d of $A < B$ as

$$v(A < B) = d \Leftrightarrow v(A \leq B) = 1 \quad \text{and} \quad v(B \leq A) = 1 - d \tag{2}$$

If $v(B \leq A) = 1$ the truth-value d of $B < A$ is

$$v(B < A) = d \Leftrightarrow v(B \leq A) = 1 \quad \text{and} \quad v(A \leq B) = 1 - d \tag{3}$$

So, the truth-value η of $A \approx B$ is defined as

$$v(A \approx B) = \eta \Leftrightarrow v(A \leq B) = 1 \quad \text{and} \quad v(B \leq A) = \eta$$

$$\text{or} \quad v(B \le A) = 1 \quad \text{and} \quad v(A \le B) = \eta \qquad (4)$$

In the case of the triangular shaped fuzzy numbers of Fig. 1, which appears often in fuzzy hypotheses tests, we can see that:

$v(A \le B) = 1$, according to (1) since the core of A lies to the left of the core of B,

$v(B \le A) = y_0$, according to (1), where y_0 the truth level of the point of intersection of the right part of A and the left part of B.

Thus according to Eq. (2) and (4),

$$v(A < B) = 1 - y_0 \quad \text{and} \quad v(A \approx B) = y_0$$

2 Hypothesis Test for the Mean of a Normal Distribution with Known Variance

We test in significance level γ the null hypothesis that the mean value μ of a random variable X that follows normal distribution with known variance σ is equal to μ_0

$$H_0 : \ \mu = \mu_0$$

with alternative the (two sided test)

$$H_1 : \ \mu \neq \mu_0$$

using a random sample of observations of X of size n. In the crisp case we test H_0 using the statistic

$$Z = \frac{\bar{x} - \mu_0}{\frac{\sigma}{\sqrt{n}}} \qquad (5)$$

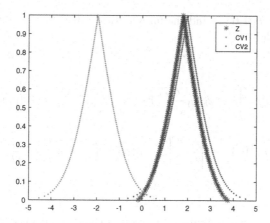

Fig. 2. The fuzzy statistic Z and the fuzzy critical values for the test of $H_0 : \ \mu = 5$ from a sample with $\bar{x}_1 = 5.4$

where $\bar{x}$ is the sample mean value.

H_0 is rejected if $z < -z_c$ or $z > z_c$ where z_c the critical value of the test

$$z_c = z_{\gamma/2} = \Phi^{-1}\left(1 - \frac{\gamma}{2}\right) \tag{6}$$

and Φ^{-1} the inverse distribution function of the standard normal distribution. While, if $-z_c < z < z_c$, then H_0 is not rejected.

In the fuzzy case for the test of H_0 we use the fuzzy statistic of Buckley [5]

$$\overline{Z} = \frac{\hat{\mu} - \mu_0}{\frac{\sigma}{\sqrt{n}}} \tag{7}$$

that is generated by (5) using a fuzzy estimator $\hat{\mu}$ of the mean value.

We use the non-asymptotic fuzzy estimator $\hat{\mu}$ the α-cuts of which are [11]

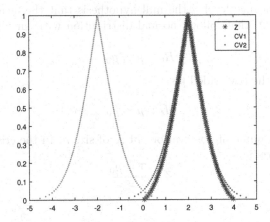

Fig. 3. The fuzzy statistic Z and the fuzzy critical values for the test of $H_0 : \mu = 5$ from a sample with $\bar{x}_1 = 5.45$

$$\hat{\mu}[\alpha] = \left[\bar{x} - z_{h(\alpha)}\frac{\sigma}{\sqrt{n}}, \ \bar{x} + z_{h(\alpha)}\frac{\sigma}{\sqrt{n}}\right], \qquad \alpha \in (0, 1] \tag{8}$$

where

$$h : (0, 1] \rightarrow \left[\frac{\gamma}{2}, \frac{1}{2}\right] \qquad h(\alpha) = \left(\frac{1}{2} - \frac{\gamma}{2}\right)\alpha + \frac{\gamma}{2} \tag{9}$$

and

$$z_{h(\alpha)} = \Phi^{-1}(1 - h(\alpha)) \tag{10}$$

where Φ^{-1} the inverse distribution function of the standard normal distribution.

From (7) and (8) follows that the α-cuts of the fuzzy statistic $\overline{Z}$ are

$$\overline{Z}[\alpha] = \left[z_0 - z_{h(\alpha)}, \ z_0 + z_{h(\alpha)}\right], \qquad \alpha \in (0, 1] \tag{11}$$

where

$$z_0 = \frac{\bar{x} - \mu_0}{\frac{\sigma}{\sqrt{n}}} \tag{12}$$

Since the test statistic is fuzzy, the critical values as described in [5] are the fuzzy numbers $\overline{CV}_2$ and $\overline{CV}_1 = -CV_2$, the α-cuts of which are

$$\overline{CV}_1[\alpha] = \left[z_{\gamma/2} - z_{\alpha/2},\ z_{\gamma/2} + z_{\alpha/2}\right] \tag{13}$$

$$\overline{CV}_2[\alpha] = \left[-z_{\gamma/2} - z_{\alpha/2},\ -z_{\gamma/2} + z_{\alpha/2}\right] \quad \alpha \in (0, 1] \tag{14}$$

Having the fuzzy test statistic $\overline{Z}$ and the fuzzy critical values $\overline{CV}_i$, our decision for rejecting H_0 or not in significance level γ, depends on the comparison of the fuzzy numbers $\overline{Z}$ and $\overline{CV}_i$, as given in Sect. 1.1. So:

if $\max\left(v(\overline{Z} > \overline{CV}_2), v(\overline{Z} < \overline{CV}_1)\right) = d$, then H_0 is rejected with a degree of confidence less or equal to d,
if $\min\left(\overline{Z} > \overline{CV}_1), v(\overline{Z} < \overline{CV}_2)\right) = d$, then H_0 is not rejected with a degree of confidence less or equal to d,
if $\max\left(v(\overline{Z} \approx \overline{CV}_2), v(\overline{Z} \approx \overline{CV}_1)\right) = 1 - d$, then we cannot make a decision on rejecting or not H_0 with degree of confidence greater or equal to d.

Example 1 Comparison of fuzzy and crisp hypothesis test. In the crisp test in significance level $\gamma = 0.05$ of the null hypothesis

$$H_0 : \mu = 5$$

with alternative the (two sided test)

$$H_1 : \mu \neq 5$$

for the mean value μ of a random variable X that follows normal distribution with standard deviation $s = 2$ using a sample of 80 observations with sample mean $\bar{x}_1 = 5.4$, we evaluate the value of the statistic (5)

$$z_0 = \frac{5.4 - 5}{\frac{2}{\sqrt{80}}} = 1.799$$

Since

$$z_0 = 1.799 < z_{0.05/2} = z_{0.025} = 1.96$$

H_0 is not rejected.

For a second sample with sample mean $\bar{x}_2 = 5.45$ the value of the statistic Eq. (5) is

$$z_0 = \frac{5.45 - 5}{\frac{2}{\sqrt{80}}} = 2.012$$

So, since

$$z_0 = 2.012 > z_{0.05/2} = z_{0.025} = 1.96$$

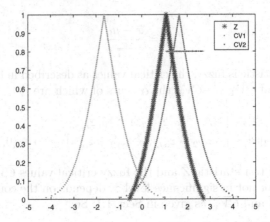

Fig. 4. Fuzzy statistic $\overline{Z}$ and fuzzy critical values $\overline{CV}_i$ for the test of $H_0 : \mu = 1$ using non-asymptotic fuzzy estimator

H_0 is rejected.

Applying the above described fuzzy test of H_0 for the first sample ($\overline{x} = 5.4$) we get Fig. 2, where the point of intersection of the fuzzy numbers $\overline{Z}$ and $\overline{CV}_2$ is $y_0 = 0.93$. So according to Eq. (2), $v(\overline{Z} < \overline{CV}_2) = 1 - 0.93 = 0.07$. Therefore, from this sample we cannot make a decision on rejecting or not H_0 with degree of confidence $d \geq 0.07$.

For the second sample ($\overline{x} = 5.45$) we plot $\overline{Z}$ and $\overline{CV}_2$ in Fig. 3, where we can see that the point of intersection of the fuzzy numbers $\overline{Z}$ and $\overline{CV}_2$ is $y_0 = 1$. So according to Eq. (4), $v(\overline{Z} \approx \overline{CV}_2) = 1$. Therefore, from this sample we cannot make a decision on rejecting H_0 or not with any degree of confidence d.

Example 2. We test in significance level 0.1 the hypothesis

$$H_0 : \mu = 1$$

with alternative the

$$H_1 : \mu \neq 1$$

for the mean value μ of a random variable X that follows normal distribution with standard deviation $s = 2$ using a sample of 100 observations with sample mean value $\overline{x}_1 = 1.24$.

Applying the fuzzy test of H_0 (using the non-asymptotic fuzzy estimator of mean value) we get Fig. 4, where the point of intersection of the fuzzy statistics Z and CV_2 is below 0.8. So, the truth-value of $Z < CV_2$ is greater than $1 - 0.8 = 0.2$. Therefore, H_0 is not rejected in degree of confidence $d = 0.2$. While, applying

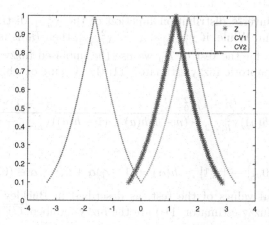

Fig. 5. Fuzzy statistic $\overline{Z}$ and fuzzy critical values $\overline{CV}_i$ for the test of $H_0 :\ \mu = 1$ using fuzzy estimator of Buckley [5]

the fuzzy test of H_0 using the fuzzy estimator of mean value of Buckley [5] we get Fig. 5, in which the point of intersection of the fuzzy test statistics $\overline{Z}$ and $\overline{CV}_2$ is above 0.8. So, $v(\overline{Z} < \overline{CV}_2) < 0.2$. Therefore, according to this test, we cannot make a decision on the rejection or not of H_0 with degree of confidence $d = 0.2$.

3 Hypothesis Test for the Variance of a Normal Distribution

We test in significance level γ the null hypothesis that the variance σ^2 of a random variable X that follows normal distribution is equal to σ_0^2

$$H_0 :\ \sigma^2 = \sigma_0^2$$

with alternative the (two sided test)

$$H_1 :\ \sigma^2 \neq \sigma_0^2$$

using a random sample of observations of X of size n.
In the crisp case we test H_0 using the test statistic

$$\chi_0^2 = \frac{(n-1)s^2}{\sigma_0^2} \tag{15}$$

where s^2 is the sample variance.
H_0 is rejected if $\chi_0^2 < \chi_{L;\gamma/2}^2$ or $\chi_0^2 > \chi_{R;\gamma/2}^2$ where $\chi_{L;\gamma/2}^2$ and $\chi_{R;\gamma/2}^2$ the critical values of the test,

$$\chi_{L;\gamma/2}^2 = F^{-1}\left(\frac{\gamma}{2}\right) \quad \text{and} \quad \chi_{R;\gamma/2}^2 = F^{-1}\left(1 - \frac{\gamma}{2}\right) \tag{16}$$

where F^{-1} the inverse distribution function of the χ^2_{n-1} distribution with n degrees of freedom. While, if $\chi^2_{L;\gamma/2} < \chi^2 < \chi^2_{R;\gamma/2}$, then H_0 is not rejected.

In the fuzzy case for the test of H_0 we use the unbiased fuzzy statistic $\overline{\chi}^2$ [5] with the non-asymptotic fuzzy estimator [11] the α–cuts of which are

$$\overline{\chi}^2[a] = \left[\frac{(n-1)\chi_0^2}{(1-h(a))\chi^2_{R;\gamma/2} + (n-1)h(a)}, \ \frac{(n-1)\chi_0^2}{(1-h(a))\chi^2_{L;\gamma/2} + (n-1)h(a)} \right] \tag{17}$$

where

$$h : (0,1] \to [\gamma, 1] \quad h(\alpha) = (1-\gamma)\alpha + \gamma, \quad \alpha \in (0,1] \tag{18}$$

The fuzzy critical values of the test, as described in Buckley [5], (using the non-asymptotic fuzzy estimator [11]) are the fuzzy numbers $\overline{CV}_1$ and $\overline{CV}_2$, the α–cuts of which are

$$\overline{CV}_1[a] = \left[\frac{(n-1)\chi^2_{L;\gamma/2}}{(1-h(a))\chi^2_{R;\gamma/2} + (n-1)h(a)}, \ \frac{(n-1)\chi^2_{L;\gamma/2}}{(1-h(a))\chi^2_{L;\gamma/2} + (n-1)h(a)} \right]$$

$$\overline{CV}_2[a] = \left[\frac{(n-1)\chi^2_{R;\gamma/2}}{(1-h(a))\chi^2_{R;\gamma/2} + (n-1)h(a)}, \ \frac{(n-1)\chi^2_{R;\gamma/2}}{(1-h(a))\chi^2_{L;\gamma/2} + (n-1)h(a)} \right]$$

Example 3. We test in significance level 0.1 the hypothesis

$$H_0 : \sigma^2 = 2$$

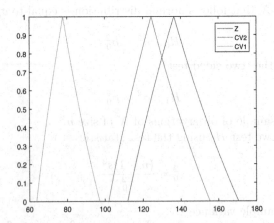

Fig. 6. The fuzzy statistic $\overline{Z}$ and the fuzzy critical values for the test of $H_0 : \sigma^2 = 2$ from a sample with variance $s^2 = 2.73$

with alternative the (two sided test)

$$H_1 : \sigma^2 \neq 2$$

using a random sample of 100 observations with sample variance $s^2 = 2.73$.

Applying the above described fuzzy test of H_0 we get the result of Fig. 6, where the point of intersection of the fuzzy numbers $\overline{Z}$ and $\overline{CV}_2$ is $y_0 = 0.79$. So, according to Sect. 1.1 $v(\overline{Z} > \overline{CV}_2) = 1 - 0.79 = 0.21$. Therefore, H_0 is rejected in degree of confidence $d \leq 0.21$.

4 Conclusions

If the value of the test statistic of a hypothesis is close to the critical values of the test, then the crisp hypothesis test is unstable, since a small change in the sample mean value (addition or removal or a change of one observation) may lead from rejection to no rejection of H_0 or vice-versa. While, the fuzzy hypothesis testing in such cases gives a very low degree of confidence of the rejection or not of the null hypothesis, as shown in Example 1.

Comparing with a computer program (in Matlab) the results of the fuzzy test with non-asymptotic fuzzy estimator with the respective fuzzy test with fuzzy estimator of Buckley [5] we see a slight difference, which in some cases leads to a different decision, as shown in the Example 2.

Our approach that uses non-asymptotic fuzzy estimators for the construction of the fuzzy statistics and a degree of confidence for the rejection or not of a hypothesis gives better results than the existing ones, since it gives us the possibility to make a decision on a partial rejection of it in a certain degree of confidence.

Acknowledgement. We would like to express our gratitude to the referees for their valuable comments.

References

1. Buckley, J.J.: On the algebra of interactive fuzzy numbers: the continuous case. Fuzzy Sets Syst. **37**, 317–326 (1990)
2. Buckley, J.J.: Fuzzy probabilities: new approach and applications. Physica-Verlag, Heidelberg (2003). https://doi.org/10.1007/978-3-642-86786-6
3. Buckley, J.J., Eslami, E.: Uncertain probabilities I: the discrete case. Soft Comput. **7**, 500–505 (2003)
4. Buckley, J.J., Eslami, E.: Uncertain probabilities II: the continuous case. Soft Comput **8**, 193–199 (2004)
5. Bucley, J.J.: Fuzzy statistic. Springer, Berlin (2004). https://doi.org/10.1007/978-3-540-39919-3
6. Dubois, D., Prade, H.: Ranking of fuzzy numbers in the setting of possibility theory. Inf. Sci. **30**, 183–224 (1983)
7. Dubois, D., Prade, H.: Fuzzy Sets and Systems. Academic Press, Cambridge (1980)

8. Klir, G., Yuan, B.: Fuzzy Sets and Fuzzy Logic: Theory and Applications. Prentice Hall, Upper Saddle River (1995)

9. Lee, H., Lee, J.-H.: A method for ranking fuzzy numbers and its application to a decision-making. IEEE Trans. Fuzzy Syst. **7**, 677–685 (1999)

10. Lee S., Lee H., Lee D.: Ranking the sequences of fuzzy values. Inf. Sci. (2003)

11. Sfiris, D., Papadopoulos, B.: Non-asymptotic fuzzy estimators based on confidence intervals. Inf. Sci. (2010)

12. Taheri, M., Arefi, M.: Testing fuzzy hypotheses based on fuzzy test statistic. Soft Comput. **13**, 617–625 (2009). https://doi.org/10.1007/s00500-008-0339-3

13. Taheri, M., Hesamian, G.: Non-parametric statistical tests for fuzzy observations: fuzzy test statistic approach. LJFIS **17**(3), 145–153 (2017)

Preservation of the Exchange Principle via Lattice Operations on (S,N)– Implications

Dimitrios S. Grammatikopoulos[iD] and Basil K. Papadopoulos[✉][iD]

Section of Mathematics and Informatics, Department of Civil Engineering,
School of Engineering, Democritus University of Thrace, 67100 Kimeria, Greece
{dimigram2,papadob}@civil.duth.gr

Abstract. In this paper, we investigate a special case of an open problem that is related to the exchange principle, a property of fuzzy implications. We focus on the cases of (S,N)– implications and the preservation of the exchange principle via lattice operations. We present and prove some sufficient conditions such that the exchange principle is preserved under the join and meet operations if we use (S,N)– implications.

Keywords: (S,N)– implication · Exchange principle

1 Introduction

The applications of fuzzy logic were the uprising of the automata theory, robotics, approximate reasoning, image processing, pattern recognition, artificial intelligence and many other scientific and applicable areas. The transition of the two valued $\{0, 1\}$– logic to the close $[0, 1]$– logic was the total revolution from the second half of the twentieth century to the present and as it appears in the future. Many definitions in fuzzy logic are generalizations from the classical ones. Fuzzy implications, their properties and some of their construction methods are also such generalizations (see [1]).

Some of them are (S,N)– implications, which are based on the following classical tautology

$$(p \Rightarrow q) \equiv (p' \vee q) \tag{1}$$

and the property exchange principle of fuzzy implications, which is based on the following classical tautology

$$[p \Rightarrow (q \Rightarrow r)] \equiv [q \Rightarrow (p \Rightarrow r)] \tag{2}$$

In this paper we investigate a special case of an open problem, that was addressed by Baczyński and Jayaram in 2008 (see [1] Remark 6.1.5). It was also formulated by Mesiar and Stupňanová in 2015 (see [6] Problem 3.1). This is the following:

© IFIP International Federation for Information Processing 2020
Published by Springer Nature Switzerland AG 2020
I. Maglogiannis et al. (Eds.): AIAI 2020, IFIP AICT 584, pp. 167–179, 2020.
https://doi.org/10.1007/978-3-030-49186-4_15

Characterize the subfamily of all fuzzy implications ((S,N)– implications, R-implications, etc.) which preserve the (EP) for lattice operations.

Although this problem has been investigated by Vemuri and Jayaram in [8,9], it is not fully solved. They investigated it, in general for any two fuzzy implications. In their results are contained some sufficient conditions for two fuzzy implications such that, (EP) is preserved via the lattice operations (see Section 3 in [8,9]). They also found and necessary ones, but under some conditions (see Section 4 in [8,9]). Furthermore, in [5] a generation of fuzzy implications with specific properties was succeeded, but not for the property (EP). All of these were the motivation for this paper. All efforts have been done for any fuzzy implications, but not extensively for a subfamily of them. Thus, we investigate it again only for (S,N)–implications. The main reason of the choice of this subfamily is that (S,N)–implications satisfy (EP). The following work will appear new conditions for the preservation of the exchange principle in the case we use (S,N)–implications.

2   Preliminaries

Definition 1. *[1, 3, 4]. A decreasing function $N : [0,1] \to [0,1]$ is called fuzzy negation, if $N(0) = 1$ and $N(1) = 0$.*

Remark 1. (i) (Example 1.4.4 in [1]) The so called, classical fuzzy negation is

$$N_C(x) = 1 - x \tag{3}$$

(ii) Moreover, the crisp fuzzy negations (see Remark 2.1 in [2]) are

$$N^\alpha(x) = \begin{cases} 0, & \text{if } x \geq \alpha \\ 1, & \text{if } x < \alpha \end{cases}, \text{ where } \alpha \in (0,1] \text{ and} \tag{4}$$

$$N_\alpha(x) = \begin{cases} 0, & \text{if } x > \alpha \\ 1, & \text{if } x \leq \alpha \end{cases}, \text{ where } \alpha \in [0,1). \tag{5}$$

Definition 2. *(Definition 2.1.1 in [1]). A function $T : [0,1]^2 \to [0,1]$ is called a triangular norm (shortly t-norm) if it satisfies, for all $x, y, z \in [0,1]$, the following conditions (Table 1)*

$$T(x,y) = T(y,x), \tag{T1}$$

$$T(x,T(y,z)) = T(T(x,y),z), \tag{T2}$$

$$\text{if } y \leq z, \text{ then } T(x,y) \leq T(x,z), \text{ i.e., } T(x,\cdot) \text{ is increasing,} \tag{T3}$$

$$T(x,1) = x. \tag{T4}$$

Definition 3. *(Definition 2.2.1 in [1]). A function $S : [0,1]^2 \rightarrow [0,1]$ is called a triangular conorm (shortly t-conorm), if it satisfies, for all $x, y, z \in [0,1]$, the following conditions*

$$S(x, y) = S(y, x), \tag{S1}$$

$$S(x, S(y, z)) = S(S(x, y), z), \tag{S2}$$

$$if\ y \leq z,\ then\ S(x, y) \leq S(x, z),\ i.e.,\ S(x, \cdot)\ is\ increasing, \tag{S3}$$

$$S(x, 0) = x. \tag{S4}$$

Table 1. Examples of t-conorms (Table 2.2 in [1]).

Name	Formula
Maximum	$S_M = max\{x, y\}$
Probor	$S_P = x + y - x \cdot y$
Nilpotent maximum	$S_{nM} = \begin{cases} 1, & if\ x + y \geq 1 \\ max\{x, y\}, & otherwise \end{cases}$

Remark 2. (Table 2.2 in [1]). In this work we use the minimum t-norm, which has the following formula

$$T_M = min\{x, y\}, x, y \in [0, 1]. \tag{6}$$

Definition 4. *(Definition 1.1.1 in [1]). A function $I : [0,1]^2 \rightarrow [0,1]$ is called a fuzzy implication if*

$$I\ is\ decreasing\ with\ respect\ to\ the\ first\ variable, \tag{I1}$$

$$I\ is\ increasing\ with\ respect\ to\ the\ second\ variable, \tag{I2}$$

$$I(0, 0) = 1, \tag{I3}$$

$$I(1, 1) = 1, \tag{I4}$$

$$I(1, 0) = 0. \tag{I5}$$

Definition 5. *(Definition 1.3.1(ii) in [1]). A fuzzy implication I is said to satisfy the exchange principle, if*

$$I(x, I(y, z)) = I(y, I(x, z)), x, y, z \in [0, 1]. \tag{EP}$$

Definition 6. *(Definition 2.4.1 in [1]). A function $I : [0, 1]^2 \to [0, 1]$ is called an (S,N)– implication if there exist a t-conorm S and a fuzzy negation N such that*

$$I(x, y) = S(N(x), y), x, y \in [0, 1] \tag{7}$$

Moreover, if I is an (S,N)– implication generated from S and N, then we will often denote it by $I_{S,N}$.

Proposition 1. *(Proposition 2.4.3(i) in [1]). If $I_{S,N}$ is an (S,N)– implication, then it satisfies (EP) (Table 2).*

Table 2. Examples of (S,N)– implications (Tables 1.3 and 2.4 in [1]).

Implication's name	S	N	(S,N)– implication
Kleene-Dienes	S_M	N_C	$I_{KD}(x, y) = max\{1 - x, y\}$
Reichenbach	S_P	N_C	$I_{RC}(x, y) = 1 - x + x \cdot y$
Fodor	S_{nM}	N_C	$I_{FD}(x, y) = \begin{cases} 1, & \text{if } x \le y \\ max\{1 - x, y\}, & \text{if } x > y \end{cases}$

Moreover, for any t-conorm S there are the following (S,N)– implications

$$I_{S,N^\alpha}(x, y) = \begin{cases} y, & \text{if } x \ge \alpha \\ 1, & \text{if } x < \alpha \end{cases}, \text{ where } \alpha \in (0, 1] \text{ and} \tag{8}$$

$$I_{S,N_\alpha}(x) = \begin{cases} y, & \text{if } x > \alpha \\ 1, & \text{if } x \le \alpha \end{cases}, \text{ where } \alpha \in [0, 1). \tag{9}$$

The lattice theory is well known by the literature. In this work we only need the lattice operations (join and meet) that are defined by Baczyński and Jayaram in Theorem 6.1.1 in [1]. Although, we will not deal with the lattice theory, we must present the preliminaries to make the problem just understandable. Let us consider as $\mathcal{FI}$ be the family of all fuzzy implications and the partial order $\le$ induced from the unit interval $[0, 1]$.

Theorem 1. *(Theorem 6.1.1 in [1]). The family $(\mathcal{FI}, \le)$ is a complete, completely distributive lattice with the lattice operations*

$$(I \bigvee J)(x, y) = max\{I(x, y), J(x, y)\} = S_M(I(x, y), J(x, y)), x, y \in [0, 1], \tag{10}$$

$$(I \bigwedge J)(x, y) = min\{I(x, y), J(x, y)\} = T_M(I(x, y), J(x, y)), x, y \in [0, 1], \tag{11}$$

where $I, J \in \mathcal{FI}$.

3   Main Results

As we have mentioned before, we focus in the Problem 3.1 in [6], in the specific case we use (S,N)– implications. According to Proposition 1, it is known that (S,N)– implications always satisfy (EP). So, the problem we investigate is the following:

Problem 3.1 (a special case): Let $I_{S,N}$ and $J_{S,N}$ be two (S,N)– implications, not necessarily generated from the same t-conorm S and fuzzy negation N. Are $I_{S,N} \bigvee J_{S,N}$ and $I_{S,N} \bigwedge J_{S,N}$ satisfy (EP)? If not, what are the conditions of the preservation of (EP)?

Firstly we present the following lemmas.

Lemma 1. *(Example 2.1.(i) in [8] and Example 2.2.(i) in [9]). Let I, J be two comparable fuzzy implications that satisfy (EP), then $I \bigvee J$ and $I \bigwedge J$ satisfy (EP).*

Proof. Without loss of generality, we assume that $I \leq J$. The proof is obvious, since $I \bigvee J = J$ and $I \bigwedge J = I$.

Lemma 2. *Let I_{S_1,N_1} and I_{S_2,N_2} be two (S,N)– implications generated from comparable t-conorms and fuzzy negations, such that $S_1 \leq S_2$ and $N_1 \leq N_2$. Then, the fuzzy implications $I_{S_1,N_1} \bigvee I_{S_2,N_2}$ and $I_{S_1,N_1} \bigwedge I_{S_2,N_2}$ satisfy (EP).*

Proof. Just notify that $I_{S_1,N_1} \leq I_{S_2,N_2}$ and the proof is deduced by Lemma 1.

A special case of Lemma 2 is the following.

Lemma 3. *Let $I_{S_1,N}$ and $I_{S_2,N}$ be two (S,N)– implications generated from the same fuzzy negation N and two comparable t-conorms. Then, the fuzzy implications $I_{S_1,N} \bigvee I_{S_2,N}$ and $I_{S_1,N} \bigwedge I_{S_2,N}$ satisfy (EP).*

Proof. It is deduced by Lemma 2.

Lemma 4. *For all $x, y, z, w \in [0,1]$ it is*

$$S_M(S_M(x,y), S_M(z,w)) = max\{x, y, z, w\} \text{ and} \tag{12}$$

$$T_M(T_M(x,y), T_M(z,w)) = min\{x, y, z, w\}. \tag{13}$$

Proof. The proof is omitted due to its simplicity.

Let us study the case that, $I_{S,N}$ and $J_{S,N}$ generated from the same t-conorm S.

Theorem 2. *Let I_{S,N_1} and I_{S,N_2} be two (S,N)– implications generated from the same t-conorm S. Then, the fuzzy implications $I_{S,N_1} \bigvee I_{S,N_2}$ and $I_{S,N_1} \bigwedge I_{S,N_2}$ satisfy (EP).*

Proof. For all $x, y, z \in [0,1]$ it is

$$I_{S,N_1}(x, (I_{S,N_1} \bigvee I_{S,N_2})(y, z))$$
$$= I_{S,N_1}(x, S_M(I_{S,N_1}(y, z), I_{S,N_2}(y, z)))$$
$$= I_{S,N_1}(x, max\{I_{S,N_1}(y, z), I_{S,N_2}(y, z)\})$$
$$\overset{(I2)}{=} max\{I_{S,N_1}(x, I_{S,N_1}(y, z)), I_{S,N_1}(x, I_{S,N_2}(y, z))\}$$
$$= S_M(I_{S,N_1}(x, I_{S,N_1}(y, z)), I_{S,N_1}(x, I_{S,N_2}(y, z))).$$

Moreover,

$$I_{S,N_1}(x, I_{S,N_2}(y, z)) = S(N_1(x), S(N_2(y), z))$$
$$\overset{(S1)}{=} S(N_1(x), S(z, N_2(y)))$$
$$\overset{(S2)}{=} S(S(N_1(x), z), N_2(y))$$
$$\overset{(S1)}{=} S(N_2(y), S(N_1(x), z))$$
$$= I_{S,N_2}(y, I_{S,N_1}(x, z)).$$

Furthermore, I_{S,N_1} satisfies (EP). So, we conclude that,

$$I_{S,N_1}(x, (I_{S,N_1} \bigvee I_{S,N_2})(y, z))$$
$$= S_M(I_{S,N_1}(x, I_{S,N_1}(y, z)), I_{S,N_1}(x, I_{S,N_2}(y, z)))$$
$$= S_M(I_{S,N_1}(y, I_{S,N_1}(x, z)), I_{S,N_2}(y, I_{S,N_1}(x, z)))$$

By swapping N_1 and N_2 it turns out the following equation

$$I_{S,N_2}(x, (I_{S,N_2} \bigvee I_{S,N_1})(y, z)$$
$$= S_M(I_{S,N_2}(y, I_{S,N_2}(x, z)), I_{S,N_1}(y, I_{S,N_2}(x, z))).$$

It is obvious by (S1) that,

$$I_{S,N_2}(x, (I_{S,N_1} \bigvee I_{S,N_2})(y, z)) = I_{S,N_2}(x, (I_{S,N_2} \bigvee I_{S,N_1})(y, z)).$$

So,

$$(I_{S,N_1} \bigvee I_{S,N_2})(x, (I_{S,N_1} \bigvee I_{S,N_2})(y, z))$$

$$= S_M(I_{S,N_1}(x, (I_{S,N_1} \bigvee I_{S,N_2})(y, z)), I_{S,N_2}(x, (I_{S,N_1} \bigvee I_{S,N_2})(y, z)))$$

$$= S_M(S_M(I_{S,N_1}(y, I_{S,N_1}(x, z)), I_{S,N_2}(y, I_{S,N_1}(x, z))),$$
$$\quad S_M(I_{S,N_2}(y, I_{S,N_2}(x, z)), I_{S,N_1}(y, I_{S,N_2}(x, z))))$$

$$\overset{(12)}{=} max\{I_{S,N_1}(y, I_{S,N_1}(x, z)), I_{S,N_2}(y, I_{S,N_1}(x, z)), I_{S,N_2}(y, I_{S,N_2}(x, z)),$$
$$I_{S,N_1}(y, I_{S,N_2}(x, z))\}$$

$$= max\{I_{S,N_1}(y, I_{S,N_1}(x, z)), I_{S,N_1}(y, I_{S,N_2}(x, z)), I_{S,N_2}(y, I_{S,N_2}(x, z)),$$
$$I_{S,N_2}(y, I_{S,N_1}(x, z))\}$$

$$\overset{(12)}{=} S_M(S_M(I_{S,N_1}(y, I_{S,N_1}(x, z)), I_{S,N_1}(y, I_{S,N_2}(x, z))),$$
$$\quad S_M(I_{S,N_2}(y, I_{S,N_2}(x, z)), I_{S,N_2}(y, I_{S,N_1}(x, z))))$$

$$\overset{(12)}{=} S_M(I_{S,N_1}(y, S_M(I_{S,N_1}(x, z)I_{S,N_2}(x, z))),$$
$$\quad I_{S,N_2}(y, S_M(I_{S,N_2}(x, z), I_{S,N_1}(x, z))))$$

$$= S_M(I_{S,N_1}(y, (I_{S,N_1} \bigvee I_{S,N_2})(x, z)), I_{S,N_2}(y, (I_{S,N_1} \bigvee I_{S,N_2})(x, z)))$$

$$= (I_{S,N_1} \bigvee I_{S,N_2})(y, (I_{S,N_1} \bigvee I_{S,N_2})(x, z))$$

thus, $I_{S,N_1} \bigvee I_{S,N_2}$ satisfies (EP).

The proof for the meet is similar, therefore it is omitted.

The same result does not hold in general, when $I_{S,N}$ and $J_{S,N}$ generated from the same fuzzy negation N. In the case we use two comparable t-conorms the lattice operations preserve (EP), according to Lemma 3. On the other hand, the preservation of (EP) is not ensured if we use two not comparable t-conorms. The proof is the following counterexample.

Example 1. Consider the fuzzy implications $I_{S_{nM},N_C} = I_{FD}$ and $I_{S_P,N_C} = I_{RC}$. The t-conorms S_{nM} and S_P are not comparable since

$$S_{nM}(0.2, 0.3) = 0.3 < 0.44 = S_P(0.2, 0.3) \text{ and}$$
$$S_{nM}(0.2, 0.9) = 1 > 0.92 = S_P(0.2, 0.9).$$

Moreover, it is

$$(I_{FD} \bigvee I_{RC})(x, y) = S_M(I_{FD}(x, y), I_{RC}(x, y))$$
$$= \begin{cases} 1, & \text{if } x \leq y \\ max\{1 - x, y, 1 - x + x \cdot y\}, & \text{if } x > y \end{cases}$$
$$= \begin{cases} 1, & \text{if } x \leq y \\ 1 - x + x \cdot y, & \text{if } x > y \end{cases}$$

and

$$(I_{FD} \bigwedge I_{RC})(x,y) = T_M(I_{FD}(x,y), I_{RC}(x,y))$$
$$= min\{I_{FD}(x,y), I_{RC}(x,y)\}$$
$$= \begin{cases} 1 - x + x \cdot y, & \text{if } x \le y \\ min\{max\{1-x,y\}, 1 - x + x \cdot y\}, & \text{if } x > y \end{cases}$$
$$= \begin{cases} 1 - x + x \cdot y, & \text{if } x \le y \\ 1 - x, & \text{if } y < x < 1 - y \\ y, & \text{if } 1 - x < y < x \end{cases}$$

Both, $I = I_{FD} \bigvee I_{RC}$ and $J = I_{FD} \bigwedge I_{RC}$ violate (EP), since

$$I(0.8, I(0.3, 0.2)) = 0.808 \ne 1 = I(0.3, I(0.8, 0.2)) \text{ and}$$
$$J(0.2, J(0.8, 0.3)) = 0.86 \ne 0.888 = J(0.8, J(0.2, 0.3)).$$

Moreover, when $I_{S,N}$ and $J_{S,N}$ generated from two comparable t-conorms and two comparable fuzzy negations the lattice operations preserve (EP), according to Lemma 3. On the other hand, the preservation of (EP) is not ensured if we use two not comparable t-conorms or fuzzy negations respectively. This will be proved in the following counterexamples.

Example 2. Consider the fuzzy implications $I_{S_{nM}, N_C} = I_{FD}$ and

$$I_{S_P, N_K} = S_P(N_K(x), y)$$
$$= N_K(x) + y - N_K(x) \cdot y$$
$$= (1 - x^2) + y - (1 - x^2) \cdot y$$
$$= 1 - x^2 + x^2 \cdot y,$$

where $N_K(x) = 1 - x^2$ (see [1] Table 1.6). It is obvious that $N_C \le N_K$. On the other hand, the t-conorms S_{nM} and S_P are not comparable (see Example 1). Moreover, it is

$$(I_{FD} \bigvee I_{S_P, N_K})(x,y) = S_M(I_{FD}(x,y), I_{S_P, N_K}(x,y))$$
$$= \begin{cases} 1, & \text{if } x \le y \\ max\{1 - x, y, 1 - x^2 + x^2 \cdot y\}, & \text{if } x > y \end{cases}$$
$$= \begin{cases} 1, & \text{if } x \le y \\ 1 - x^2 + x^2 \cdot y, & \text{if } x > y \end{cases}$$

and

$$(I_{FD} \bigwedge I_{S_P, N_K})(x,y) = T_M(I_{FD}(x,y), I_{S_P, N_K}(x,y))$$
$$= min\{I_{FD}(x,y), I_{S_P, N_K}(x,y)\}$$
$$= \begin{cases} 1 - x^2 + x^2 \cdot y, & \text{if } x \le y \\ min\{max\{1-x,y\}, 1 - x^2 + x^2 \cdot y\}, & \text{if } x > y \end{cases}$$

$$= \begin{cases} 1 - x^2 + x^2 \cdot y, & \text{if } x \le y \\ 1 - x, & \text{if } y < x < 1 - y \\ min\{y, 1 - x^2 + x^2 \cdot y\}, & \text{if } 1 - x < y < x \end{cases}$$

$$= \begin{cases} 1 - x^2 + x^2 \cdot y, & \text{if } x \le y \\ 1 - x, & \text{if } y < x < 1 - y \\ y, & \text{if } 1 - x < y < x \end{cases}$$

Both, $I = I_{FD} \bigvee I_{SP,N_K}$ and $J = I_{FD} \bigwedge I_{SP,N_K}$ violate (EP), since

$$I(0.8, I(0.5, 0.2)) = 1 \neq 0.872 = I(0.5, I(0.8, 0.2)) \text{ and}$$
$$J(0.2, J(0.8, 0.3)) = 0.972 \neq 0.98208 = J(0.8, J(0.2, 0.3)).$$

Example 3. Consider the fuzzy implication $I_{S_M,N_C} = I_{KD}$. Let the fuzzy negation

$$N_{Ex3}(x) = \begin{cases} 1, & \text{if } x \in [0, 0.5) \\ 0.3, & \text{if } x \in [0.5, 1) \\ 0, & \text{if } x = 1 \end{cases}$$

and the corresponding (S,N)– implication

$$I_{SP,N_{Ex3}} = S_P(N_{Ex3}(x), y)$$
$$= N_{Ex3}(x) + y - N_{Ex3}(x) \cdot y$$
$$= \begin{cases} 1, & \text{if } x \in [0, 0.5) \\ 0.3 + 0.7 \cdot y, & \text{if } x \in [0.5, 1) \\ y, & \text{if } x = 1 \end{cases}$$

It is known that $S_M \le S_P$ (see [1] Remark 2.2.5 (viii)). On the other hand the fuzzy negations N_C and N_{Ex3} are not comparable. Moreover, it is

$$(I_{KD} \bigvee I_{SP,N_{Ex3}})(x, y) = S_M(I_{KD}(x, y), I_{SP,N_{Ex3}}(x, y))$$
$$= \begin{cases} max\{1 - x, y, 0.3 + 0.7 \cdot y\}, & \text{if } x \in [0.5, 1) \\ y, & \text{if } x = 1 \\ 1, & \text{otherwise} \end{cases}$$
$$= \begin{cases} max\{1 - x, 0.3 + 0.7 \cdot y\}, & \text{if } x \in [0.5, 1) \\ y, & \text{if } x = 1 \\ 1, & \text{otherwise} \end{cases}$$

and

$$(I_{KD} \bigwedge I_{SP,N_{Ex3}})(x, y) = T_M(I_{KD}(x, y), I_{SP,N_{Ex3}}(x, y))$$
$$= \begin{cases} min\{max\{1 - x, y\}, 1\}, & \text{if } x \in [0, 0.5) \\ min\{max\{1 - x, y\}, 0.3 + 0.7 \cdot y\}, & \text{if } x \in [0.5, 1) \\ min\{max\{1 - x, y\}, y\}, & \text{if } x = 1 \end{cases}$$
$$= \begin{cases} max\{1 - x, y\}, & \text{if } x \in [0, 0.5) \\ min\{max\{1 - x, y\}, 0.3 + 0.7 \cdot y\}, & \text{if } x \in [0.5, 1) \\ y, & \text{if } x = 1 \end{cases}$$

Both, $I = I_{KD} \bigvee I_{SP,N_{Ex3}}$ and $J = I_{KD} \bigwedge I_{SP,N_{Ex3}}$ violate (EP), since

$$I(0.8, I(0.5, 0.2)) = 0.65 \neq 0.608 = I(0.5, I(0.8, 0.2)) \text{ and}$$
$$J(0.8, J(0.5, 0.1)) = 0.37 \neq 0.44 = J(0.5, J(0.8, 0.1)).$$

Example 4. Consider the fuzzy implications $I_{S_{nM},N_C} = I_{FD}$ and

$$I_{SP,N_{Ex3}} = \begin{cases} 1, & \text{if } x \in [0, 0.5) \\ 0.3 + 0.7 \cdot y, & \text{if } x \in [0.5, 1) \\ y, & \text{if } x = 1 \end{cases}$$

The t-conorms S_{nM} and S_P are not comparable (see Example 1). The same holds for the fuzzy negations N_C and N_{Ex3}. Moreover, it is

$$(I_{FD} \bigvee I_{SP,N_{Ex3}})(x, y) = S_M(I_{FD}(x, y), I_{SP,N_{Ex3}}(x, y))$$

$$= \begin{cases} max\{1 - x, y, 0.3 + 0.7 \cdot y\}, & \text{if } x \in [0.5, 1) \text{ and } x > y \\ y, & \text{if } x = 1 \\ 1, & \text{otherwise} \end{cases}$$

$$= \begin{cases} max\{1 - x, 0.3 + 0.7 \cdot y\}, & \text{if } x \in [0.5, 1) \text{ and } x > y \\ y, & \text{if } x = 1 \\ 1, & \text{otherwise} \end{cases}$$

and

$$(I_{FD} \bigwedge I_{SP,N_{Ex3}})(x, y) = T_M(I_{FD}(x, y), I_{SP,N_{Ex3}}(x, y))$$

$$= \begin{cases} 1, & \text{if } x \in [0, 0.5) \text{ and } x \leq y \\ min\{1, max\{1 - x, y\}\}, & \text{if } x \in [0, 0.5) \text{ and } x > y \\ min\{1, 0.3 + 0.7 \cdot y\}, & \text{if } x \in [0.5, 1) \text{ and } x \leq y \\ min\{max\{1 - x, y\}, 0.3 + 0.7 \cdot y\}, & \text{if } x \in [0.5, 1) \text{ and } x > y \\ y, & \text{if } x = 1 \end{cases}$$

$$= \begin{cases} max\{1 - x, y\}, & \text{if } x \in [0, 0.5) \text{ and } x > y \\ 0.3 + 0.7 \cdot y, & \text{if } x \in [0.5, 1) \text{ and } x \leq y \\ min\{max\{1 - x, y\}, 0.3 + 0.7 \cdot y\}, & \text{if } x \in [0.5, 1) \text{ and } x > y \\ y, & \text{if } x = 1 \\ 1, & \text{otherwise} \end{cases}$$

Both, $I = I_{FD} \bigvee I_{SP,N_{Ex3}}$ and $J = I_{FD} \bigwedge I_{SP,N_{Ex3}}$ violate (EP), since

$$I(0.8, I(0.5, 0.2)) = 0.65 \neq 0.608 = I(0.5, I(0.8, 0.2)) \text{ and}$$
$$J(0.4, J(0.5, 0.3)) = 1 \neq 0.72 = J(0.5, J(0.4, 0.3)).$$

Despite of the above counterexamples, there are some special cases, where the lattice operations preserve (EP). These are the cases we use at least one fuzzy negation N which has trivial range, i.e., $N(x) \in \{0, 1\}$ for all $x \in [0, 1]$.

Proposition 2. *Let I_{S_1,N_1} and I_{S_2,N^α} be two (S,N)– implications. Then, the fuzzy implications $I_{S_1,N_1} \bigvee I_{S_2,N^\alpha}$ and $I_{S_1,N_1} \bigwedge I_{S_2,N^\alpha}$ satisfy (EP).*

Proof. For all $x, y, z \in [0,1]$ it is

$$(I_{S_1,N_1} \bigvee I_{S_2,N^\alpha})(y,z) = S_M(I_{S_1,N_1}(y,z), I_{S_2,N^\alpha}(y,z))$$

$$\overset{(8)}{=} \begin{cases} max\{S_1(N_1(y),z), z\}, \text{ if } y \geq \alpha \\ max\{I_{S_1,N_1}(y,z), 1\}, \text{ if } y < \alpha \end{cases}$$

$$= \begin{cases} S_1(N_1(y),z), \text{ if } y \geq \alpha \text{ (see [5] Proposition 9)} \\ 1, \qquad\qquad \text{ if } y < \alpha \end{cases}$$

Thus,

$$(I_{S_1,N_1} \bigvee I_{S_2,N^\alpha})(x, (I_{S_1,N_1} \bigvee I_{S_2,N^\alpha})(y,z))$$

$$= \begin{cases} (I_{S_1,N_1} \bigvee I_{S_2,N^\alpha})(x, S_1(N_1(y),z)), \text{ if } y \geq \alpha \\ (I_{S_1,N_1} \bigvee I_{S_2,N^\alpha})(x, 1), \qquad\qquad \text{ if } y < \alpha \end{cases}$$

$$= \begin{cases} S_1(N_1(x), S_1(N_1(y),z)), \qquad\qquad \text{ if } x \geq \alpha \text{ and } y \geq \alpha \\ 1, \qquad\qquad\qquad\qquad\qquad\qquad \text{ if } x < \alpha \text{ and } y \geq \alpha \\ S_M(S_1(N_1(x),1), S_2(N^\alpha(x),1)), \text{ if } y < \alpha \end{cases}$$

$$= \begin{cases} S_1(N_1(x), S_1(N_1(y),z)), \text{ if } x \geq \alpha \text{ and } y \geq \alpha \\ 1, \qquad\qquad\qquad\qquad \text{ if } x < \alpha \text{ and } y \geq \alpha \\ 1, \qquad\qquad\qquad\qquad \text{ if } y < \alpha \end{cases}$$

$$= \begin{cases} S_1(N_1(x), S_1(N_1(y),z)), \text{ if } x \geq \alpha \text{ and } y \geq \alpha \\ 1, \qquad\qquad\qquad\qquad \text{ otherwise} \end{cases}$$

By swapping x and y we have

$$(I_{S_1,N_1} \bigvee I_{S_2,N^\alpha})(y, (I_{S_1,N_1} \bigvee I_{S_2,N^\alpha})(x,z))$$

$$= \begin{cases} S_1(N_1(y), S_1(N_1(x),z)), \text{ if } y \geq \alpha \text{ and } x \geq \alpha \\ 1, \qquad\qquad\qquad\qquad \text{ otherwise} \end{cases}$$

$$\overset{(S1)}{=} \begin{cases} S_1(N_1(y), S_1(z, N_1(x))), \text{ if } y \geq \alpha \text{ and } x \geq \alpha \\ 1, \qquad\qquad\qquad\qquad \text{ otherwise} \end{cases}$$

$$\overset{(S2)}{=} \begin{cases} S_1(S_1(N_1(y),z), N_1(x)), \text{ if } y \geq \alpha \text{ and } x \geq \alpha \\ 1, \qquad\qquad\qquad\qquad \text{ otherwise} \end{cases}$$

$$\overset{(S1)}{=} \begin{cases} S_1(N_1(x), S_1(N_1(y),z)), \text{ if } x \geq \alpha \text{ and } y \geq \alpha \\ 1, \qquad\qquad\qquad\qquad \text{ otherwise} \end{cases}$$

$$= (I_{S_1,N_1} \bigvee I_{S_2,N^\alpha})(x, (I_{S_1,N_1} \bigvee I_{S_2,N^\alpha})(y,z)).$$

Thus, $I_{S_1,N_1} \bigvee I_{S_2,N^\alpha}$ satisfies (EP).

Similarly for the meet, it is

$$(I_{S_1,N_1} \bigwedge I_{S_2,N^\alpha})(y,z) = T_M(I_{S_1,N_1}(y,z), I_{S_2,N^\alpha}(y,z))$$

$$\overset{(8)}{=} \begin{cases} min\{S_1(N_1(y),z), z\}, \text{ if } y \geq \alpha \\ min\{S_1(N_1(y),z), 1\}, \text{ if } y < \alpha \end{cases}$$

$$= \begin{cases} z, \qquad\qquad \text{ if } y \geq \alpha \text{ (see [5] Proposition 9)} \\ S_1(N_1(y),z), \text{ if } y < \alpha \end{cases}$$

Thus,

$$(I_{S_1,N_1} \bigwedge I_{S_2,N^\alpha})(x, (I_{S_1,N_1} \bigwedge I_{S_2,N^\alpha})(y, z))$$

$$= \begin{cases} (I_{S_1,N_1} \bigwedge I_{S_2,N^\alpha})(x, z), & \text{if } y \geq \alpha \\ (I_{S_1,N_1} \bigwedge I_{S_2,N^\alpha})(x, S_1(N_1(y), z)), & \text{if } y < \alpha \end{cases}$$

$$= \begin{cases} z, & \text{if } x \geq \alpha \text{ and } y \geq \alpha \\ S_1(N_1(x), z), & \text{if } x < \alpha \text{ and } y \geq \alpha \\ S_1(N_1(y), z), & \text{if } x \geq \alpha \text{ and } y < \alpha \\ S_1(N_1(x), S_1(N_1(y), z)), & \text{if } x < \alpha \text{ and } y < \alpha \end{cases}$$

By swapping x and y we have

$$(I_{S_1,N_1} \bigwedge I_{S_2,N^\alpha})(y, (I_{S_1,N_1} \bigwedge I_{S_2,N^\alpha})(x, z))$$

$$= \begin{cases} z, & \text{if } y \geq \alpha \text{ and } x \geq \alpha \\ S_1(N_1(y), z), & \text{if } y < \alpha \text{ and } x \geq \alpha \\ S_1(N_1(x), z), & \text{if } y \geq \alpha \text{ and } x < \alpha \\ S_1(N_1(y), S_1(N_1(x), z)), & \text{if } y < \alpha \text{ and } x < \alpha \end{cases}$$

$$\overset{(S1)}{=} \begin{cases} z, & \text{if } x \geq \alpha \text{ and } y \geq \alpha \\ S_1(N_1(y), z), & \text{if } x \geq \alpha \text{ and } y < \alpha \\ S_1(N_1(x), z), & \text{if } x < \alpha \text{ and } y \geq \alpha \\ S_1(N_1(y), S_1(z, N_1(x))), & \text{if } x < \alpha \text{ and } y < \alpha \end{cases}$$

$$\overset{(S2)}{=} \begin{cases} z, & \text{if } x \geq \alpha \text{ and } y \geq \alpha \\ S_1(N_1(x), z), & \text{if } x < \alpha \text{ and } y \geq \alpha \\ S_1(N_1(y), z), & \text{if } x \geq \alpha \text{ and } y < \alpha \\ S_1(S_1(N_1(y), z), N_1(x)), & \text{if } x < \alpha \text{ and } y < \alpha \end{cases}$$

$$\overset{(S1)}{=} \begin{cases} z, & \text{if } x \geq \alpha \text{ and } y \geq \alpha \\ S_1(N_1(x), z), & \text{if } x < \alpha \text{ and } y \geq \alpha \\ S_1(N_1(y), z), & \text{if } x \geq \alpha \text{ and } y < \alpha \\ S_1(N_1(x), S_1(N_1(y), z)), & \text{if } x < \alpha \text{ and } y < \alpha \end{cases}$$

$$= (I_{S_1,N_1} \bigwedge I_{S_2,N^\alpha})(x, (I_{S_1,N_1} \bigwedge I_{S_2,N^\alpha})(y, z)).$$

Thus, $I_{S_1,N_1} \bigwedge I_{S_2,N^\alpha}$ satisfies (EP).

Proposition 3. *Let I_{S_1,N_1} and I_{S_2,N_α} be two (S,N)– implications. Then, the fuzzy implications $I_{S_1,N_1} \bigvee I_{S_2,N_\alpha}$ and $I_{S_1,N_1} \bigwedge I_{S_2,N_\alpha}$ satisfy (EP).*

Proof. The proof is omitted because it is similar to the proof of Proposition 2.

Although, the proofs of the Propositions 2 and 3 could also been deduced by Proposition 4.1 in [7] (see also [8] Proposition 5.2, [9] Proposition 5.1), an alternative proof has been presented. That is because the induced formula of lattice operations (join and meet) should also been mentioned. Moreover, Propositions 2 and 3 hold for an (S,N)– implication, whose negation N has trivial range and any other fuzzy implication I [7–9], but this is out of the purpose of this paper, since we study only (S,N)– implications.

4   Conclusions

In this paper, we have investigated the solution of a specific case of an open problem (see [1] Remark 6.1.5, [6] Problem 3.1) that is related to the preservation of the exchange principle (EP) of fuzzy implications via lattice operations. We have investigated these solutions with the use of only (S,N)– implications. We have presented sufficient, but not necessary conditions for this preservation. More specific, the conclusions are the following: It is ensured that (EP) is preserved via the lattice operations, when two (S,N)– implications generated from the same t-conorm S or at least one of them generated from a fuzzy negation N, which has trivial range. Moreover, the same result holds, when two (S,N)– implications, I_{S_1,N_1} and I_{S_2,N_2}, generated from comparable t-conorms and fuzzy negations, such that $S_1 \leq S_2$ and $N_1 \leq N_2$. On the other hand, we have presented counterexamples (Examples 1, 2, 3 and 4) that proved the violation of (EP) in general, in many other cases. However, this problem needs more investigation. Our intention is to study this problem in detail in the near future.

References

1. Baczyński, M., Jayaram, B.: STUDFUZZ. Studies in Fuzziness and Soft Computing, vol. 231. Springer, Heidelberg (2008). https://doi.org/10.1007/978-3-540-69082-5
2. Dimuro, G.P., Bedregal, B., Bustince, H., Jurio, A., Baczyński, M., Mis, K.: QL-operations and QL-implication functions constructed from triples (O, G, N) and the generation of fuzzy subsethood and entropy measures. Int. J. Approx. Reason. **82**, 170–192 (2017)
3. Drewniak, J.: Invariant fuzzy implications. Soft. Comput. **10**, 506–513 (2006). https://doi.org/10.1007/s00500-005-0526-4
4. Fodor, J.C., Roubens, M.: Fuzzy Preference Modeling and Multicriteria Decision Support. Kluwer, Dordrecht (1994)
5. Grammatikopoulos, D.S., Papadopoulos, B.K.: A method of generating fuzzy implications with specific properties. Symmetry **12**(1), 155–170 (2020)
6. Mesiar, R., Stupňanová, A.: Open problems from the 12th international conference on fuzzy set theory and its applications. Fuzzy Sets Syst. **261**, 112–123 (2015)
7. Vemuri, N.R.: Mutually exchangeable fuzzy implications. Inf. Sci. **317**, 1–24 (2015)
8. Vemuri, N.R., Jayaram, B.: Preservation of the exchange principle under lattice operations on fuzzy implications. In: 8th International Summer School on Aggregation Operators AGOP 2015, pp. 227–232. University of Silesia, Katowice (2015)
9. Vemuri, N.R., Jayaram, B.: Lattice operations on fuzzy implications and the preservation of the exchange principle. Fuzzy Sets Syst. **301**, 64–78 (2016)

Versatile Internet of Things for Agriculture: An eXplainable AI Approach

Nikolaos L. Tsakiridis[1]([✉]), Themistoklis Diamantopoulos[1],
Andreas L. Symeonidis[1], John B. Theocharis[1], Athanasios Iossifides[2],
Periklis Chatzimisios[2], George Pratos[3], and Dimitris Kouvas[4]

[1] Electrical and Computer Engineering Department,
Aristotle University of Thessaloniki, Thessaloniki, Greece
`tsakirin@ece.auth.gr, thdiaman@issel.ee.auth.gr,`
`{asymeon,theochar}@eng.auth.gr`
[2] Department of Information and Electronic Engineering,
International Hellenic University, Thessaloniki, Greece
`aiosifidis@el.teithe.gr, peris@it.teithe.gr`
[3] Infinite Informatics Ltd., Thessaloniki, Greece
`pratos@indinf.gr`
[4] ScientAct S.A., Thessaloniki, Greece
`dgk@scientact.com.gr`

Abstract. The increase of the adoption of IoT devices and the contemporary problem of food production have given rise to numerous applications of IoT in agriculture. These applications typically comprise a set of sensors that are installed in open fields and measure metrics, such as temperature or humidity, which are used for irrigation control systems. Though useful, most contemporary systems have high installation and maintenance costs, and they do not offer automated control or, if they do, they are usually not interpretable, and thus cannot be trusted for such critical applications. In this work, we design Vital, a system that incorporates a set of low-cost sensors, a robust data store, and most importantly an explainable AI decision support system. Our system outputs a fuzzy rule-base, which is interpretable and allows fully automating the irrigation of the fields. Upon evaluating Vital in two pilot cases, we conclude that it can be effective for monitoring open-field installations.

Keywords: Precision irrigation · Internet of Things · eXplainable AI

1 Introduction

The number of devices connected to the Internet, known as *IoT (Internet of Things) devices*, has been continuously growing. Indicatively, in 2008 their number exceeded the global population, while in 2019 there were more than 26 billion

© IFIP International Federation for Information Processing 2020
Published by Springer Nature Switzerland AG 2020
I. Maglogiannis et al. (Eds.): AIAI 2020, IFIP AICT 584, pp. 180–191, 2020.
https://doi.org/10.1007/978-3-030-49186-4_16

connected devices[1]. This number is expected to reach 75 billion by 2025[2]. Apart from home/consumer use, IoT today is largely used in different industrial sectors, including manufacturing, security, transportation, and agriculture [20].

As the global population is expected to rise[3], there is a demand for augmenting food production coupled with increased use of arable land water resources [7]. The challenge of achieving sustainable agriculture and increasing food production using technologies such as Artificial Intelligence (AI) and IoT has become evident ever since the beginning of the 21st century, when Zhang et al. [30] coined the term *precision agriculture*. The goal of this area is to maximize production while mitigating the relevant costs, i.e. minimizing water/energy consumption and optimizing the use of pest-control/growth chemicals.

Research in precision agriculture is broad, focusing on all three layers of IoT: the perception layer (i.e. sensors that are placed at fields), the network layer (i.e. communication protocols between sensors and applications), and the application layer (i.e. data storage and decision-making) [28]. Several challenges still exist: the data transfer cost, the need for uninterrupted power supply (that undermines their applicability), the lack of efficient and versatile data storage, and most importantly the need for automated control systems that are smart and can support precision agriculture with minimal human intervention.

The advent of increasingly opaque decision systems, such as deep and ensemble learning techniques, highlights a key issue: when decisions derived from such systems affect the livelihood of humans, as in agriculture, there is an increasing need to understand how the AI methods derive them. The consideration of interpretability as an additional design driver can improve the models by: i) facilitating the robustness by highlighting the potential perturbations, ii) ensuring impartiality in decision-making, and iii) guaranteeing that an underlying truthful causality (linking the cause with the effect) exists in model reasoning.

In this work, we present a system overcoming the aforementioned limitations. Our system, named *Vital*, employs IoT devices with low installation, use, and maintenance costs, capable of supporting open-field irrigation by measuring all relevant metrics (temperature, humidity, etc.). Furthermore, our system employs a distributed and scalable data management infrastructure, and a smart fuzzy rule-based system that fully automates the irrigation of open fields. Our system lies in the broad category of eXplainable AI (XAI) [3,9] systems, therefore it produces interpretable high-efficiency results, while enabling humans to inspect the model reasoning and thus understand, trust, and manage the AI technologies.

---

[1] https://safeatlast.co/blog/iot-statistics/.

[2] https://www.statista.com/statistics/802690/worldwide-connected-devices-by-access-technology/.

[3] https://www.un.org/development/desa/publications/world-population-prospects-2019-highlights.html.

2   Background and Related Work

IoT has already been applied in several areas relevant to agriculture [28]. In this section we focus on IoT for precision agriculture and specifically on what is referred to as *open-field agriculture*. Open-field deployments typically comprise sensors that measure climate conditions as well as soil sensors, and their main purpose is to optimize irrigation and enhance crop production.

One of the first systems to support such functionality [13] is based on three nodes: one measuring soil moisture and temperature, another for weather parameters (air temperature and humidity, wind speed, and luminosity), and one for irrigation control. The first two nodes send data to a base station and the user can manually activate the valve connected to the third node to irrigate the field. A similar system was proposed in [21], which also monitors these metrics and further integrates a solar-power node to reduce the energy needs of the system. Project SWAMP [12] takes water control one step further, by integrating not only water consumption but also water reserve and distribution metrics. All of these systems, however, are not automated, so an expert agronomist has to continuously monitor them and take any decisions concerning irrigation.

An interesting alternative proposed in [18] employs a set of rules to irrigate automatically based on soil moisture. These are set by the administrator (usually an agronomist) based on past data, while the farmer may override them and manually irrigate. A similar algorithm based on sensor and weather data is proposed in [16]. Though useful, these approaches still require expert monitoring. A more advanced method is proposed in [11], employing Support Vector Regression to predict soil moisture for a range of upcoming days, which is then used to determine whether and when irrigation should start. The main drawback in this case lies with interpretability; the inability to review the underlying model is a risky practice, considering errors may prove critical for the crop.

As a result, several research efforts have been directed towards fuzzy rule-based systems (FRBSs), which are interpretable and generalize well in cases when data are derived from sensors and thus may be too fine-grained. The involvement of fuzzy logic is justified by the following: i) expert knowledge is fuzzy and not precise and can thus be easier integrated, ii) data from sensors in the real world are noisy and are reported with a degree of fuzziness, iii) fuzzy consequents enable a smoother output response than traditional crisp systems [3,9]. One of the first approaches [10] includes a fuzzy logic model that receives as input soil humidity values from a sensor and determines the irrigation duration using four linguistic variables: zero, short-term, middle-term, and large-time. A more complex model is proposed in [29], which is based on temperature (low, moderate, high), soil moisture (dry, moderate, wet), and humidity (less, medium, high) and results in a set of rules capable of determining pump operation and its duration. In [17] five triangular membership functions are used as input corresponding to sunlight intensity, wind speed, humidity, temperature, and soil moisture, while the output indicates the necessary irrigation quantity. Similarly, the model proposed in [15] uses Gaussian membership functions and receives input from ten sensors, two air sensors (for humidity and temperature) and eight ground moisture sensors placed in different parts of the area, to provide more accurate recommendations.

Although the aforementioned systems are quite effective in certain scenarios, they are usually not applicable to multiple generic precision irrigation scenarios. They usually require human intervention in order to adapt to different fields and they are largely based on specific sensors that may not be available in all cases (e.g. certain systems may require a very large number of sensors [15]).

In this work, we propose a flexible IoT system operating on a set of sensors that is upgradeable and further employs a data management infrastructure able to handle large amounts of values from these sensors. We design an interpretable FRBS which functions automatically, thus not requiring any input from the farmer/agronomist. As mentioned, our system can be classified into the XAI domain. Within this domain, FRBSs have demonstrated their potential to attain both interpretability and accuracy in a variety of application areas [8,23,24].

3   Vital: An IoT System for Precision Agriculture

The Vital system comprises the following components (depicted in Fig. 1):

- The slave devices (end nodes) connected physically with the devices (e.g. weather stations, sensors, valves etc.);
- The IoT gateway that wirelessly connects to multiple slave devices and exchanges data and other information;
- The IoT platform which collects the data from multiple gateways, transmitted through 3G networks, and acts as the data focal point;
- The web application acting as the front-end of the system and relaying the output of the rule-based system via SMS to the relevant gateway; and
- The rule-based system which reads the data stored in the IoT platform and collected by the web application, and infers data-driven decisions.

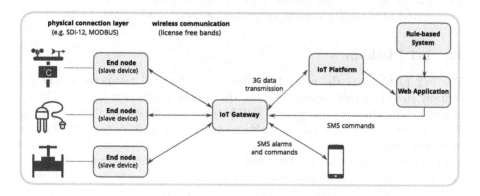

Fig. 1. Overview of the Vital system.

3.1 Wireless Slaves

The wireless slaves are ultra low-power, battery-powered end nodes in the network. The sensor nodes are frequently synchronized by the gateway in order to achieve real-time data collection with a fixed update rate of 5 min. Two types of slave devices were considered, namely ADS-200 and ADS-210. ADS-200 is the end-node used to receive input data from sensors, while ADS-210 is used to drive 12 V latching valves and relays. Their D-size Lithium Thionyl battery can provide autonomous operation for over 10 years. ADS-200 incorporates 1 configurable input (digital, pulse counter or analog), and multiple excitation options for powering transducers. The device supports acquisition of up to 16 measurement channels, based on the popular SDI-12 communication protocol and 8 channels, based on the MODBUS protocol. ADS-210 on the other hand, incorporates 2 digital inputs, 1 with counting capability, 1 analog input, and a full bridge output for driving 12 V latching valves and relays.

3.2 IoT Gateway

The IoT gateway (ADU-700) communicates wirelessly in license-free bands with the slaves and transmits their data to the IoT platform. The gateways can both receive the data (telemetry) from the slaves (upstream component) and relay the commands received through SMS text messages (downstream component), to perform specific tasks, like opening or closing a valve. Able to withstand ambient conditions, they are low-power IP66 certified and capable of securely connecting to multiple slave devices. The wireless network coverage reaches a radius of 1–6 km (line of sight), enabling the use of one gateway across multiple fields. The network uses a proprietary protocol with bidirectional communications. Modes of operation include the autonomous battery operation and an external power supply operation with uninterruptible transition to battery operation during power outage. The ADU-700 incorporates a 3G, GSM/GPRS modem and supports periodical data transmission via FTP and alerting via SMS. The data are transmitted to the IoT platform at regular intervals, configurable by the user.

3.3 IoT Platform

The IoT platform is the central datahub, i.e. where all telemetry values are securely stored. In this implementation, the Cenote platform is used [4]. Cenote is a Big Data Management System (BDMS) with analytics capabilities for the Web of Things, providing (near) real-time analytics capacities to event streams stemming from heterogeneous sources of information. Moreover, it is an open source platform following a component-based development approach. Deployed in a distributed and scalable manner, it acts as a general-scope, out-of-the-shelf, BDMS, supporting analytics out-of-the-box in many scenarios that involve event stream processing. Its overall architecture is presented in Fig. 2.

3.4 Web Application

The web front-end of the platform was developed in Javascript and is an interface to visualize telemetry data. It further has an API connecting to an external SMS service, enabling the platform to task a gateway with actions via SMS commands (e.g. for irrigation this concerns the opening of a valve for a specified number of minutes). Moreover, it enables the management of data (e.g. translating gateway ids to actual fields, using proper units, etc.), manages the access of users, and enables users to inspect and tweak the rule-based system. This latter component is very crucial for XAI; by enabling users to visualize the inner manifestations of the rule-based system the model is transparent and thus trustworthy.

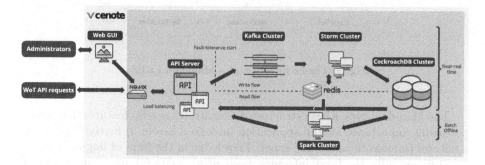

Fig. 2. Overview of the Cenote IoT platform depicting its components.

3.5 Rule-Based System

Fuzzy systems are an important tool in the domains of Computational Intelligence and Soft Computing, having been applied to a plethora of different tasks. Most commonly, a model structure in the form of an FRBS is considered. These systems have displayed their superiority to classical decision trees and crisp rule-based systems and their robustness in different AI applications including classification [2,6,25], big data analysis [6,27], and regression [19,26]. FRBSs allow for the definition of simple and verbally formulated rules over imprecise domains which can be combined to generate precise yet understandable results. They are an extension to classical rule-based systems, since they involve "IF-THEN" type of rules, but the antecedents and consequents comprising each rule are composed of fuzzy logic statements, instead of classical ones. The use of fuzzy sets with linguistic labels is particularly noteworthy, as the output system has a significant interpretability degree for the expert to understand the reasoning mechanisms of the model and the inner details of the problem characteristics.

These types of systems are called Mamdani-type FRBSs [5] and have two main components (illustrated in Fig. 3): i) the fuzzy inference system, which implements the fuzzy reasoning process to be applied on the input to infer the output, and ii) the fuzzy knowledge base (KB), representing the knowledge about the problem being solved. The KB contains the fuzzy rules composed of linguistic variables that take values from a descriptive term set with a real-world meaning. This distinction between the fuzzy rule structures and their meaning allows us to define two different sub-components, namely the fuzzy rule base (RB) containing the collection of fuzzy rules, and the data base (DB) containing the membership functions of the fuzzy partitions associated to the linguistic variables.

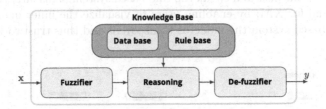

Fig. 3. The architecture of a Mamdani FRBS.

The Mamdani-type FRBS structure demonstrates several features that are of particular importance for the application presented herein: i) firstly, it provides a natural framework to include expert knowledge in the form of linguistic fuzzy rules; ii) secondly, the fuzzy inference mechanism makes full use of the power of fuzzy logic-based reasoning; and iii) thirdly, it composes a highly flexible means to formulate knowledge, while at the same time it remains interpretable.

More formally, given a dataset $E = \{x_1^p, \ldots, x_N^p, y^p\}_{p=1}^{Q}$, which is composed of Q patterns, with each pattern consisting of an N-dimensional predictor $\mathbf{x}$ and one response (or target) variable y, the FRBS models the input-output relationship using a total of R fuzzy rules of the form:

$$\mathcal{R}^k : \text{IF } x_1 \text{ is } \tilde{A}_1^k \text{ AND } \ldots \text{ AND } x_N \text{ is } \tilde{A}_N^k \text{ THEN } y \text{ is } B^k \text{ with } CF^k$$

where $\mathcal{R}^k$ is the k-th fuzzy rule, $\tilde{A}_i^k$ is the set of interpretable linguistic terms for feature i, B^k is the linguistic term for the output, and CF^k is the certainty factor (or degree of importance) for B^k.

The input space and the output variable are partitioned using uniformly distributed triangular membership functions into ℓ_{in} and ℓ_{out} fuzzy sets respectively. Thus, they constitute a strong (or proper) fuzzy partition [14], implying that in a given sample space $\mathcal{X}$:

$$\forall x : \sum_{q=1}^{\ell} \mu_{L^q}(x) = 1 \tag{1}$$

where L^q is the q-th linguistic fuzzy set of space $\mathcal{X}$. The DNF (disjunctive normal form) approach [5] is used, where each input variable x_i of $\mathcal{R}^k$ takes a set of consecutive linguistic terms as a value $\tilde{A}_i^k = \{L_i^1 \text{ or } \ldots \text{ or } L_i^{\ell_{in}}\}$, joined

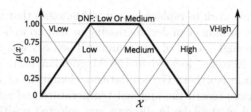

Fig. 4. A strong fuzzy partition of the space $\mathcal{X}$ using triangular membership functions. Highlighted is the fuzzy set $\tilde{A} = \{L_2 \text{ or } L_3\}$ following the DNF approach.

by the disjunctive operator OR. Figure 4 depicts the DNF approach in a strong fuzzy partition using triangular membership functions for a given space $\mathcal{X}$.

The matching degree between a pattern p and a rule $\mathcal{R}^k$ is:

$$\mu_{\mathcal{A}}^k(p) = \bigwedge_{i=1}^{N} \tilde{A}_i^k = \bigwedge_{i=1}^{N} \left\{ \bigvee_{q=1}^{\ell_{in}} \mu_{L_i^q}(x_i^p) \right\} \tag{2}$$

where $\mu_{L_i^q}(x_i^p)$ is the membership grade of each linguistic term, while $\wedge$ and $\vee$ denote the AND and OR operators, respectively. The AND operator is implemented using the minimum operator, whereas the OR operator is implemented using the bounded sum, defined as boundedSum$(a, b) = \min{(1, a + b)}$. Given an input vector $\mathbf{x}^p$, the output of the system is calculated in the following way:

$$\hat{y}^p = \sum_{k=1}^{R} \{\mu_{\mathcal{A}}^k(p) \cdot CF^k \cdot \text{CoG}(B^k)\} / \{\sum_{k=1}^{R} \mu_{\mathcal{A}}^k(p) \cdot CF^k\} \tag{3}$$

The DB parameters were $\ell_{in} = 5$ (i.e. five uniformly distributed triangular sets for the input variables) and $\ell_{out} = 7$. The larger granularity in the output space allows for more fine-grained irrigation control. The learning of the RB through the dataset takes place by following the learning algorithm presented in [26].

The rule-based system is continuously operating and monitoring the raw telemetry; once new data are received the rules are checked. If the conditions are met (i.e. the proper rules are activated by the input) the relevant action is determined from the output of the rule base and sent to the gateway as an SMS command. If no rules are activated, then no irrigation command is transmitted.

4   Pilot Cases

To demonstrate the efficacy of our approach we considered two independent pilot cases which are presented below.

4.1   Integration with Existing Telemetry Network

In this pilot case, the goal was to demonstrate the potential of the proposed system to integrate data of existing (legacy) network stations. This demonstrates the suitability of this work to replace legacy technologies that were installed but were either malfunctioning or underfunctioning or had no web-based integration capabilities. Accordingly, a set of environmental sensors was selected that was operating in lake Koronia, located in the Mygdonian basin of the Region of Central Macedonia in Northern Greece (latitude/longitude of 40.68/23.15). The environmental sensors record the ambient temperature and relativity humidity, as well as the level of the lake in discrete and selected positions. The sensors were connected via the SDI-12 interface using three separate slave devices, while a single gateway acted as the local data collector. The integration was successful, indicating that our system can integrate and/or replace legacy installations.

4.2   Precision Irrigation

In this pilot application we focused on a pilot area of young olive trees within the premises of the Farm School the Aristotle University of Thessaloniki in Northern Greece (at a latitude/longitude of 40.54033/23.00083), with the field covering an area of 12 acres and consisting of 1780 individual trees, using a drip irrigation system. The field was then divided into 4 equal areas (hereafter referred to as plots), each controlled individually by a separate irrigation valve. One of the plots was used as control (i.e. operated manually by an expert), another was operated by the expert using remote commands (i.e. via SMS commands), while the other two were irrigated by the automated system. An all-in-one weather station (ATMOS 41) was placed in the center of the field to supply the meteorological data (i.e. solar irradiance, precipitation, wind speed and direction, temperature and humidity). In the three experimental plots a GS3 sensor by Decagon was installed in the soil that measures the soil's temperature, conductivity and dielectric permittivity. All sensors and irrigation valves were connected to the slave devices that relayed the telemetry to the gateway and thence it was transmitted to the Cenote platform over TLS. The raw telemetry data were checked for abnormal values (e.g. if the data were below or above predefined minima and maxima values) before they were stored to the database.

In addition to the raw telemetry values, the soil moisture content (calculated via the Topp equation [22] using as input the soil's dielectric permittivity) and the reference crop evapotranspiration ETo (according to the FAO Penman–Monteith equation [1]) were calculated on-the-fly and served through the web-based platform to the FRBS as inputs. Both soil moisture content and ETo were potential inputs and could be taken into account by the FRBS in order to infer the best time to irrigate as well as the amount of irrigation necessary.

In the two plots controlled by the FRBS two independent courses of actions were followed. In both cases, the fuzzy KB was preselected by an expert (i.e. an agronomist) using the linguistic description of the input variables. This was done to demonstrate the ability of fuzzy systems to incorporate the expert knowledge in a straightforward way, enabling people not familiar with machine learning models to effortlessly develop an AI-based model. In one of these two plots however, the rule base was allowed to evolve to automatically tune the components of the fuzzy KB. This was deemed important as the irrigation mode of operation represents an open-loop type of control, where a decision to irrigate is sent to the gateway and thence to the relevant slave device and valve, without immediate feedback—feedback is received once the gateway synchronizes its data with the IoT platform at its predetermined time interval. The FRBS can then use this input-output relationship from prior actions and improve its structure and components by tweaking the necessary parameters, in accordance to the delayed feedback it receives, so that it can improve in the future. Additionally, and because the system automatically redefines its components to address the needs, the rule base can accordingly be more robust to perturbations and changes (e.g. in ambient conditions). Hence, this smart AI-driven system can be more precise in the management of irrigation and thus *ipso facto* help conserve water and resources. This is demonstrated in Fig. 5 for the month of January (which was dry and had the most irrigation needs). The crop needs are the calculated ETo minus the precipitation; the FRBS-auto is the closest to the theoretical values with an RMSE of 0.63 mm, compared to 2.10 mm for FRBS-expert and 6.47 mm for control, which was the most wasteful of all.

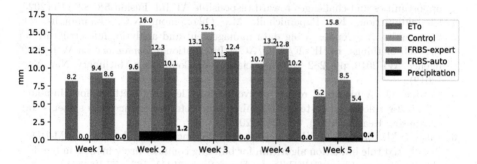

Fig. 5. Comparison among the different irrigation techniques for January of 2020.

5   Conclusion

In this paper, we have presented Vital, a system that employs IoT devices in order to monitor sensor values of fields and automate irrigation. Vital integrates a set of IoT devices (sensors) with low installation and maintenance costs, as well as a robust distributed data management infrastructure. Furthermore, we design a fuzzy-rule based system that effectively takes decisions about the irrigation of the fields. Compared to other open-field agriculture systems, our system encompasses an eXplainable AI approach, thus ensuring maximum efficiency as well as increased interpretability, by enabling humans to inspect the model, which is highly desirable in such critical applications.

Concerning future work, one may consider further adapting our system by adding more types of sensors (e.g. for digital imaging) and/or even attaching other types of valves apart from irrigation (e.g. for pesticide control). Finally, concerning the smart system, we plan to evaluate it against scenarios with missing data and assess whether it can be used effectively in such cases, as well as perform comparisons against other approaches in this area.

Acknowledgments. This research has been co-financed by the European Regional Development Fund of the European Union and Greek national funds through the Operational Program Competitiveness, Entrepreneurship and Innovation, under the call RESEARCH - CREATE - INNOVATE (project code: T1EDK-02296).

References

1. Allen, R.G., Pereira, L.S., Raes, D., Smith, M., et al.: Crop evapotranspiration-Guidelines for computing crop water requirements-FAO Irrigation and drainage paper 56. FAO, Rome **300**(9), D05109 (1998)
2. Antonelli, M., Bernardo, D., Hagras, H., Marcelloni, F.: Multiobjective evolutionary optimization of type-2 fuzzy rule-based systems for financial data classification. IEEE Trans. Fuzzy Syst. **25**(2), 249–264 (2017)
3. Arrieta, A.B., et al.: Explainable artificial intelligence (XAI): concepts, taxonomies, opportunities and challenges toward responsible AI. Inf. Fusion **58**, 82–115 (2020)
4. Chatzidimitriou, K., Papamichail, M., Oikonomou, N.C., Lampoudis, D., Symeonidis, A.: Cenote: a big data management and analytics infrastructure for the web of things. In: IEEE/WIC/ACM International Conference on Web Intelligence, WI 2019, pp. 282–285. Association for Computing Machinery, New York (2019)
5. Cordón, O.: A historical review of evolutionary learning methods for Mamdani-type fuzzy rule-based systems: designing interpretable genetic fuzzy systems. Int. J. Approx. Reason. **52**(6), 894–913 (2011)
6. Elkano, M., Sanz, J.A., Barrenechea, E., Bustince, H., Galar, M.: CFM-BD: a distributed rule induction algorithm for building compact fuzzy models in big data classification problems. IEEE Trans. Fuzzy Syst. **28**(1), 163–177 (2020)
7. FAO, IFAD, UNICEF, WFP and WHO: The State of Food Security and Nutrition in the World 2019. Safeguarding against economic slowdowns and downturns. FAO, Rome, Italy (2019)
8. Fernández, A., Carmona, C.J., del Jesus, M.J., Herrera, F.: A view on fuzzy systems for big data: progress and opportunities. Int. J. Comput. Intell. Syst. **9**(Suppl. 1), 69–80 (2016)
9. Fernandez, A., Herrera, F., Cordon, O., del Jesus, M.J., Marcelloni, F.: Evolutionary fuzzy systems for explainable artificial intelligence: why, when, what for, and where to? IEEE Comput. Intell. Mag. **14**(1), 69–81 (2019)
10. Gao, L., Zhang, M., Chen, G.: An intelligent irrigation system based on wireless sensor network and fuzzy control. J. Netw. **8**(5), 1080–1087 (2013)
11. Goap, A., Sharma, D., Shukla, A., Krishna, C.R.: An IoT based smart irrigation management system using Machine learning and open source technologies. Comput. Electron. Agric. **155**, 41–49 (2018)
12. Kamienski, C., et al.: SWAMP: an IoT-based smart water management platform for precision irrigation in agriculture. In: 2018 Global Internet of Things Summit (GIoTS), pp. 1–6 (2018)
13. Khriji, S., Houssaini, D.E., Jmal, M.W., Viehweger, C., Abid, M., Kanoun, O.: Precision irrigation based on wireless sensor network. IET Sci. Meas. Technol. **8**(3), 98–106 (2014)

14. Klir, G.J., Yuan, B.: Fuzzy Sets and Fuzzy Logic: Theory and Applications. Prentice-Hall Inc., Upper Saddle River (1995)
15. Kokkonis, G., Kontogiannis, S., Tomtsis, D.: A smart IoT fuzzy irrigation system. IOSR J. Eng. **07**(06), 15–21 (2017)
16. Mohanraj, I., Ashokumar, K., Naren, J.: Field monitoring and automation using IOT in agriculture domain. Procedia Comput. Sci. **93**, 931–939 (2016)
17. Mohapatra, A.G., Lenka, S.K.: Neural network pattern classification and weather dependent fuzzy logic model for irrigation control in WSN based precision agriculture. Procedia Comput. Sci. **78**(C), 499–506 (2016)
18. Muangprathub, J., Boonnam, N., Kajornkasirat, S., Lekbangpong, N., Wanichsombat, A., Nillaor, P.: IoT and agriculture data analysis for smart farm. Comput. Electron. Agric. **156**, 467–474 (2019)
19. Rodríguez-Fdez, I., Mucientes, M., Bugarín, A.: S-FRULER: scalable fuzzy rule learning through evolution for regression. Knowl.-Based Syst. **110**, 255–266 (2016)
20. Sharma, N., Shamkuwar, M., Singh, I.: The history, present and future with IoT. In: Balas, V.E., Solanki, V.K., Kumar, R., Khari, M. (eds.) Internet of Things and Big Data Analytics for Smart Generation. ISRL, vol. 154, pp. 27–51. Springer, Cham (2019). https://doi.org/10.1007/978-3-030-04203-5_3
21. Shuwen, W., Changli, Z.: Study on farmland irrigation remote monitoring system based on ZigBee. In: 2015 International Conference on Computer and Computational Sciences (ICCCS), Noida, India, pp. 193–197 (2015)
22. Topp, G.C., Davis, J.L., Annan, A.P.: Electromagnetic determination of soil water content: measurements in coaxial transmission lines. Water Resour. Res. **16**(3), 574–582 (1980)
23. Tsakiridis, N., Theocharis, J., Ben-Dor, E., Zalidis, G.: Using interpretable fuzzy rule-based models for the estimation of soil organic carbon from VNIR/SWIR spectra and soil texture. Chemometr. Intell. Lab. Syst. **189**, 39–55 (2019)
24. Tsakiridis, N., Theocharis, J., Panagos, P., Zalidis, G.: An evolutionary fuzzy rule-based system applied to the prediction of soil organic carbon from soil spectral libraries. Appl. Soft Comput. J. **81**, 105504 (2019)
25. Tsakiridis, N., Theocharis, J., Zalidis, G.: DECO$_3$R: a differential evolution-based algorithm for generating compact fuzzy rule-based classification systems. Knowl.-Based Syst. **105**, 160–174 (2016)
26. Tsakiridis, N., Theocharis, J., Zalidis, G.: DECO$_3$RUM: a differential evolution learning approach for generating compact Mamdani fuzzy rule-based models. Expert Syst. Appl. **83**, 257–272 (2017)
27. Tsakiridis, N., Theocharis, J., Zalidis, G.: An evolutionary fuzzy rule-based system applied to real-world Big Data - the GEO-CRADLE and LUCAS soil spectral libraries. In: IEEE International Conference on Fuzzy Systems (2018)
28. Tzounis, A., Katsoulas, N., Bartzanas, T., Kittas, C.: Internet of Things in agriculture, recent advances and future challenges. Biosyst. Eng. **164**, 31–48 (2017)
29. Yadav, R., Daniel, A.K.: Fuzzy based smart farming using wireless sensor network. In: 2018 5th IEEE Uttar Pradesh Section International Conference on Electrical, Electronics and Computer Engineering (UPCON), pp. 1–6 (2018)
30. Zhang, N., Wang, M., Wang, N.: Precision agriculture - a worldwide overview. Comput. Electron. Agric. **36**(2), 113–132 (2002)

Machine Learning

Acoustic Resonance Testing of Glass IV Bottles

Ivan Kraljevski[1]([✉])[iD], Frank Duckhorn[1][iD], Yong Chul Ju[1][iD],
Constanze Tschoepe[1][iD], and Matthias Wolff[2][iD]

[1] Fraunhofer Institute for Ceramic Technologies and Systems IKTS,
Dresden, Germany
{ivan.kraljevski,frank.duckhorn,yong.chul.ju,
constanze.tschoepe}@ikts.fraunhofer.de
[2] Chair of Communications Engineering, Brandenburg University of Technology
(BTU) Cottbus-Senftenberg, Cottbus, Germany
matthias.wolff@b-tu.de

Abstract. In this paper, acoustic resonance testing on glass intravenous (IV) bottles is presented. Different machine learning methods were applied to distinguish acoustic observations of bottles with defects from the intact ones. Due to the very limited amount of available specimens, the question arises whether the deep learning methods can achieve similar or even better detection performance compared with traditional methods.

The results from the binary classification experiments are presented and compared in terms of Balanced Accuracy Rate, F1-score, Area Under the Receiver Operating Characteristic Curve and Matthews Correlation Coefficient metrics.

The presented feature analysis and the employed classifiers achieved solid results, despite the rather small and imbalanced dataset with a highly inconsistent class population.

Keywords: Acoustic resonance testing · Machine learning · Glass IV bottles · Non-destructive testing

1 Introduction

Glass materials are ubiquitous in many areas of everyday life, in the home, industry, medicine, vehicles, etc. The production quality of glass is of the highest importance, however, defects in glass materials may occur in the form of cracks, spots, bubbles and inclusions, holes and abrasions [19].

Different defects will have different impacts on the usage of glass products, while cracks may have little effect on the ordinary household glass, even hairline or micro-sized cracks will have large effects on the glassware used in the pharmaceutical and medical industries.

© IFIP International Federation for Information Processing 2020
Published by Springer Nature Switzerland AG 2020
I. Maglogiannis et al. (Eds.): AIAI 2020, IFIP AICT 584, pp. 195–206, 2020.
https://doi.org/10.1007/978-3-030-49186-4_17

Therefore, it is necessary to detect the presence of any defects in glass materials by employing non-destructive testing (NDT) in fabrication and in-service inspections to ensure product integrity and reliability.

Common approaches to NDT of glass materials are X-ray Computer Tomography (XCT), optical systems (OS) and acoustic emission testing (AE) [1]. In production, machine vision-based systems are providing reliable defect detection [18], while in-service it would be difficult to deploy equipment and trained operators to track defects and subjectively interpret results. Many studies present such systems [14, 21, 26, 32], where image processing algorithms are used to detect glassware defects.

To the best of our knowledge, machine learning (ML) approaches have not been extensively applied in the field of NDT on glassware and there are fewer studies on it. In one of them [15], the authors applied Convolutional Neural Networks (CNNs) on images to detect defects in the mouth, body and the bottom of glass bottles achieving an average accuracy rate of 98.4%.

Alternatives to image-based systems are Resonance Acoustic Method (RAM) [25], known as Acoustic Resonance Testing (ART) [7] NDT systems.
The mechanical vibrations in the structure are produced by impact and transmitted as audible signals carrying information about the object's material, structure and geometry in their entirety, but alone it will generally not diagnose the location, size, or type of defect.

ART was successfully applied in many areas, such as in automotive production lines [12]. Artificial Neural Networks (ANN) were used for classifying automotive components as intact, without damage or defective with results that completely prevented false positives and only 2.61% of good parts detected as defective [23]. Classification of magnetic tiles (qualified/unqualified) by their acoustic resonance after impact with a metal block was presented in [30].

Impact tests to determine defects in glassware using Fourier analysis of the resonance frequency are presented in [11]. Wavelet packet transforms and ANNs were used for the same problem in [10], achieving an accuracy rate of 96.6% on 3 classes (no glass, good, defective). Smart signal processing in ART is crucial since every defect will produce a corresponding dynamic response depending on its size and the feature analysis should be able to reliably capture the deviations.

In this paper, acoustic resonance testing of glass IV (intravenous) bottles is presented. We collected a small database in trials where a hammer impact was used as excitation. The observation signals were labeled, pre-processed and feature analysis was performed. Different ML approaches were applied, and the results of the binary classification were presented and compared. Since the amount of the collected data is small, the question arises whether deep learning could provide comparable or even better performance against traditional ML methods.

The paper is organized as follows: Sect. 2 describes the selection of glass IV bottles used in the experiments, as well as, the data collection and their organization in a dataset and the employed feature analysis algorithms.

In Sect. 3, we first present the experimental setup and then each of the used ML approaches with their specific parameter configurations. In Sect. 4 we present the achieved results in the classification experiments across different machine learning and feature analysis approaches. We close this paper with concluding remarks in Sect. 5.

(a) Specimen B2

(b) Specimen B3

Fig. 1. Recording setup with naturally damaged bottles

2 Materials and Methods

The selection of glass IV bottles consists of used and empty ones which are: intact (120), artificially (10) and naturally (3) damaged.

To provide representative examples of defects, different types of damages were artificially introduced to intact bottles in the form of smaller and larger cracks, abrasions and holes. In this case, each specimen has characteristic damage of different sizes and placement.

Bottles with defects that occurred naturally have small (the specimens B1 and B2) to middle size (the specimen B3) hairline cracks on the glass surface.

The bottles have the used rubber cap still on, and there was some intravenous fluid still present in most of them, which influenced their physical structure and consequently defined a unique set of characteristic spectral and temporal features, e.g. vibration properties.

2.1   Data Collection and Organization

The recording sessions were carried out in a soundproof room, where the acoustic responses of the bottles were measured by a microphone array [28].

A human operator was knocking the bottles using a modal hammer on three different locations, a force sensor was used as a recording trigger, ensuring consistent acoustic signal onset across the recordings. The recording setup with two different damaged bottles with marked cracks is presented in Fig. 1.

The acoustic responses were acquired by four high-performance Microtech Gefell MK 301 E microphone cartridges placed horizontally at: −125, −25, 25 and 125 mm of the field center, and 125 mm vertically from the base. The bottles were placed at a distance of 30 cm and a height of 165 mm, the holder was aligned by a perforated base plate with the microphone field in the center. The hammer impacts were applied at the positions (20 times on each): 0, 120 and 240 degrees counter-clockwise rotation, the 0 degree position is where the bottle label is aligned with a marked holder.

The dataset contains 8002 observations labeled according to the condition of the IV bottle specimens. The originally damaged bottles were labeled with "B" (broken) - 181 (2.3%), the artificially damaged with "D" (defective) - 602 (7.5%) and the undamaged bottles were labeled with "V" (valid) 7219 (90.2%) observations.

2.2   Feature Analysis

Acoustic response observations were obtained by simple delay-and-sum beam-forming of individual microphone signals, providing more acoustic information than separate ones, at the same time reducing the computational costs of the feature analysis and the classification.

The acoustic resonance observations were transformed into three different feature sets:

1. The raw signal (SIG) observations, where the analog signals were A/D converted to 48 kHz, 16 bit, little-endian PCM and summed into one single observation with a fixed duration of one second, the dimension of the features is 48000×1 elements.
2. The primary feature analysis (PFA) was done over the SIG features, by short-time Fourier transformation with a Blackman window of 512, and MEL filter-bank with a triangular transfer function. The feature vector dimension per observation is 298×30.
3. Secondary feature analysis (SFA) was performed over the PFA features by their standardization to zero mean and unit variance, computing the delta features and performing Principal Component Analysis (PCA) for size reduction, producing feature vectors with a dimension of 298×24.

3   Experiments and Results

The choice of an appropriate machine learning approach should also be based on the following factors:

- tolerance of high dimensionality,
- the capability of exploiting a small dataset, and
- handling of imbalanced datasets.

The objective is to find out which ML approach is best suited for the given dataset: Hidden Markov Models (HMMs), Support Vector Machine (SVM) models or Deep Learning methods.

Stratified 5-fold Cross-Validation (CV) was employed in all experiments, ensuring proper representations of the classes in the train and the test folds. The 95% confidence intervals (CI) for balanced accuracy rate (BAR) [3], the macro average (μ) and standard deviation (σ) of the F1 score were used as performance metrics. The Area Under the Curve (AUC) and the Matthews Correlation Coefficient (MCC) were calculated, as well. The class weights were applied accordingly in the training procedure to anticipate the highly imbalanced dataset during training.

The objective of Experiment 1 is to classify acoustic responses of intact and artificial damaged bottles. The dataset folds were created by random sampling at the same time ensuring that the train and the test sets do not contain the same signal observations. The class distribution was also taken into account and the classes were accordingly represented in both sets.

In the second experiment (Experiment 2), the training was performed by leaving out a specific group of bottles from the k-fold training runs. They were used as an additional dataset and tested on the models trained for each fold.

This represents a real use-case, where few damaged specimens are available. In the case when no damaged specimens are available at all, other ML approaches for one-class or anomaly (outliers) detection have to be considered.

Here, the dataset was divided into disjoint groups: V1, V2, and V3 (each with different 40 bottles from class V), B1, B2, and B3 (each with one bottle from class B). The training was performed in the following combinations:

- training V1, V2, B1, B2, and test V3, B3;
- training V1, V3, B1, B3, and test V2, B2;
- training V2, V3, B2, B3, and test V1, B1.

The false negative rates (FNR) of the observations of natural damaged bottles ("B") with the 95% CI were calculated. Miss-classifying observations of damaged bottles (positive identification) as observations of intact ones is more critical than the miss-classifying observations of intact bottles as damaged (false positives).

3.1   Hidden Markov Models

Hidden Markov Models (HMM), model the signals as Markov processes where the states emit the observations by a Gaussian probability density function.

Apart from other areas of application (speech and handwriting recognition, bioinformatics, etc.) they were also successfully applied in NDT [27].

In the experiments, the HMMs were created using the dLabPro software [13,29]. To represent the subtle temporal features, the classes were modeled by five-state forward connected HMMs with exactly one Gaussian probability density function (PDF) and full covariance matrix per state. The model parameters are iteratively estimated by the Viterbi training algorithm alternated with Gaussian splitting, with maximum of 5 splits and a different number of iterations per split.

The similarity measure describes how well a model fits an observation and is estimated by computing neg-log likelihoods (NLLs). The model which yields the highest likelihood defined predicted class label.

3.2   Support Vector Machines

SVMs predict class labels by finding the best hyperplane that separates samples of one class from those of the other class. They are very effective in the case where the dimension of the features is much higher than the number of observations in the dataset. The choice of kernel parameters plays an important role in achieving acceptable results for binary classification with SVMs [8].

To train the SVM models, we used the python scikit-learn [20] interface to the well known LIBSVM library [4]. The Radial Basis Function (RBF) was chosen as a kernel because of its good general performance and the SVMs were tuned over a range of the cost (10^{-4} to 10^1) and the gamma (10^{-9} to 10^1) parameters.

3.3   Deep Neural Networks

Deep Feed-forward (DFFN) and Convolutional Neural Network (CNN) were trained over the feature sets with Keras [5] and Tensorflow [17]. For all employed classifiers random search was performed over the hyper-parameter space in the pre-tests to estimate appropriate architectures and parameter values.

The DFFN architectures consist of an input layer corresponding to the feature vectors, output layer with one unit and softmax activation. They have two fully connected hidden layers: the first with 256 (128 for SFA features) units and the second with the half of units from the first layer, respectively.

The Leaky Rectified Linear Unit (LeakyReLU) [16] was used as the activation function, batch normalization and dropout layers [24] with Adaptive Gradient Algorithm (Adagrad) [9] as the optimization function.

For PFA and SFA features, the CNN architectures have two 2D convolutional layers with 32 filters of size 5×5, Exponential Linear Unit activation (ELU) [6], followed by batch normalization, dropout, and MaxPooling layers.

The output of the convolutional layers was flattened and feed to a fully connected layer with 90 units and ReLU activation and Adadelta [31] as optimization function.

One dimensional CNN based on the Soundnet architecture [2] was used for the SIG features (Table 1), with Rectified Linear Unit (ReLU) activation, batch normalization, dropout and MaxPooling layers (MPL).

Table 1. The configuration of the convolutional and pooling layers applied to raw audio signals.

	Layers							
	CL1	MPL1	CL2	MPL2	CL3	CL4	CL5	MPL3
Dim.	24000	3000	1500	188	24	12	6	2
# Filters	16	16	32	32	64	128	256	256
Filter size	64	8	32	8	16	8	4	4
Stride	2	8	2	8	2	2	2	4

The output of the convolution layers (with a dimension of 512) was flattened into two fully connected layers with 64 and 32 units respectively, providing a rather small size of trainable parameters (283217) well suited for small datasets. For all of them, the output layer has a softmax activation and binary-cross entropy as a loss function.

To avoid over-fitting which is even more emphasized in the case of a small dataset and larger feature dimensions, proper parameters for optimizer algorithms, training batch size, dropout rate, and early stopping criteria, were discovered during hyperparameter optimization.

The maximum number of epochs was set to 200 with the condition of 10 (DFFNs) or 25 (CNNs) epochs with no improvement (patience parameter) after which the training was stopped. The patience parameter and the high dropout rate allow longer training and better generalization.

4   Results and Discussion

The inconsistency of the dataset samples across classes was the main issue as already described in Sect. 2. This influenced the overall resonance frequencies and could mask those related to the condition of the bottle.

Also, the hammer impacts made by a human operator gave inconsistent acoustic resonance responses per specimen. On the other hand, this represents the real use-case, where inexperienced personnel should assess the glassware condition in a non-ideal environment.

4.1   Experiment 1

Table 2 presents the results achieved in Experiment 1, in terms of balanced accuracy rate (BAR) with the 95% confidence intervals and the mean (μ) of the F1 scores across folds with standard deviation (σ), across all features and classifiers.

Table 2. Balanced Accuracy Rate (%) and F1-score.

Experiment 1	SIG	PFA	SFA
95% CI of BAR			
HMM	N/A	$91.17^{+0.86}_{-0.94}$	$93.90^{+0.67}_{-0.78}$
SVM	$89.45^{+0.93}_{-1.02}$	$90.85^{+1.27}_{-1.39}$	$84.75^{+1.58}_{-1.62}$
DFFN	$60.66^{+1.51}_{-1.49}$	$82.36^{+1.62}_{-1.68}$	$80.80^{+1.68}_{-1.72}$
CNN	$86.60^{+1.47}_{-1.53}$	$93.05^{+1.12}_{-1.23}$	$88.50^{+1.43}_{-1.47}$
F1-score: μ (σ)			
HMM	N/A	0.77 (0.01)	0.83 (0.02)
SVM	0.74 (0.01)	0.93 (0.01)	0.89 (0.01)
DFFN	0.63 (0.04)	0.86 (0.02)	0.84 (0.02)
CNN	0.87 (0.02)	0.94 (0.01)	0.90 (0.02)

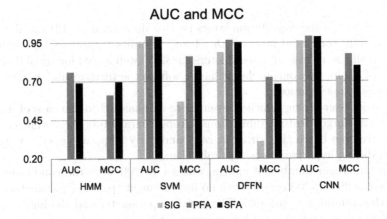

Fig. 2. AUC and MCC across features and classifiers.

The SIG/HMM combination was omitted from the experiments as infeasible due to the dimensionality of the features.

It can be seen that CNN models performed well across all feature types in terms of BAR and better than any other in terms of F1-score, despite small and imbalanced dataset. The models were able to discover patterns that can be interpreted as detectors of dynamic events in the time domain (SIG), as well as, in the time-frequency domain (PFA and SFA). The PFA features have the edge over the raw signals for all classifiers except HMMs, mainly because of the dimensionality reduction which compensates the smaller observation counts.

In the pre-trials with even a smaller dataset (600 samples), the performance gap between the raw signal-based features and the other two for each classifier was even larger. Given more training data, it is expected that CNNs outperform the other approaches on raw signals, where more optimal that the handcrafted features will be discovered.

Figure 2 presents the AUC and the MCC metrics for the binary classification, where the damaged bottle label was chosen as positive, since detecting those observations correctly is more important than the intact ones.

The CNNs are performing equally well as the SVMs which, in general, are more suitable choice for binary classification given the size and nature (imbalanced data) of the dataset.

4.2 Experiment 2

Because they are of higher importance, Table 3 presents only the results for false negative rates on class "B" obtained in Experiment 2. It can be seen that most of the classifier/feature models failed to reliably detect unseen damaged specimens.

Table 3. False Negative Rate (%) for class "B".

Experiment 2	SIG	PFA	SFA
Train: V1V2B1B2		*Test: V3B3*	
HMM	N/A	$10.67^{+4.06}_{-3.26}$	$20.33^{+5.00}_{-4.41}$
SVM	$58.33^{+5.64}_{-5.80}$	$29.00^{+5.49}_{-5.07}$	$62.67^{+5.49}_{-5.74}$
DFFN	$82.33^{+4.14}_{-4.80}$	$32.33^{+5.61}_{-5.26}$	$37.00^{+5.74}_{-5.48}$
CNN	$32.00^{+5.60}_{-5.24}$	$13.00^{+4.34}_{-3.59}$	$21.33^{+5.08}_{-4.50}$
Train: V1V3B1B3		*Test: V2B2*	
HMM	N/A	$4.00^{+2.88}_{-1.92}$	$46.33^{+5.82}_{-5.75}$
SVM	$79.00^{+4.47}_{-5.05}$	$17.33^{+4.77}_{-4.11}$	$49.67^{+5.80}_{-5.80}$
DFFN	$97.67^{+1.39}_{-2.41}$	$26.33^{+5.37}_{-4.89}$	$43.67^{+5.82}_{-5.69}$
CNN	$31.67^{+5.59}_{-5.23}$	$2.33^{+2.41}_{-1.39}$	$7.67^{+3.62}_{-2.74}$
Train: V2V3B2B3		*Test: V1B1*	
HMM	N/A	$18.69^{+4.84}_{-4.22}$	$50.49^{+5.75}_{-5.76}$
SVM	$87.87^{+3.44}_{-4.20}$	$37.38^{+5.70}_{-5.45}$	$83.61^{+3.97}_{-4.64}$
DFFN	$87.54^{+3.49}_{-4.24}$	$20.98^{+5.01}_{-4.43}$	$56.07^{+5.65}_{-5.77}$
CNN	$24.59^{+5.23}_{-4.73}$	$16.07^{+4.61}_{-3.94}$	$14.10^{+4.42}_{-3.70}$

The reason is the small amount of the damaged bottles samples (only from two specimens) in the training set and the large variation in the acoustic resonance responses of the intact bottles.

Although class weighting was applied, the characteristic features of damaged bottles were not properly generalized for the unseen specimens.

McNemar's test [22] showed that there are significant differences in FNR ($p \leq 0.05$) across all classifier-feature combinations, except in the cases: CNN/PFA against HMM/PFA and DFFN/PFA against CNN/SIG.

In general, the CNNs along with HMMs performed better in predicting observations from damaged bottles than other models, particularly with PFA features. In comparison with others, they managed to reliably capture the subtle differences between the damaged and intact bottles.

The CNN/PFA model achieved an overall BAR (%) of $94.94^{+0.68}_{-0.72}$ and FNR (%) of $9.97^{+1.48}_{-1.34}$, while the HMM/PFA achieved overall BAR (%) of $94.36^{+0.67}_{-0.73}$ and FNR (%) of $9.81^{+1.47}_{-1.33}$.

5   Conclusions

Different machine learning approaches in acoustic resonance testing of glass IV bottles are presented. Despite the rather small and imbalanced dataset with highly inconsistent classes, the presented feature analysis, and the employed classifiers achieved solid results.

The HMM and CNN models in combination with handcrafted features achieved the best overall performance in both experiments. Providing a larger amount of data, the CNNs have the potential of improving the prediction of raw signals.

In practical application, when a high detection accuracy is of critical importance, ensemble modeling should be considered. Here multiple models are created on different training sets, and the prediction is performed by aggregating their classification results.

Alternatively, higher detection performance can be achieved using one trained model and combining the predictions from multiple impacts on different positions on a tested specimen. Moreover, the classifier system should be able to detect and adapt to changing environmental conditions and data properties.

References

1. Aastroem, T.: From fifteen to two hundred NDT-methods in fifty years. In: 17th World Conference on Nondestructive Testing, pp. 25–28 (2008)
2. Abdoli, S., Cardinal, P., Koerich, A.L.: End-to-end environmental sound classification using a 1D convolutional neural network. Expert Syst. Appl. **136**, 252–263 (2019)
3. Brodersen, K., et al.: The balanced accuracy and its posterior distribution. In: 20th International Conference on Pattern Recognition, pp. 3121–3124. IEEE (2010)
4. Chang, C., Lin, C.: LIBSVM: a library for support vector machines. ACM Trans. Intell. Syst. Technol. (TIST) **2**(3), 27 (2011)
5. Chollet, F., et al.: Keras (2015). https://github.com/fchollet/keras
6. Clevert, D.A., Unterthiner, T., Hochreiter, S.: Fast and accurate deep network learning by exponential linear units (ELUs). arXiv preprint arXiv:1511.07289 (2015)
7. Coffey, E.: Acoustic resonance testing. In: 2012 Future of Instrumentation International Workshop (FIIW) Proceedings, pp. 1–2. IEEE (2012)
8. Cortes, C., Vapnik, V.: Support-vector networks. Mach. Learn. **20**(3), 273–297 (1995). https://doi.org/10.1007/BF00994018

9. Duchi, J., Hazan, E., Singer, Y.: Adaptive subgradient methods for online learning and stochastic optimization. J. Mach. Learn. Res. **12**(Jul), 2121–2159 (2011)
10. Gokmen, G.: The defect detection in glass materials by using discrete wavelet packet transform and artificial neural network. J. Vibroeng. **16**(3), 1434–1443 (2014)
11. Gunathilaka, G.: Using Fourier analysis of the resonance frequency in glassware to identify defects. In: Research Symposium on Pure and Applied Sciences. Faculty of Science, University of Kelaniya, Sri Lanka (2018)
12. Hertlin, I., Schultze, D.: Acoustic resonance testing: the upcoming volume-oriented NDT method. In: III Pan-American Conference for Nondestructive Testing (2003)
13. Hoffmann, R., Eichner, M., Wolff, M.: Analysis of verbal and nonverbal acoustic signals with the Dresden UASR system. In: Esposito, A., Faundez-Zanuy, M., Keller, E., Marinaro, M. (eds.) Verbal and Nonverbal Communication Behaviours. LNCS (LNAI), vol. 4775, pp. 200–218. Springer, Heidelberg (2007). https://doi.org/10.1007/978-3-540-76442-7_18
14. Huang, B., Ma, S., Wang, P., Wang, H., Yang, J., Guo, X., Zhang, W., Wang, H.: Research and implementation of machine vision technologies for empty bottle inspection systems. Eng. Sci. Technol. Int. J. **21**(1), 159–169 (2018)
15. Liang, Q., Xiang, S., Long, J., Sun, W., Wang, Y., Zhang, D.: Real-time comprehensive glass container inspection system based on deep learning framework. Electron. Lett. **55**(3), 131–132 (2019). https://doi.org/10.1049/el.2018.6934
16. Maas, A.L., Hannun, A.Y., Ng, A.Y.: Rectifier nonlinearities improve neural network acoustic models. In: Proceedings of the ICML, vol. 30, no. 1, p. 3 (2013)
17. Martín, M., et al.: Tensorflow: a system for large-scale machine learning. In: 12th USENIX Symposium on Operating Systems Design and Implementation (OSDI 16), pp. 265–283 (2016)
18. Mery, D., Medina, O.: Automated visual inspection of glass bottles using adapted median filtering. In: Campilho, A., Kamel, M. (eds.) Image Analysis and Recognition. Lecture Notes in Computer Science, vol. 3212, pp. 818–825. Springer, Heidelberg (2004). https://doi.org/10.1007/978-3-540-30126-4_99
19. Rosli, N.S., Fauadi, M., Awang, N., Noor, A.: Vision-based defects detection for glass production based on improved image processing method. J. Adv. Manuf. Technol. (JAMT) **12**(1 (1)), 203–212 (2018)
20. Pedregosa, F., et al.: Scikit-learn: machine learning in Python. J. Mach. Learn. Res. **12**, 2825–2830 (2011)
21. Peng, X., Li, X.: An online glass medicine bottle defect inspection method based on machine vision. Glass Technol. Eur. J. Glass Sci. Technol. A **56**(3), 88–94 (2015)
22. Salzberg, S.L.: On comparing classifiers: Pitfalls to avoid and a recommended approach. Data mining and knowledge discovery **1**(3), 317–328 (1997). https://doi.org/10.1023/A:1009752403260
23. Sankaran, V.: Low cost inline NDT system for internal defect detection in automotive components using Acoustic Resonance Testing. In: Proceedings of the National Seminar and Exhibition on Non Destructive Evaluation, pp. 237–239 (2011)
24. Srivastava, N., Hinton, G., Krizhevsky, A., Sutskever, I., Salakhutdinov, R.: Dropout: a simple way to prevent neural networks from overfitting. J. Mach. Learn. Res. **15**(1), 1929–1958 (2014)
25. Stultz, G., Bono, R., Schiefer, M.: Fundamentals of resonant acoustic method NDT. Adv. Powder Metall. Part. Mater. **3**, 11 (2005)
26. Tai, K.: The application of digital image processing technology in glass bottle crack detection system. Acta Technica **62**(1A), 381–390 (2017)

27. Tschoepe, C., Wolff, M.: Statistical classifiers for structural health monitoring. IEEE Sens. J. **9**(11), 1567–1576 (2009)

28. Wawra, J.: Experiments in acoustic pattern recognition. B.Sc. thesis, Brandenburgische Technische Universität, Cottbus - Senftenberg, in German (2015)

29. Wolff, M.: dLabPro: a signal processing and acoustic pattern recognition toolbox (2014). https://github.com/matthias-wolff/dLabPro

30. Xie, L., et al.: Internal defect inspection in magnetic tile by using acoustic resonance technology. J. Sound Vib. **383**, 108–123 (2016)

31. Zeiler, M.D.: ADADELTA: an adaptive learning rate method. arXiv preprint arXiv:1212.5701 (2012)

32. Zhou, X., et al.: A surface defect detection framework for glass bottle bottom using visual attention model and wavelet transform. IEEE Trans. Industr. Inf., 1 (2019). https://doi.org/10.1109/TII.2019.2935153

AI Based Real-Time Signal Reconstruction for Wind Farm with SCADA Sensor Failure

Nadia Masood Khan[1]([✉]), Gul Muhammad Khan[2]([✉]), and Peter Matthews[1]([✉])

[1] Durham University, Durham, England, UK
{nadia.m.khan,p.c.matthews}@durham.ac.uk
[2] National Center of Artificial Intelligence (NCAI), University of Engineering and Technology Peshawar, Peshawar, Pakistan
gk502@uetpeshawar.edu.pk

Abstract. Supervisory Control and Data Acquisition (SCADA) systems used in wind turbines for monitoring the health and performance of a wind farm can suffer from data loss due to sensor failure, transmission link breakdown or network congestion. Sensory data is used for important control decisions and such data loss can make the failures harder to detect. This work proposes various solutions to reconstruct the lost information of important SCADA parameters using Linear and non-linear Artificial Intelligence (AI) algorithms. It comprises of three major contributions; (1) signal reconstruction from other available SCADA parameters, (2) comparison of linear and non-linear AI models, and (3) generalization of the AI algorithms between turbines. Experimental results demonstrate the effectiveness of the developed methodologies for reconstruction of the lost information for valuable planning decisions.

Keywords: Signal reconstruction · SCADA data · Condition monitoring · Linear regression · Random forest · Neural network

1 Introduction

Wind energy generation is an ideal source of green energy due to which the capacity of wind farms has been increased 30 times with a 17% cumulative growth in the last few years. Wind energy is expected to supply 12% of worldwide energy demand by 2020 [22]. Due to growth in the wind industry, wind turbines are most likely to be installed in diverse climatic conditions, onshore and offshore which would need continuous monitoring. These systems are monitored using a Supervisory Control and Data Acquisition (SCADA) system for their operation and performance. Unexpected failures of wind turbine components cause

© IFIP International Federation for Information Processing 2020
Published by Springer Nature Switzerland AG 2020
I. Maglogiannis et al. (Eds.): AIAI 2020, IFIP AICT 584, pp. 207–218, 2020.
https://doi.org/10.1007/978-3-030-49186-4_18

an increase in machine downtime, repair cost and subsequently cost of energy. Condition monitoring is often used to monitor health parameters, e.g. temperature or vibration that shows the condition of a machinery and any significant change in its pattern is indicative of developing failure [4]. One of the most significant problems arises when a communication link or a sensor fails causing faulty or no data to be received for timely probable control decisions [13]. This research develops a system for reconstruction of the lost signal from low correlated parameters when one of the SCADA sensors fails to send data. Linear and non-linear AI algorithms have been analysed to find a generalized model which will be robust and perform better for all wind turbines in a wind farm rather than one turbine. The wind power curve defines the relationship between wind speed and power. It is frequently used to monitor the health of a wind turbine using SCADA data received from other system parameters. This study will assume that wind power, being the most important parameter to monitor the performance and health of a wind turbine, is lost or corrupted in transmission. Artificial intelligence (AI) based models are extensively used in detecting failures and predicting wind power from a SCADA system of a wind farm [17].

Research in the literature is focused mostly on signal reconstruction from historical data. Signal reconstruction from other available parameters was never considered as an option. The motivation of this research is to reconstruct a signal when one of the SCADA sensors fails to send data either due to a sensor failure or a communication breakdown for longer than expected. Also, when the highly correlated variables and historical data is not available for a very long time, signal reconstruction becomes vital for optimum operation of the plant. Electrical power generated from a wind turbine is considered to be the corrupted/lost signal in this case, since it is the most important parameter describing the normal operation of a wind turbine and is hard to predict due to its high degree of fluctuations and randomness. AI algorithms have the ability to learn and model the non-stationary behaviours. We have explored two AI methods: random forest and Cartesian Genetic programming evolved Artificial Neural Network (CGPANN) and then compared these results with a linear regression model to find out the best performing model for accurate estimation of the failed sensor data. Training and test results on the same turbine demonstrate random forest performing much better than its counterparts. Its performance degrades when tested on data from other wind turbines in the same wind farm. The CGPANN model having multilayered feed-forward architecture arranged in Cartesian format show remarkable generalization and continue to perform better in diverse data conditions [6].

2 Background

SCADA collects data from a machine and send it to a central processing unit for proactive measures. Data collected and stored from SCADA comprises of information regarding every aspect of a wind farm which can be used to infer overall health of the wind turbine in real-time. SCADA systems are often at risk due to various factors such as the sensor failure or network congestion, limited power or equipment abnormality resulting in data loss [12]. Sensor failure means it might be sending abnormal data or not sending data at all, this work is focused on the latter one. Missing/lost data is a challenge faced in engineering and industry specifically in applications employing sensing technologies implementing intelligent real-time monitoring and control such as an offshore wind farm SCADA data and wireless sensor networks [12]. A framework for the effective data management and dealing with issues of missing and corrupted samples in the acquired data is developed in [15] for effective fatigue assessment of offshore wind turbines. Yang et al. [23] proposed a machine learning based reconstruction model for real-time condition monitoring and fault detection. They performed correlation analysis to select input parameters and then used Support Vector Regression (SVR) for building a reconstruction model. Their focus was on failures caused due to high temperature, so signals relevant to temperature faults are selected as input features to estimate generator drive end temperature. Singh [19] used wind power curve to identify the abnormal operation of a wind turbine. Wind power being considered the vulnerable parameter, since deviation of power from its normal operational values helps in identifying probable failures in advance. Establishing a generalised model to reconstruct the lost/corrupted SCADA signal of wind power is the focus of this work. Lind et al. [11] have explored a stochastic approach to reconstruct the tower top acceleration signals from a single external variable, i.e. wind speed and previous values of the tower top acceleration. Their finding was that signal reconstruction can be used to monitor and detect abnormal behaviour. De-noising auto-encoder (DAE) is proposed in [1] for reconstruction of original sensor measurements from a corrupted SCADA system due to covert cyber-deception attack (CCDA). An ad-hoc method is presented in [14] to reconstruct the long bursts of data lost by SCADA due to sensor or communication failure. Lamrini et al. [10] applied self-organizing map (SOM)-based methods for the reconstruction of data from a water treatment system to deal with data loss due to sensor failure or corrupted input data. A number of statistical and artificial intelligence models have been proposed for wind power estimation to prevent damage to wind turbines and ensure stability of the power system. Sun et al. [21] proposed a hybrid model of deep belief networks (DBN) and random forest for short term wind power forecasting.

3 Dataset

SCADA is widely used in different areas for monitoring and control in real time. The SCADA data used in this study is acquired from La Haute Borne wind farm (ENGIE Green)[1] which consists of four wind turbines from Senvion MM82 technology located in the Grand East region in north-eastern France. The SCADA system collected data from 31 parameters along with their statistics such as average, maximum, minimum and standard deviation of each parameter. Since the average value of these parameters captures most of the information, only the

---

[1] ENGIE Green, https://opendata-renewables.engie.com/explore/dataset/la-haute-borne-data-2013-2016/.

average value of each parameter has been used for the experiments. Each data point is sampled at 10 min interval. Nominal power for each turbine at the La Haute Borne wind farm is 2050 kW, with a rotor diameter of 82 m and a hub height of 80 m. Cut in wind speed of 3.5 m/s, nominal wind speed of 14.5 m/s and cut-out wind speed of 25 m/s. Some of the parameters from the dataset are, active power, reactive power, vane position, wind speed, nacelle angle, gearbox bearing temperature, generator bearing temperature, pitch angle, torque and converter torque. The layout of the wind farm with latitude/longitude and inter-turbine distances in meters is shown in Fig. 1. The wind farm comprises of four wind turbines: R80711, R80780, R80721 and R80736.

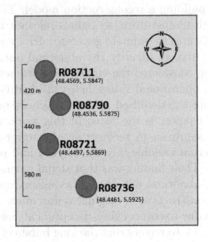

Fig. 1. Wind farm layout.

4   Methodology

The objective of this work is to develop a reliable signal reconstruction model for wind power prediction from other SCADA parameters. This is necessary because the power produced from a wind turbine depicts its health statistics and eventually help in monitoring the condition of a wind turbine. Accurate wind power prediction has a significant economical and technical impact for a reliable large-scale wind power integration and important energy management planning decisions. There has been ample work carried out to estimate power generated from wind speed and an acceptable accuracy is reported in literature [2]. The challenging part is to estimate wind power when the closely correlated parameters such as wind speed, torque, apparent power, rotor speed etc. are not available, and the system has to predict highly fluctuating wind power from very low correlated parameters with non-linear and non-stationary characteristics. This work is focused on restoring the lost information from the low correlated parameters, assuming that all the highly correlated parameters are missing. Due to the non-linear nature of the problem, an adequately accurate algorithm needs to be developed to model this complex relationship. Wind power is estimated in absence of highly correlated variables. Any deviation of a power curve from its normal operational values is indicative of a number of faults such as blade pitch angle failure, blade damage, pitch control failure and blades affected by ice or dirt [20].

The idea is to develop a model in which the output has low correlation with input signals and input signals have low cross-correlation with one another [18]. As part of the methodology, a correlation matrix has been generated and input signals having an absolute cross-correlation coefficient greater than 0.8 are removed from input when developing wind power signal reconstruction model in this study. One year (2014) data is divided into 3 segments based on wind speed. Figure 2 shows the performance curve for each region. The first segment contains the data when the wind speed is below cut-in and the turbine blades are trying to overcome friction. There is no power produced in this region. Instead the wind turbine takes power from grid to keep the turbine blades moving to prevent damage to blades due to ice and dust. The second segment is the most important to model, as it contains data above cut-in speed to rated wind speed; there is a rapid growth in power produced. In segment 3, constant power is produced until the wind speed reaches the cut-off speed and the turbine is turned off to prevent damage due to excessive wind speed. These three segments represent the various key modes of wind turbine operation. Linear regression, random forest and CGPANN are implemented on each of the data segment and then on the data without segmentation. The data from wind turbine R80721 between 01/01/2014 to 31/12/2014 is selected for training and testing following the split strategy 75% training, 25% testing [9]. Testing has been performed on the same turbine and on data from the other three wind turbines in the wind farm. The reconstruction algorithm performance is tested on all three segments as well as on the whole dataset. Similarly, data from the other three wind turbines is divided into segments and evaluated to find the best signal reconstruction algorithm.

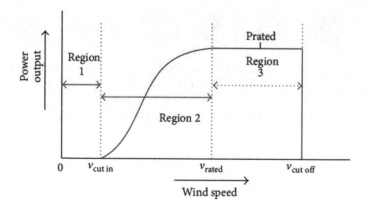

Fig. 2. Typical power curve for a wind turbine.

4.1 Multiple Linear Regression (MLR)

Linear regression is a statistical method used for prediction of a dependent variable from a single independent variable. It is termed as Multiple linear regression (MLR) [5] when the dependent variable Y is predicted from a set of independent variables X_k where k is an index of the predictor variables.

$$Y = \beta_0 + \beta_1 X_1 + \ldots + \beta_k X_k + \epsilon \tag{1}$$

The model parameters β_0 (intercept), β_1 to β_k (regression coefficients) are learned from the data and ϵ is the residual standard deviation in Y. Python's Scikit-learn library [16] has implemented a number of algorithms. The linear regression is implemented using the Scikit-learn library in this study on the transformed dataset.

4.2 Random Forest (RF)

Random forest [3] is a supervised machine learning algorithm which consists of an ensemble of decision trees. More trees means more robust performance. The ensemble decision trees are trained using a bagging method. This trains a number of learning models and its combination increases the overall results. RF can be represented as an ensemble of C number of trees $T_1(X), T_2(X), ..., T_C(X)$ having m inputs given by $X = x_1, x_2, ...x_m$. The resulting ensemble produces C outputs $\hat{Y}_1 = T_1(X), \hat{Y}_2 = T_2(X), ..., \hat{Y}_C = T_C(X)$ and the mean is calculated from the output of these randomly generated trees to give final prediction $\hat{Y}$ of random forest regression tree.

Random forest comes under the umbrella of artificial intelligence models which develops decision trees with the root node having the most important feature. Random forest is implemented using Scikit-learn library [16] and is used with the default parameter setting. Hyper-parameters are not optimized since it performed well with default parameters.

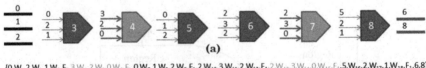

(a)

$\{0,W_1,2,W_2,1,W_3,F_1,3,W_4,2,W_5,0,W_6,F_1,0,W_7,1,W_8,2,W_9,F_1,2,W_{10},3,W_{11},2,W_{12},F_1,2,W_{13},3,W_{14},0,W_{15},F_1,5,W_{16},2,W_{17},1,W_{18},F_1,6,8\}$

(b)

Fig. 3. a. CGPANN phenotype, b. CGPANN genotype

4.3 Cartesian Genetic Programming Evolved Artificial Neural Network

Cartesian genetic programming evolved Artificial Neural Network (CGPANN) was first proposed in 2010 [8]. The study conducted in [7] shows that CGPANN has performed best when compared with Hidden Markov model (HMM), Auto Regressive Integrated Moving Average (ARIMA), regression model, Classification and Regression tree (CART), and neural network model for time series prediction. The essence of CGPANN is to tune the hyper parameters using evolutionary programming called Cartesian genetic programming (CGP) instead of traditional gradient based methods. All neurons are arranged in Cartesian format in CGPANN. Recently, a single row format has been used, since it provides more flexibility of generating infinite graphs as shown in Fig. 3. Figure 3 shows the CGPANN phenotype with corresponding genotype in part A and B. Not all neurons are connected in CGPANN while creating graphs from output to inputs. This ability of CGPANN provides the flexibility of generating arbitrary graphs. The numbers in genotype can change during evolutionary training using mutation operator, causing the network topology to change accordingly producing novel solutions to the problem. Direct encoding scheme is used to encode connection weights, connection type, and the topology of the network. The $1 + \lambda(\lambda = 9)$ evolutionary strategy is followed to produce population of probable solutions. In this work, the log sigmoid function is used as the activation function as represented by Eq. 3.

$$h_\theta(x) = \frac{1}{1 + e^{-\theta^T x}} \tag{2}$$

4.4 Evaluation of Model Performance

The following two performance metrics are applied to evaluate the abilities of three models to reconstruct the wind power from loosely correlated parameters.

$$Root Mean Square Error (RMSE) = \sqrt{(\frac{1}{N}) \sum_{i=1}^{n}(y_i - \hat{y}_i)^2} \tag{3}$$

$$Mean Absolute Error (MAE) = \frac{1}{N} \sum_{i=1}^{n} |y_i - \hat{y}_i|. \tag{4}$$

where y_i is the actual value, $\hat{y}_i$ is the estimated value and N is the total number of observations.

5 Experimental Setup

The proposed wind power reconstruction model is trained on the parameters available in the SCADA dataset under consideration. The framework for the

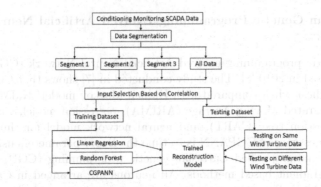

Fig. 4. Framework of wind power data reconstruction model

experimental setup is shown in Fig. 4. Conditioning monitoring data from the SCADA system is segmented based on different wind speeds, it results in four sub-dataset. Each segment is analysed separately to select inputs for training. Figure 5 shows the correlation heat map for each segment. The heat map is used for two purposes in this research, first to verify the different modes of operation of a wind turbine by showing the difference in heat map of each segment. It demonstrates that correlation between different variables varies across various segments. Second, to locate the variables having an absolute cross-correlation greater than 0.8 and removing these from the input. The main aim of this research is to reconstruct the important parameter i.e. wind power which shows the health of a wind turbine from poorly correlated input parameters. Pitch angle *Ba* has very low correlation with power in segment 1, but noticeable negative correlation in segment 2 and 3. Blade pitch angle keeps on changing to capture most of the wind energy. When the power production gets low due to change in wind direction in segment 2, pitch angle is changed to increase the power production. In segment 3, when power reaches its rated maximum value, pitch angle is adjusted to stop production and prevent turbine blades from damage. Similarly, wind speed is recorded by three sensors (Ws, Ws1, Ws2) at different locations in a wind turbine. Wind speed has highest correlation with power in segment 2, since power produced relies on wind speed in this region. In segments 1 and 3, correlation between wind speed and power produced is low as there is constant power produced due to wind energy in these two regions. The heat map demonstrates various operational modes in each segment to exploit wind turbine physics. After the input selection, each sub-dataset is split into training and testing data. Training data is then used to develop three reconstruction models using Linear Regression, Random Forest and CGPANN. The trained reconstruction models are then tested on two types of testing datasets, that include testing data from same wind turbine it is trained on and the data from other three wind turbines in the same wind farm.

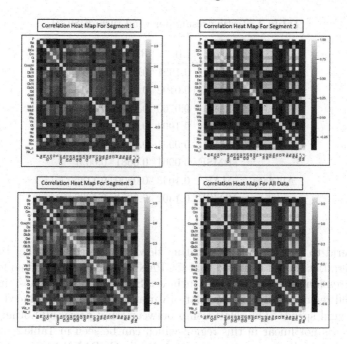

Fig. 5. Heat map showing cross-correlation between input parameters in each segment.

5.1 Results and Analysis

To validate the performance of linear and non-linear methods for wind power reconstruction, experiments have been carried out on four wind turbines' data for the year 2014. Table 1 shows the results obtained by testing the trained algorithms' performance on different segments and all year (2014) data of the turbine R80721. Random forest (RF) performs exceptionally well in all three segments as well as on the overall data without segmentation when tested on same wind turbine data (R80721) it has been trained on. Random forest performs well on segments 1 and 2 giving mean absolute error (MAE) of 0.02. On segment 3 MAE is 0.06 and on data without segmentation MAE is 0.05. High values of error on complete data shows that the data segmentation can be helpful in accurately reconstructing the power production of a wind turbine from other SCADA parameters.

Trained models have been evaluated for their performance on other wind turbines from the same wind farm to check if the developed models had learned the power variation and fluctuation pattern in wind farms in general rather then learned a single wind turbine's operation. Table 2 shows the performance of the three trained reconstruction models on different wind turbines in the same wind farm. Random forest (RF) still performs well in segment 1 due to the fact that segment 1 does not have a lot of variations and power produced in this region is constant. Ideally, the wind turbine does not produce any power below cut in wind speed, but it is not the case in real time as the wind turbine takes power from

Table 1. Testing results for wind turbine 80721

	Error	LR	RF	CGPANN
Segment 1	MAE	0.1321	0.0251	0.1020
	RMSE	0.0192	0.0071	0.0513
Segment 2	MAE	0.0760	0.0274	0.1163
	RMSE	0.1004	0.0488	0.1838
Segment 3	MAE	0.0961	0.0669	0.0092
	RMSE	0.0961	0.0681	0.0097
All data	MAE	0.1043	0.0500	0.1135
	RMSE	0.1302	0.0851	0.1838

national grid to keep the blades rotating slowly in the cold weather to prevent icing. Overall, it should be noted that CGPANN outperforms random forest and linear regression in all three wind turbines. Segment 2 is the most important operational region of all as it has a lot of variations in power produced based on different wind speeds. The relationship between wind power and other SCADA parameters is not linear in this region which can be seen in Table 2 that linear regression has highest error in segment 2. While CGPANN has lowest MAE of approximately 0.14 and the random forest having MAE ≈0.6 which verifies that a neural network has the property of transferability and able to learn the power patterns in a wind farm.

Table 2. Performance evaluation in terms of MAE and RMSE for various turbines

	Error	R80711			R80736			R80790		
		LR	RF	CGPANN	LR	RF	CGPANN	LR	RF	CGPANN
Segment 1	MAE	0.1321	0.0251	0.0209	0.3686	0.0562	0.1368	0.3064	0.0681	0.1020
	RMSE	0.1777	0.0563	0.0503	0.3842	0.0853	0.1417	0.3246	0.0834	0.1077
Segment 2	MAE	2.2599	0.6258	0.1432	2.7365	0.6469	0.1455	2.6814	0.6462	0.1337
	RMSE	2.3981	0.6559	0.1980	2.8903	0.6737	0.2285	2.8701	0.6728	0.1951
Segment 3	MAE	0.0900	0.1537	0.0197	0.0945	0.1909	0.0668	0.0689	0.2737	0.0193
	RMSE	0.0987	0.1797	0.0448	0.1131	0.2156	0.0754	0.0848	0.3037	0.0249
All data	MAE	0.2391	0.1302	0.1313	0.2336	0.1292	0.1306	0.2317	0.1216	0.1220
	RMSE	0.3350	0.2100	0.1903	0.3277	0.2196	0.2116	0.3134	0.1762	0.1860

In Table 2, the low error on all data without segmentation when tested on different wind turbines is because segments 1 and 3 have low errors as compared to segment 2. Table 3 depicts the size of data in each segment. This emphasizes the importance of segmentation as the model might show good performance on all data but not well on segment 2 with high variations as shown in Table 2.

Table 3. Wind turbines training and testing data size

	R80721		R80736
	Training	Testing	Testing
Segment 1	7959	2654	10530
Segment 2	39420	13140	40520
Segment 3	8	3	22
All data	47387	15797	53560

6 Conclusion and Future Work

This paper presents a methodology to deal with the challenge of wind power estimation from low correlation data and proposes a signal reconstruction model for wind power in case of SCADA sensor failure in a wind farm. The proposed model can be used to monitor the wind power signal continuously even when the power sensor fails and the signals of parameters that are highly correlated with wind power are also not received either. Linear regression, random forest and CGP evolved ANN is used for real time prediction of electric power produced from a wind turbine. Data segmentation based on wind speeds help in the accurate estimation of wind power and emphasizes the importance of segment 2. Although Random Forest testing results are better for a specific wind turbine, CGPANN generalizes better when exposed to wind turbines other than it is trained on. Accurate and timely prediction of these important parameters are able to help in important decisions for a wind farm. This work can be extended to the cases where more than one sensor fails. In future, the reconstructed signal will be used to identify faults in the wind turbine so that an alarm can be generated before the actual failure.

References

1. Ahmed, S., Lee, Y., Hyun, S.H., Koo, I.: Mitigating the impacts of covert cyber attacks in smart grids via reconstruction of measurement data utilizing deep denoising autoencoders. Energies **12**(16), 3091 (2019)
2. Barbosa de Alencar, D., de Mattos Affonso, C., Limão de Oliveira, R.C., Moya Rodriguez, J.L., Leite, J.C., Reston Filho, J.C.: Different models for forecasting wind power generation: Case study. Energies **10**(12), 1976 (2017)
3. Breiman, L.: Random forests. Mach. Learn. **45**(1), 5–32 (2001)
4. Da Silva, R.R., Costa, E.D.S., De Oliveira, R.C., Mesquita, A.L.: Fault diagnosis in rotating machine using full spectrum of vibration and fuzzy logic. J. Eng. Sci. Technol. **12**(11), 2952–2964 (2017)
5. Freedman, D.A.: Statistical Models: Theory and Practice. Cambridge University Press, Cambridge (2009)
6. Hassanzadeh, H., Nguyen, A., Karimi, S., Chu, K.: Transferability of artificial neural networks for clinical document classification across hospitals: a case study on abnormality detection from radiology reports. J. Biomed. Inform. **85**, 68–79 (2018)

7. Khan, G.M., Ahmad, A.M.: Breaking the stereotypical dogma of artificial neural networks with cartesian genetic programming. In: Stepney, S., Adamatzky, A. (eds.) Inspired by Nature. ECC, vol. 28, pp. 213–233. Springer, Cham (2018). https://doi.org/10.1007/978-3-319-67997-6_10

8. Khan, M.M., Khan, G.M., Miller, J.F.: Evolution of neural networks using cartesian genetic programming. In: IEEE Congress on Evolutionary Computation, pp. 1–8. IEEE (2010)

9. Kusiak, A., Zhang, Z.: Short-horizon prediction of wind power: a data-driven approach. IEEE Trans. Energy Convers. **25**(4), 1112–1122 (2010)

10. Lamrini, B., Lakhal, E.K., Le Lann, M.V., Wehenkel, L.: Data validation and missing data reconstruction using self-organizing map for water treatment. Neural Comput. Appl. **20**(4), 575–588 (2011)

11. Lind, P.G., Vera-Tudela, L., Wächter, M., Kühn, M., Peinke, J.: Normal behaviour models for wind turbine vibrations: comparison of neural networks and a stochastic approach. Energies **10**(12), 1944 (2017)

12. Liu, X., Zheng, Z., Zhang, Z., Cao, Z.: A statistical learning framework for the intelligent imputation of offshore wind farm missing SCADA data (2019)

13. Marti-Puig, P., Martí-Sarri, A., Serra-Serra, M.: Different approaches to scada data completion in water networks. Water **11**(5), 1023 (2019)

14. Marti-Puig, P., Martí-Sarri, A., Serra-Serra, M.: Double tensor-decomposition for scada data completion in water networks. Water **12**(1), 80 (2020)

15. Martinez-Luengo, M., Shafiee, M., Kolios, A.: Data management for structural integrity assessment of offshore wind turbine support structures: data cleansing and missing data imputation. Ocean Eng. **173**, 867–883 (2019)

16. Pedregosa, F., et al.: Scikit-learn: machine learning in Python. J. Mach. Learn. Res. **12**, 2825–2830 (2011)

17. Qian, P., Tian, X., Kanfoud, J., Lee, J.L.Y., Gan, T.H.: A novel condition monitoring method of wind turbines based on long short-term memory neural network. Energies **12**(18), 3411 (2019)

18. Schlechtingen, M., Santos, I.F.: Comparative analysis of neural network and regression based condition monitoring approaches for wind turbine fault detection. Mech. Syst. Signal Process. **25**(5), 1849–1875 (2011)

19. Singh, P.: Analytical techniques of SCADA data to assess operational wind turbine performance (2013)

20. Stetco, A., et al.: Machine learning methods for wind turbine condition monitoring: a review. Renew. Energy **133**, 620–635 (2019)

21. Sun, Z., Sun, H., Zhang, J.: Multistep wind speed and wind power prediction based on a predictive deep belief network and an optimized random forest. Math. Probl. Eng. **2018** (2018)

22. Wan, C., Lin, J., Wang, J., Song, Y., Dong, Z.Y.: Direct quantile regression for nonparametric probabilistic forecasting of wind power generation. IEEE Trans. Power Syst. **32**(4), 2767–2778 (2016)

23. Yang, C., Liu, J., Zeng, Y., Xie, G.: Real-time condition monitoring and fault detection of components based on machine-learning reconstruction model. Renew. Energy **133**, 433–441 (2019)

Autonomous Navigation for Drone Swarms in GPS-Denied Environments Using Structured Learning

William Power$^{(\boxtimes)}$, Martin Pavlovski, Daniel Saranovic, Ivan Stojkovic, and Zoran Obradovic

Center for Data Analytics and Biomedical Informatics, Temple University, Philadelphia, PA, USA
{william.power,martin.pavlovski,daniel.saranovic,
ivan.stojkovic,zoran.obradovic}@temple.edu

Abstract. Drone swarms are becoming a new tool for many tasks including surveillance, search, rescue, construction, and defense related activities. As their usage increases, so does the possibility of adversarial attacks on their contribution to these use cases. One possible avenue, whether deliberate or not, is to deny access to the position feedback offered by the Global Positioning System (GPS). Operating in these 'GPS denied' environments poses a new challenge; both in navigation, and in collision avoidance. This study proposes two novel concepts; a structural model of environmental deviance to aid in autonomous navigation, and a method to use the output of said model to implement a collision avoidance system. Both of these concepts are developed and tested in the framework of a simulated environment that mimics a GPS-denied scenario. Using data from hundreds of simulated swarm flights, this work shows structured learning can improve navigational accuracy without the need for externally provided position feedback.

Keywords: Machine learning · Adaptive control · Swarm intelligence · Decision-making

1 Introduction

Drone swarms are collections of remotely or autonomously controlled aerial vehicles which maintain some form of internal structure between the individuals. These swarms have utility in many domains and tasks; surveillance (coverage), search and rescue, construction, and defense related activities [9,16]. Swarms typically rely on some kind of position feedback to maintain the prescribed structure of the swarm. Referred to as localization, this task is typically achieved by

© IFIP International Federation for Information Processing 2020
Published by Springer Nature Switzerland AG 2020
I. Maglogiannis et al. (Eds.): AIAI 2020, IFIP AICT 584, pp. 219–231, 2020.
https://doi.org/10.1007/978-3-030-49186-4_19

utilizing GPS data and other sensor modalities [10, 20]. Without position feedback, maintaining a safe, collision-free flight within a crowded airspace becomes an increasingly difficult task [1, 17].

In domains where adversaries or technical issues might disable or impede the efficacy of position feedback, drone swarms are at risk of losing the position information prerequisites that allow for safe, collision-free flight. In these GPS-denied environments, other methods must be used to provide position information to the drones, or to provide other navigation schemes that allow for safe flight. This paper seeks to investigate a method for coping with a GPS-denied environment without incurring these systemic costs of complex sensor fusion and navigational calculation. Given a set of drones with the most basic of intra-drone communication, and with a limited computational budget, are there still navigation methods that can cope with the loss of GPS provided position feedback? One of the simplest methods for dealing with the lack direct positional feedback operate is the technique of dead reckoning. In such an approach, the current estimate of a drones position is taken as the 'true' position, and new control signals are calculated based on the relative location of this position to a desired position.

Applying the method of dead reckoning to a drone swarm introduces a new source of error that might lead to collisions of drones within a swarm due to environmental deviations. Dead reckoning assumes the position estimate at the prior time-step is accurate, and only influenced by the action of the drone. It lacks a model of the influences of the environment. However, additional sensors, if available in the GPS-denied environment, provide intra-drone communication and distance keeping. Radio frequency sensors, or simple optical systems may provide such feedback [7, 12]. While this is useful for collision avoidance, ultimately, they cannot also solve the navigation issue. This work assumes the existence of a simple inter-drone communication channel. This can be used to craft a nearest neighbor graph of the swarm. The history of drone locations over time, coupled with this graph, provides a spatio-temporal data-set.

This work hypothesizes that such a spatio-temporal data-set can be used to train a model which will improve upon the efficacy of dead reckoning in GPS-denied environments by predicting future environmental deviations based on changes in the structure of the swarm network. We hypothesize that such a network encodes some information about the environmental deviations that introduce the accumulated errors of dead reckoning. By using appropriate structural machine learning tools, this information could be leveraged to correct the estimated position of each drone within the swarm. In turn, this will reduce the accumulation of error, and lead to position estimates that are closer to their true value when the model is used versus pure dead reckoning.

The novelty of the proposed procedure is that it provides a framework to safely navigate a swarm of drones in a GPS-denied environment with a purposefully simplified set of sensor modalities and computational capabilities. This framework is shown to provide course maintenance and collision avoidance via structural learning of environmental deviance's in an online manner.

To support this, in this study we answer the following research questions: *1) Can a prediction-based approach properly estimate a drones trajectory after communication loss? 2) If so, do structure-based models improve this prediction capability? 3) Can a method for avoiding collisions between swarms be implemented, using the variance of the structured model as input?*

2   Prior Work

The past decade has seen substantial work on the issues of collision avoidance, localization, and mapping [6]. The prevalence of these swarms has incentivized the creation of high fidelity models of both individual drones [8] and of swarms of these individuals [2,18].

Due to their prevalence, the security and consistency of communication within a swarm becomes paramount [3]. The manner of communication that enables the localization of drones within the swarms becomes an attack surface. In systems where GPS is the communication channel for localization, entering a 'GPS-denied' environment poses a threat to the swarm. Work has been done to produce systems that leverage visual input to accommodate localization in such settings. On board visual sensors have been utilized to localize drones with respect to observations of visual indicators applied to each drone [14]. The bearing of the swarm were combined with observed landmarks to implement a sufficient simultaneous localization and mapping (SLAM) framework. It is also shown that other modalities can be utilized [15]. Radio frequency ranging has been investigate. This can provide a means of inter-drone localization within a swarm by observing the received signal strength of RF signals from neighboring drones [12].

When position feedback is lost, the technique of dead reckoning can be used. This manner of control simply assumes the current position estimate of the drone location is correct, and flies accordingly. Such a system is prone to quick accumulation of error when used in isolation. However, work has been done to improve upon these estimates. The combination of dead reckoning, and observations using a Kalman filter can still allow for improved trajectory tracking [21]. Dead reckoning can also be combined with other localization methods to improve its accuracy. Such a method is proposed in which the received signal strength of neighboring drone broadcasts is combined with a dead reckoning scheme to enable accurate localization [19].

3   Method

3.1   Problem Statement

Assume a swarm of D drones that is heading towards a certain destination in a stochastic environment. Given that global communication (e.g., GPS communication) is lost on the way towards the destination, the objective is to predict the positions of all drones at all subsequent time-steps.

Formally, a swarm $\mathcal{S}^{[T]} = \{\mathcal{D}_1^{[T]}, \ldots, \mathcal{D}_D^{[T]}\}$ is defined as a collection of D individual drones, where $\mathcal{D}_i^{[T]} = \{\mathbf{p}_i^1, \ldots, \mathbf{p}_i^T\}$ represents the trajectory of the i-th drone containing its positions $\mathbf{p}_i^t \in \mathbb{R}^K$ for each time-step $t = 1, \ldots, T$. Over the course of a swarm's flight during which *global communication is lost* at a certain time-step t_c, the *objective* is to learn a swarm-level function $F_S = (f_1, \ldots, f_D) : \mathbb{R}^{D \times \omega K} \to \mathbb{R}^{D \times K}$. Each $f_i : \mathbb{R}^{\omega K} \to \mathbb{R}^K$ is a drone-level function that, when at any time-step $t > t_c$, *predicts* the position of the i-th drone at the next time-step $\mathbf{p}_i^{t+1}$ based on a ω-sized *window* of past positions $\boldsymbol{\omega}_i^t = [\mathbf{p}_i^{t-\omega+1}, \ldots, \mathbf{p}_i^t]$.

The above statement assumes that none of the individual drones is aware of its true position upon communication loss since any position feedback will be unavailable (e.g., after a GPS-denied regime was initiated). Nevertheless, an intra-swarm communication method can be leveraged to allow the drones to share their true and predicted positions before and after communication loss, respectively. This assumption is supported by the fact that low cost and complexity methods are available for enabling such communication [7,12]. In addition, the task of distributing this information falls within a well researched field [4,5], which is out of the scope of this paper.

3.2   Structured Learning for Trajectory Prediction

Given a swarm $S^{[t_c]}$ that lost global communication at time-step t_c, we propose to use a Multi-Target Gaussian Conditional Random Field (MT-GCRF) model to predict the drones' positions at each subsequent time-step $t > t_c$ by leveraging the available positional data from all time-steps prior to t_c. To that end, an input matrix $\mathbf{W} = [\mathbf{W}^1, \ldots, \mathbf{W}^{t_c - \omega}]^\top$ is constructed for each previous time-step t. For a specific time-step t, $\mathbf{W}^t$ contains the ω-sized windows of all D drones before t, i.e. $\mathbf{W}^t = [\boldsymbol{\omega}_1^{t+\omega-1}, \ldots, \boldsymbol{\omega}_D^{t+\omega-1}]^\top$. Similarly, a target matrix $\mathbf{P} = [\mathbf{P}^1, \ldots, \mathbf{P}^{t_c-\omega}]^\top$ is obtained, where $\mathbf{P}^t = [\mathbf{p}_1^{t+\omega}, \ldots, \mathbf{p}_D^{t+\omega}]^\top$ contains the positions of all D drones at time-step t. Moreover, to capture the between-target similarities, a similarity matrix $\mathbf{S}$ was additionally constructed as $S_{ij}^t = 1 - \frac{d_{ij}^t}{\max_{i,j} d_{ij}^t}$, where d_{ij}^t is the Euclidean distance between $\mathbf{p}_i^t$ and $\mathbf{p}_j^t$. The resulting $\mathbf{S}^1, \ldots, \mathbf{S}^{t_c-\omega}$ are diagonally placed in a single supra matrix $\mathbf{S}$.

The input and target matrices $\mathbf{W}$ and $\mathbf{P}$, along with the similarity matrix $\mathbf{S}$, are then used to learn the parameters of an MT-GCRF model by maximizing the probability of the drones' target positions conditioned on their corresponding windows of previous points, $P(\mathbf{P}|\mathbf{W})$. Finally, the learned MT-GCRF model is used to estimate the positions for all drones at each time-step following t_c.

MT-GCRF Overview. Let $\mathbf{X} = [\mathbf{x}_1, \ldots, \mathbf{x}_N]^\top$ and $\mathbf{Y} = [\mathbf{y}_1, \ldots, \mathbf{y}_N]^\top$ denote an input matrix of explanatory variables and a target vector of response variables, respectively, such that $\mathbf{y}_i \in \mathbb{R}^K$ is associated with $\mathbf{x}_i \in \mathbb{R}^M$, for each

$i = 1, \ldots, N$. A conventional Continuous Conditional Random Field [11] models the conditional probability over all target vectors $\mathbf{Y}$ given $\mathbf{X}$ as

$$P(\mathbf{Y}|\mathbf{X}) = \frac{1}{Z} \exp\left\{ -\sum_{k=1}^{K}\left[\alpha \sum_{i=1}^{N}(y_{i,k} - f_k(\mathbf{x}_i))^2 + \beta \sum_{i \sim j} S_{ij}(y_{i,k} - y_{j,k})^2 \right] \right\},$$
(1)

where $[\alpha, \beta]$ are the model parameters, f_k is a parameterized function that outputs an estimate of the k-th dimension of $\mathbf{y}_i$ given the i-th input vector, and Z is a normalization constant. The first term in the exponent is the association potential that models the pairwise relationship between the k-th target of $\mathbf{y}_i$ and its estimate. In the seconds term, S_{ij} describes the similarity between $\mathbf{y}_i$ and $\mathbf{y}_j$. The entire similarity matrix $\mathbf{S} = [S_{ij}]_{N \times N}$ can be thought of as an adjacency matrix of a weighted similarity graph. The structure of such a graph reflects the similarities among its nodes.

MT-GCRF Learning. To allow for efficient parameter learning, the conditional probability in Eq. 1 can be transformed to a Multivariate Gaussian form [13]:

$$P(\mathbf{Y}|\mathbf{X}) = \frac{1}{(2\pi)^{N/2}|\mathbf{\Sigma}|^{1/2}} \exp\left\{ -\frac{1}{2}(\mathbf{Y} - \boldsymbol{\mu})^\top \mathbf{\Sigma}(\mathbf{Y} - \boldsymbol{\mu}) \right\}.$$
(2)

The above Gaussian-form model is referred to as a Multi-Target Conditional Random Field (MT-GCRF). The parameters of MT-GCRF are determined by maximizing the conditional likelihood of Eq. (2).

MT-GCRF Inference. Given an input matrix $\mathbf{X} \in \mathbb{R}^{J \times M}$, inference is carried out by calculating $\boldsymbol{\mu} = \alpha \mathbf{\Sigma}[\mathbf{b}_1, \ldots, \mathbf{b}_J]^\top$, where $\mathbf{b}_i = [f_1(\mathbf{x}_i), \ldots, f_K(\mathbf{x}_i)]^\top$.

3.3   Variance Estimation and Collision Avoidance

To address the third research question, a method for swarm-based collision avoidance was developed. "Swarm-based" refers to the fact that this is a procedure that allows a set of swarms to model each others environmental deviation. In the proposed approach each swarm will manage and train its own structure-based model. Then, recalling the assumption of an existing means of distributed communication, we extend the assumption to imply swarms will be close enough to share models.

Simply sharing the models does not provide a collision avoidance system. To leverage the models in creation of one, this work chooses to interpret the uncertainty of predicted future locations as output variance of the structured model, and in turn uses this to set the radius of sphere representing the 'possible collision' volume for the swarm. Since the swarms will have a copy of each others' models, this radius can be deduced for each swarm. More specifically, we take the mean of the predicted positions of all drones in a swarm to be the swarm's estimated 'center'. Consequently, we define a swarm's variance as the distance of the swarm's center to the predicted position of the furthest drone, plus the maximum per-dimension variance of that drone.

During inference, each swarm uses the shared models to determine if their radius intersects that of another swarm since in such a situation a possible collision might occur soon. To avoid collision, the swarms update the target positions of their control systems such that their revised targets are at positions opposite one-another along the line connecting the centers of the swarms variance radius. The trajectories update distance along this line is set to keep the radii of the possibly colliding swarms from intersecting while simultaneously minimizing change from the intended mission plan. This is done by setting the distance from the center points along the connecting line for each swarm to be half the value of the overlap of the swarm radii. Once these new safe positions have been reached, the swarms reset their targets to the original location.

4   Experiments

We have conducted two sets of experiments. The first set is aimed to characterize the efficacy of the model, as measured by its ability to correct the errors induced by environmental deviations when communication is lost and the swarm utilizes dead reckoning without position feedback. In these experiments, the distance from a ground truth flight path (where no feedback is lost) is used to compare the errors from pure dead reckoning, dead reckoning corrected by an unstructured model, and dead reckoning corrected by the proposed structural MT-GCRF model. Parameters of the models were learned from simulated training data gathered while drones locations were available and the prediction were utilized to augment location estimates in a GPS-denied environment. In the second set of experiments, the qualitative efficacy of the proposed collision avoidance mechanism is tested.

4.1   Drone-Level Trajectory Prediction

To compare the efficacy of dead reckoning (DR) with no model, an unstructured regression model (UR), and a structured model (MT-GCRF in this case); four sets of data were collected for each simulated flight path. Each of these data sets is collected from a simulation that shares the same set of randomly generated gusts. This is accomplished by managing the random state for the simulations, so that they are all initialized with the same random seed.

The first simulation for each 'run' of the experiment is the baseline, or ground truth run. In this, no communication loss occurs, and the swarm proceeds in navigating with full position feedback for the entire flight path. This provides the basis of comparison for the three methods.

The second, third, and fourth simulations make use of dead reckoning, the unstructured model UR, and the structured model MT-GCRF, respectively. In the case of DR, once communication is lost the navigation of the drone swarm becomes based on an estimate of the current position. Without feedback, this estimate is updated by adding the current instantaneous velocity of each drone to its previous position estimate. In the third simulation, the drones' positions after communication loss are predicted by UR, which are later used by MT-GCRF. In a sense, this is as if we set the β parameter of MT-GCRF to 0. In the fourth simulation, the full MT-GCRF model is used, which learns the relevance of the UR's outputs and the swarm's structure, represented by α and β, respectively.

In both sets of trajectory prediction experiments, the models were ran on simulated flight paths consisting of 150 time steps, with communication loss occurring at the 100-th timestep; meaning that 100 timestep were used for training, while the remaining 50 were used to evaluate the models. All prediction-based models used a ω-sized window, with ω being 10, to construct their training data. The UR model (and therefore the underlying model in MT-GCRF) was a Multi-Target Neural Network with a single hidden layer containing 30 hidden units, trained using the L-BFGS optimization method.

The root mean squared error (RMSE) between the true and predicted drones' positions, along a certain dimension, was used. For a swarm $\mathcal{S}$ of D drones, over a flight-path of T time-steps, the error along the k dimension is computed as $RMSE(\mathcal{S}, k) = \frac{1}{T-t_c} \sum_{t=t_c+1}^{T} \sqrt{\frac{1}{D} \sum_i^D (p_{i,k}^t - \hat{p}_{i,k}^t)^2}$ where $\hat{p}_{i,k}^t$ is the predicted position of the i-th drone at timestep t after and $p_{i,k}^t$ is its true position at timestep t.

This set of experiments was split into two groups. One to investigate the lower-complexity swarms, and another to investigate higher-complexity swarm topologies. This was done to aid in highlighting the effect of structure in the efficacy of the prediction. More complex layouts provide more interesting structure, and in turn show a different response to the structured learning method.

Experiments for Lower-Complexity Swarms. In the lower-complexity swarms, the layouts was a four-drone planar swarm, and an eight-drone cubic swarm. In each case, the swarms start flights at one corner of a volume, and move towards the diagonal corner, with a set critical time in the center of the flight path. Each flight had at least one gust present in the training and inference stages. A set of 100 random simulations were conducted, collecting the previously described data.

Experiments for Higher-Complexity Swarms. The higher-complexity swarm experiments were conducted in the same manner as the lower-complexity swarm experiments, but using nine-drone planar, and 27-drone cubic layouts. This is of particular interest because these configurations contain 'internal' drones, whereas the low complexity configurations do not. This is hypothesized to increase the influence of the structure (the similarity matrix S) within the structural model, improving its efficacy over the unstructured model when compared to the lower complexity layouts.

4.2 Multi Swarm Collision Avoidance Experiments

To explore the utility of the proposed collision avoidance procedure, a simulated mission was constructed that places two swarms at opposite corners of a volume, with target flight paths that intersect. The critical time for GPS denial was set prior to the ostensible collision time of the swarms. Each swarm uses a separate model that will be shared with the other swarm. Each swarm's model was bagged five times using bootstrap aggregation to obtain its uncertainty (prediction variance). All other experimental settings remain the same as those described in Sect. 4.1. The resulting behaviour of the swarms is gathered from observation of animations of the flights, as well as from calculating the number (if any) of collisions.

4.3   Simulation Design

To test the hypothesis that a structural model can aid in navigation, a simulator must be able to model the behaviour of the drones in a deterministically reproducible way, under a reproducible set of environmental deviations. This must be done while tracking the spatio-temporal data set encoded in the nearest-neighbor graph of the drones. To accomplish this, a simulator was written in Python 3.6.9.

To simulate the use of GPS, and the subsequent loss of GPS feedback, the program tracks an actual and estimated position for each member of the swarm. When in a GPS-enabled regime, the actual position is used, while in a GPS-denied regime the estimated position is used.

The simulation introduces environmental deviation by use of a simple gusting wind model. A gust of wind is represented as a single vector of a random magnitude, pointing a randomly selected direction on the plane. For the duration of the gust (also selected from a random distribution), this vector is used to move the drone position within the simulated environment.

The wind deviation vector is not applied to each drones position identically. To introduce more realistic behaviour, and to partially encode some information about the deviation within the network structure, a model of 'drafting' was introduced. In this model, the relative location of each drone to others, along the axis in which the gust is pointing, will decrease the magnitude of the deviation applied to that drone. That is, drones 'behind' other drones along this gust vector will feel less deviation. To achieve this, a convex hull is calculated, and used to quickly determine the influence of drones on the hull onto those inside, with respect to the wind direction. This process is done recursively until the drafting effect of each drone on each other is calculated, yielding a scaled wind deviation for each drone.

5   Results

5.1   Drone-Level Trajectory Prediction

For the lower and higher complexity trajectory prediction tasks described in Sect. 4.1, the error is reported using the RMSE measure. The behaviour typical of all models in a single run is illustrated in Fig. 1. The unstructured model has a decreased stability compared to the structured model (2 of the 4 drones have erratic flight paths). When the structured model is used, the similarity matrix, and the learned association potentials, can effectively 'smooth out' the impact of the instability of the underlying regressors.

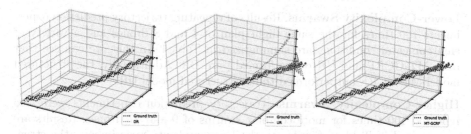

Fig. 1. Example of a simulated flight trajectories of a 4 drone planar swarm. Subfigures contain the ground truth trajectory and the predicted trajectories by DR (left), UR (middle), and MT-GCRF (right) upon communication loss.

Table 1. Trajectory prediction for lower-complexity swarms (4 and 8 drones) by DR, UR and MT-GCRF. RMSE and its 90% confidence interval in three dimensions (x, y, z) is reported for 100 random flights over 50 inference timesteps in a GPS-denied environment.

Model	RMSE		
	x	y	z
4 Drone Planar			
DR	2.4290 ± 0.0298	1.4102 ± 0.0141	0.3813 ± 0.0230
UR	0.5854 ± 0.0677	0.3095 ± 0.0333	0.3327 ± 0.0426
MT-GCRF	$\mathbf{0.4280 \pm 0.0447}$	$\mathbf{0.2531 \pm 0.0233}$	$\mathbf{0.2264 \pm 0.0292}$
8 Drone Cubic			
DR	2.4204 ± 0.0265	1.4094 ± 0.0141	0.3401 ± 0.0224
UR	0.5994 ± 0.0713	0.3155 ± 0.0210	0.3582 ± 0.0264
MT-GCRF	$\mathbf{0.4228 \pm 0.0374}$	$\mathbf{0.2395 \pm 0.0130}$	$\mathbf{0.2875 \pm 0.0198}$

Table 2. Trajectory prediction for higher-complexity swarms (9 and 27 drones) by DR, UR and MT-GCRF. RMSE and its 90% confidence interval in three dimensions (x, y, z) is reported for 100 random flights over 50 inference timesteps in a GPS-denied environment.

Model	RMSE		
	x	y	z
9 Drone Planar			
DR	2.3408 ± 0.0245	1.3720 ± 0.0145	0.4002 ± 0.0239
UR	0.7208 ± 0.1019	0.3826 ± 0.0480	0.4850 ± 0.0709
MT-GCRF	$\mathbf{0.5171 \pm 0.0628}$	$\mathbf{0.3476 \pm 0.0370}$	$\mathbf{0.2753 \pm 0.0449}$
27 Drone Cubic			
DR	2.2934 ± 0.0224	1.3701 ± 0.0127	$\mathbf{0.3748 \pm 0.0212}$
UR	0.6932 ± 0.0591	0.4127 ± 0.0385	0.4810 ± 0.0425
MT-GCRF	$\mathbf{0.5014 \pm 0.0282}$	$\mathbf{0.3637 \pm 0.0160}$	0.3965 ± 0.0238

Lower-Complexity Swarms. Results of repeating trajectory prediction experiments 100 times for the same flight plan of 4 and 8 drones and a variety of environmental conditions are summarized in Table 1. In this, we can see that both MT-GCRF and UR outperformed pure DR. Additionally, the structured model MT-GCRF outperformed the unstructured model UR.

Higher-Complexity Swarms. Trajectory prediction experiments were also repeated 100 times for more complex swarms of 9 and 27 drones. Results are summarized in Table 2. Similar to the low-complexity experiments, the structured model (MT-GCRF) outperforms the unstructured regressors (UR), which in turn outperforms dead reckoning (DR), in most instances.

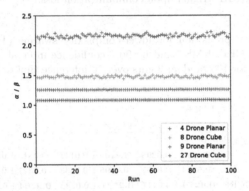

Fig. 2. The ratio of α to β learned for each swarm size and shape, across all runs. Recall that α and β represent the importances of the unstructured predictor and the swarm structure, respectively (refer to Eq. (1) for more details).

Parameter Analysis. The parameters learned by MT-GCRF, over all 100 runs, for both low and higher-complexity swarms are presented in Fig. 2. There are two main insights that the figure suggests. First, the α/β ratio decreases for more complex swarms, suggesting that the larger/complex a swarm is, the more relevant the structure is to the model (i.e. the structural term in (1) has a higher relative value). Moreover, the ratio of the structural terms exhibits a higher stability over multiple runs for larger/complex swarms (i.e. MT-GCRF is more stable w.r.t. its parameters).

5.2   Multi Swarm Collision Avoidance

In the trajectory prediction experiments it was observed that MT-GCRF was more accurate than its alternatives, therefore, MT-GCRF was selected as the prediction model for each of two swarms. Each swarm's MT-GCRF was ran on 150-timestep flight path, with t_c set at the 100-th timestep.

   To investigate the efficacy of the avoidance procedure, 100 such simulations were performed and 0 collisions were observed. As a reference, the states of a two-swarm collision procedure, from a single simulation, before, during, and after collision avoidance was triggered, are illustrated in Fig. 3.

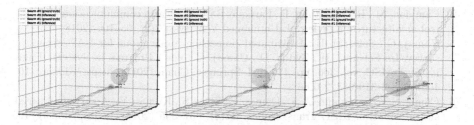

Fig. 3. Multi-swarm collision avoidance. Positions of each swarm's drones as well as the swarm's estimated variances are displayed for a time-step: Left: Prior to triggering the collision avoidance mechanism. Middle: During collision avoidance. Right: After the collision avoidance mechanism has terminated.

6 Conclusions

The results of our study provide evidence that the prediction-based approaches can sufficiently model deviations in drone swarms trajectories after communication loss without using additional sensors. In hundreds of conducted experiments both unstructured and structured predictive models showed significantly improved swarm-level RMSE over dead reckoning when compared to the ground truth. Spatio-temporal patterns of environmental deviations observed before GPS signal was lost were exploited the most effectively by the proposed MT-GCRF structured learning method for autonomous navigation in both the lower and higher complexity swarm configurations. Finally, the proposed structured regression based autonomous navigation method was able to be leveraged to form a collision avoidance framework that successfully managed and avoided a collision between two swarms in GPS-denied environment.

Acknowledgements. This research was supported in part by the NSF grants IIS-1842183, SES-1659998 and AFRL Contract No. FA8750-18-C-0041, subcontract 555027-78056.

References

1. Barbeau, M., Garcia-Alfaro, J., Kranakis, E.: Geocaching-inspired resilient path planning for drone swarms. In: IEEE Conference on Computer Communications Workshops (INFOCOM WKSHPS), IEEE INFOCOM 2019, Paris, France, pp. 620–625. IEEE, April 2019. https://doi.org/10.1109/INFCOMW.2019.8845318
2. Belkhouche, F., Vadhva, S., Vaziri, M.: Modeling and controlling 3D formations and flocking behavior of unmanned air vehicles. In: 2011 IEEE International Conference on Information Reuse & Integration, Las Vegas, NV, USA, pp. 449–454. IEEE, August 2011. https://doi.org/10.1109/IRI.2011.6009590
3. Brust, M.R., Danoy, G., Bouvry, P., Gashi, D., Pathak, H., Gonçalves, M.P.: Defending against intrusion of malicious UAVs with networked UAV defense swarms. In: 2017 IEEE 42nd Conference on Local Computer Networks Workshops (LCN Workshops), pp. 103–111, October 2017. https://doi.org/10.1109/LCN.Workshops.2017.71. http://arxiv.org/abs/1808.06900

4. Campion, M., Ranganathan, P., Faruque, S.: UAV swarm communication and control architectures: a review. J. Unmanned Veh. Syst. **7**(2), 93–106 (2018)
5. Chmaj, G., Selvaraj, H.: Distributed processing applications for UAV/drones: a survey. In: Selvaraj, H., Zydek, D., Chmaj, G. (eds.) Progress in Systems Engineering. AISC, vol. 366, pp. 449–454. Springer, Cham (2015). https://doi.org/10.1007/978-3-319-08422-0_66
6. Chung, S.J., Paranjape, A.A., Dames, P., Shen, S., Kumar, V.: A survey on aerial swarm robotics. IEEE Trans. Robot. **34**(4), 837–855 (2018). https://doi.org/10.1109/TRO.2018.2857475. https://ieeexplore.ieee.org/document/8424838/
7. Faigl, J., Krajník, T., Chudoba, J., Přeučil, L., Saska, M.: Low-cost embedded system for relative localization in robotic swarms. In: 2013 IEEE International Conference on Robotics and Automation, pp. 993–998. IEEE (2013)
8. Fernando, H.C.T.E., De Silva, A.T.A., De Zoysa, M.D.C., Dilshan, K.A.D.C., Munasinghe, S.R.: Modelling, simulation and implementation of a quadrotor UAV. In: 2013 IEEE 8th International Conference on Industrial and Information Systems, Peradeniya, Sri Lanka, pp. 207–212. IEEE, December 2013. https://doi.org/10.1109/ICIInfS.2013.6731982. http://ieeexplore.ieee.org/document/6731982/
9. Hayat, S., Yanmaz, E., Muzaffar, R.: Survey on unmanned aerial vehicle networks for civil applications: a communications viewpoint. IEEE Commun. Surv. Tutor. **18**(4), 2624–2661 (2016). https://doi.org/10.1109/COMST.2016.2560343
10. Kokić, P., Tomaš, B.: Enhanced drone swarm localization using GPS and trilateration based on RF propagation model. In: Central European Conference on Information and Intelligent Systems, pp. 259–264. Faculty of Organization and Informatics Varazdin (2017)
11. Lafferty, J., McCallum, A., Pereira, F.C.: Conditional random fields: probabilistic models for segmenting and labeling sequence data (2001)
12. Lockspeiser, J.R., Don, M.L., Hamaoui, M.: Radio frequency ranging for swarm relative localization. Technical ARL-TR-8194, US Army Research Laboratory (2017)
13. Radosavljevic, V., Vucetic, S., Obradovic, Z.: Continuous conditional random fields for regression in remote sensing. In: ECAI, pp. 809–814 (2010)
14. Saska, M., et al.: System for deployment of groups of unmanned micro aerial vehicles in GPS-denied environments using onboard visual relative localization. Auton. Robots **41**(4), 919–944 (2016). https://doi.org/10.1007/s10514-016-9567-z
15. Singh, S., Sujit, P.: Landmarks based path planning for UAVs in GPS-denied areas. IFAC-PapersOnLine **49**(1), 396–400 (2016). https://doi.org/10.1016/j.ifacol.2016.03.086
16. Tahir, A., Böling, J., Haghbayan, M.H., Toivonen, H.T., Plosila, J.: Swarms of unmanned aerial vehicles – a survey. J. Ind. Inf. Integr. **16**, 100106 (2019). https://doi.org/10.1016/j.jii.2019.100106
17. Wang, C.-L., Wang, T.-M., Liang, J.-H., Zhang, Y.-C., Zhou, Y.: Bearing-only visual SLAM for small unmanned aerial vehicles in GPS-denied environments. Int. J. Autom. Comput. **10**(5), 387–396 (2013). https://doi.org/10.1007/s11633-013-0735-8
18. Watson, N., John, N., Crowther, W.: Simulation of unmanned air vehicle flocking. In: Proceedings of Theory and Practice of Computer Graphics, Birmingham, UK, pp. 130–137. IEEE Computer Society (2003). https://doi.org/10.1109/TPCG.2003.1206940. http://ieeexplore.ieee.org/document/1206940/
19. Wu, H., Qu, S., Xu, D., Chen, C.: Precise localization and formation control of swarm robots via wireless sensor networks. Math. Prob. Eng. **2014**, 1–12 (2014). https://doi.org/10.1155/2014/942306

20. Yanmaz, E., Yahyanejad, S., Rinner, B., Hellwagner, H., Bettstetter, C.: Drone networks: communications, coordination, and sensing. Ad Hoc Netw. **68**, 1–15 (2018). https://doi.org/10.1016/j.adhoc.2017.09.001

21. Zhou, Q.L., Zhang, Y., Qu, Y.H., Rabbath, C.A.: Dead reckoning and Kalman filter design for trajectory tracking of a quadrotor UAV. In: Proceedings of 2010 IEEE/ASME International Conference on Mechatronic and Embedded Systems and Applications, QingDao, China, pp. 119–124. IEEE, July 2010. https://doi.org/10.1109/MESA.2010.5552088

Chemical Laboratories 4.0: A Two-Stage Machine Learning System for Predicting the Arrival of Samples

António João Silva[1,2] ⓘ, Paulo Cortez[2(✉)] ⓘ, and André Pilastri[1] ⓘ

[1] Centro de Computação Gráfica, 4804-533 Guimarães, Portugal
{antonio.silva,andre.pilastri}@ccg.pt
[2] ALGORITMI Center, Department of Information Systems, University of Minho, 4804-533 Guimarães, Portugal
pcortez@dsi.uminho.pt

Abstract. This paper presents a two-stage Machine Learning (ML) model to predict the arrival time of In-Process Control (IPC) samples at the quality testing laboratories of a chemical company. The model was developed using three iterations of the CRoss-Industry Standard Process for Data Mining (CRISP-DM) methodology, each focusing on a different regression approach. To reduce the ML analyst effort, an Automated Machine Learning (AutoML) was adopted during the modeling stage of CRISP-DM. The AutoML was set to select the best among six distinct state-of-the-art regression algorithms. Using recent real-world data, the three main regression approaches were compared, showing that the proposed two-stage ML model is competitive and provides interesting predictions to support the laboratory management decisions (e.g., preparation of testing instruments). In particular, the proposed method can accurately predict 70% of the examples under a tolerance of 4 time units.

Keywords: Automated Machine Learning · Industry 4.0 · Regression

1 Introduction

The Industry 4.0 concept assumes a high usage of Artificial Intelligence (AI), where industrial physical processes generate data that can be analyzed by Business Analytics, namely Data Mining (DM) and Machine Learning (ML) techniques, aiming to improve the factory efficiency (e.g., reduce costs, enhance production levels) [21]. This concept is transforming the Chemical industry, which has a large impact in the world economy (e.g., petrochemicals, pharmaceuticals).

In this work, we address a relevant Business Analytics need of a chemical company, which is adopting a Industry 4.0 transformation. To ensure the

© IFIP International Federation for Information Processing 2020
Published by Springer Nature Switzerland AG 2020
I. Maglogiannis et al. (Eds.): AIAI 2020, IFIP AICT 584, pp. 232–243, 2020.
https://doi.org/10.1007/978-3-030-49186-4_20

quality of the products being manufactured, samples taken from the company production processes need to be tested in laboratories. The tests assure that the products are compliant with quality standards, allowing their usage by the company clients. Under this context, predicting the arrival of production samples at the laboratory is a key issue, since it helps in the allocation of equipment and human resources. Aiming to solve this task, this paper presents a novel two-stage ML prediction system, which was developed during the implementation of a CRoss-Industry Standard Process for DM (CRISP-DM) [25] project that included three iterations, each focusing on a distinct regression strategy. During the modeling stage of the three CRISP-DM iterations, an Automated ML (AutoML) [12] procedure was adopted, allowing to compare and configure six state-of-the-art ML algorithms.

The paper is structured as follows. Section 2 describes the related work. The business task, data and proposed approach are presented in Sect. 3. The obtained results are shown in Sect. 4. Finally, Sect. 5 concludes the paper.

2 Related Work

In recent years, there has been an increased interest in the field of AI, due to the rise of data, computational power and sophisticated learning algorithms (e.g., Deep Learning) [9]. Following the Industry 4.0 revolution [21], many factories now are generating data that can be analyzed by DM and ML techniques in order to support managerial decision-making. Yet, several real-world DM projects tend to fail due to a misalignment between business needs and ML analyses [10]. The CRISP-DM is an open standard and robust methodology that was specifically developed to reduce this misalignment and increase the success of DM projects [25]. CRISP-DM includes six stages that are executed through several iterations and that involve both business and ML experts: business understanding, data understanding, data preparation, modeling, evaluation and deployment. CRISP-DM is a popular methodology. For instance, it has been applied to the Banking [18] and Health Care [3] domains.

Regarding the analyzed chemical industry, the quality testing laboratories are mostly managed manually, with the usage of Information Technology (IT) being more focused on storing the test values rather than the process [16,22]. Moreover, the data is typically spread through different databases what work as information silos (e.g., production, laboratory testing), thus it is difficult to have an easy access to all data under a single version of the truth. By adopting the Industry 4.0 concept, which assumes a better usage of IT, there is a potential gain to optimize the management of the chemical laboratories. In this work, we describe one aspect of the Industry 4.0 transformation that is being conducted by a chemical company. It corresponds to the result of implementing a CRISP-DM project that uses both production and laboratory testing databases.

In terms of Predictive Analytics applied to the industry, most studies target predictive maintenance via several ML algorithms, such as Random Forest (RF) [4], Neural Networks (NN) [23] and Gradient Boosting Machines (GBM) [17].

There are also studies about non maintenance prediction applications, such as: the classification of quality products produced by injection molding processes via Boosting, RF and NN models [5]; and estimation of endpoint temperature and chemical concentration of a furnace when producing low-carbon steel using RF and ridge regression algorithms [19]. All these studies require the selection and configuration of the right ML algorithm, which often depends on the ML expert knowledge and that involves the usage of heuristics or trial-and-error experiments [14]. In order to avoid this time-consuming procedure (in terms of the ML expert effort), we adopt an AutoML [12] during the modeling stage of the CRISP-DM. This systematization and automation the ML model selection provides two main advantages. First, it alleviates the effort of the ML analyst, allowing to focus on other ML aspects in order to provide a better business value. In particular, in this paper, it allowed to implement more iterations of the CRISP-DM methodology, which was helpful to design the proposed two-stage ML model. Second, it reduces the ML maintenance effort, since the ML can be retrained automatically, as new data arrives, which is advantageous for the analyzed company.

3   Materials and Methods

3.1   Business Task

The analyzed chemical company produces several products, in batches. During the production-batch execution process, a sequence of samples, called In-Process Control (IPC), are selected for quality laboratory inspection, in order to ensure that the production process is running as expected. In terms of the chemical laboratories, the IPC samples have the highest priority, because the production process can not continue without their approval. A fixed amount of IPC samples are selected from each production-batch ($s \in \{1, ..., IPC_{\max}\}$). The production information system registers several attributes related to the IPC sample production, including its initial production time, denoted here as IPC production time PT_s. One by one, the IPC samples arrive at the laboratory at time LT_s, under irregular intervals that are difficult to be estimated in advance.

The business goal is thus the non-trivial task of predicting of arrival time for each IPC sample at the chemical laboratories. Solving this task efficiently allows a better management of the laboratory equipment and human resources. For instance, some IPC quality tests require a setup time, in which the analysts need to prepare in advance the laboratory testing instruments. The business goal was addressed as a regression task, under two main target goals. In the first CRISP-DM iteration, we only used laboratory temporal data and the target goal was defined as predict $y_1 = LT_{s+1} - LT_s$, which corresponds to the time lag between the next IPC sample arrival (LT_{s+1}) and the current (already known) laboratory sample arrival (LT_s). In the second and third CRISP-DM iterations, we explored production temporal data, predicting $y_2 = LT_s - PT_s$, where the laboratory arrival time can be immediately estimated once the IPC sample starts its production.

3.2 Data Understanding and Preparation

We used an Extract, Transform, load (ETL) procedure to merge the relevant data from two main databases related with the production and laboratory testing information systems, populating an integrated and business oriented data warehouse system. The ETL resulted in a raw file with 226,929 rows and 33 columns regarding all laboratory samples that were analyzed during a three-year time period. The data warehouse was further filtered in order to contain rows related with IPC samples and with complete values in terms of the input and output attributes (Table 1), leading to a dataset with 26,611 instances. The input variables were manually selected and defined from the filtered raw file using expert domain knowledge, obtained by interacting with the chemistry experts. Due the complexity of the chemical factory processes and information system integration issues, it was not possible to have access to a more richer set of data features (e.g., which components and machines were used to produce the samples). Thus, the resulting set of 8 inputs is rather small, which makes more challenging the prediction task. Both output targets were computed using a particular time unit, which is not disclosed here due to business privacy issues.

Table 1. Summary of the data attributes.

Name	Description	Range
Input attributes		
Day	Day of the week when the production-batch started	$\{1, ..., 7\}$
Month	Month when the production-batch started	$\{1, ..., 12\}$
Product	Product type (nominal code)	155 levels
Version	Version of the product (numeric)	$\{1, ..., 108\}$
Grade	Product grade (nominal, related with the lab tests)	15 levels
Stage	Product stage (nominal, related with the lab tests)	1,272 levels
Batch	Batch identification of the product (nominal)	925 levels
s	Sequence number of the sample ($s \in \{1, ..., IPC_{\max}\}$)	$\{1, ..., 169\}$
Output targets		
y_1	Time lag arrival of two consecutive samples	$[0.2, 5315.3]$
y_2	Time lag between PT_s and LT_s	$[0.0, 3270.0]$

3.3 Machine Learning Models

In terms of computational environment, we adopted the R tool and its `rminer` package [8] for data manipulation and ML result evaluation, while the AutoML adopts the H2O implementation [7]. The AutoML procedure was configured to select the regression model and its hyperparameters based on the best Root Mean Squared Error (RMSE) computed using a validation set that is obtained by applying an internal 10-fold cross-validation method over the training data. All computational experiments were executed on the same personal computer and each individual ML model was trained up to a maximum running time of 3,600 s. Once a ML model is selected, the model was retrained with all training data. As in [11], the AutoML was configured to include a total of 6 distinct regression algorithms: RF, Extremely Randomized Trees (XRT), Generalized Linear Model (GLM), GBM, XGBoost (XG) and a Stacked Ensemble (SE). The RF is a popular ensemble method that combines a large number of decision trees based on bagging and random selection of input features [15]. The XRT algorithm extends the RF approach by randomly selecting the decision thresholds of the tree nodes [13]. GLM estimates regression models for exponential distributions (e.g., Gaussian, Poisson, gamma) [15]. The GBM algorithm is a based on a generalization of tree boosting, sequentially building regression trees for all data features [15]. XG is another ensemble tree method that uses boosting to enhance the prediction results [6]. The SE method, also known as stacked regression [2], combines the predictions of different base learners by using a second-level ML algorithm. The H2O implementation [7] uses the following AutoML setup: RF and XRT – set with the default hyperparameters; GLM – grid search used to set one hyperparameter (*alpha*, a regularization parameter); GBM and XG – grid search used to tune nine and ten hyperparameters (e.g., number of trees, maximum depth, minimum rows); SE – all five algorithms (RF, XRT, GLM, GBM, XG) are used as base learners and the individual predictions are weighted by using a second-level GLM learner. For the ML algorithms that require numeric inputs (e.g., GLM), the nominal inputs (e.g., product, grade) are previously transformed by using the standard one-hot encoding, which assigns one boolean input per categorical level. For instance, a categorical feature with three levels ($\{a,b,c\}$) is encoded as: $a = (1, 0, 0)$, $b = (0, 1, 0)$ and $c = (0, 0, 1)$.

A total of three CRISP-DM iterations were executed, aiming to improve the regression results and the potential value of the ML models. The first CRISP-DM iteration targeted the y_1 output, while the second and third CRISP-DM iterations approached y_2, under two variants. The y_1 target is assumes that at least one IPC sample from the production-batch as arrived at the laboratory. The trained ML model can be used each time new sample arrives, allowing to estimate when the next sample will be delivered ($\widehat{y}_1$). A different perspective is adopted by the y_2 target, since the fitted ML model can be applied to predict the laboratory sample arrival once an IPC sample production has started. The model employed in the second CRISP-DM iteration uses a simple regression with a single ML model ($\widehat{y}_2$). During the evaluation stage of the second CRISP-DM iteration, we identified that there were some high prediction errors, in particular when predicting the arrival times for the first sample of the production-batch ($s = 1$). In order to check if we could improve these results, a third CRISP-DM iteration was executed, in which we specialize two distinct ML models (α and β). The first ML model (α) is trained using only the first product-batch sample examples ($s = 1$) and thus the fitted model includes only seven input attributes ($\{$day, month, product, version, grade, stage, batch$\}$). The second model (β) is only activated when producing the other product-batch IPC samples ($s > 1$). Similarly to the second CRISP-DM iteration model, this ML model is trained with all eight inputs (including s, the sample sequence number). The proposed two-stage model ($\widehat{y}_{2\alpha\beta}$) is shown in Fig. 1.

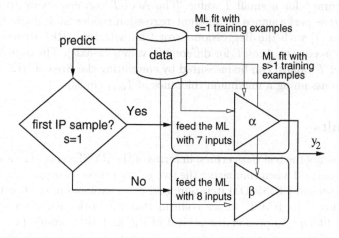

Fig. 1. Schematic of the proposed two-stage ML prediction model ($\widehat{y}_{2\alpha\beta}$).

3.4   Evaluation

The collected data was divided into three main sets, by using a chronological order. The last 20 weeks of data (total of 5,110 examples) was kept out of the initial ML experiments. The goal is apply this additional unseen data in a more realistic evaluation, provided by a Rolling Window (RW) validation [24] that is executed for the best ML regression approach. The remaining and oldest 21,501 examples (not used as test set by the RW) were further divided into training and test sets (holdout split) [20]. The time ordered Holdout Split (HS) was used to compare the three distinct main regression approaches (from the CRISP-DM iterations). The training data included the oldest 15,050 examples (around 70%). As for the HS test set, it included 6,451 instances.

Regarding the RW, it was set using a fixed training window with six months of data and a weekly testing of the ML models, in a total of 20 iterations. In the first iteration, at the first Sunday, the ML was trained with the last six months of historical data. Then, the model was used to perform sample arrival predictions for the incoming week (fixed test size of seven days). In the second iteration, executed at the second Sunday, the training window was updated by discarding one week of the oldest data and adding the previous week examples, allowing to update (retrain) the ML model, which then predicted the next week sample arrival times, and so on.

In this work, we adopt two popular regression error measures: RMSE and Mean Absolute Error (MAE). We also use the Acc@T metric, which is more easily understood by the business analysts, since it measures the percentage of examples accurately predicted when assuming an absolute error tolerance of T. A quality regression model should provide low RMSE and MAE values and also a high accuracy for a small T value. The Acc@T concept allows to compare the predictive performance of different regression modes in a single graph, as proposed in [1] with the Regression Error Characteristic (REC) curves, which plot in the y-axis the Acc@T for different T values (x-axis). The overall quality (for distinct T values) can be measured by computing the Area of REC (AREC) curve when assuming a maximum tolerance of $T_{\max}$ (in %).

4   Results

Table 2 presents the test data errors, in terms of the RMSE error measure, for the HS evaluation and when comparing the two y_2 prediction strategies: $\widehat{y}_2$, executed during the second CRISP-DM iteration; and $\widehat{y}_{2\alpha\beta}$, explored in the third CRISP-DM iteration. The RMSE values confirm that for both prediction strategies, it is more difficult to predict the arrival of the first IPC sample ($s = 1$) than the arrival of the remaining samples ($s > 1$). It is interesting to notice that by specializing a learning model for each of these IPC sample types, as executed in the third CRISP-DM iteration ($\widehat{y}_{2\alpha\beta}$), a substantial error reduction is achieved for both sample types ($s = 1$ and $s > 1$).

Table 2. Test data holdout results for $s = 1$ and $s > 1$ IPC sample arrival (best values in **bold**).

Approach	RMSE	
	$s = 1$	$s > 1$
$\widehat{y}_2$	209.9	188.9
$\widehat{y}_{2\alpha\beta}$	**124.8**	**41.3**

The full comparison of the aggregated HS results, assuming all IPC samples, is shown in Table 3, which contains: the evaluation method used (**Eval.**); the best model selected using the AutoML procedure (**Model**); and several predictive performance measures. The AREC was computed by using a maximum tolerance of $T_{\max=16}$ time units. All performance measures confirm that the best predictive model was achieved by $\widehat{y}_{2\alpha\beta}$, while $\widehat{y}_1$ obtained better results than $\widehat{y}_2$. When compared with $\widehat{y}_1$, $\widehat{y}_{2\alpha\beta}$ achieved a substantial predictive improvement: RMSE – reduction of 46.8 points; MAE – difference of 14.1 points; and AREC – increase of 10% points. As for the ML algorithms, the AutoML selected GBM and SE as the best performing models when using the 10-fold internal cross-validation (applied over training data). The $\widehat{y}_{2\alpha\beta}$ uses GBM for predicting the arrival times of the $s = 1$ samples and SE for the other ones.

Figure 2 complements the HS results by showing the respective REC curves for the three main regression approaches. The plot confirms that for most of the low tolerance range (x-axis), $\widehat{y}_{2\alpha\beta}$ provides a higher classification accuracy, resulting in an overall higher AREC. Indeed, the proposed two-stage ML model can predict correctly 37%, 59% and 70% of the samples for low tolerance values of $T = 1$, $T = 2$ and $T = 4$, a value that increases to 85% when the tolerance is increased to $T = 16$ time units.

Table 3. Test data results (best HO values in **bold**).

Approach	Eval.	Model	RMSE	MAE	AREC	Acc@T				
						$T = 1$	$T = 2$	$T = 4$	$T = 8$	$T = 16$
$\widehat{y}_1$	HO	GBM	98.0	27.0	61%	28%	45%	56%	66%	76%
$\widehat{y}_2$		SE	190.3	112.1	6%	1%	1%	3%	5%	12%
$\widehat{y}_{2\alpha\beta}$		α:GBM; β:SE	**51.2**	**12.9**	**71%**	**37%**	**59%**	**70%**	**77%**	**84%**
$\widehat{y}_{2\alpha\beta}$	RW	α:GBM; β:SE	37.5	11.4	71%	38%	56%	69%	76%	85%

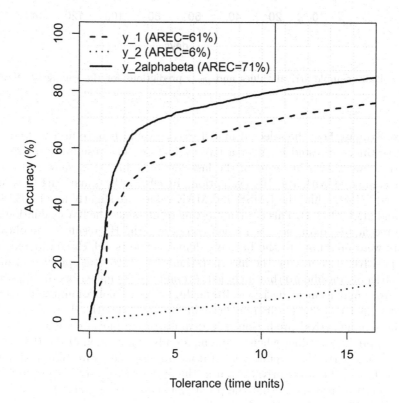

Fig. 2. Holdout REC curves for the three regression approaches.

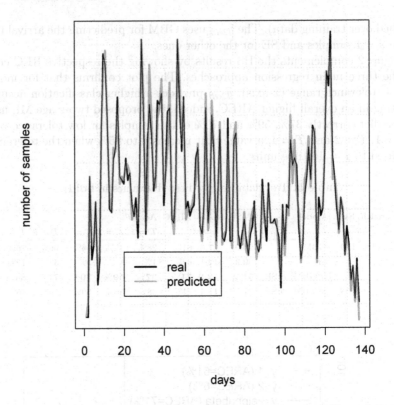

Fig. 3. Daily sample arrival values and $\hat{y}_{2\alpha\beta}$ predictions for the rolling window test data.

To estimate how the selected model $(\hat{y}_{2\alpha\beta})$ would behave in a real environment setting, we tested it under a RW evaluation. The results for all 20 week iterations are shown in terms of the last row of Table 3 and show consistency when compared with the HS evaluation. In effect, the same AREC value is achieved (71%), while the RMSE and MAE values are slightly lower (RMSE of 37.5 and MAE of 11.4). This is an interesting result, since the RW evaluation used more recent test data, not seen when comparing the HS results. The obtained results were presented to the business domain experts, which considered them very positive, encouraging the incorporation of the two-stage prediction model into a friendly dashboard that included several business indicators to support the laboratory management decisions. To facilitate the visualization, the dashboard was designed to provide different granularity levels (hourly, daily or monthly) for the sample arrival prediction. For demonstrative purposes, Fig. 3 plots the real and predicted values when assuming a daily aggregation of the IPC sample arrival for a particular chemical laboratory and for the entire RW testing time period. Due to business privacy issues, the scale of the y-axis is omitted from the graph. Figure 3 shows that the predictions are very close to the real values, denoting a high quality fit of the prediction model.

5   Conclusions

This paper addresses the non-trivial task of predicting the arrival of In-Process Control (IPC) samples at chemical laboratories for quality testing. To solve this task, we implemented the CRoss-Industry Standard Process for Data Mining (CRISP-DM) methodology, under three iterations, each focusing on a different regression approach. During the data understanding and preparation CRISP-DM stages, we collected recent data from a chemical company, resulting in 26,611 sample arrival examples related with a three-year time period. As for the modeling stage of CRISP-DM, we employed an Automated Machine Learning (AutoML) procedure, to automatically select and configure the best model when exploring six state-of-the-art ML algorithms.

Several experiments were held. Using a time ordered Holdout Split (HS), we compared the three main regression approaches: $\hat{y}_1$ - predict the time lag between the arrival of two consecutive samples (y_1), executed in the first CRISP-DM iteration; $\hat{y}_2$ - predict the time lag between starting the production of the sample and its arrival to the laboratory (y_2), explored in the second CRISP-DM iteration; and $\hat{y}_{2\alpha\beta}$ - a two-stage ML model to predict y_2, developed in the third CRISP-DM iteration. For all predictive performance measures, the best results were achieved at the two-stage ML model, which obtained interesting results (e.g., it can accurately predict 70% of the examples under a tolerance of $T = 4$ time units). The selected two-stage ML model ($\hat{y}_{2\alpha\beta}$) was further evaluated using a realistic Rolling Window (RW) procedure, which considered 20 weeks of unseen data. A similar predictive performance was achieved, when compared with the HS results, showing that the proposed two-stage ML model is robust for the analyzed chemical company. In effect, the ML model was incorporated into a friendly dashboard prototype, obtaining a valuable feedback from the chemical laboratory managers.

In future work, we intend to apply the two-stage model to predict the arrival of other types of samples (e.g., raw material). Moreover, we intend to further explore the deployment stage of CRISP-DM, by better integrating the proposed model in a decision support system tool. For instance, by using the predictions to directly optimize the laboratory human resources and instruments.

Acknowledgments. This work has been supported by FCT – Fundação para a Ciência e Tecnologia within the R&D Units Project Scope: UIDB/00319/2020. The authors also wish to thank the chemical company staff involved with this project for providing the data and also the valuable domain feedback.

References

1. Bi, J., Bennett, K.P.: Regression error characteristic curves. In: Fawcett, T., Mishra, N. (eds.) Machine Learning, Proceedings of the Twentieth International Conference (ICML 2003), Washington, DC, USA, 21–24 August 2003, pp. 43–50. AAAI Press (2003). http://www.aaai.org/Library/ICML/2003/icml03-009.php

2. Breiman, L.: Stacked regressions. Mach. Learn. **24**(1), 49–64 (1996). https://doi.org/10.1007/BF00117832
3. Caetano, N., Cortez, P., Laureano, R.M.S.: Using data mining for prediction of hospital length of stay: an application of the CRISP-DM methodology. In: Cordeiro, J., Hammoudi, S., Maciaszek, L., Camp, O., Filipe, J. (eds.) ICEIS 2014. LNBIP, vol. 227, pp. 149–166. Springer, Cham (2015). https://doi.org/10.1007/978-3-319-22348-3_9
4. Canizo, M., Onieva, E., Conde, A., Charramendieta, S., Trujillo, S.: Real-time predictive maintenance for wind turbines using big data frameworks. In: 2017 IEEE International Conference on Prognostics and Health Management, ICPHM 2017, Dallas, TX, USA, 19–21 June 2017, pp. 70–77. IEEE (2017). https://doi.org/10.1109/ICPHM.2017.7998308
5. Charest, M., Finn, R., Dubay, R.: Integration of artificial intelligence in an injection molding process for on-line process parameter adjustment. In: 2018 Annual IEEE International Systems Conference, SysCon 2018, Vancouver, BC, Canada, 23–26 April 2018, pp. 1–6. IEEE (2018). https://doi.org/10.1109/SYSCON.2018.8369500
6. Chen, T., Guestrin, C.: XGBoost: a scalable tree boosting system. In: Krishnapuram, B., Shah, M., Smola, A.J., Aggarwal, C.C., Shen, D., Rastogi, R. (eds.) Proceedings of the 22nd ACM SIGKDD International Conference on Knowledge Discovery and Data Mining, San Francisco, CA, USA, 13–17 August 2016, pp. 785–794. ACM (2016). https://doi.org/10.1145/2939672.2939785
7. Cook, D.: Practical Machine Learning with H2O: Powerful, Scalable Techniques for Deep Learning and AI. O'Reilly Media Inc., Newton (2016)
8. Cortez, P.: Modern Optimization with R. Springer, Heidelberg (2014). https://doi.org/10.1007/978-3-319-08263-9
9. Darwiche, A.: Human-level intelligence or animal-like abilities? Commun. ACM **61**(10), 56–67 (2018). https://doi.org/10.1145/3271625
10. Deal, J.: The ten most common data mining business mistakes, June 2013. https://www.elderresearch.com/most-common-data-science-business-mistakes
11. Ferreira, L., Pilastri, A., Martins, C., Santos, P., Cortez, P.: An automated and distributed machine learning framework for telecommunications risk management. In: Proceedings of the 12th International Conference on Agents and Artificial Intelligence, ICAART 2020, Valletta, Malta, February, vol. 2, pp. 99–107. SciTePress (2020)
12. Feurer, M., Klein, A., Eggensperger, K., Springenberg, J.T., Blum, M., Hutter, F.: Efficient and robust automated machine learning. In: Cortes, C., Lawrence, N.D., Lee, D.D., Sugiyama, M., Garnett, R. (eds.) Advances in Neural Information Processing Systems 28: Annual Conference on Neural Information Processing Systems 2015, Montreal, Quebec, Canada, 7–12 December 2015, pp. 2962–2970 (2015)
13. Geurts, P., Ernst, D., Wehenkel, L.: Extremely randomized trees. Mach. Learn. **63**(1), 3–42 (2006). https://doi.org/10.1007/s10994-006-6226-1
14. Gibert, K., Izquierdo, J., Sànchez-Marrè, M., Hamilton, S.H., Rodríguez-Roda, I., Holmes, G.: Which method to use? An assessment of data mining methods in environmental data science. Environ. Model. Softw. **110**, 3–27 (2018)
15. Hastie, T., Tibshirani, R., Friedman, J.: The Elements of Statistical Learning: Data Mining, Inference, and Prediction. Springer, Heidelberg (2009). https://doi.org/10.1007/978-0-387-84858-7
16. Kammergruber, R., Robold, S., Karlič, J., Durner, J.: The future of the laboratory information system-what are the requirements for a powerful system for a laboratory data management? Clin. Chem. Lab. Med. (CCLM) **52**(11), 225–230 (2014)

17. Liulys, K.: Machine learning application in predictive maintenance. In: 2019 Open Conference of Electrical, Electronic and Information Sciences (eStream), pp. 1–4. IEEE (2019)

18. Moro, S., Laureano, R., Cortez, P.: Using data mining for bank direct marketing: an application of the crisp-DM methodology. In: Proceedings of European Simulation and Modelling Conference-ESM 2011, pp. 117–121. EUROSIS-ETI (2011)

19. Sala, D.A., Jalalvand, A., Deyne, A.V.Y., Mannens, E.: Multivariate time series for data-driven endpoint prediction in the basic oxygen furnace. In: Wani, M.A., Kantardzic, M.M., Mouchaweh, M.S., Gama, J., Lughofer, E. (eds.) 17th IEEE International Conference on Machine Learning and Applications, ICMLA 2018, Orlando, FL, USA, 17–20 December 2018, pp. 1419–1426. IEEE (2018). https://doi.org/10.1109/ICMLA.2018.00231

20. Schorfheide, F., Wolpin, K.I.: On the use of holdout samples for model selection. Am. Econ. Rev. **102**(3), 477–81 (2012)

21. Shrouf, F., Ordieres, J., Miragliotta, G.: Smart factories in industry 4.0: a review of the concept and of energy management approached in production based on the internet of things paradigm. In: 2014 IEEE International Conference on Industrial Engineering and Engineering Management, pp. 697–701, December 2014. https://doi.org/10.1109/IEEM.2014.7058728

22. Skobelev, D., Zaytseva, T., Kozlov, A., Perepelitsa, V., Makarova, A.: Laboratory information management systems in the work of the analytic laboratory. Meas. Tech. **53**(10), 1182–1189 (2011)

23. Spendla, L., Kebisek, M., Tanuska, P., Hrcka, L.: Concept of predictive maintenance of production systems in accordance with industry 4.0. In: 2017 IEEE 15th International Symposium on Applied Machine Intelligence and Informatics (SAMI), pp. 000405–000410. IEEE (2017)

24. Tashman, L.J.: Out-of-sample tests of forecasting accuracy: an analysis and review. Int. J. Forecast. **16**(4), 437–450 (2000). https://doi.org/10.1016/S0169-2070(00)00065-0

25. Wirth, R., Hipp, J.: Crisp-DM: towards a standard process model for data mining. In: Proceedings of the 4th International Conference on the Practical Applications of Knowledge Discovery and Data Mining, pp. 29–39 (2000)

Predicting Physical Properties of Woven Fabrics via Automated Machine Learning and Textile Design and Finishing Features

Rui Ribeiro[1,4], André Pilastri[1], Carla Moura[2], Filipe Rodrigues[3],
Rita Rocha[3], José Morgado[3], and Paulo Cortez[4(✉)]

[1] EPMQ - IT Engineering Maturity and Quality Lab, CCG ZGDV Institute,
Guimarães, Portugal
{rui.ribeiro,andre.pilastri}@ccg.pt
[2] Riopele, Pousada de Saramagos, Portugal
[3] CITEVE - Centro Tecnológico das Indústrias Têxtil e do Vestuário de Portugal,
Familicão, Portugal
[4] ALGORITMI Centre, Department of Information Systems, University of Minho,
Guimarães, Portugal
pcortez@dsi.uminho.pt

Abstract. This paper presents a novel Machine Learning (ML) approach to support the creation of woven fabrics. Using data from a textile company, two CRoss-Industry Standard Process for Data Mining (CRISP-DM) iterations were executed, aiming to compare three input feature representation strategies related with fabric design and finishing processes. During the modeling stage of CRISP-DM, an Automated ML (AutoML) procedure was used to select the best regression model among six distinct state-of-the-art ML algorithms. A total of nine textile physical properties were modeled (e.g., abrasion, elasticity, pilling). Overall, the simpler yarn representation strategy obtained better predictive results. Moreover, for eight fabric properties (e.g., elasticity, pilling) the addition of finishing features improved the quality of the predictions. The best ML models obtained low predictive errors (from 2% to 7%) and are potentially valuable for the textile company, since they can be used to reduce the number of production attempts (saving time and costs).

Keywords: Textile fabrics · Regression · Machine Learning

1 Introduction

The introduction of the Industry 4.0 concept is transforming diverse industry sectors due to the adoption of Information Technology (IT), such as Internet of Things (IoT), Big Data, Cloud Computing and Artificial Intelligence (AI) [10,23]. In particular, the Industry 4.0 transformation can enhance the textile industry by improving the production efficiency (e.g., reducing costs) and assisting in the design of woven fabrics.

I. Maglogiannis et al. (Eds.): AIAI 2020, IFIP AICT 584, pp. 244–255, 2020.
https://doi.org/10.1007/978-3-030-49186-4_21

In this work, we address a textile company that is being transformed by the Industry 4.0. The company produces custom made woven fabrics for diverse clients. Currently, the fabric design is mostly based on the designer experience and intuition, which results in the execution of several trial-and-error production experiments that require resources (production materials, machines, human labour) and time. Each new fabric production attempt also requires laboratory quality tests, to verify if the produced fabric complies with quality standards and the client requirements. If a fabric is not approved, a new design attempt is set, resulting in an additional production time and costs. All these production steps generate data that can be explored by AI tools, namely Data Mining (DM) and Machine Learning (ML), to support the design of new woven fabrics.

In this paper, we report the implementation of a CRoss-Industry Standard Process for DM (CRISP-DM) [24] project for the prediction of the final fabric physical properties, as measured by nine laboratory quality tests (e.g., abrasion, pilling). The goal is to use a ML model as an "oracle", providing estimates of the fabric real physical properties for several input design options, thus aiding the textile design experts and reducing the number of fabric production attempts. To better focus on input feature selection and transformation, we adopt an Automated Machine Learning (AutoML) procedure during the modeling stage of CRISP-DM, allowing to automatically select and tune the hyperparameters of the predictive ML models [19]. In particular, we focus on input variables that can be set during the textile design phase, namely based on fabric design (e.g., composition, amount of finished threads) and finishing (e.g., washing, drying, singeing) features. In total, we executed two major CRISP-DM iterations, in which we explored different input feature engineering strategies.

This paper is organized as follows: Sect. 2 introduces the related work; Sect. 3 presents the two CRISP-DM iterations; Sect. 4 details the obtained results; and finally Sect. 5 presents the main conclusions.

2 Related Work

The creation of a new woven fabric is composed of several phases (e.g., design, production, testing). In particular, fabric testing has a crucial role in assessment product quality and performance, ensuring regulatory compliance and it provides information about the properties of the fabrics [18]. The overall process of new fabric creation generates large amounts of data, which under the Industry 4.0 concept can be used by AI tools (DM and ML) to extract valuable knowledge [17]. Following the increasing interest in DM, the CRISP-DM was proposed as a standard methodology to support the execution of real DM projects [24]. The methodology involves interactions between business domain and DM/ML experts and several iterations that can include up to six main phases: business understanding, data understanding, data preparation, modeling, evaluation and deployment. Regarding the textile domain, use of DM techniques is more recent, involving mainly classification tasks, such as defect detection and estimating the quality of yarns [26].

In what concerns a data-driven modeling of textile quality tests, the research is more scarce. Fan and Hunter [12] used a backpropagation Neural Network (NN) with one single hidden layer with 30 inputs based on fibre, yarn, and fabric constructional parameters to predict nine fabric properties (e.g., abrasion, seam slippage). A similar NN model was adopted in [1] to predict the pilling propensity of fabrics. In other study, Support Vector Machines (SVM) with 17 input features related to fiber and yarn were used to predict 8 different rates of pilling [25]. In the same study, backpropagation NN were also proposed to predict other textile properties, such as seam strength and elongation. In a more recent study [11], a simple multiple regression model was used to estimate the relationship between fabric tear strength and other input variables, such as yarn tensile strength, yarn count and fabric linear density. In previous work [22], we performed an initial exploration of ML algorithms, such as Random Forests (RF) and Gradient Boosting Machines (GBM), to predict two fabric properties (tear strength in warp and weft directions). While interesting results were achieved, the study explored a very limited set of inputs (e.g., no finishing data was used).

The performance of a ML algorithm is dependent on a correct feature engineering and ML model selection [9]. Most of the mentioned state-of-the-art works adopt a simple and fixed set of input variables (defined *a priori*) when predicting the textile physical properties. Moreover, the prediction models were obtained by using an empirical trial-and-error process that often requires a substantial effort from the ML expert [15]. In contrast with the related works (e.g., [1,12,25]), we employ in this paper an AutoML that automatically selects the best among several state-of-the-art ML algorithms. The adoption of the AutoML procedure allowed us to better focus on feature engineering, which is a non-trivial task in this domain. For instance, finishing features were considering a future challenge and thus were excluded from the predictive study performed in [12]. Woven fabric feature engineering is a complex task due to two main reasons. First, the design and finishing processes of the fabric creation includes a variable number of input features (e.g., yarns, finishing operations) that can influence the targeted textile physical properties. Second, most of these features are nominal and often present a high cardinality. Since most regression ML algorithms work only work with numeric values, a nominal to numeric transform is needed.

3   CRISP-DM Methodology

3.1   First CRISP-DM Iteration

Business Understanding: The creation of a new woven fabric starts with the definition desired characteristics. The fabric developer uses its experience and intuition, taking into account the textile requirements and starts to analyse the most similar fabrics already produced. Then, a several design elements are initially set, such as the type and number of fibers and the pick count. Some of these design elements involve a single value per fabric (e.g., number of picks), while others involve a variable number of choices (e.g., which and how many yarns to use). Next, a physical sample is produced using several materials (e.g., yarns) and machines (e.g., loom). The final production stage includes a variable sequence of finishing operations (e.g., washing, drying). Then, the produced fabric is tested via laboratory instruments, allowing to infer the physical properties and check if it meets the desired characteristics. If the fabric does not comply with the quality standards or client requirements, then the whole fabric creation process is repeated. In practice, several iterations are executed until a quality fabric is achieved, which results in additional production time and costs. The analyzed textile company expressed the need to get a fast and cheap estimate of the true fabric physical properties by adopting a ML approach. The goal is to use the predictive ML models as "oracles", quickly checking some fabric design and finishing alternative choices, thus reducing the number of attempts necessary to produce a woven fabric.

In total, the company identified nine target properties: abrasion, seam slippage (warp and weft directions), elasticity (warp and weft directions), dimensional stability to steam (warp and weft directions), bias distortion and pilling. All these nine properties are measured using numeric values. In this work, each property is measured as a separate regression task.

Data Understanding and Preparation: The data was collected from two main data sources: the company Enterprise Resource Planning (ERP), with fabric production records, and the laboratory testing database, with fabric quality tests performed between February 2012 to March 2019. The ERP data included 90,034 examples with 2,391 features per row. Using a manual analysis and domain expert knowledge, the ERP features were filtered into a total of 805 potentially relevant attributes. The laboratory dataset had 149,388 examples with the results for the nine selected physical tests. To aggregate all data, a Data Warehouse system was implemented, using an Extraction, Transform, Load (ETL) process to merge and preprocess the two data sources. During the ETL, some records were discarded since they had missing features (e.g., no yarns or no composition values).

When analysing the obtained historical data, we identified a small fraction of laboratory database entries (around 1%) that included slight different physical test values for the same fabric. After consulting the laboratory analysts, it become clear that the differences were due to the execution of laboratory tests at different fabric finishing procedures (e.g., before or after drying). Since the laboratory database did not include when such tests were executed, we opted to compute average values, in order to get a single number per fabric and test.

The initial set of input features explored in this CRISP-DM iteration is presented in Table 1. Figure 1 exemplifies how some of these features are related with the textile fabric. The first 11 rows of the table are related with a fixed set of design attributes that are defined for all fabrics. Each fabric is composed by two main elements warp and weft, each including a variable mixture of yarns, from 1 (minimum) to a maximum of 21 (in our database). Moreover, each yarn has four main characterizing features plus the number of its repetitions in the warp or weft (these features are shown in the last five rows of Table 1). Thus, the proper preprocessing of yarn data, to feed the regression models, is a non-trivial issue. In this work, we propose the following yarn representation. For each fabric, we use a sequence that has a maximum of max_y yarns for warp and then another sequence of max_y yarns for weft. Each sequence is thus composed by the elements $<y_1, ..., y_{max_y}>$, where y_i denotes the i-th yarn representation data. In this work, we adopt the threshold of $max_y = 6$ yarns per warp and weft. This value allows the representation of 99.7% of the fabrics without any information loss, while using a larger threshold would increase the sparseness of the input space, increasing the complexity of the predictive models. When a fabric does not have 6 yarns, we use a zero padding to fill the "empty" yarn values, which is a popular text preprocessing technique that adds null values (e.g., 0) to non-existent features. Finally, we explore two yarn alternative representations: **A**, use of the code (unique value) and its number of repetitions, where y_i is set as the tuple $(code_i, repetitions_i)$; and **B**, use of all yarn characterizing elements except the code, where y_i is set as $(composition_i, folds_i, count_i, repetitions_i)$. In total, the **A** representation assumes 35 input variables $(11 + 2 \times 6 \times 2)$, while the **B** encoding results in 59 input features $(11 + 2 \times 6 \times 5)$. Before feeding the ML algorithms, all the numeric inputs were standardized to a zero mean and one standard deviation. As for the nominal variables, several of them contain a high cardinality. For instance, the analyzed database includes 6,265 distinct types of yarns. A popular nominal to numeric transform is the one-hot encoding, which assigns one boolean value per nominal level. However, this transformation would highly increase the input space, resulting in a very sparse representation that would prejudice the learning of the regression models, also enlarging the computational memory and effort. To handle this issue, in this work we transform all nominal attributes with the Inverse Document Frequency (IDF) function:

$$IDF(l) = ln(n/n_l) \tag{1}$$

where n is the total number of examples in the training set and n_l is the number of examples that contain the level l in the analyzed attribute [5]. The advantage of this transform is that is encodes a nominal attribute into a single numeric value, with the most frequent levels being set near the zero (but with a larger "space" between them), and the less frequent ones being more close to each other and near a $IDF(l)$ maximum value.

Table 2 presents the nine output targets. The last column (**Range**) shows the admissible range values for each target, as defined by the textile company. All examples outside such range were considered outliers (e.g., uncommon military fabrics) and thus removed from the dataset. Since different quality tests can be assigned to different fabrics (depending on the client requirements), a variable number of examples is presented for each output (column **Examples**).

Modeling: The experiments were conducted in a personal computer using two different computational environments: the R statistical tool and its rminer package for data manipulation and evaluation of ML algorithms [8], and H20 software which implements a AutoML procedure [7]. As previously discussed, during the

Table 1. Initial list of fabric design input features.

Attribute	Description (data type)	Range
composition	Composition of the fabric (nominal code)	660 levels
t_pol	Number of finished threads per centimeter (numeric)	[18, 1321]
p_pol	Number of finished picks per centimeter (numeric)	[7, 510]
weight/m^2	Weight (in grams) per square meter (numeric)	[22, 1690]
finished width	Width in centimeters (numeric)	[122, 168]
weave design	Weave pattern of the fabric (nominal code)	20 levels
reed width	Width of the reed in centimeters (numeric)	[30, 242]
denting	Number of the reed dents per centimeter (numeric)	[5, 252]
ends/dent	Number of yarns per dent (numeric)	{0, 1, ..., 8}
n_picks	Number of picks on loom per centimeter (numeric)	[0, 3450]
warp total ends	Total number of threads on the warp (numeric)	[1026, 6862]
yarn code	Identification code of the yarn (nominal code)	6,265 levels
yarn composition	Composition of the yarn (nominal code)	88 levels
n_folds	Number of single yarns twisted (numeric)	{0, 1, ..., 12}
yarn count	Mass per unit length of the yarn (numeric)	[0, 268]
yarn repetitions	Number of yarn repetitions in warp or weft (numeric)	{0, 1, ..., 8}

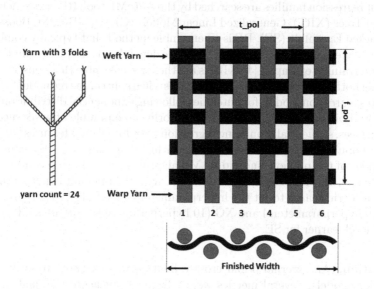

Fig. 1. Visualization of some woven fabric features.

first iteration of CRISP-DM we explored the issue of yarn representation, thus two main strategies as compared: **A** and **B**. During this modeling stage, to find the best ML algorithm we adopt an AutoML procedure.

The AutoML was configured to automatically select the regression model and its hyperparameters based on the best Mean Absolute Error (MAE), using a internal 5-fold cross-validation applied over the training data. We adopted the same

Table 2. List of output target variables.

Test	Examples	Range
Abrasion	456	[5000, 30000]
Seam slippage (warp)	10,605	[1, 20]
Seam slippage (weft)	10,279	[1, 20]
Elasticity (warp)	7,901	[5, 55]
Elasticity (weft)	12,698	[5, 70]
Dimensional stability to steam (warp)	8,773	[−4, 2]
Dimensional stability to steam (weft)	8,871	[−4, 2]
Bias distortion	15,141	[1, 14]
Pilling	11,912	[1, 4.5]

AutoML configuration executed in [13]. The computational experiments were executed on a desktop computer and each ML algorithm was trained using a maximum running time of 3,600 s. After selecting the best ML algorithm, its best set of hyperparameters are fixed and the ML is retrained with all training data. A total of 6 different regression families are searched by the AutoML tool: RF, Extremely Randomized Trees (XRT), Generalized Linear Models (GLM), GBM, XGBoost (XG) and Stacked Ensemble (SE). RF is an ensemble method that typically combines a large set of tree predictors, such that each tree depends on a random sample of features and training examples [4]. XRT is another tree ensemble that consists of randomizing both attribute and cut-point choices when splitting a tree node [14]. GLM estimates regression models for outcomes following exponential distributions (e.g., Gaussian, Poisson, gamma) [21]. GBM performs an ensemble of weak successive decision trees, sequentially building regression trees for all data features [20]. XG is another popular boosting decision tree algorithm [6]. Finally, the SE combines the predictions of the previous individual ML algorithms by using a second-level ML algorithm [3]. The H20 tool sets RF and XRT with their default hyperparameters, performs a grid search to set the hyperparameters for GLM (1 hyperparameter), GBM (9 hyperparameters) and XG (10 hyperparameters), and uses GLM as the second-level learner for SE.

Evaluation: An external 3-fold cross-validation was executed to evaluate the regression models. Several metrics were selected to measure the quality of the predictions: MAE, Normalized MAE (NMAE), Adjusted R2 (Adj.R^2) and classification accuracy for a given tolerance T (Acc@T). Regarding MAE and NMAE, the lower the values, the better are the predictions. The NMAE measure normalizes the MAE by the range of the output target on the test set, thus it provides a percentage that is easy to interpret and is scale independent. In the case of Adj.R^2 and Acc@T (from 0 to 1), higher values indicate better predictions. The Acc@T value is based on the Regression Error Characteristic (REC) curves and it measures the percentage of correctly classified examples when assuming a fixed absolute error tolerance (T) [2]. In this work we use $T \in \{5\%, 10\%, 20\%\}$. We note that the percentage of error tolerance is computed by considering the range of the target values. The first CRISP-DM iteration results are discussed in Sect. 4.

3.2 Second CRISP-DM Iteration

After showing the Sect. 4 results to the textile experts, it was decided to perform a second CRISP-DM iteration to check the utility of finishing features. During a new business understanding phase, it become clear that the finishing process should influence the final fabric properties. The finishing consists of a predefined sequence of operations that are applied to a fabric with the goal to increase the attractiveness or serviceability of the textile product [16].

At the new data understanding and preparation stages, we identified that the company had a total of 61 different types of finishing operations. Moreover, the sequence of finishing operations can be different for each fabric and it can include repetitions of the operations (e.g., several wash and dry cycles). In the analyzed database, the number of executed finishing operations ranged from 1 (minimum) to 39 (maximum), with an average of 6.82. Table 3 presents the top ten most used types of finishing and the respective number of usages (column **Examples**). In the table, we added the special value "Others" to represent a merge of distinct finishing operations for which there was no description data.

Table 3. Ten most used fabric finishing operations.

Rank	Finishing	Examples	Rank	Finishing	Examples
1	Dry	145,008	6	Finish_Fixate	53,717
2	Wash	122,513	7	Dyeing	44,162
3	Sanforization	73,401	8	Others	31,704
4	Finish	67,774	9	Shear_Right	21,600
5	Singeing	58,875	10	Decatizing	20,685

Similarly to the yarn encoding strategy, in this work we will assume a sequence with a maximum of max_f finishing operations to represent the finishing process: $<f_1, ..., f_{max_f}>$, where f_i denotes the i-th finishing operation. In this work, we set $max_f = 10$ as a reasonable value that represents around 85% of all fabrics without information loss, helping to reduce the number of inputs that are fed into the ML models. To encode each finishing operation (nominal attribute) we adopt the same IDF transform (Eq. 1). In the modeling phase, the best previous input encoding (**A**) is compared with the new encoding **C** that merges all **A** inputs with the finishing features, resulting in 45 (35 + 10) input variables. The evaluation phase was executed similarly to the first CRISP-DM iteration.

Table 4. AutoML predictive results (average NMAE test set values, in %; best results per CRISP-DM iteration in **bold**).

Target	First iteration		Second iteration	
	A	B	A	C
Abrasion	**4.81**	5.17	**4.81**	4.93
Seam slippage (warp)	**4.54**	5.00	**4.54**	4.43
Seam slippage (weft)	3.09	**2.62**	3.09	**2.56**
Elasticity (warp)	**2.94**	3.42	**2.94**	2.59
Elasticity (weft)	**2.39**	2.87	**2.39**	2.15
Dimensional stability to steam (warp)	**6.60**	7.52	**6.60**	6.17
Dimensional stability to steam (weft)	**4.27**	6.58	**4.27**	4.12
Bias distortion	**4.16**	4.33	**4.16**	3.79
Pilling	**6.80**	8.14	**6.80**	6.70
Average	**4.41**	5.07	**4.41**	**4.16**

4   Results

In all experiments performed the AutoML always selected the GBM or SE algorithms. GBM provided the best overall results (lowest NMAE averaged over the external 3 cross-validation iterations), while for some targets and specific folds (e.g., Bias Distortion and third fold experiment), the selected model was SE. To compare the feature strategy results, we always assume the best algorithm (GBM or SE) per external fold validation, denoting this as the AutoML model.

Table 4 summarizes the predictive performance results, in terms of the 3-fold average NMAE values for the best AutoML model that were obtained during the first and second CRISP-DM iterations. For the first iteration, it becomes clear that **A** is the best yarn representation strategy. It provides the lowest NMAE results for eight of the nine fabric targets and it also obtains the lowest average value, over all output tasks (difference of 0.66% points when compared with **B**). Moreover, **A** has the additional advantage of producing less inputs (35 and not 59), leading to predictive models that require less computational memory and fitting effort. Following these results, we adopted the **A** encoding to represent the yarns. In the second iteration, the usage of fabric finishing features (**C**) improves the prediction results for eight of the nine targets. Overall, **C** provides the lowest average NMAE, with a 0.25% point improvement when compared with **A**. Table 5 complements the results by showing the other predictive measures for **A** and **C** (represented in column **Str.**). In general, when **C** obtains the lowest MAE error, it also outperforms the **A** strategy for the other measures (Acc@T and Adj.R^2). For demonstrative purposes, Fig. 2 shows the AutoML elasticity (warp) predictions (x-axis) versus the target values for a particular external 3-fold iteration. The plot includes the tolerance ranges of the $T = 5\%$ and $T = 10\%$, showing that an interesting percentage of the values are correctly predicted within those ranges (e.g., 43% of accuracy for $T = 10\%$).

Table 5. AutoML predictive results (other regression measures; best results in **bold**).

Test	Str.	Regression metrics				
		MAE	Acc@5%	Acc@10%	Acc@20%	Adj.R^2
Abrasion	A	**924.03**	1%	1%	2%	**0.76**
	C	948.60	0%	0%	1%	0.75
Seam slippage (warp)	A	0.83	24%	35%	47%	0.79
	C	**0.81**	**28%**	**42%**	**54%**	**0.80**
Seam slippage (weft)	A	0.55	28%	52%	65%	0.83
	C	**0.46**	**61%**	**67%**	**73%**	**0.84**
Elasticity (warp)	A	1.46	11%	17%	24%	**0.92**
	C	**1.29**	**26%**	**34%**	**42%**	**0.92**
Elasticity (weft)	A	1.25	10%	17%	25%	**0.92**
	C	**1.12**	**13%**	**20%**	**30%**	**0.92**
Dimensional stability to steam (warp)	A	0.40	17%	31%	50%	0.63
	C	**0.37**	**30%**	**40%**	**53%**	**0.65**
Dimensional stability to steam (weft)	A	0.25	55%	61%	67%	0.73
	C	0.25	55%	**62%**	**68%**	**0.75**
Bias distortion	A	0.50	13%	25%	40%	0.53
	C	**0.46**	**14%**	**26%**	**44%**	**0.59**
Pilling	A	0.24	**35%**	**48%**	**63%**	0.76
	C	**0.23**	25%	37%	54%	**0.78**

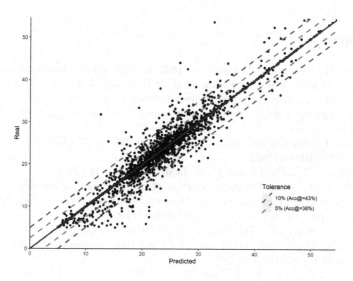

Fig. 2. Predicted versus real elasticity (warp) values.

5   Conclusions

This paper addresses a textile company that is being transformed under the Industry 4.0 concept and that identified the need to reduce the number of production attempts when designing new woven fabrics by using a Machine Learning (ML) approach. To handle this goal, we implemented two iterations of the CRoss-Industry Standard Process for Data Mining (CRISP-DM) methodology. Each iteration focused on a feature engineering task, aiming to check the value of input fabric yarn and finishing feature representations. During the modeling stage of CRISP-DM, an Automated ML (AutoML) was used to select the best among six state-of-the-art ML algorithms. The best results were achieved by an input set of features that includes a fixed sequence with a simple yarn code representation and another fixed sequence with fabric finishing operations (strategy **C**). Interesting predictive results were achieved for nine targeted fabric properties, with an average NMAE error that ranges from 2% to 7%. The results were shown to the textile company, which considered them valuable to reduce the number of fabric creation attempts, thus having a potential to save the production time and costs. In future work, we intend to apply a similar approach in the prediction of other fabric quality tests, such as residual extension and traction.

Acknowledgments. This work was carried out within the project "TexBoost: less Commodities more Specialities" reference POCI-01-0247-FEDER-024523, co-funded by *Fundo Europeu de Desenvolvimento Regional* (FEDER), through Portugal 2020 (P2020).

References

1. Beltran, R., Wang, L., Wang, X.: Predicting the pilling propensity of fabrics through artificial neural network modeling. Text. Res. J. **75**(7), 557–561 (2005)
2. Bi, J., Bennett, K.P.: Regression error characteristic curves. In: Proceedings of the 20th International Conference on Machine Learning (ICML 2003), pp. 43–50 (2003)
3. Breiman, L.: Stacked regressions. Mach. Learn. **24**(1), 49–64 (1996). https://doi.org/10.1007/BF00117832
4. Breiman, L.: Random forests. Mach. Learn. **45**(1), 5–32 (2001)
5. Campos, G.O., et al.: On the evaluation of unsupervised outlier detection: measures, datasets, and an empirical study. Data Min. Knowl. Discov. **30**(4), 891–927 (2016). https://doi.org/10.1007/s10618-015-0444-8
6. Chen, T., Guestrin, C.: XGBoost: a scalable tree boosting system. In: Proceedings of the 22nd ACM SIGKDD International Conference on Knowledge Discovery and Data Mining, pp. 785–794 (2016)
7. Cook, D.: Practical Machine Learning with H2O: Powerful, Scalable Techniques for Deep Learning and AI. O'Reilly Media Inc., Sebastopol (2016)
8. Cortez, P.: Modern Optimization with R. Springer, Cham (2014). https://doi.org/10.1007/978-3-319-08263-9
9. Domingos, P.: A few useful things to know about machine learning. Commun. ACM **55**(10), 78–87 (2012)

10. Drath, R., Horch, A.: Industrie 4.0: Hit or hype? [Industry Forum] (2014). https://doi.org/10.1109/MIE.2014.2312079
11. Eltayib, H.E., Ali, A.H., Ishag, I.A.: The prediction of tear strength of plain weave fabric using linear regression models. Int. J. Adv. Eng. Res. Sci. **3**(11), 151–154 (2016)
12. Fan, J., Hunter, L.: A worsted fabric expert system: Part II: an artificial neural network model for predicting the properties of worsted fabrics. Text. Res. J. **68**(10), 763–771 (1998). https://doi.org/10.1177/004051759806801010
13. Ferreira, L., Pilastri, A., Martins, C., Santos, P., Cortez, P.: An automated and distributed machine learning framework for telecommunications risk management. In: Proceedings of the 12th International Conference on Agents and Artificial Intelligence, ICAART 2020, Volume 2, Valletta, Malta, February, pp. 99–107. SciTePress (2020)
14. Geurts, P., Ernst, D., Wehenkel, L.: Extremely randomized trees. Mach. Learn. **63**(1), 3–42 (2006)
15. Gibert, K., Izquierdo, J., Sànchez-Marrè, M., Hamilton, S.H., Rodríguez-Roda, I., Holmes, G.: Which method to use? An assessment of data mining methods in environmental data science. Environ. Model. Softw. **110**, 3–27 (2018)
16. Hall, M.E.: Finishing of technical textiles. In: Handbook of Technical Textiles, p. 152 (2000)
17. Han, J., Kamber, M., Pei, J.: Data Mining: Concepts and Techniques. Elsevier Inc., Amsterdam (2012). https://doi.org/10.1016/C2009-0-61819-5
18. Hu, J.: Fabric Testing. Elsevier Ltd., Amsterdam (2008). https://doi.org/10.1533/9781845695064
19. Le, T.T., Fu, W., Moore, J.H.: Scaling tree-based automated machine learning to biomedical big data with a feature set selector. Bioinformatics (2019). https://doi.org/10.1093/bioinformatics/btz470
20. Natekin, A., Knoll, A.: Gradient boosting machines, a tutorial. Front. Neurorobotics **7**, 21 (2013)
21. Nelder, J.A., Wedderburn, R.W.: Generalized linear models. J. R. Stat. Soc. Ser. (Gen.) **135**(3), 370–384 (1972)
22. Ribeiro, R., Pilastri, A., Moura, C., Rodrigues, F., Rocha, R., Cortez, P.: Predicting the tear strength of woven fabrics via automated machine learning: an application of the CRISP-DM methodology. In: Proceedings of the 22th International Conference on Enterprise Information Systems – ICEIS2020, Prague, Czech Republic. SciTePress, May 2020
23. Wang, S., Wan, J., Li, D., Zhang, C.: Implementing smart factory of Industrie 4.0: an outlook. Int. J. Distrib. Sens. Netw. **2016** (2016). https://doi.org/10.1155/2016/3159805
24. Wirth, R., Hipp, J.: CRISP-DM: towards a standard process model for data mining. In: Proceedings of the 4th International Conference on the Practical Applications Of Knowledge Discovery and Data Mining, pp. 29–39. Springer, London (2000)
25. Yap, P.H., Wang, X., Wang, L., Ong, K.L.: Prediction of wool knitwear pilling propensity using support vector machines. Text. Res. J. **80**(1), 77–83 (2010). https://doi.org/10.1177/0040517509102226
26. Yildirim, P., Birant, D., Alpyildiz, T.: Data mining and machine learning in textile industry. Wiley Interdiscip. Rev. Data Min. Knowl. Discov. **8**(1), e1228 (2018)

Real-Time Surf Manoeuvres' Detection Using Smartphones' Inertial Sensors

Dinis Moreira[1]([⊠])(iD), Diana Gomes[1](iD), Ricardo Graça[1](iD), Dániel Bányay[2],
and Patrícia Ferreira[3]

[1] Fraunhofer Portugal AICOS, Porto, Portugal
{dinis.moreira,diana.gomes,ricardo.graca}@fraunhofer.pt
[2] KTH Royal Institute of Technology, Stockholm, Sweden
dbanyay@kth.se
[3] Damel - Confecção de Vestuário Lda, Amorim, Portugal
patricia@damel.pt

Abstract. Surfing is currently one of the most popular water sports in the world, both for recreational and competitive level surfers. Surf session analysis is often performed with commercially available devices. However, most of them seem insufficient considering the surfers' needs, by displaying a low number of features, being inaccurate, invasive or not adequate for all surfer levels. Despite the fact that performing manoeuvres is the ultimate goal of surfing, there are no available solutions that enable the identification and characterization of such events. In this work, we propose a novel method to detect manoeuvre events during wave riding periods resorting solely to the inertial sensors embedded in smartphones. The proposed method was able to correctly identify over 95% of all the manoeuvres in the dataset (172 annotated manoeuvres), while achieving a precision of up to 80%, using a session-independent validation approach. These findings demonstrate the suitability and validity of the proposed solution for identification of manoeuvre events in real-world conditions, evidencing a high market potential.

Keywords: Surf · Manoeuvre detection · Inertial sensors · Monitoring system · Machine learning · Smartphone · Sports performance

1 Introduction

Surf has been increasing its popularity worldwide, for both competitive and recreational levels. Minimal training and equipment makes this an appealing water sport for everyone [20]. Paddle, stationary, wave riding and some miscellaneous events are the four main activities of a surf session [20,21]. Despite the exponential increase in the field of sports trackers, there have not been significant

Financial support from project TexBoost - less Commodities more Specialities (POCI-01-0247-FEDER-024523), co-funded by Portugal 2020, framed under the COMPETE 2020 (Operational Program Competitiveness and Internationalization) and European Regional Development Fund (ERDF) from European Union (EU).

© IFIP International Federation for Information Processing 2020
Published by Springer Nature Switzerland AG 2020
I. Maglogiannis et al. (Eds.): AIAI 2020, IFIP AICT 584, pp. 256–267, 2020.
https://doi.org/10.1007/978-3-030-49186-4_22

developments in the specific surf area [13]. Although performing manoeuvres is the ultimate goal in surfing, there are few solutions available for surfers' assistance, especially concerning the characterization of rotational movements and performed manoeuvres.

Tools that can track and measure surfers' progress over time are quite appealing and needed for surf practitioners, specially when no external guidance is provided [11,20]. However, the analysis of the events that occur during wave riding periods can be a challenge. Moreover, even if a great part of surf session analytics (e.g. paddle duration, wave counting) are currently being performed by some commercial solutions, these still lack manoeuvre identification and evaluation, which is a relatively new and unexplored domain [11,20].

Most of these surf monitoring systems generate biofeedback based on sensor data retrieved during the training session, and are capable of providing some additional information about the executed movements [9,11,19]. Currently, most of these solutions can only count the number of waves and paddle time, estimate speed, distance and movements pattern during the session using Global Positioning System (GPS) and/or Inertial Measurement Units (IMU) measurements [1,2,10,13]. However, they lack detail during the most important surf event – wave riding [11]. Moreover, and specifically for manoeuvre detection purposes, there is a low number of available studies and real-world validations [13,25]. As such, there is a gap in the market concerning surf manoeuvre detection.

Surf manoeuvres may be difficult to distinguish, especially for non-experts [7,26]. Thus, the automatic analysis of these events may be considered a difficult task. Differences between two manoeuvres are often only evident in the "dynamics" or "elegance" in their execution. Even in surf competitions, the evaluation process made by the judges is usually subjective and based on the average of their opinions, highlighting even more the importance of having a solution capable of thorough identification, characterisation and evaluation all of the performed manoeuvres [7].

This context motivated the development of a new algorithm for the detection and characterisation of the performed manoeuvres during wave riding periods in real-world conditions, solely resorting to the inertial sensors embedded in a smartphone. In this sense, this manuscript proposes a novel approach for the segmentation and identification of the performed manoeuvres for moderate to experienced surfers.

The remainder of the paper is organised as follows: Sect. 2 describes prior work conducted in this field; Sect. 3 describes the dataset and proposed methodology of the study; Sects. 4 and 5 report and discuss the main findings of this study, respectively; Sect. 6 highlights the main conclusions and points out possible directions for future work.

2   Related Work

There are only a few commercially available solutions for surf monitoring purposes, and even less if we only consider manoeuvre detection and analysis.

Most of the current commercially available solutions are only based in GPS data. Thus, these solutions can only extract top-level information about the surf session, such as wave count, travelled distance or wave speed. Detecting more complex surfing movements, such as in-wave manoeuvres, is hard or even impossible to perform with this type of solutions given the single source of available data. One example of this type of commercially available solutions is the *Rip Curl Search GPS* watch. Additionally, some mobile applications have also been developed such as *Surf Track* [3], *Dawn Patrol* [5] and *WavesTracker* [4], but very little information regarding their functioning, validity or system setup is provided.

The combined use of GPS and inertial sensor data, i.e. accelerometer, gyroscope and magnetometer sensors, has been widely reported in the literature to effectively increase overall robustness and precision of human activity recognition applications [13]. Therefore, other solutions that make use both of GPS and inertial sensor data, such as the *GlassyPro* wristband [1] or *Xensr Air* [6] surfboard-mounting device, are some of the commercially available solutions that currently combine several sources of data. However, they may not be entirely suitable for manoeuvre detection purposes. Wrist-worn devices, like the *GlassyPro*, may be practical and easy to use, but due to their positioning on the body, may be insufficient for manoeuvre detection and characterization purposes. Torso and board rotations are important metrics for wave performance analysis and extremely important when trying to identify and evaluate certain manoeuvres, and wrist-mounted sensors may not be suitable for this task. There are also devices that were designed to be mounted in the surfboard, like *Xensr Air* and *Trace Up*. However, to the best of the authors' knowledge, these devices are currently not available for purchase. Moreover, these can be associated with some safety issues due to its size and attachment to the surfboard, and they do not track the surfer's actual movements.

Besides the commercially available solutions, some research studies were also conducted, aiming surf monitoring and/or performance analysis. Madureira *et al.* [18] proposed an algorithm comparing the use of GPS sensor alone and together with inertial sensors data for wave detection. Similarly, Hoettinger *et al.* [14] proposed a machine learning based approach for differentiating wave from non-wave events, using sliding windows of 2.0 s with 75% overlap, also achieving accurate results. However, none of these studies performed any type of analysis related to manoeuvre identification or characterisation, highlighting the novelty of this study.

3 Methodology

3.1 Data Annotation

The dataset reported in [13] was used to conduct this study, featuring raw data from the accelerometer, gyroscope and magnetometer during all of the recorded surf sessions. This dataset had already been annotated for several surf events, including wave riding periods. We selected the sessions of the 5 advanced-level surfers

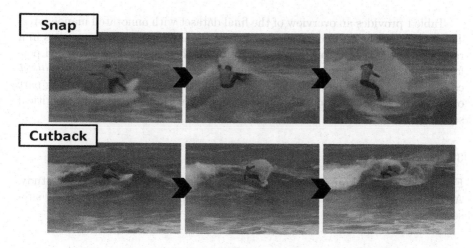

Fig. 1. Snapshot frames of the execution of the detected manoeuvres: *snap* (top) and *cutback* (bottom).

in the dataset, since appropriately performing manoeuvres during wave riding is a demanding task in the sport, characteristic of experienced surfers (Fig. 1).

A dual-source data annotation tool enabling synchronized signal and video visualization was created to finely annotate manoeuvre periods in each wave. Manoeuvre annotation was a challenging process, due to the similarity between some manoeuvres, the particular technique of each surfer, and lack of clear view of the surfer in some situations (e.g. agitated sea, distance). Moreover, annotations were performed by non-specialists in this area and therefore, can be subjective and prone to human error. To minimize its influence, we opted not to consider events in which the annotator was unsure of the label to assign or the temporal limits of the movement, and unsuccessful manoeuvres, i.e. ending in a fall.

Table 1. Dataset description.

No. sessions	8		
No. users	5		
No. waves	165		
Wave duration (avg ± std)	9.73 ± 5.43 s		
No. manoeuvres	172	*Snap*	71
		Cutback	94
		Other	7
Manoeuvre duration (avg ± std)	2.45 ± 0.75 s		

Table 1 provides an overview of the final dataset with annotated manoeuvres. The most represented manoeuvres were *cutback* and *snap*. A *snap* consists in a radical change of trajectory in the pocket or on the top of the wave; when performed abruptly, it produces spectacular and flashy buckets of spray. A *cutback* consists in riding up the wave shoulder, turning back towards the breaking part of the wave without losing speed and ending up with a re-entry in the critical section of the wave.

3.2   Data Processing

Figure 2 presents a graphical overview of the processing steps used in this study. A thorough explanation of each of these steps is provided in the following subsections.

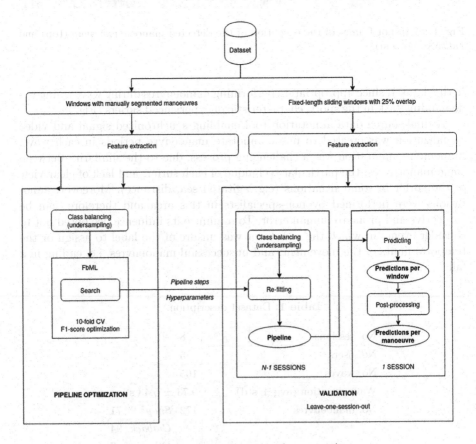

Fig. 2. Data processing operations overview.

Data Stream Segmentation. An overlapping sliding window approach was implemented for data stream segmentation (Fig. 3). Since the selection of window size may be considered an empirical and task-oriented problem, highly related with the duration of the event of interest and/or the maximum expected recognition latency, different window sizes were evaluated, considering the distribution of manoeuvre duration in our dataset, presented in Table 1: 2.0, 2.5, 3.0, and 3.5 s, with fixed overlap of 25%. Each time-window with over 70% of match with a certain annotated manoeuvre was assigned a positive class label. All remaining samples were assigned a negative class label.

Annotated ground truth manoeuvres' windows were also considered for the pipeline optimization step.

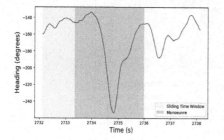 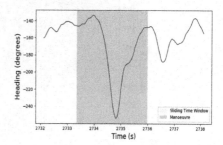

Fig. 3. Example of two consecutive sliding windows.

Feature Extraction. A set of generic and domain-specific features were extracted from each of the 3-axis orientation components (heading, pitch, roll) obtained after sensor fusion using the gradient descent-based algorithm of [17], the magnitude of the 3-axis linear acceleration, and the magnitude of the XY components of the linear acceleration for each time-window.

Time-domain features such as minimum, maximum, variance, skewness, kurtosis, mean cross ratio, waveform length [16] and interquartile range values were calculated for each time window.

Features based on the wavelet transformation of the input signals were also calculated. Wavelet transform decomposes a signal according to the frequency, representing the frequency distribution in the time domain [22]. We used the implementation of the Wavelet Packet Transform of the PyWt library [15], with *Daubechies 2* (db2) as mother wavelet and a maximum decomposition level of *3*, for performing a time-frequency analysis of the linear acceleration and orientation signals, respectively. For each resolution level, the relative wavelet energy was calculated for each associated frequency band [23,24]. A metric based on the statistical variance of the wavelet coefficients was also calculated for each resolution level and associated frequency bands [27].

A set of other features which intended to describe the rotational dynamics of the movements were also extracted from the heading orientation signal. All local *extremas* were identified to generate *minima-maxima-minima* and *minima-maxima-minima-maxima* sequences (Fig. 4). These domain-specific features consisted in extracting the sum, average and maximum angular displacement and velocity values for each min-max/max-min sequence.

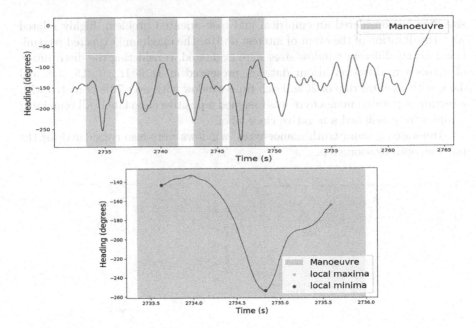

Fig. 4. Heading evolution over time for a wave riding period with several annotated manoeuvres (top) and a manoeuvre time-window with local *extrema* detection (bottom).

The feature extraction step resulted in a total of 199 features for each time window.

Pipeline Optimization. A domain-specific resampling strategy was implemented to handle dataset imbalance, since the number of instances from the positive class was always inferior to the number of instances from the negative class for all tested window sizes. This strategy consisted in discarding non-manoeuvre segments with the highest percentages of manoeuvre match successively until the classes were balanced.

The optimization and selection of the learning pipeline was performed using a tool called Feature-based Machine Learning (FbML), created at Fraunhofer AICOS. This tool is based on the open-source project *auto-sklearn* [12], and allows a search space initialization via meta-learning (search similar datasets and initialize hyper-parameter optimization algorithm with the found configuration) while providing a vast list of options for data pre-processing (balancing, imputation of missing values, re-scaling), feature transformation, and feature and classifier selection. As such, we explored pipelines generated with the following combinations of methods:

1. **Scalers:** Standardization (zero mean and unit variance); Min-Max Scaling; Normalization to unit length; Robust Scaler; Quantile Transformer; None.

2. **Feature Transformation/Selection:** Principal component analysis (PCA); Univariate Feature Selection; Classification Based Selection (Extremely Randomized Trees and L1-regularized Linear SVM); None.
3. **Classifiers:** Gaussian Naive Bayes; K-Nearest Neighbors; Linear and Nonlinear Support Vector Machines; Decision Trees; Random Forest; Adaboost.
4. **Validation Strategy:** 10-Fold Cross Validation.
5. **Optimization Metric:** F1-score.

At each new test, the results and parameters of the 5 best classification pipelines were stored for further evaluation.

Leave-One-Session-Out Validation. In order to study model generalization for different acquisition conditions (i.e. sessions), we implemented a leave-one-session-out validation approach. At each iteration i, where $i \in [1, N]$ and N represents the total number of different sessions, all of the instances from the surf session S_i were selected for testing while the remaining ones were used for re-fitting the pipeline. The pipeline was defined by the best combination of methods and hyperparameters which resulted from the FbML optimization with cross-validation.

While this approach allows us to assess the performance of the method under session-independent conditions, it will still not be enough to fully assess performance in real-world conditions, since consecutive windows containing data from a same manoeuvre should be merged in order to deliver a proper count and useful information to the user. As such, we created a methodology (*post-processing* step of Fig. 2) which merges consecutive positive predictions in pairs, setting them to correspond to a same manoeuvre M. If, after two windows are already merged, a new and single positive prediction occurs, this window may also be considered to belong to M if its classification probability is greater than that of the previous window. Otherwise, it will be set to correspond to a new manoeuvre.

Manoeuvre detection results were computed considering the predictions per window (for selection of the best window size and an overall assessment of pipelines' performance) and after considering the post-processing needed for utilization in real-world conditions (predictions per manoeuvre), as Fig. 2 indicates.

4 Results

Table 2 combines the results of the best set of pipelines generated by the FbML after the leave-one-session-out validation considering the predictions per window, for different window sizes with a fixed overlap of 25%. These results support that the best overall performance was attained with windows of 2 s, associated with a F1-score (optimization metric) of 0.91. This segmentation approach was thus selected and used in all further experiments.

Table 3 exhibits the final manoeuvre detection results, after the post-processing step. True positives (TP) correspond to detected manoeuvres' time segments which overlap with annotated manoeuvres' periods. False positives

Table 2. Average performance of the top 5 classification pipelines obtained by the FbML considering the predictions per window, using leave-one-session-out validation.

Window size (s)	Average ± standard deviation			
	F1-score	Precision	Recall	Accuracy
2	0.91 ± 0.01	0.92 ± 0.02	0.90 ± 0.01	0.91 ± 0.01
2.5	0.86 ± 0.02	0.88 ± 0.01	0.84 ± 0.03	0.86 ± 0.02
3.0	0.88 ± 0.01	0.89 ± 0.01	0.86 ± 0.01	0.88 ± 0.01
3.5	0.84 ± 0.02	0.85 ± 0.02	0.83 ± 0.03	0.83 ± 0.02

(FP) correspond to detected manoeuvres' periods which do not overlap with annotated manoeuvres. False negatives (FN) correspond to annotated manoeuvres which were not detected (do not overlap) with any positive prediction segment.

Table 3. Manoeuvre detection performance after application of the time-windows' merging criteria (real-world conditions) for each of the top 5 pipelines derived from the FbML optimization process.

Pipeline	Scaler	Selection	Classifier	TP	FP	FN	Recall	Precision	F1-score
1	Min-Max	–		168	54	4	0.98	0.76	0.85
2	Robust	–		164	45	8	0.95	0.78	0.86
3	Min-Max	Univariate	Adaboost	167	42	5	0.97	0.80	0.88
4	Min-Max	–		164	41	8	0.95	0.80	0.87
5	Min-Max	Univariate		165	40	7	0.96	0.80	0.88

All of the top 5 pipelines derived from the FbML optimization process relied on an Adaboost classifier, and performed a feature scaling step. The pipelines which implemented univariate feature selection are associated with the highest F1-scores (0.88). All pipelines were able to detect at least 95% of the annotated manoeuvres, despite the demanding circumstances under which the tests took place, since session-independence was preserved. The impact of false positives was also taken into consideration: the lowest precision was 0.76.

5   Discussion of Results

The results reported in Table 2 enabled a reasoned selection of the best segmentation approach and an overall understanding of the performance of the method. Our results support that, out of the experimented window sizes, windows of 2 s are the most appropriate for manoeuvre segmentation and classification. Windows of 2 s led to the highest number of samples in the dataset. Moreover, considering the average manoeuvre duration in our dataset, it is reasonable to conclude that keeping segments of 2 s and an overlap of 25% between consecutive samples guarantees that most manoeuvre periods are contained in 1 window or 2 consecutive windows. This approach was, thus, selected as segmentation method.

Overall, the results obtained using the predictions per window were considered very promising towards adequate manoeuvre detection using our method. However, the performance metrics exhibited in Table 3 are the most critical to understand if the method generates reliable and intelligible information for the surfer in terms of manoeuvre detection in real-world conditions.

We were able to correctly identify over 95% of all annotated manoeuvres in the dataset, while achieving a precision of up to 80%. False positive occurrences were mostly related with the following situations: 1) finishing wave riding with a failed manoeuvre, ending in a fall; 2) segments which most likely corresponded to manoeuvres, but corresponding to times when the annotator did not have clear sight of the surfer; 3) conservative annotation process which only considered periods of absolute certainty as ground truth manoeuvres. Optimizing the trade-off between false positive and negative predictions is a well-known challenge of machine learning problems. As such, and considering the aforementioned situations, we consider that the reported results are appropriate and support the adequate performance of the method in real-world conditions.

Another important detail of this study is that it maintained a session-independent validation approach. Thus, the attained results support the appropriate generalization of the method for different users and different acquisition conditions, including sea level and agitation variation and slightly different positioning of the smartphone in the users' back.

5.1   Challenges and Limitations of the Study

Despite the comprehensive amount of collected sessions available in the dataset of [13], we were only able to use the sessions from 5 surfers, since these were the only users with the necessary level of expertise to perform in-wave manoeuvres. Our dataset was finally mainly composed of *cutback* and *snap* manoeuvres, with very little representation of the remaining ones. This can be considered a limitation of the study, as there is no certainty of the performance of the method for unseen manoeuvres.

Another limitation is related with the fine time limits for ground truth manoeuvre annotation, and the challenges of the annotation process (discussed above), which may impair a full reliable quantification of the performance of the method [8]. A second annotation round with surf experts may be an adequate approach to tackle this limitation, followed by a comparative analysis of the expected improvements.

The labelling criteria for each time-window may also be a source of error of the method, as the definition of these criteria was empirically performed. It would also be interesting to implement and test dynamic data stream segmentation techniques, and assess if it would be possible to achieve improved fits of the generated time segments with annotated manoeuvre periods using such techniques.

6   Conclusions

This manuscript details the development of a surf manoeuvre detection algorithm, using data from the smartphone's inertial sensors and a machine learning pipeline optimized for the problem in hands. Several time-window sizes were tested, and windows of 2 s with 25% overlap delivered the best results. Manoeuvres were detected with up to 88% F1-score under our real-world conditions validation, which is very promising for a real-world application and should have a high market potential.

As future work, we intend to combine the outcome of this study with the work of [13] to create a full surf monitoring solution which simultaneously detects surf session events (namely, waves), and further segments these periods to deliver more performance metrics to the surfer concerning fine events, i.e. in-wave manoeuvres.

References

1. Glassy pro (2018). https://glassy.pro/index.html. Accessed 18 July 2018
2. Rip curl search gps (2018). https://searchgps.ripcurl.com/welcome/the-watch.php. Accessed 18 July 2018
3. Surf track app (2018). https://itunes.apple.com/us/app/surf-track/id1020948920? mt=8. Accessed 18 July 2018
4. Waves tracker (2018). http://www.wavestracker.com/. Accessed 18 July 2018
5. Dawn patrol (2019). https://itunes.apple.com/us/app/dawn-patrol/id11610141-79#?platform=iphone. Accessed 24 May 2019
6. Xensr air (2019). http://xensr.com/products/. Accessed 29 May 2019
7. Judges training manual (2020). http://www.surfclubs.org/documents.html. Accessed 20 Feb 2020
8. Alsallakh, B., et al.: A visual analytics approach to segmenting and labeling multivariate time series data. In: EuroVA@ EuroVis (2014)
9. Chambers, R., Gabbett, T.J., Cole, M.H., Beard, A.: The use of wearable microsensors to quantify sport-specific movements. Sport. Med. **45**(7), 1065–1081 (2015)
10. Coutts, A.J., Duffield, R.: Validity and reliability of GPS devices for measuring movement demands of team sports. J. Sci. Med. Sport. **13**(1), 133–135 (2010)
11. Farley, O.R., Abbiss, C.R., Sheppard, J.M.: Performance analysis of surfing: a review. J. Strength Cond. Res. **31**(1), 260–271 (2017)
12. Feurer, M., Klein, A., Eggensperger, K., Springenberg, J., Blum, M., Hutter, F.: Efficient and robust automated machine learning. In: Cortes, C., Lawrence, N.D., Lee, D.D., Sugiyama, M., Garnett, R. (eds.) Advances in Neural Information Processing Systems, vol. 28, pp. 2962–2970. Curran Associates, Inc. (2015). http://papers.nips.cc/paper/5872-efficient-and-robust-automated-machine-learning.pdf
13. Gomes, D., Moreira, D., Costa, J., Graça, R., Madureira, J.: Surf session events' profiling using smartphones' embedded sensors. Sensors **19**(14), 3138 (2019)
14. Hoettinger, H., Mally, F., Sabo, A.: Activity recognition in surfing-a comparative study between hidden markov model and support vector machine. Procedia Eng. **147**, 912–917 (2016)
15. Lee, G., Gommers, R., Waselewski, F., Wohlfahrt, K., O'Leary, A.: PyWavelets: a python package for wavelet analysis. J. Open Source Softw. **4**(36), 1237 (2019)

16. Lotte, F.: A new feature and associated optimal spatial filter for EEG signal classification: waveform length. In: Proceedings of the 21st International Conference on Pattern Recognition (ICPR2012), pp. 1302–1305. IEEE (2012)

17. Madgwick, S.O., Harrison, A.J., Vaidyanathan, R.: Estimation of IMU and MARG orientation using a gradient descent algorithm. In: 2011 IEEE International Conference on Rehabilitation Robotics (ICORR), pp. 1–7. IEEE (2011)

18. Madureira, J., Lagido, R., Sousa, I.: Comparison of number of waves surfed and duration using global positioning system and inertial sensors. World Acad. Sci. Eng. Technol. Int. J. Med. Health Biomed. Bioeng. Pharm. Eng. 9(5), 444–448 (2015)

19. Mendes Jr, J.J.A., Vieira, M.E.M., Pires, M.B., Stevan Jr., S.L.: Sensor fusion and smart sensor in sports and biomedical applications. Sensors 16(10), 1569 (2016)

20. Mendez-Villanueva, A., Bishop, D.: Physiological aspects of surfboard riding performance. Sport. Med. 35(1), 55–70 (2005)

21. Minghelli, B.: Time-motion analysis in surf: benefits (2018)

22. Rhif, M., Ben Abbes, A., Farah, I.R., Martínez, B., Sang, Y.: Wavelet transform application for/in non-stationary time-series analysis: a review. Appl. Sci. 9(7), 1345 (2019)

23. Rosso, O.A., Blanco, S., Yordanova, J., Kolev, V., Figliola, A., Schürmann, M., Başar, E.: Wavelet entropy: a new tool for analysis of short duration brain electrical signals. J. Neurosci. Methods 105(1), 65–75 (2001)

24. Salwani, M., Jasmy, Y.: Relative wavelet energy as a tool to select suitable wavelet for artifact removal in EEG. In: 2005 1st International Conference on Computers, Communications, & Signal Processing with Special Track on Biomedical Engineering, pp. 282–287. IEEE (2005)

25. dos Santos, M.B.: Surf biomechanics and bioenergetics (2018)

26. Scarfe, B., Elwany, M., Mead, S., Black, K.: The science of surfing waves and surfing breaks-a review (2003)

27. Ziaja, A., Antoniadou, I., Barszcz, T., Staszewski, W.J., Worden, K.: Fault detection in rolling element bearings using wavelet-based variance analysis and novelty detection. J. Vib. Control. 22(2), 396–411 (2016)

SDN-Enabled IoT Anomaly Detection Using Ensemble Learning

Enkhtur Tsogbaatar[1](), Monowar H. Bhuyan[1,2], Yuzo Taenaka[1],
Doudou Fall[1], Khishigjargal Gonchigsumlaa[3], Erik Elmroth[2],
and Youki Kadobayashi[1]

[1] Laboratory for Cyber Resilience, NAIST, Nara 630 0192, Japan
{tsogbaatar.enkhtur.ta4,yuzo,doudou-f,youki-k}@is.naist.jp
[2] Computing Science Department, Umeå University, Umeå, Sweden
{monowar,elmroth}@cs.umu.se
[3] SICT, MUST, 133 43 Ulaanbaatar, Mongolia

Abstract. Internet of Things (IoT) devices are inherently vulnerable due
to insecure design, implementation, and configuration. Aggressive behavior change, due to increased attacker's sophistication, and the heterogeneity of the data in IoT have proven that securing IoT devices is a making challenge. To detect intensive attacks and increase device uptime, we
propose a novel ensemble learning model for IoT anomaly detection using
software-defined networks (SDN). We use a deep auto-encoder to extract
handy features for stacking into an ensemble learning model. The learned
model is deployed in the SDN controller to detect anomalies or dynamic
attacks in IoT by addressing the class imbalance problem. We validate the
model with real-time testbed and benchmark datasets. The initial results
show that our model has a better and more reliable performance than the
competing models showcased in the relevant related work.

Keywords: Internet of Things (IoT) · Anomaly detection ·
Autoencoder · Probabilistic Neural Networks (PNN) · Software defined
network (SDN) · Ensemble learning

1 Introduction

The rapid evolution of the Internet of Things has brought billions of internet-
enabled devices into our daily life to make it smarter by bridging the gap between
the physical and the virtual world. Frost and Sullivan [7] have predicted that
the number of connected devices will increase by up to 45.41 billion by 2023.
Autonomous decision making, information to end-users, machine-to-machine and
machine-to-user interaction have boosted the acceptance of IoT as a critical
asset in the service chain. IoT systems open up several opportunities in areas of
autonomous transportation and industrial automation. As manufacturers hastily
produce new IoT devices without basic security and privacy checks; thus, allowing attackers to easily and swiftly identify vulnerabilities that allow them to

© IFIP International Federation for Information Processing 2020
Published by Springer Nature Switzerland AG 2020
I. Maglogiannis et al. (Eds.): AIAI 2020, IFIP AICT 584, pp. 268–280, 2020.
https://doi.org/10.1007/978-3-030-49186-4_23

evade, manipulate, and take over IoT networks. The failure to implement proper security and privacy measures have already resulted in dire consequences for certain IoT manufacturers and service providers in terms of reputation and financial penalties [5]. Hence, security is becoming crucial to protect IoT devices and applications from cyberattacks in both small and large-scale networks comprised of physical and virtual infrastructures.

SDN [10] attracts both academia and industry due to appealing features such as flexibility, dynamicity in network operations and resource management. By leveraging SDN, we can provide several attractive benefits for IoT security. Unfortunately, managing heterogeneous IoT networks and detecting dynamic attacks become challenging due to changing behaviour, constraint resources, and limited scalability. Hence, the ensemble learning model is developing a detection module within the SDN framework by: (i) isolating compromised devices, (ii) doing early detection and mitigation, (iii) reducing resource wastage, (iv) improving accuracy by deploying sophisticated algorithms. Mostly, detection models suffer class imbalance problems that incur significant performance loss for detecting attacks in IoT. This work takes the benefits of SDN controller at switches by grabbing the features of data back-and-forth into IoT devices.

To address these challenges, we present an SDN-enabled ensemble learning model for IoT anomaly detection that utilizes the appealing features of deep auto-encoders and probabilistic neural networks. Our proposal uses both device and network switching level features to build an efficient detection system, which will ensure the increase of IoT device uptime and performance. We make the following contributions.

- A novel SDN-enabled IoT anomaly detection using ensemble learning:
  - It utilizes the principles of deep autoencoders and probabilistic neural networks.
  - It can detect dynamic attacks and handle the class imbalance problem.
- Systematic and extensive experimental analysis using testbed and benchmark datasets, showing the proposed method is superior to competitors in terms of $F1$ measure.

2 Related Work

Several significant amounts of works [2,12–14] have been proposed to address the machine learning-driven IoT anomaly detection problems with and without the SDN. Bhunia et al. [2] present a machine learning-driven anomaly detection and mitigation method for IoT using the SDN. This method deploys the SVM model at the SDN controller to monitor and learn the behavior of IoT devices over time and detect attacks. The evaluation was ensured based on Mininet-based emulation by considering multiple attack scenarios and claimed that the method can detect and mitigate attacks within a few seconds. Zolanvari et al. [14] demonstrate the efficiency of Artificial Neural Network (ANN) in detecting anomalies with Industrial IoT (IIoT) testbed datasets having class imbalance. This method leverages Synthetic Minority Over-Sampling Technique (SMOTE) to address the class imbalance problem in IIoT datasets to achieve expected ANN performance. Recently, Thien et al. [13] propose an autonomous self-learning distributed system for detecting compromised IoT devices using GRU (Gated Recurrent Units) known as DÏoT. This method was evaluated with 33 IoT devices and demonstrates True Positive Rate (TPR) is 95.6% and can detect compromised devices within 257 ms.

However, still, several opportunities are left to address IoT security problems in small and large-scale industries with a comprehensive solution focus either centralized or distributed fashion. Some problems including increase uptime of devices that deployed in the edge of the network, device heterogeneity, reliability, and early detection of attacks irrespective of class imbalance issues. We are motivated to address two primary aspects including early detection of dynamic attacks and class imbalance problems using the data-driven mechanism.

3   System Model

We present a novel ensemble learning-based anomaly detection model for IoT using SDN primarily to detect anomalies or dynamic attacks for increasing device uptime and detection efficiency. This model aims to provide security of IoT devices by monitoring traffic and system metrics in SDN switches. Also, it can handle the class imbalance problem where attack classes are rarer than legitimate classes. An SDN-enabled IoT anomaly detection framework is given in Fig. 1. This framework has three primary components, including an SDN controller, SDN switches, and IoT devices. Further, the proposed framework has data collection and preprocessing, a learning module, a detection module, a flow management module, and the maintenance of a status table. The network operators employ the SDN-enabled framework to isolate the services, increase reachability, improve service-oriented policies at the switch level. The SDN controller disintegrates policies into service-specific rules and colonizes into flow tables of SDN-switches through the standard channel like OpenFlow [11]. Each packet is

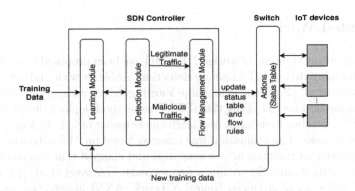

Fig. 1. SDN-enabled framework for IoT anomaly detection.

forwarded based on the enabled rules in the flow table. Each rule have three most common fields including *matching field, actions,* and *flow counters.* If a packet header matches a rule, then the controller must take actions (e.g., forwarding to a specific device) and update the counters immediately. We assume that the controller has complete information of network topology and can make a request for each counter rule from switches [6]. A new rule is installed or updated reactively when new flows come to the network without any matching rules. In the following subsections, we discuss the anomaly detection, flow management and maintenance of the status table.

3.1   Anomaly Detection

The proposed model aims to detect anomalies in IoT based on the dynamic observation of both packet and flow level traffic instances that pass through SDN switches as well as system metrics. We deploy the detection module that can monitor traffic and system metrics of deployed devices as well as applications for anomaly detection.

Learning Module: Anomaly detection models have a vital requirement to have aggregated data either at an endpoint or from multiple sources. The model adopts packet level, flow level, and system metrics data to detect anomalies in IoT. Because the attackers primarily target different performance or system metrics. We validate this model using both testbed and benchmark datasets. For testbed data, we set up a real-time testbed comprising multilevel hierarchical architecture from physical infrastructures to IoT devices. The preprocessing function extracts significant features that the attackers typically utilize and labels them based on the behavioral analysis. For the benchmark, we use a recent dataset, called N-BaIoT [12] for our experimentation. We provide extended explanations of each dataset in Sect. 4.

To make an efficient and better representation of features, we use autoencoder and deep feature representation by a non-linear transformation of features set before feeding data into the learning model. This module employs both legitimate and anomalous features or system metrics to learn the model for anomaly detection in IoT. Let's assume that $X = \{x_1, x_2, \cdots, x_n\}$ is the input data, $X' = \{x'_1, x'_2, \cdots, x'_n\}$ is the encoded output, $F = \{f_1, f_2, \cdots, f_n\}$ is the features set, and h is the hidden layers.

Autoencoder and deep feature representation are multilayer neural networks having multiple hidden layers, h, to encode the input and to reconstruct the output as similar as possible to the input. The network has two parts: an encoder, $h = f(x)$ and a decoder $r = g(h)$ [9]. We employ deep autoencoder (DAE) to extract and represent the features set obtained from the preprocessing function. The primary advantage of using deep autoencoder is having multiple hidden layers. More hidden layers incur better feature representation, which is advantageous for an anomaly detection model [1]. The under-complete autoencoders have a lower number of nodes in hidden layers.

Detection Module: This module employs the outcome of a learning module for detecting anomalies in IoT within an SDN-enabled framework. We explain the components of the detection module below.

Probabilistic Neural Networks (PNN) is a multilayered feedforward network with four primary layers, including input, pattern, summation, and output layers. The PNN is represented as a Kernel Discriminant Analysis (KDA), which is a generalization of Linear Discriminant Analysis (LDA) to find a linear combination of features that separates classes. A PNN consists of several sub-networks that estimate the Parzen window probability density function of each class using the samples of the training set. Each node of the network calculates the probability density function for each training sample according to the Eq. (1).

$$p(x) = \frac{1}{\sigma}\omega(\frac{x - x_i}{\sigma}) \tag{1}$$

where x_i is the ith sample and x is the input instance (unknown), $\omega()$ is the weighting function and σ is the smoothing parameter. The nodes are grouped according to the classes of the training sample in the pattern layer, and each group sums up for the next layer to get the class-wise probability. In summation layer, the lth nodes aggregate the values from the pattern layer of lth classes. This summation is estimated based on mixed Gaussian or Parzen window estimator as defined in Eq. (2).

$$f_l(x) = \frac{1}{n\sigma}\omega\sum_{n=1}^{n_l}(\frac{x - x_i}{\sigma}). \tag{2}$$

where n_l is the number of sample in l^{th} classes. Hence, the summation layer maps the lth nodes to the lth classes. For new samples, $f_l(x)$ can be estimated without retraining. The PNN needs more samples to achieve a high probability of mapping score from input instances to underlying classes where $f_l(x)$ has a maximum posterior probability of a class.

The weight function $\omega()$ is chosen as a kernel function (e.g., Radial Basis Function (RBF)) to compute the distance between the known and unknown sample points. If the distance is nearest, then it has more influence on the end class. The use of PNN provides multiple benefits, including insensitive to an outlier in the data, new input patterns stores in the network, and the smoothing parameter σ. Additionally, it reduces the retraining of the network if the training samples become large. Each sub-network of PNN implies a Parzen density estimator for a particular class. These features boost the detection of exceptional events in the data. However, it has observed that PNN alone cannot provide better results. Hence, we have used deep autoencoder probabilistic neural networks to detect anomalies in IoT.

Deep autoencoder probabilistic neural networks (DAE-PNN) is an autoencoder integrated with probabilistic neural networks. It encodes input using multilayer neural networks instead of stacked layers. In anomaly detection, anomalous classes are rare, whereas legitimate classes are frequent. Hence, binary classifier gets more biased performance [8]. Such classifiers can be used to refine the decision boundary between the rare and frequent classes. In the proposed method, we have used deep autoencoder PNN for encoding the input and feed them into PNN for classification. These inputs include samples from both rare and frequent classes. In anomaly detection, we have to make the trade-off between generalization and specialization to refine the decision boundary for achieving high accuracy. Most anomaly detection models are not specialized except just giving a bias to rare classes. The proposed model mitigates these drawbacks and gains substantial performance improvement thanks to ensemble learning. In PNN, the smoothing parameter σ determines the spread of RBF where it reaches a peak in the center of weighting. Selecting optimal values of σ implies a better spread of RBF in PNN. A shallow value of σ causes model to over-fit whereas an extremely high value may cause model to under-fit. However, both factors are essential to address the class imbalance problem during anomaly detection in IoT. These problems motivate us to develop an ensemble learning model, which we explain below.

Ensemble Learning employs multiple PNNs as weak classifiers to address the biases by fine-tuning the smoothing parameter σ. This model takes inputs from the encoded features of the deep autoencoder PNN to construct inputs for the next layer. Let's assume that $A = \{x_{a1}, x_{a2}, \cdots, x_{an}\}$ is the set of anomalous instances, $L = \{x_{l1}, x_{l2}, \cdots, x_{ln}\}$ is the legitimate instances, S is the training dataset, Y is the test dataset, tr and te indicate training and testing instances, then we can get S and Y as in Eq. (3):

$$S = A_{tr} \cup L_{tr}$$
$$Y = A_{te} \cup L_{te}$$

$$(3)$$

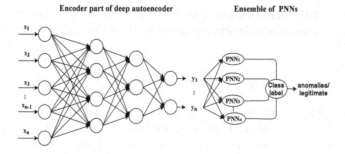

Fig. 2. The proposed model: an integration of deep autoencoder and ensemble of PNNs.

As the legitimate instances are more than attack instances, we divided the legitimate instances into N subsets and we kept anomalous instances as one class for training. We used N number of PNNs for ensemble learning, where i^{th} PNN is trained with i^{th} subset of datasets. Hence, we chose i^{th} PNN for training with $(i+1)^{th}$ PNN for binary classification. However, we chose N^{th} PNNs for the majority of the classes and one more PNN for an anomalous class, specifically for the multiclass problem. For this, we used deep autoencoder, since we know that autoencoder provides the non-linear transformation of input data to reduce features set. A pictorial representation of the proposed model is given in Fig. 2. Let f be a non-linear activation function with weight Ω and bias b then the deep autoencoder is formulated as in Eq. (4):

$$\xi = \left(f(\Omega x + b) \right)$$
$$DAE(x) = \left(\xi_1(\xi_2(\xi_3(\cdots \xi_h(x)))) \right)$$

(4)

where $\xi()$ is the encoding function whereas $\xi_i()$ is the ith deep encoder, h is the number of hidden layers, each feature vector x is transformed to $\hat{x}$ using $DAE(x)$ defined in Eq. (5).

$$z_c^n(\hat{x}) = f_c^n \left(DAE(x) \right)$$

(5)

Each input instance x is assigned to C classes based on intermediate RBF score received from the nth PNNs. However, if there are N PNNs for the ensemble, then each instance, x is assigned to C classes based on the Eq. (6). The classification score is computed using Eq. (7) for each instance x that belongs to a specific class.

$$z_c(x) = \sum_{n=1}^{N} (z_c^n(\hat{x}))$$

(6)

$$s_c(x) = \frac{z_c(x) - min(z_c(x))}{max(z_c(x)) - min(z_c(x))}$$

(7)

Once we get the classification score for each instance, we label the unknown instance x as anomalous or legitimate based on the node's maximum probability, $max(s_c(x))$.

3.2 Flow Management

This module is enabled to prepare appropriate rules and to dynamically update the switch as a set of actions. If the network flow is marked as legitimate, it can then pass the flow across the controller without any interruption. Otherwise, the controller blacklists the flow and investigates further to verify the types of attacks.

3.3   Maintenance of Status Table

This module takes input from the flow management module and generates profiles for each IoT device with the outcome of metrics such as packet, flow, and system. This table is comprised of three columns, including time, device ID, and status. The status of each IoT device is marked either as 0 (legitimate) or as 1 (anomalies) for each time point. The system manager utilizes this feature to easily handle large-scale IoT devices as well as their services to the end-users.

4   Performance Evaluation

This section reports and explains the intensive experimental results obtained from a real-time testbed and benchmark datasets. We begin with datasets description and proceed with experimental results.

4.1   Datasets

The proposed method is evaluated using two datasets: (i) real-time testbed data, and (ii) benchmark data. The real-time testbed data is generated with a significant amount of attacks on IoT devices and applications.

Real-Time Testbed Data: The experiment is performed in a virtualized environment with a hierarchical deployment of IoT devices to physical servers in the real-time testbed. Figure 3 illustrates the architecture of the real-time testbed setup. The testbed is comprised of multiple servers and applications at a physical level, virtualized level, IoT devices and IoT applications. We consider multiple VMs, one of the VMs is a target of attackers. We generate both DoS and DDoS attacks using the Targa2[1] attack generation tool, the D-ITG internet traffic generator tool [4], the BoNeSi botnet simulator[2], and the stress-ng[3] system resources load generator. We generate multiscale attacks [3] to the IoT devices and applications when they are deployed in the virtualized environment. We collect multiple metrics (e.g., packet, flow, system metrics, device uptime, etc.) from devices to physical infrastructures for learning the proposed ensemble model. POX[4] is an OpenFlow controller used to deploy the learning algorithm to detect anomalies in IoT devices based on dynamic policy updating within the SDN framework.

Benchmark Data: Due to the non-availability of benchmark datasets, we used the N-BaIoT [12] dataset for our experiments. This dataset was prepared using two attack tools, i.e., Mirai (scan, ACK flooding, SYN flooding, UDP flooding, UDPplain attacks) and Bashlite (scan, junk, UDP flooding, TCP flooding,

---

[1] http://packetstormsecurity.com/.

[2] https://github.com/Markus-Go/bonesi.

[3] https://kernel.ubuntu.com/~cking/stress-ng/.

[4] https://github.com/noxrepo/pox.

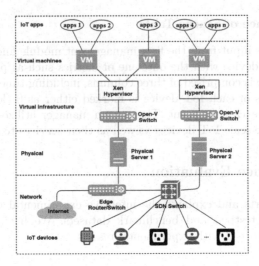

Fig. 3. Real-time testbed setup.

COMBO attacks), with 9 commercial IoT devices. There are 5 Bashlite, 5 Mirai, and 1 legitimate datasets, having 115 features in each of them. We created an imbalanced dataset by separating 98514 legitimate and 9850 anomalous instances with a combination of 10 different kinds of Mirai and Bashlite attacks.

4.2   Results

The proposed framework is evaluated using both real-time testbed and benchmark datasets. We present the results based on real-time testbed datasets that consider multi-level metrics such as packet, flow, and system metrics. To observe the behavior of both real-time testbed and benchmark datasets, we estimate the cumulative density for legitimate and attack instances, as shown in Fig. 4. From the Fig. 4, it is clear that the distribution of legitimate instances differs from attack instances.

In this experiment, we used deep autoencoder with an ensemble of PNNs with the 'adadelta' optimizer and 'binary-crossentropy' as a loss function. The model was pre-trained with 100 epoch, 100 batch size, and 'relu' activation function by considering four hidden layers with the compositions 17-16-16-15 for testbed data and 115-100-50-25 for benchmark data. These compositions indicate 17 features at the first layer and finish with 15 the encoded outcome from the deep autoencoder. Regarding the testbed and benchmark datasets, we have used non-overlapping 70% for training and 30% for testing, so that the class imbalance problem remains the same in both the training and the test set.

Dependency Analysis on σ: The proposed framework analyzes the dependency of σ to identify suitable values that illustrate the maximum detection

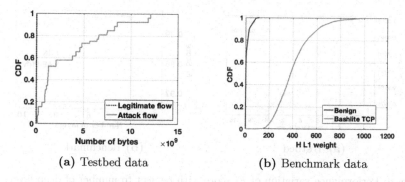

Fig. 4. Characterization of data: cumulative density for legitimate vs. attack instances.

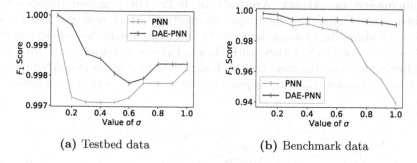

Fig. 5. Performance variation of F_1 score with respect to σ of PNN.

rate. The performance of PNN depends on the suitable values of σ. Notably, the lower values of σ provide a lower false-negative rate and higher values of σ yields a lower false-positive rate. We have heuristically identified the range of values for σ as 0.1–0.3 for real-time data and 0.1–0.3 for benchmark data, respectively. These experiments were conducted by considering imbalanced datasets. Figure 5 illustrates the performance in the real-time testbed and benchmark datasets.

Dependency Analysis on Deep Layers: We used deep autoencoder to extract features from both real-time testbed and benchmark datasets. As we know, an increasing number of hidden layers in the autoencoder excels the performance of the model. However, finding the optimal depth of hidden layers for a specific domain and address the class imbalance problem in parallel is still tricky. Figure 6 provides the empirical investigation on the number of deep layers (3 for real-time testbed and 4 for benchmark datasets) to acquire the best performance of the model. Our model recommends that deep neural networks improve the model performance significantly, but a vast number of layers may overfit the model.

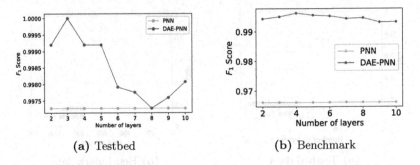

(a) Testbed   (b) Benchmark

Fig. 6. Performance variation of F_1 score with respect to number of deep layers.

Performance on Attack Analysis in IoT: The proposed framework is evaluated with intensive experiments for multiple attack scenarios using real-time testbed and benchmark datasets. In the testbed data, we consider multi-scale attacks (i.e., DoS, DDoS) with class imbalance scenarios. Also, we found multiple attack scenarios from the benchmark dataset. Figures 7a and b illustrate the improved performance of the proposed framework for detecting anomalies in IoT.

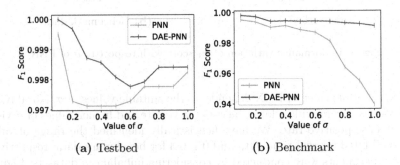

(a) Testbed   (b) Benchmark

Fig. 7. F_1 score variation with respect to σ using ensemble model for detecting anomalies in IoT.

4.3   Comparison with Competing Methods

To assess the efficiency of the proposed framework for detecting anomalies in IoT, we compared it with existing methods including SoftThings [2], network-based IoT anomaly detection [12], and DÏoT [13]. The SoftThings [2] can detect and mitigate dynamic attacks in IoT using SDN. The system achieves 98% precision. It employs Support Vector Machine (SVM) when detecting IoT attacks at the SDN controller in a mininet simulation environment. The network-based IoT anomaly detection [12] employs the deep autoencoders to detect anomalies in IoT. The system has been evaluated using testbed datasets and it achieves 100% detection rate. The DÏoT [13] reports an autonomous self-learning anomaly detection system for IoT. The system provides evidence to detect anomalies with 95.6% detection rate.

Although the existing methods have positive performance, the majority of existing methods or systems were evaluated either as testbed data or in simulated environments. Additionally, our framework shows its superiority in terms of the following points: (i) we detect anomalies with 99.8% detection rate for testbed and 99.9% detection rate for benchmark datasets, (ii) we address the class imbalance problem by using ensemble learning, (iii) we increase IoT device uptime.

5 Conclusion and Future Work

This work presents a novel ensemble learning model for IoT anomaly detection using SDN while addressing the class imbalance problems. Our model integrates the deep autoencoder to encode the features set and impedes them to an ensemble of PNNs for performance improvement of the baseline learning model. The proposed method is shown to perform better than the existing techniques for binary classification anomalies in IoT by employing the appealing features of SDN. We demonstrated superior performance as compared to existing methods in the real-time testbed and benchmark datasets with 99.8%.

In the future, we would like to extend our work to design and implement the deep ensemble learning model for multi-class anomaly detection in IoT. The deep ensemble learning model will consider the maximum composition of multi-scale attacks in IoT within SDN framework.

Acknowledgment. Part of this study was funded by the ICS-CoE Core Human Resources Development Program. Additional support was provided by the JST CREST Grant Number JPMJCR1783, Japan.

References

1. Berman, D.S., Buczak, A.L., Chavis, J.S., Corbett, C.L.: A survey of deep learning methods for cyber security. Information **10**, 122 (2019)
2. Bhunia, S.S., Gurusamy, M.: Dynamic attack detection and mitigation in IoT using SDN. In: 2017 27th International Telecommunication Networks and Applications Conference (ITNAC), pp. 1–6. IEEE (2017)
3. Bhuyan, M.H., Elmroth, E.: Multi-scale low-rate DDoS attack detection using the generalized total variation metric. In: 17th IEEE International Conference on Machine Learning and Applications, Orlando, Florida, USA, 17–20 December 2018 pp. 1040–1047. IEEE SMC (2018)
4. Botta, A., Dainotti, A., Pescapè, A.: A tool for the generation of realistic network workload for emerging networking scenarios. Comput. Netw. **56**(15), 3531–3547 (2012)

5. Farris, I., Taleb, T., Khettab, Y., Song, J.: A survey on emerging and NFV security mechanisms for IoT systems. IEEE Commun. Surv. Tutor. **21**(1), 812–837 (2018)
6. Foundation, O.N.: OpenFlow switch specification. Report ONF TS-023, Open Networking Foundation (2015)
7. Frost and Sullivan: IoT Security Market Watch-Key Market Needs and Solution Providers in the IoT Landscape, June 2017. https://store.frost.com/iot-security-market-watch-key-market-needs-and-solution-providers-in-the-iot-landscape.html
8. Ghosh, A.: Big data and its utility. Consult. Ahead **10**(1), 52–68 (2016)
9. Goodfellow, I., Bengio, Y., Courville, A.: Deep Learning, pp. 493–495. MIT Press, Cambridge (2017)
10. He, M., Alba, A.M., Basta, A., Blenk, A., Kellerer, W.: Flexibility in softwarized networks: classifications and research challenges. IEEE Commun. Surv. Tutor. **21**(3), 2600–2636 (2019)
11. McKeown, N., et al.: OpenFlow: enabling innovation in campus networks. SIGCOMM Comput. Commun. Rev. **38**(2), 69–74 (2008)
12. Meidan, Y., et al.: N-BaIoT - network-based detection of IoT botnet attacks using deep autoencoders. IEEE Pervasive Comput. **17**(3), 12–22 (2018)
13. Nguyen, T.D., Marchal, S., Miettinen, M., Fereidooni, H., Asokan, N., Sadeghi, A.R.: DÏoT: a federated self-learning anomaly detection system for IoT. In: IEEE 39th International Conference on Distributed Computing Systems, Dallas, Texas, USA, 7–9 July 2019, pp. 756–767 (2019)
14. Zolanvari, M., Teixeira, M.A., Jain, R.: Effect of imbalanced datasets on security of industrial iot using machine learning. In: 2018 IEEE International Conference on Intelligence and Security Informatics (ISI), pp. 112–117, November 2018. https://doi.org/10.1109/ISI.2018.8587389

Harnessing Social Interactions on Twitter for Smart Transportation Using Machine Learning

Narayan Chaturvedi(✉), Durga Toshniwal, and Manoranjan Parida

Centre for Transportation Systems, Indian Institute of Technology, Roorkee, India
narayanchaturvedi@gmail.com, {durga.toshniwal,m.parida}@ce.iitr.ac.in

Abstract. Twitter is generating a large amount of real-time data in the form of microblogs that has potential knowledge for various applications like traffic incident analysis and urban planning. Social media data represents the unbiased actual insights of citizens' concerns that may be mined in making cities smarter. In this study, a computational framework has been proposed using word embedding and machine learning model to detect traffic incidents using social media data. The study includes the feasibility of using machine learning algorithms with different feature extraction and representation models for the identification of traffic incidents from the Twitter interactions. The comprehensive proposed approach is the combination of following four steps. In the first phase, a dictionary of traffic-related keywords is formed. Secondly, real-time Twitter data has been collected using the dictionary of identified traffic related keywords. In the third step, collected tweets have been pre-processed, and the feature generation model is applied to convert the dataset eligible for a machine learning classifier. Further, a machine learning model is trained and tested to identify the tweets containing traffic incidents. The results of the study show that machine learning models built on top of right feature extraction strategy is very promising to identify the tweets containing traffic incidents from micro-blogs.

Keywords: Machine learning · Twitter data analysis · Traffic incident detection

1 Introduction

Rapid urbanization and an increased number of social media users attract the researchers for harnessing social media data to resolve the problems raised due to rapid urbanization. Recently, the government of developing countries like India has also been focusing on the need for smart cities to resolve the problems of urbanization. Smart cities also demand smart transportation and utilization of continuously generating social media data to improve city transportation. Detection of transport services disruptions, travel complains resolutions, seasonal messages to commuters, and more importantly transport event or incident detection

© IFIP International Federation for Information Processing 2020
Published by Springer Nature Switzerland AG 2020
I. Maglogiannis et al. (Eds.): AIAI 2020, IFIP AICT 584, pp. 281–290, 2020.
https://doi.org/10.1007/978-3-030-49186-4_24

are the typical applications of social media data harnessing for the field of transportation. In [17], the authors evaluated the potential of Twitter data for transit customer satisfaction and in [10], the authors proposed an approach for traffic incidents detection using social media data.

People generally express their daily activities using different social media platforms. However, the microblogging platform, like twitter, is most popular among all present microblogging websites. Microblogging service, Twitter has 328 million active users every month [16] who are tweeting at 230000 messages per minute [1]. Twitter is a cost-effective solution for generating a vast amount of real-time data. Relevant information extraction from Twitter data is the primary requirement to characterize the traffic events. In text mining, words of short messages are the tokens which are the basic unit of feature, and these tokens are the key to knowledge discovery from short messages. Since written tweet text cannot be directly utilized in machine learning models for tweet discrimination. Therefore, features of the tweets need to be converted into a numeric sparse matrix representation. The selection of an effective approach for feature extraction will improve the overall accuracy of the traffic event identification task.

The main objective of this study is to present and compare the text feature representation models and state of the art embedding to built a powerful machine learning based framework for the detection of traffic incidents. On the basis of the results acquired, we have proposed a comprehensive approach that works on social media data to extract patterns containing traffic incidents. The main contributions of our work are summarized as following: (1) To collect traffic-related social media data, a dictionary containing prominent traffic related keywords that are popular in social media and day to day communication of citizens, has been created. (2) Keyword filtration and manual approaches are applied to label the collected tweets for the training and testing of machine learning models. The paper proposes a computational framework based on a machine learning model to detect non-recurrent traffic causes.

Rest of the paper is organized as follows: Sect. 2 discusses the work done related to social media and transportation. Section 3 covers the details of data collection and labeling processes, and Sect. 4 explains the steps taken for preprocessing of collected data. Section 5 presents the methodology proposed and Sect. 6 contains the results of the study. Finally, the paper is concluded in Sect. 7.

2   Literature Review

Recently, several studies in the literature have been come out that utilizes social media data in various applications. Studies applied machine learning techniques on top of various non-numeric representation of text features and numeric sparse matrix feature representation to classify social media data containing traffic incidence/traffic-related information and data not related to traffic. A tweet contains many attributes like coordinates, creation time, language, place, timestamp, tweet text, etc. In [4], the authors analyzed not only tweet text but also other attributes of the tweet to demonstrate user activities and their moving

pattern. Along with other attributes, the tweet text has been given more importance in twitter-based studies. The traffic related microblogging messages were
retrieved using Support Vector Machine in [3]. The studies in [6,7,14] utilized
machine learning models for the detection of tweets containing traffic-related
information.

Several studies have been performed to classify the tweets containing traffic
information using machine learning (ML) models with numeric feature representation. To know the effect of feature extraction on ML models, there is a need
for this study which focuses on a feasibility study of bag-of-words, TF-IDF and
word2vec all three on three different machine learning models.

3   Data Collection

Domain-specific data collection is the first challenge in social media data based
studies in order to identify relevant tweets. The objective of labeling is to allocate
a class identity to every user posted tweet, as related to traffic incident or not.
Real-time publicly published tweets are collected using the Twitter streaming
API. The streaming API collects and filter tweets based on language, hash-tags,
keywords, and geographic bounding box.

In this study, we have collected twitter data based on geographical location
and identified general keywords matching specific hashtags such as road, accident, injury, potholes, congestion, jam related to different transportation services
for 26 March 2018 to 26 April 2018. The keywords have chosen from the news-
paper and research articles and also validated with transportation experts. Our
work focuses on tweets related to transportation services with the geographic
location of densely populated capital region Delhi of India for our study.

3.1   Dictionary Formation and Labeling

In order to collect traffic incident related tweets using Twitter streaming Api, a
dictionary of related keywords has been created. In the first phase of dictionary
formation, traffic incident related keywords have been collected manually from
the related literature, research papers and news articles. The manually collected
keywords have been validated by transport experts. Further, one synonym of
every keyword has been fetched from the wordnet dictionary database [12] using
python library. Adding synonyms doubles the number of keywords in the dictionary. The keyword dictionary is used to collect Twitter data.

To train a machine learning model, we need to label each tweet with a class
name. In this study, we have collected geo-tagged tweets based on traffic related
keyword but still, there are tweets collected that do not contain any traffic related
information. For example, some keywords like *accident* is a popular keyword for
road accident but many times people may use such keywords in other references
also. Because of this use of the same keyword in several different references,
those tweets that contain the keywords, but still are irrelevant for our study
get collected. For example, tweet: *jiocare is taking follows up from last2 years.*

But no improvement in connectivity and congestion at m... is irrelevant to our study but collected due to keyword *congestion*. To train a machine learning model to classify such tweets, tweets are manually labelled in to two classes: i. t - tweets that contain traffic event/incident ii. n - non-traffic tweets. Traffic related and non-traffic tweet classes are abbreviated as t and n, respectively.

The interesting factor of this study is the data collection strategy in which a dictionary of traffic-related keywords is formed and used to collect geo-tagged tweets to train the classification model. This approach of data collection represents that collected Twitter data-set contains both types of tweets: traffic-related and non-traffic related but most of the non-traffic tweets have different reference/domains containing some of the similar keywords of other class.

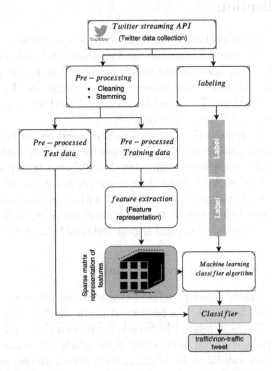

Fig. 1. Detailed steps of the proposed methodology.

4 Preprocessing

The main objective of the pre-processing is to make the tweets eligible for the classification task. Special characters like punctuation and stop words are frequently used in tweets and these stop words do not have any meaning. Therefore, the removal of such special characters and hyperlinks are the primary step of pre-processing. We have removed the # symbol, "@" symbol from the tweets. Tweets have been split into keywords (tokens) using Natural language Toolkit library. This process is known as tokenization of tweets. Further, upper case keywords could not be interpreted same so all keywords have been changed to lowercase. We performed Stemming to reduce the words to their stem. Thus, we have been changed each tweet into a bag of tokens eligible for text mining task.

5   Methodology

K Nearest Neighbour, Naive Bayes and Support Vector Machine are popular machine learning techniques that have been applied by researchers in text mining task [3,6,8,14]. The comprehensive methodology for detecting tweets consist of traffic events have shown in Fig. 1 with all the steps taken, starting from tweets crawling to built classifier. The Bag-of-words, TF-IDF and Word2Vec embedding have been applied in conjunction with machine learning models to construct an efficient framework for traffic incident detection.

5.1   Feature Extraction Model

The section gives the brief idea about the feature extraction model and word embedding applied in the study.

Bag-of-words (BOW) [2] is a simple technique to represent written text document with machine learning algorithms. BOW keeps the frequency of terms present in a text document and this term frequency is used to represent text in numeric form. We have used CountVectorizer from sklearn library to count the occurrences of words and further represent them in form of sparse matrix for machine learning technique implementation.

Term frequency-inverse document frequency (TF-IDF) is a statistical measuring technique to calculate the weight which decides the significance of the word in the document. Term frequency of a word w in a document of tweets T_n, $TF(w, T_n)$ can be calculated as per Eq. 1. IDF score of a word indicates the rareness of word in a written text document. IDF score of a word w which is present in T number of tweets out of total T_n collected tweets can be calculated as per Eq. 2. Further, TF-IDF weight of a word w is calculated to know the final normalized significance value as per Eq. 3.

$$TF(w, T_n) = \frac{count(w)}{word\_count(T_n)} \tag{1}$$

where count(w) is the frequency of word w in text document and $word\_count(T_n)$ is the total word count in text document.

$$IDF(w, T_n) = log_e \frac{T}{T_n} \tag{2}$$

$$TF - IDF(w, T_n) = TF(w, T_n) * IDF(w, T_n) \tag{3}$$

Word2Vec (W2V) is neural network-based word embedding tool in natural language processing to create a n-dimensional vector corresponding to every word of a text sentence. Continuous bag of words (CBOW) and skip-gram models are the two main approaches to train W2V model [11]. In the process of W2V model training, a vocabulary from tokenized training data creates in first phase. In second phase, similar text features are grouped based on cosine similarity distance. In this study, we have created separate W2V word vectors from the training and test dataset to train and set the machine learning classifier. To implement the feature extraction model and generate the numeric feature vector, gensim [15] python library has been used.

5.2   Machine Learning Model

The section covers the brief idea about the machine learning classifiers used in the study.

K-nearest neighbors is a non-parametric machine learning classifier which scores its nearest neighbors in training data and k-top-scored neighbors' class is used to classify the new input data [5,9]. The decision rule can be written as Eq. 4.

$$Score(t, c) = \sum_{t_n \epsilon KNN(t)} similarity(t, t_n) D(t_n, c) \tag{4}$$

$D(t_n, c)$ is the categorization for tweet t_n with respect to category c that is defined as Eq. 5.

$$D(t_n, c) = \begin{cases} 1, & \text{if } t_n \in c \\ 0, & \text{if } t_n \notin c \end{cases} \tag{5}$$

Naive Bayes (NB) classifier is based on an assumption that probability of one feature to be in a class is independent to the presence of any other feature in the class. Naive Bayes model is based on Bayes' theorem and performs well on big datasets. It is used in text classification [13] that uses a method of estimating the possibility of different classes based on different features of a written text document. MultinomialNB has been used to implement the classifier with BOW and TF-IDF model while the Gaussian distribution NB has been applied with W2V vectors because W2V vectors contain negative values and Gaussian NB works well with such values.

Support Vector Machine (SVM) is the supervised classification technique which categorizes the data. The basic SVM model training is performed as follows: 1. Plot features in a dimensional space where each plotted point co-ordinates are known as support vectors. 2. A separating line needs to be found in such a way so that each point should be farthest from the separating line. 3. In the testing phase, depending on the test, where the data is found on either side of the line, we can categorize new data.

6   Results and Analysis

This section evaluates the performance of machine learning classifiers with the combinations of feature representation models. The machine learning model training and the confusion matrix are also investigated.

6.1   Performance Metrics

In this study, mainly two popular classification metrics have been used to measure the performance of the classifier.

- Accuracy - the percentage/fraction of rightly classified twitter messages is known as the accuracy of trained model classifier.

$$Accuracy = \frac{TN + TP}{TN + TP + FN + FP} \tag{6}$$

Where TN is true negative, TP is true positive, FN is false positive and FP is false positive.
- Precision and Recall - Precision is the percentage/fraction of relevant tweets while recall is the percentage of actual positives rightly identified.

$$Precision = \frac{TP}{TP + FP} \tag{7}$$

$$Recall = \frac{TP}{TP + FN} \tag{8}$$

- F-measure - harmonic average of precision and recall. F-measure is the function of precision P and recall R and can be defined as in Eq. 9.

$$F - measure = \left(\frac{P^{-1} + R^{-1}}{2} \right)^{-1} \tag{9}$$

6.2   Classifier Performance

Table 1 shows the accuracy of various combinations of machine learning models with Bag-of-words, TF-IDF and W2V feature extraction models to categorize tweets containing traffic incidents. To find the value of number of neighbors K for the KNN classifier, the variation between the number of neighbors and misclassification error has been calculated. The value of K at which minimum value of misclassification error has been obtained is assumed to be the optimal value of K. The variation of Misclassification error and K has been shown in Fig. 2 for W2V. The optimal value of K = 3 in case of BOW, K = 5 in case of TF-IDF and K = 9 in case of W2V has been used to train the KNN classifier. The accuracy results in Table 1 clearly depicts the competative results of all three machine learning models. However, SVM with the combination of W2V

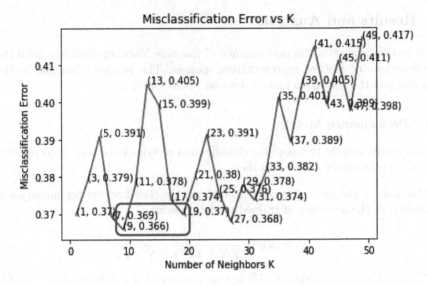

Fig. 2. optimal value of K in KNN classifier with W2V.

embedding technique outperforms other combinations of feature extraction and machine learning models by almost 3% and It successfully brings about the Precision, recall and F-measure of 77%, 84% and 80% for the class of tweets containing traffic incidents, respectively.

Since the SVM model with the W2V has achieved the best accuracy, this combination is treated as the representative of traffic incident detection approach from social interactions.

Table 1. Accuracy of the machine learning models (ML models), trained and tested on top of bag-of-words, TF-IDF and W2V techniques

ML model	Accuracy (%)		
	Bag-of-Words	TF-IDF	Word2Vec
KNN Classifier	62	64	73
Naive Bayes	61	64	71
Support vector machine	69	74	76

6.3   Real World Case Assessment and Confusion Matrix

In this study, we have collected geo-tagged twitter data for the capital region Delhi of India to evaluate the performance of machine learning based incident detection framework. However, our assessment has not been taken the case of real world in to consideration. The final objective of this study is to identify traffic incidents from the citizen's posted messages on micro-blogging platform

Table 2. Classification report(Y_{test}, *prediction*)

	Precision	Recall	F1-score
n	.75	.68	.71
t	.77	.84	.80
Micro avg.	.76	.76	.76
Macro avg.	.76	.76	.76
Weighted avg.	.76	.76	.76

like twitter. In order to detect traffic incidents from twitter streams in real time, the classifiers need to trained on labeled historical tweets and then future tweets can be analyzed for identifying traffic events. Since real time evaluation is a need for such studies, 70% of labeled tweets have been selected to train the model while remaining labeled tweets worked as test dataset. The classifier's performance is examined on the test data-set to conclude the optimal comprehensive approach.

Confusion matrix is the best way to analyse the predictions of both classes of tweets. Table 2 represents the precision, recall, F1-score, and confusion matrix for the tweets test set. Further, the effectiveness of the classifying model is directly proportional to precision, recall, and F1-score for both the classes. Some studies reported the lower precision and recall for traffic class tweets. The reason behind this lower value may be a lower number of traffic class tweets. However, in our case, we have different data collection strategy as described in Sect. 3 due to that we have collected a high number of traffic class tweets t in comparison to non-traffic class tweets number. Therefore, as depicted in Table 2, the performance metrics for non-traffic class n is obtained lower.

7   Conclusion

The study proposes a computational framework that includes a machine learning model built with word embedding technique to detect tweets that contain information related to traffic incidents. The method uses a text feature extraction and its transformation into numeric sparse matrix representation to implement text features on top of machine learning classifiers. Further, a dictionary of traffic-related keywords has been formed which is used to collect the tweets containing traffic incidents. Collected tweets are analyzed, pre-processed, and labeled to model a classifier for detecting tweets containing traffic incidents.

The results of the study present that neural network-based W2V shows slightly better results than TF-IDF and BOW for traffic incident detection. However, as one of the limitations of the W2V model, training of the model is much more complex which requires more processing and memory resource in comparison to TF-IDF model. A tradeoff occurs in choosing TF-IDF and W2V model in different applications. The study clearly identifies that the combination of W2V model with SVM classifier gives a best fit computational framework to detect traffic incidents from social media data.

References

1. Ashtari, O.: The super tweets of# sb47. Twitter. com Blog (2013)
2. Berry, M.W., Castellanos, M.: Survey of text mining. Comput. Rev. **45**(9), 548 (2004). https://doi.org/10.1007/978-1-4757-4305
3. de Carvalho, S.F.L., et al.: Real-time sensing of traffic information in twitter messages (2010)
4. Chaniotakis, E., Antoniou, C.: Use of geotagged social media in urban settings: Empirical evidence on its potential from twitter. In: 2015 IEEE 18th International Conference on Intelligent Transportation Systems, pp. 214–219. IEEE (2015)
5. Cover, T.M., Hart, P.E., et al.: Nearest neighbor pattern classification. IEEE Trans. Inf. Theory **13**(1), 21–27 (1967)
6. D'Andrea, E., Ducange, P., Lazzerini, B., Marcelloni, F.: Real-time detection of traffic from twitter stream analysis. IEEE Trans. Intell. Transp. Syst. **16**(4), 2269–2283 (2015)
7. Fu, K., Nune, R., Tao, J.X.: Social media data analysis for traffic incident detection and management. Technical report (2015)
8. Gu, Y., Qian, Z.S., Chen, F.: From twitter to detector: real-time traffic incident detection using social media data. Transp. Res. Part C Emerg. Technol. **67**, 321–342 (2016)
9. Han, X., Liu, J., Shen, Z., Miao, C.: An optimized k-nearest neighbor algorithm for large scale hierarchical text classification. In: Joint ECML/PKDD PASCAL Workshop on Large-Scale Hierarchical Classification, pp. 2–12 (2011)
10. Mai, E., Hranac, R.: Twitter interactions as a data source for transportation incidents. Technical report (2013)
11. Mikolov, T., Sutskever, I., Chen, K., Corrado, G.S., Dean, J.: Distributed representations of words and phrases and their compositionality. In: Advances in neural information processing systems, pp. 3111–3119 (2013)
12. Miller, G.A., Beckwith, R., Fellbaum, C., Gross, D., Miller, K.J.: Introduction to wordnet: An on-line lexical database. Int. J. Lexicogr. **3**(4), 235–244 (1990)
13. Mitchell, T.: Machine learning. mccraw hill, 1996. 93 d. moniere et d. labbé. essai de stylistique quantitative. In: JADT, pp. 561–569 (2002)
14. Pereira, J., Pasquali, A., Saleiro, P., Rossetti, R.: Transportation in social media: an automatic classifier for travel-related tweets. In: Oliveira, E., Gama, J., Vale, Z., Lopes Cardoso, H. (eds.) EPIA 2017. LNCS (LNAI), vol. 10423, pp. 355–366. Springer, Cham (2017). https://doi.org/10.1007/978-3-319-65340-2_30
15. Rehurek, R., Sojka, P.: Software framework for topic modelling with large corpora. In Proceedings of the LREC 2010 Workshop on New Challenges for NLP Frameworks. Citeseer (2010)
16. Sadam, R.: Twitter reports 6 pct increase in monthly active users. https://www.reuters.com/article/twitter-results/twitter-reports-6-pct-increase-in-monthly-active-users-idUSL4N1HY48L. Accessed 23 Jun 2019
17. Wu, B., Idris, A.O.: Measuring and visualizing transit customers' satisfaction using twitter data. Technical report (2018)

Medical-Health Systems

An Intelligent Cloud-Based Platform for Effective Monitoring of Patients with Psychotic Disorders

Ilias Maglogiannis[1]([envelope]), Athanasia Zlatintsi[2], Andreas Menychtas[1],
Dennis Papadimatos[1], Panayiotis P. Filntsis[2], Niki Efthymiou[2],
George Retsinas[2], Panayiotis Tsanakas[2], and Petros Maragos[2]

[1] Department of Digital Systems, University of Piraeus, Piraeus, Greece
{imaglo,amenychtas}@unipi.gr, dpapadimatos@gmail.com
[2] School of ECE, National Technical University of Athens, 15773 Athens, Greece
{nzlat,panag,maragos}@cs.ntua.gr,
{filby,nefthymiou,gretsinas}@central.ntua.gr

Abstract. The therapy of patients with psychotic disorders (i.e., bipolar disorder and schizophrenia) could benefit from the constant monitoring of their physiological and motor parameters. In this paper, we present an innovative and advanced cloud based platform that facilitates the effective monitoring of such patients. A commodity smartwatch is used for biosignal and motion data collection at a 24/7 basis. The paper describes the technical details of the implemented application both on the smartwatch and the cloud server side. Technical challenges regarding the upload, the storage and the battery constraints of the smartwatch are also discussed, along with the initial results regarding data visualization and processing.

Keywords: Healthcare platforms · Biosignal collection · Patient monitoring · Psychotic disorders · Motion analysis

1 Introduction

A key issue in the therapy of patients with psychotic disorders (i.e., bipolar disorder and schizophrenia) is the ability to constant monitor their physiological and mental status, even 24/7 if that is possible. In this research work, we present an innovative and advanced computer-based platform that facilitates the effective monitoring and can support the relapse prevention in such patients. This

This research has been co-financed by the European Regional Development Fund of the European Union and Greek national funds through the Operational Program Competitiveness, Entrepreneurship and Innovation, under the call RESEARCH – CREATE – INNOVATE (project code: T1EDK-02890, acronym: e-Prevention).

The original version of this chapter was revised: an author's first name was corrected to "Niki." The correction to this chapter is available at https://doi.org/10.1007/978-3-030-49186-4_38

I. Maglogiannis et al. (Eds.): AIAI 2020, IFIP AICT 584, pp. 293–307, 2020.
https://doi.org/10.1007/978-3-030-49186-4_25

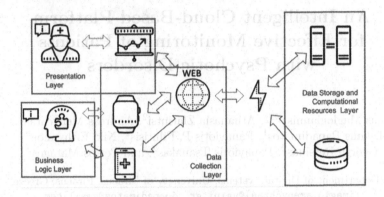

Fig. 1. The overall architecture of the proposed system.

research work is done within the e-Prevention research and development project that is coordinated by the National Technical University of Athens (https:// eprevention.gr/) and it is co-funded by the European Commission and Greek National funds.

The innovative key offerings of the envisioned system, are as follows: 1) long-term continuous recordings of biometric signals and motion data through simple wearable sensors (i.e., smartwatch), 2) a portable device (tablet) installed in the patient's residence, which records short-term audio-visual videos of the patient while communicating with the doctor; and by using affective computing methodologies is able to understand the emotional status of the patient, and 3) automatic and systematic storing and management of the captured data in a Cloud Infrastructure. The stored data are then processed through machine learning and signal processing techniques in order to detect changes and patterns, to facilitate the prediction of clinical symptoms and side effects of the patients medication.

The final user of the platform, who is the attending physician, is able to continuously monitor and optionally also annotate the data and receive information about the patient's daily mental and physiological status by observing the data from the wearable sensors and the emotion analysis of the short videos. Additionally, the computational intelligence algorithms incorporated in the platform enable big data processing and statistical analysis and prediction of salient events regarding the patient's daily routine along with the corresponding visualizations. The overall architecture of the proposed e-Prevention system is illustrated in Fig. 1.

2 Related Work

The evolution and rapid dissemination of mobile and wearable consumer technology, such as smartwatches and fitness trackers, has created unique opportunities for personalized data collection in an unobtrusive and even affordable way. Furthermore, the enormous technological advances have enabled the reliable recording and quantification of a large number of behavioral and biometric indexes through their sensors [23, 25]. Such sensors usually include accelerometers, gyroscopes and heart rate monitors among others for measuring and detecting the user's motion/kinetic activity, sleep patterns and autonomic function [9, 26], opening this way the possibility for non-intrusive acquisition of activity data.

Over the last years, wearable technology due to the inclusion of these sensors has in fact become more proximal to human activity, having as an eminent result to become increasingly acceptable even in healthcare. As early as 2006, the Institute of Medicine recognized the potential to transform mental health services by providing more continuous and precise information on patient-specific behavior, symptoms, and medication side effects. In order to do that, novel methods are demanded to be developed, to transform raw data into knowledge and then use this in turn to support personalized interventions [6] so as to transform hospital-centered healthcare practice to proactive, individualized care. Behavioral and biometric indexes have been already used in general medicine and sports and nowadays the evidence shows that they could be introduced into clinical psychiatry [6], as well. Specifically, psychosis is a spectrum of disorders that underlie different etiopathogenic mechanisms acting on the Central Nervous System (CNS), leading to common symptoms [31]. Despite extensive research over the last 60 years in neurobiology and neurophysiology of psychotic disorders, their cause remains unclear and reliable biometric indexes for the diagnosis and prediction of the course of the psychotic symptomatology have not yet been found. Based on the fact that the process of psychosis is continuous and relapse is a "biological" process that evolves over time [13, 21, 32], it would be expected to observe changes in such indexes that are related and likely precede the onset and/or worsening of psychotic conditions. To mention a few, some of the typical early warning signs of mental illnesses include parkinsonian type symptoms, i.e., rigidity, tremor, jerking arm movements, or involuntary movements of the limps, awkward gait, unusual gestures or postures, decreased physical activity and social interactions, abnormal sleep patterns [1, 5, 20]. The continuous monitoring assessed through such wearables, opens the way for more precise and personalized digital interventions, and complement by enabling detection of early signs of illness relapse, medication adherence or even treatment efficacy and thus may help increase the number of positive clinical outcomes in mental healthcare [10, 28].

Major research areas in psychiatry [7, 8, 17, 22] have suggested employing technological advancement for accurate and continuous monitoring of patients to reduce the impact of mental illness on a patient's daily activities, for early diagnosis and prevention of psychotic relapses so as to increase the effectiveness of treatments. For instance, sensor technology, has been used in order to understand patterns of daily behavior that may be indicative of trends or changes in factors related to mental health, including sleep, mood, and stress [15, 24], or to characterize a diverse array of psychopathology, from schizophrenia and depression to general mental health [11]. In [30] a combination of accelerometer and location sensors could detect relapse in depressive symptoms before patients reported such changes. It has also been found that such sensor technology could predict changes in affect and behavior (for a review, please see [15]).

3   System Architecture

The availability of commodity consumer devices (i.e., smartwatches) has eased motion data collection but in the context of the e-Prevention project the requirement to collect and transmit constantly large amounts of raw biosignal data did not fit the normal use of available consumer devices. We evaluated a number of available devices and selected the Samsung Gear S3 Frontier as the user data collection device. Compared to other smartwatches that were tested, the Samsung Gear has the ability to store and send data from acceleration, angular velocity and pulse sensors, providing also the heartbeat interval period value (R-R interval), while it has a large storage capacity (4 GB) capable of storing data for a few days and a WiFi interface removing the need for a smartphone for connectivity.

In addition, the smartwatch has the ability to run specific applications to collect data and upload them to a server. An application was developed in the Tizen Studio environment using technologies to implement Tizen applications (Tizen Hybrid Application) [2]. Hybrid Application applications consist of sections implemented either with native C code to achieve efficient operation or with javascript code to implement interfaces with available Tizen Advanced UI (TAU) libraries. The application implements smartwatch data collection functions and transmission to a cloud infrastructure. Due to the large volume of data produced, various compression techniques without loss of information were considered taking into account the limited computational and energy resources of the smartwatch. Analytics were also incorporated to gather critical smartwatch operating parameters such as power levels, free storage and network availability.

3.1   Architecture of the Smartwatch Application

The architecture of the smartwatch application is illustrated in Fig. 2 and consists of the following modules:

1. The Sensor Service module that interacts with the smartwatch sensors to collect data and store them temporarily on the device. Due to performance requirements it is implemented as a Tizen native service.
2. The Data Transmission Module (Network Service) that compresses and transfers the data to the server; also implemented as a Tizen native service.
3. The Control Service that coordinates the data collection and transmits units taking into account the state of the internet connection and the smartwatch energy status. It also aggregates the smartwatch operating parameters (analytics) for sending to the server.
4. The Tizen Web App interface that provides the user with basic information on the functionality of the application. The implementation utilizes advanced web design technologies based on TAU wearable interface libraries.

3.2   Biosignal Data Collection and Energy Consumption of Smartwatch

The sensor data are provided by the Tizen operating system in various ways dependent on the characteristics of the measured parameters. Extensive testing during the pilot development of the system leaded to the optimum data sampling rate in relation to the optimum time between successive battery charges.

Table 1 shows the sampling frequencies that were selected according to the requirements of the project's medical team to ensure continuous recordings between charges of approximately 24 h.

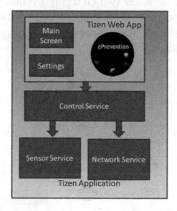

Fig. 2. Smartwatch application architecture.

Step data and sleep duration are also stored as calculated by the smartwatch, while GPS tracking was not used as it significantly increased power consumption and reduced the time required between successive loads. The actual output data amounts to about 300 MB/user/day, where the day of use includes approximately 22 h of continuous recording. The data is compressed at about 100–120 MB/user/day. About 20 users require about 2.5 GB of storage per day. In total, 24 months of compressed logging require approximately 1.7 TB of storage.

Table 1. Biosignal collection parameters.

Data parameter/hr	Sampling frequency	Data volume
Linear acceleration	20 Hz	ca. 7 MB/hr
Angular velocity	20 Hz	ca. 7 MB/hr
Pulse data	5 Hz	ca. 1.5 MB/hr

3.3 Data Collection Module

The requirements for collecting large volumes of data from smartwatch sensors require an appropriate software architecture based on efficient event-driven sensor management. Tizen uses an asynchronous programming interface (API) for sensors based on function callbacks that allow data transfer when available. The data collection module initializes the sensors required and attaches appropriate management routines to transfer the data and store it along with a timestamp in the smartwatch storage.

Data are organized into separate folders that contain files depending on the sensor type as well as timing information. Each data structure corresponds to a smartwatch usage period between two charging events. The above organization allows smartwatch data to be cached if there is no network connection or data upload problems on the server. Also independent storage in separate structures reduces the risk of data loss.

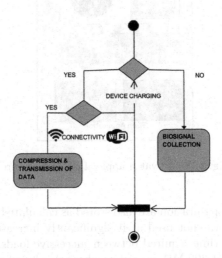

Fig. 3. Flowchart of data collection and transmission.

3.4 Data Transmission Module

The watch has advanced connectivity to local wireless WiFi networks but the volume of data requires efficient management of the data upload. The upload process is demanding on watch resources especially concerning energy, so it only takes place when the smartwatch is in the charging state.

Due to the high volume, the data collection module ensures efficient storage with minimal processing (Fig. 3). The data are stored in the smartwatch's storage in chronological order and are organized into groups corresponding to the intervals between device charging events. The available smartwatch storage capacity (approx. 4 GB) allows continuous recording and storage of measurements corresponding to 8 days. If the smartwatch is connected to a network, then as described below the data are transferred to the cloud and this time is renewed.

Transmission is organized into packets and compressed with the DEFLATE algorithm as implemented by the zlib library [3]. This library is a reliable solution with efficient use of the limited smartwatch computing resources and it is supported by a middleware on the node.js server platform. Running the decompression processes on the server concurrently allows data to be received from multiple smartwatches simultaneously.

3.5 Architecture of the Biosignal Collection Server

The server uses the '∼okeanos' national public infrastructure-as-a-service (IaaS), see Fig. 4 for the server architecture. The application is implemented with node.js technologies and uses the NoSQL MongoDB database. The application runs in an isolated Docker container and is structured as follows:

1. Authentication Controller that implements system resources access policies according to user roles.

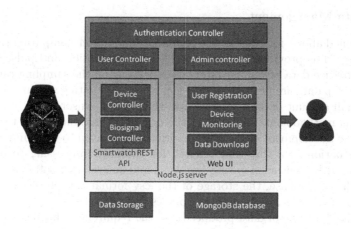

Fig. 4. Biosignal Collection server architecture.

2. Device Controller that implements the process of registering new devices in the system and gathering watch information on the functionality of the connected devices.
3. Biosignals Controller that manages the efficient transfer of biosignal data.
4. Admin Controller that implements the system management and monitoring.
5. The WEB UI module that constitutes the interface of the administrators with the application described in Sect. 4.1.
6. Web REST interface (REST interface) that routes REST calls from the devices and Web UI application to the appropriate module.

4   The System in Practice

4.1   Web UI for System Administrators

The Administrator Web UI interface (Fig. 5) allows devices to monitor and access downloaded data. It consists of a client subsystem implemented with front-end responsive frameworks such as Bootstrap and Vue.js and a server subsystem implemented with node.js. The portal features are: i) access to user log data, ii) new device registration, iii) e-mail notification system to administrators when a device has not uploaded data for more than 1 day, and iv) watch analytics.

The alert system uses information from the Device Controller module concerning device connectivity to generate alerts to administrators. The administrator interface allows access to functionality concerning activity alerts, user data and smartwatch analytics. It also provides access to data collected from the accompanying Android application used to collect activity data, which are manually annotated by the users.

4.2   Data Management

As mentioned above, for efficient use of limited smartwatch computing resources, data is stored as provided by the available sensors. Motion and pulse tracking sensors produce data at regular intervals depending on the sampling rate, while step and sleep data are provided by the smartwatch operating system comprising a very small amount of data compared to the original sensor data.

Various methodologies for storing the original sensor data were tested, but prior to the development of the system, priority was given to recording all available information in a way that allowed for thorough processing to eliminate errors caused by sensor operating problems and smartwatch software. In addition, in the pilot phase, the storage of the .csv format was chosen to save the processing time so as not to limit the processing capabilities.

The sheer volume of data produced and the requirement for flexible ways of securely accessing data create high storage requirements; thus was mitigated by the use of proven data compression techniques. Structuring the application of the

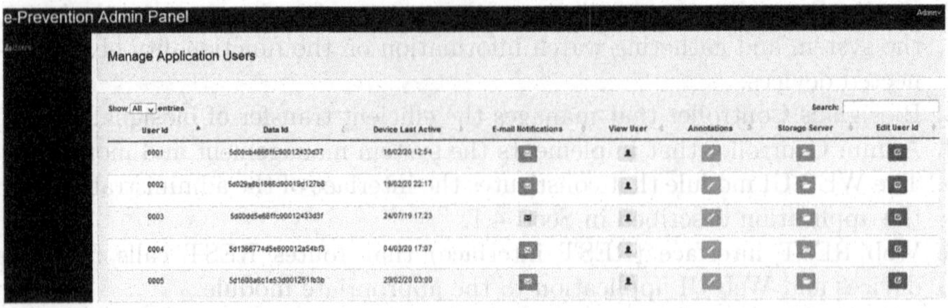

Fig. 5. Administrator Web UI.

server as a set of "micro services" for data transfer, without loss of information, entails the transfer of biosignals from the efficient processing and storage process. As part of the pilot process, the data is stored either uncompressed in the form of .csv files for immediate local processing or as compressed packets for a specific data collection time period ready for off-server transfer.

The need to store a large volume of data led to the creation of a new virtual machine in the ~okeanos infrastructure for the ultimate storage of the compressed data, and it is available for further processing through an appropriate sftp interface.

5 Initial Data Processing and Results

5.1 Qualitative Analysis

We performed an exploratory data analysis (i.e., kinetic and heart rate data), employing both traditional signal processing techniques, such as short-time analysis, as well as more sophisticated non-linear methods; i.e., multi-scale fractal and non-linear dynamics analysis, in order to extract descriptors that efficiently convey behavioral and biometric information. Such non-linear descriptors have been shown to be of importance for the modeling of a series of other 1D signals, see as for instance [16,19,27,33]). Our first preliminary results validate the efficiency of the features, thus they could be employed for building tools that can be used for more complex pattern extraction, i.e., behavioral changes, that can be correlated with mental health issues or relapses. Figure 6 shows examples

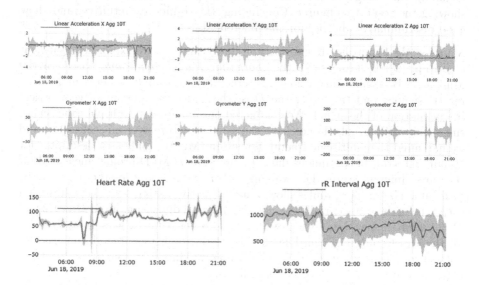

Fig. 6. Visualization of user movement and heart rate data, i.e., accelaration data (top row), gyroscope data (middle row) for all three axis (x, y, z-axis) and data collected by the heart rate monitor (i.e., heart rate and RR-interval data). (Color figure online)

of the raw data from the accelererometer (acc), gyroscope (gyr) and heart rate monitor (hrm) collected form a control user. All data is shown for a 24 h interval (where the horizontal blue line indicates when the user was asleep), so obvious differences can be observed.

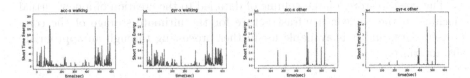

Fig. 7. Short Time Energy of the acc and gyr signals (x-axis) for two different states (walking and other, e.g., resting, standing), using a 10 s sliding window.

Short-Time Energy Analysis: We extracted the short-time energy, to explore differences when the user is walking, sleeping, or doing other activities. Figure 7 presents the calculated short-time energy, on the acc and gyr signals (x-axis), showing two 10-min. intervals corresponding to the different states of walking vs. other activity. The short-time energy is computed using a 400-sample window applied on the signal and provides insights for the signal content. We can observe that the short-time energy presents a multitude of peaks in the acc and gyr data during walking, and fewer peaks with higher values of energy when doing other activities. On the contrary, during sleep, the energy was very low, mainly including the energy of the noise produced by the sensors, thus we omitted showing the respective figures. Concluding, the results are intuitive, and show good discriminative abilities for the different states.

Multi-scale Fractals: Regarding non-linear analysis, we explored an efficient algorithm [18] that measures the *short-time fractal dimension*, based on the Minkowski-Bouligand dimension [12]. Fractal dimension D is between 1 and 2 for 1D signals; and the larger the D is, the larger the amount of geometrical fragmentation of the signal. We conducted our analysis at multiple time scales, since real-world signals do not have the same structure over all scales [18], and we measured the multiscale fractal dimension (MFD) profile as a descriptor for 10-min. intervals of continuous data. Figure 8 shows MFD profiles measured for the three different states, i.e., walking, other and sleeping for the acc and the hrm data. The various profiles show quite distinctive patterns for every state and each type of data. Specifically, we note high fractal dimension up to $D = 1.2$ and $D = 1.4$ for the acc data during walking and other, while it gets up to $D = 1.6$ for the acc-sleep data. Regarding the hrm data, the higher fractal dimension is noticed for sleeping (up to $D = 1.4$), while walking is around $D = 1$ getting higher and up to $D = 1.3$ for larger scales s.

Nonlinear Dynamics and Attractor Reconstruction: Next, we considered that the human movements, as well as the human heart rate system, constitute *separate* nonlinear dynamics systems. The two variables that must be calculated

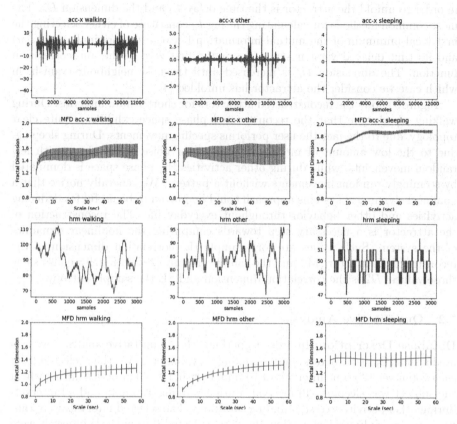

Fig. 8. Raw signals and mean and std (error bars) of the multiscale fractal dimension distribution of the acc (x-axis) and the hrm data for walking, other and sleep (the MFDs are shown for 60-s analysis windows, updated every 30 s).

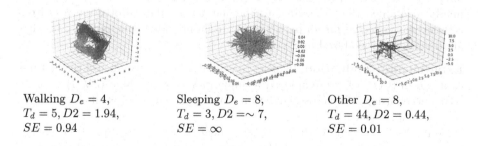

Walking $D_e = 4$, $T_d = 5, D2 = 1.94$, $SE = 0.94$

Sleeping $D_e = 8$, $T_d = 3, D2 =\sim 7$, $SE = \infty$

Other $D_e = 8$, $T_d = 44, D2 = 0.44$, $SE = 0.01$

Fig. 9. Reconstructed attractors for the three states. Each additional dimension above 3 and up to 6 is shown using different colors. D2 is the correlation dimension, and SE the sample entropy.

in order to unfold the attractor is the time delay T_d and the dimension D_e. The most common methods for calculating the lag T_d constitute of either selecting the first local minimum of the mutual information between the original signal $s(n)$ and its time-delayed version $s(n + T_d)$ or the first zero of the autocorrelation function. The dimension D_e is increased until the false neighbours vanish, in which case we consider the attractor has unfolded [4].

Figure 9 shows the reconstructed attractors for short 1-min. intervals. During walking we observe that the reconstructed phase space exhibits a quite clear topology, due to the fact the user performs specific movements. During sleeping, due to the low amount of movement, the sensor noise is prevalent, with few random movements, while during other activities the phase space is dominated by seemingly random movements without a pattern. We generally notice that a nonlinear dynamical systems approach could be proven to be useful for deducing activities and other behavior throughout everyday life. The reconstruction of the attractor is a necessary step towards computing the nonlinear dynamics of the system. Such values are for example the correlation dimension of the reconstructed attractor [14], and the sample entropy [29]. In Fig. 9 we also show those values; $D2$ is the correlation dimension and SE the sample entropy.

5.2   Quantitative Analysis

Database Description: In order to perform the quantitative analysis, we created a small dataset by randomly sampling 10 volunteers. We then selected from each volunteer 120 time intervals of 10-min. length, i.e., 40 intervals during which the subject was sleeping, 40 intervals during walking/running, and 40 intervals during other activities (rest, standing, working). The selected intervals were random, and we only made sure that the data collected during these intervals were valid, since in some cases, sensor malfunctions resulted in data loss, or the heart rate monitor was unable to accurately detect the heart rate. The analysis that follows use this dataset and attempts to identify useful cues/features, that can be used for accurately detecting the state of a person (sleeping, walking, running), so that later on can generalize to other physiological states. It should be finally mentioned that according to the the act for data protection a written consent was collected for all participating users providing the motion data that are describe and analyzed in Sect. 5.

Experimental Evaluation: In order to validate the performance of the features explored in Sect. 5.1, we employed a Support Vector Machine (SVM) with an RBF kernel, using 5-fold cross validation. Table 2 presents the results for different subset combinations.

With STE we denote the statistics (mean and std) of short-time energy analysis performed on the: 3 accelerometer (x, y, z-axis), 3 gyroscope (x, y, z-axis), and 1 heart rate signals. By taking the mean and std of the short-time energy calculated over each interval we have a total of 14 features for each interval. CH denotes the correlation dimension and the sample entropy calculated by performing delay embedding with $D_e = 4$ for the kinetic data and

Table 2. Results on the collected data for various linear and non-linear features.

Features	Acc (%)	Features	Acc (%)
STE	72.17	STE + CH	74.58
MFD-G	72.08	STE + MFD-G	76.92
MFD-A	74.16	STE + MFD-GA	**77.66**
MFD-H	60.50	STE + MFD-GAH	77.33
CH	65.30	STE + MFD-GAH + CH	77.58

$D_e = 2$ for the heart rate data. MFD-G, MFD-A, and MFD-H denote features from the multi-scale fractal analysis of the gyroscope (x-axis), the accelerometer (x-axis), and the heart rate, respectively. In order to reduce the number of features extracted by the MFD analysis (900 dimensions for kinetic data and 300 for heart rate), we perform linear sampling, picking a total of 30 points for the heart rate and 32 points for the accelerometer and gyroscope MFD analysis.

As we can see from the accuracy (%) results in Table 2, both the traditional speech processing methods, such as the STE analysis, as well as the multi-scale fractal analysis can be used in order to discriminate between the different states. The same can be observed using the CH features which combined with STE, increase the accuracy of the system. Our results show that the extracted descriptors examined in this paper can provide useful insights and can be used in order to build a tool able to assess the state of the user.

6   Conclusions

Concluding, the novelty of this work is the implementation of a cloud based platform for the extraction of key markers and indicators that can possibly assist the medical personnel in monitoring patients with mental disorders and predict psychotic symptom's relapses and adverse medicine side effects. This is done initially by long-term recordings of biosignal and motion related indexes and further processing. As this project evolves the collected data will be combined with the short-time videos and their affective analysis as well as with responses of patients to standard questionnaires. Nevertheless, the results so far prove the feasibility of our research effort in the continuous collection of biometric and motion data along with their prognostic value.

References

1. Schizophrenia.com. http://schizophrenia.com/
2. https://developer.tizen.org/ko/development/guides/native-application/location-and-sensors/device-sensors (2019)
3. http://zlib.net (2019)
4. Abarbanel, H., Frison, T.W., Tsimring, L.S.: Obtaining order in a world of chaos [signal processing]. IEEE Signal Process. Mag. **15**(3), 49–65 (1998)

5. American Psychiatric Association.: American Psychiatric Association: Desk reference to the diagnostic criteria from DSM-5®. American Psychiatric Pub, Washington, DC (2014)
6. Aung, M.H., Matthews, M., Choudhury, T.: Sensing behavioral symptoms of mental health and delivering personalized interventions using mobile technologies. Depress. Anxiety **34**(7), 603–609 (2017)
7. Bauer, M., et al.: Self-reporting software for bipolar disorder: validation of chronorecord by patients with mania. Psychiatr. Res. **159**(3), 359–366 (2008)
8. Blum, J., Magill, E.: M-psychiatry: sensor networks for psychiatric health monitoring. In: Proceedings of The 9th Annual Postgraduate Symposium The Convergence of Telecommunications, Networking and Broadcasting, Liverpool John Moores University, pp. 33–37. Citeseer (2008)
9. Boletsis, C., McCallum, S., Landmark, B.F.: The use of smartwatches for health monitoring in home-based dementia care. In: Zhou, J., Salvendy, G. (eds.) ITAP 2015. LNCS, vol. 9194, pp. 15–26. Springer, Cham (2015). https://doi.org/10.1007/978-3-319-20913-5_2
10. Collins, P.Y., et al.: Grand challenges in global mental health. Nature **475**(7354), 27–30 (2011)
11. Cornet, V.P., Holden, R.J.: Systematic review of smartphone-based passive sensing for health and wellbeing. J. Biomed. Inform. **77**, 120–132 (2018)
12. Falconer, K.: Fractal Geometry: Mathematical Foundations and Applications. Wiley, Chichester (2004)
13. Gaebel, W., et al.: Early neuroleptic intervention in schizophrenia: are prodromal symptoms valid predictors of relapse? Br. J. Psychiatr. **163**(S21), 8–12 (1993)
14. Grassberger, P., Procaccia, I.: Measuring the strangeness of strange attractors. Physica D **9**(1–2), 189–208 (1983). https://doi.org/10.1007/978-0-387-21830-4_12
15. Javelot, H., et al.: Telemonitoring with respect to mood disorders and information and communication technologies: overview and presentation of the psyche project. BioMed Res. Int. **2014**, 12 (2014)
16. Kokkinos, I., Maragos, P.: Nonlinear speech analysis using models for chaotic systems. IEEE Trans. Speech Audio Process. **13**(6), 1098–1109 (2005)
17. Koutsouleris, N., et al.: Early recognition and disease prediction in the at-risk mental states for psychosis using neurocognitive pattern classification. Schizophr. Bull. **38**(6), 1200–1215 (2011)
18. Maragos, P.: Fractal signal analysis using mathematical morphology. In: Advances in electronics and electron physics, vol. 88, pp. 199–246. Elsevier (1994)
19. Maragos, P., Potamianos, A.: Fractal dimensions of speech sounds: Computation and application to automatic speech recognition. JASA **105**(3), 1925–1932 (1999)
20. Maxhuni, A., et al.: Classification of bipolar disorder episodes based on analysis of voice and motor activity of patients. Pervasive Mobile Comput. **31**, 50–66 (2016)
21. McCandless-Glimcher, L., McKnight, S., Hamera, E., Smith, B.L., Peterson, K.A., Plumlee, A.A.: Use of symptoms by schizophrenics to monitor and regulate their illness. Psychiatr. Serv. **37**(9), 929–933 (1986)
22. McGorry, P., et al.: Biomarkers and clinical staging in psychiatry. World Psychiatr. **13**(3), 211–223 (2014)
23. Menychtas, A., Tsanakas, P., Maglogiannis, I.: Automated integration of wireless biosignal collection devices for patient-centred decision-making in point-of-care systems. Healthc. Technol. Lett. **3**(1), 34–40 (2016)
24. Mohr, D.C., Zhang, M., Schueller, S.M.: Personal sensing: understanding mental health using ubiquitous sensors and machine learning. Annu. Rev. Clin. Psychol. **13**, 23–47 (2017)

25. Panagopoulos, C., et al.: Utilizing a homecare platform for remote monitoring of patients with idiopathic pulmonary fibrosis. In: Vlamos, P. (ed.) GeNeDis 2016. AEMB, vol. 989, pp. 177–187. Springer, Cham (2017). https://doi.org/10.1007/978-3-319-57348-9_15

26. Patel, S., Park, H., Bonato, P., Chan, L., Rodgers, M.: A review of wearable sensors and systems with application in rehabilitation. J. Neuroeng. Rehabil. 9(1), 21 (2012)

27. Pitsikalis, V., Maragos, P.: Analysis and classification of speech signals by generalized fractal dimension features. Speech Commun. 51(12), 1206–1223 (2009)

28. Reinertsen, E., Clifford, G.D.: A review of physiological and behavioral monitoring with digital sensors for neuropsychiatric illnesses. Physiol. Meas. 39(5), 05TR01 (2018)

29. Richman, J.S., Moorman, J.R.: Physiological time-series analysis using approximate entropy and sample entropy. Am. J. Physiol. Heart Circ. Physiol. 278(6), 2039–2049 (2000)

30. Saeb, S., et al.: Mobile phone sensor correlates of depressive symptom severity in daily-life behavior: an exploratory study. J. Med. Internet Res. 17(7), e175 (2015)

31. Van Os, J., Kapur, S.: Schizophrenia. seminar. Lancet 374, 635–645 (2009)

32. Wiersma, D., Nienhuis, F.J., Slooff, C.J., Giel, R.: Natural course of schizophrenic disorders: a 15-year followup of a dutch incidence cohort. Schizophr. Bull. 24(1), 75–85 (1998)

33. Zlatintsi, A., Maragos, P.: Multiscale fractal analysis of musical instrument signals with application to recognition. IEEE Trans. Audio Speech Lang. Process. 21(4), 737–748 (2012)

Applying Deep Learning to Predicting Dementia and Mild Cognitive Impairment

Daniel Stamate[1,2,4(✉)], Richard Smith[1,2], Ruslan Tsygancov[1], Rostislav Vorobev[1], John Langham[1,2], Daniel Stahl[3], and David Reeves[4]

[1] Data Science & Soft Computing Lab, London, UK
d.stamate@gold.ac.uk
[2] Computing Department, Goldsmiths, University of London, London, UK
[3] Department of Biostatistics and Health Informatics, King's College London, London, UK
[4] Division of Population Health, Health Services Research and Primary Care,
University of Manchester, Manchester, UK

Abstract. Dementia has a large negative impact on the global healthcare and society. Diagnosis is rather challenging as there is no standardised test. The purpose of this paper is to conduct an analysis on ADNI data and determine its effectiveness for building classification models to differentiate the categories Cognitively Normal (CN), Mild Cognitive Impairment (MCI), and Dementia (DEM), based on tuning three Deep Learning models: two Multi-Layer Perceptron (MLP1 and MLP2) models and a Convolutional Bidirectional Long Short-Term Memory (ConvBLSTM) model. The results show that the MLP1 and MLP2 models accurately distinguish the DEM, MCI and CN classes, with accuracies as high as 0.86 (SD 0.01). The ConvBLSTM model was slightly less accurate but was explored in view of comparisons with the MLP models, and for future extensions of this work that will take advantage of time-related information. Although the performance of ConvBLSTM model was negatively impacted by a lack of visit code data, opportunities were identified for improvement, particularly in terms of pre-processing.

Keywords: Dementia prediction · Artificial Neural Networks · Deep Learning · Multi-Layer Perceptron · ConvBLSTM · ReliefF · SMOTE

1 Introduction

Dementia is a loss of cognitive function. People with this condition have problems with socialization and with their daily lives. The disorder leads to damage or loss of memory, language skills, problem-solving, visual perception, self-management, and the reduced ability to pay attention and to focus. Some people with dementia cannot control their emotions, and their personalities may change. Alzheimer's disease accounts for 60% to 80% of cases of dementia [1].

Worldwide, more than 47.5 million people suffer from dementia currently, and every year there are 7.7 million new cases. The proportion of the general population aged 60

© IFIP International Federation for Information Processing 2020
Published by Springer Nature Switzerland AG 2020
I. Maglogiannis et al. (Eds.): AIAI 2020, IFIP AICT 584, pp. 308–319, 2020.
https://doi.org/10.1007/978-3-030-49186-4_26

and over with dementia is estimated to be between 5% and 8% [1]. For comparison, dementia is currently more expensive in terms of health care and social assistance than cancer, stroke and chronic heart disease taken together [1]. It is estimated that by 2030, the population with dementia will be approximately 75 million and this condition will cost society US\$ 2 trillion [2]. There is no treatment at the present to cure dementia, however, the life of those patients can be supported and improved. The principal goals for dementia care are: providing information and long-term support to caregivers; optimising cognition, activity, physical health and well-being; detecting and treating behavioural and psychological symptoms; and early diagnosis. In terms of diagnosing dementia, the earlier a person receives a correct diagnosis, the sooner help can be provided. There is however a significant obstacle in diagnosing dementia because there is no standardized test for detecting it [3]. On the other hand, current thinking suggests that 35% of cases of dementia could be prevented if personalized risk could be estimated in advance, helping early informed decisions by clinicians and patients. As such, in addition of efforts to improve dementia diagnosis, there is lately a particular focus of researchers also in developing approaches to predicting risk of dementia [10, 19, 21].

This work proposes a machine learning approach to predicting dementia, in particular based on deep learning. The health care sector is one of the most important areas for machine learning applications. However, it is one of the most complex fields [4] and one of the most challenging, especially in the areas of diagnosis and prediction [5].

According to literature, there are hundreds of possible predictors for dementia, which can generally be categorized based on the following types of models: neuropsychological based models, health-based models, multifactorial models and genetic risk scores [19]. The applicability of these models spreads in multiple directions [16, 20, 21]. The magnetic resonance imaging (MRI), in combination with multiplex neural networks, have been used to segregate healthy brains from progressive mild cognitive impairment (pMCI), based on the structural atrophy of the brain because of Alzheimer's disease [20]. Routine primary care patient records in UK have been and are currently used to develop a risk score for the purposes of estimating how at risk an individual may be of developing dementia, by using conventional statistical methods and modern machine learning algorithms [10, 21]. Positron emission tomography (PET) scans and the regional analysis of the protein amyloid-β, have been used by a Random Forest classifier to identify patients with age-related stable MCI and pMCI [22]. In a recent EMIF-AD study [16], a machine learning methodology based on Extreme Gradient Boosting XGBoost, Random Forest and Deep Learning, has been proposed for Alzheimer's based dementia diagnosis using metabolites in the blood which were proven by the study to be as accurate predictors as the widely accepted but invasive to measure cerebrospinal fluid (CSF) biomarkers.

The present work proposes a new approach to predicting dementia using Deep Learning based on the ADNI (Alzheimer's Disease Neuroimaging Initiative) data [6, 17]. The ADNI dataset we use in this study contains the classes Cognitive Normal (CN), Mild Cognitive Impairment (MCI), and Dementia (DEM). Information gain and ReliefF methods [9, 18] were alternatively used for feature selection. The training, tuning and testing of the predictive models were performed with cross-validation using Deep Neural Network algorithms. The stability of model performances was studied using Monte Carlo simulations. In particular, this work explores the use of three Deep Learning models:

two Deep Feedforward Networks which are Multi-Layer Perceptron models (MLP1 and MLP2), and a Convolutional Bidirectional Long Short-Term Memory (ConvBLSTM) model [7, 8]. Good prediction performances were obtained with MLP1 and MLP2 for dementia: the sensitivities were 0.87 (SD 0.03) and 0.77 (SD 0.01) and the specificities were 0.97 (SD 0.01) and 0.98 (SD 0.01), respectively. The ConvBLSTM model was slightly less accurate in this study but was explored in view of comparisons with the MLP models, and also for future extensions of this work that will take advantage of time-related information in the ADNI data by using the ConvBLSTM's capabilities exploiting such information.

The remaining of the paper is organised as follows. Section 2 introduces the ADNI data used in this study, and our approach of machine learning methodology based on feature selection, missing data and data imbalance treatments, and deep learning predictive modelling. The prediction results and model performance evaluation are presented in Sect. 3. Finally a discussion and conclusion Sect. 4 ends the paper.

2 Data and Methods

2.1 ADNI Data Repository

The Alzheimer's Disease Neuroimaging Initiative (ADNI) data repository was used for this work [17]. The ADNI study was launched in 2003 by the National Institute on Aging (NIA), the National Institute of Biomedical Imaging and Bioengineering (NIBIB), the Food and Drug Administration (FDA), private pharmaceutical companies and non-profit organizations as a $60 million dollar, 5-year, public-private partnership.

There are three main goals of the ADNI study: (1) to detect Alzheimer's Disease (AD) at the earliest possible stage (pre-dementia) and identify ways to track the disease's progression with biomarkers; (2) to support advances in AD intervention, prevention and treatment through the application of new diagnostic methods at the earliest possible stages (when intervention may be most effective); and (3) to continually administer ADNI's innovative data-access policy, which make access to this data possible for scientists worldwide without any embargo.

2.2 The Dataset and the Description of Variables

The data used in this study was downloaded via the adnimerge R package [6], which merges together report forms and biomarkers from the ADNI subprojects ADNIGO, ADNI1, and ADNI2 [17]. The data consists of several different sources: clinical and genetic data, MRI data, PET data and some additional biospecimen. There is also longitudinal information as each participant in the ADNI dataset can have more than one screening visit. In this study we did not work with medical images directly, but used features that had been extracted from the data already.

Our target label consisted of three different classes: Cognitive Normal (CN), Mild Cognitive Impairment (MCI) and Dementia (DEM). The original data comprised 113 variables and 13,272 observations (visits), with multiple observations per participant. The variables extracted from the original dataset were as follows [17]:

- *Baselines Demographics*: age, gender, ethnicity, race, marital status, and education level were included as predictors.
- *Functional Activities Questionnaire (FAQ)* is a test that can be used to assess the dependency on another person that a participant requires to carry out normal daily tasks.
- *Mini-Mental State Exam (MMSE)* is used to estimate the severity and progression of cognitive impairment and to follow the course of cognitive changes in an individual over time.
- *PET measurements (FDG, PIB, AV45)* are participants' brain function measurements.
- *MRI measurements (Hippocampus, intracranial volume (ICV), MidTemp, Fusiform, Ventricles, Entorhinal and WholeBrain)* are structural measurements of participants' brain.
- *APOE4* is an integer measurement representing the appearance of epsilon 4 allele of the APOE gene.
- *ABETA, TAU, PTAU* are cerebrospinal fluid (CSF) biomarker measurements.
- *Rey's Auditory Verbal Learning Test (RAVLT)* are neurophysiological tests evaluating an individual's episodic memory.
- *Everyday cognitive evaluations (Ecog)* are questionnaires that illustrates a participant's ability to carry out everyday tasks.
- *Logical Memory – Delayed Recall Total Number of Story Units Recalled (LDELTO-TAL)* is a neuropsychological test that evaluates a person's ability to recall information after a prescribed amount of time.
- *Modified Preclinical Alzheimer Cognitive Composite (mPACC)* are tests that evaluate a person's cognition, episodic memory and timed executive function.
- *ADAS and MOCA* are generalized neuropsychological tests that evaluate a person's cognitive ability (e.g. memory, visuospatial, etc.).

The resulting data contained 1851 participants and 51 input attributes, and one output attribute. 75% of data was used as training data and 25% was used as test data to evaluate the performance of the models.

2.3 Feature Selection

To select a set of predictors, two feature selection methods were alternatively employed here. On one hand, we employed the Information Gain method and selected attributes scoring at least 0.01 [18]. One the other hand, we used ReliefF method [9, 18] which we chose to combine with a permutation test based on 500 random permutations of the labels. For instance, features with an observed Relief score corresponding to a distance of at least 1.96 standard deviations from the centre of the normal distribution built with the Relief scores, repeatedly calculated 500 times while labels were permuted randomly, were selected for further processing. This corresponds to the application of the statistical permutation test with significance level alpha $= 0.05$. Figure 1 illustrates the example of such a variable mPACCdigit.bl with an observed relief score of 0.38 which is far away from the centre of the distribution, which is an indication that the variable is predictive. The analysis in our study was performed by considering alternative values for the significance level, namely 0.05 and 0.1. However, here we report results obtained with the latter value, which involves a less stringent selection of features, so leads to a larger number of predictors.

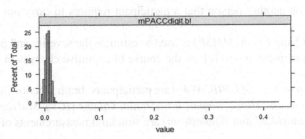

Fig. 1. mPACCdigit.bl variable with an observed relief score of 0.38 which is far away from the centre of the distribution, is considered predictive.

Table 1 illustrates the top 10 ranked features according to the two employed feature selection methods, Information Gain, and ReliefF combined with a permutation test, whose results seem to concord significantly.

Table 1. Top 10 features selected with ReliefF and information gain methods.

Rank	ReliefF	Information gain
1	mPACCdigit.bl	LDELTOTAL.bl
2	mPACCtrailsB.bl	mPACCdigit.bl
3	LDELTOTAL.bl	mPACCtrailsB.bl
4	ADAS13.bl	MMSE.bl
5	FAQ.bl	FAQ.bl
6	MMSE.bl	ADAS13.bl
7	ADAS11.bl	EcogSPTotal.bl
8	EcogSPTotal.bl	ADASQ4.bl
9	EcogSPMem.bl	ADAS11.bl
10	EcogSPPlan.bl	EcogSPMem.bl

Due to lack of space, the predictive modelling results reported in this paper correspond to the second feature selection method only, i.e. ReliefF with permutation test, as this led to better prediction results. Note that mPACCdigit.bl predictor illustrated in Fig. 1 is ranked top by this feature selection method, as illustrated by Table 1. However, the downside of the ReliefF method combined with a permutation test in this study, is that it becomes computationally more expensive.

2.4 Missing Value Imputation

For the two MLP models MLP1 and MLP2, 2-dimensional datasets were used, which did not contain longitudinal information. We imputed each visit of a participant from their first visit in order to retain the relationship between different visits and so that the final dataset consisted of only one data instance for each participant.

The ConvBLSTM model used 3-dimensional data, with longitudinal information provided via screening visit data. We replaced the date of each screening with the time difference between visits in order to include information about the sequence of visits as well as the interval between visits. The number of instances of visit codes decreases rapidly after the first visit so the GAIN imputation technique [11] was used to mitigate the lack of later visit data.

2.5 Balancing Classes

Large class imbalance has, most of the times, negative consequences on the performance of the predictive models, as machine learning algorithms tend to focus more on detecting the larger classes. The class distribution for our data is as follows: CN: 617, MCI: 886, Dementia: 348, which suggests that the minority class Dementia could be poorly detected by the predictive models in this case. This problem was mitigated using SMOTE algorithm (Synthetic Minority Over-sampling TEchnique) [12] which reduces class imbalance. SMOTE chooses a data point randomly from the minority class, determines the k nearest neighbours to that point and then uses these neighbours to generate new synthetic data points using interpolation. The synthetic data was added to the minority class to overcome the gap between the majority and minority classes in terms of number of instances. Our analysis used $k = 5$ neighbours, as a default value. This technique is to be applied only on the training set as it shouldn't affect the distribution of classes in the test set in order to obtain unbiased estimates for the performance metrics on the test/unseen data.

2.6 Tuning the Deep Learning Models

Parameter values, such as learning rate, decay value and AMSGrad option of ADAM optimiser, for the models were optimised, using Tree of Parzen Estimators (TPE) optimisation algorithm from 'hyperopt' Python package. Models were then evaluated on the test set. Figures 2 and 3 show the topologies of MLP1 and MLP2 optimised models, respectively, which used LeakyReLU activation functions in the hidden layers and a softmax activation function in the output layer. The models were tuned in a 10-fold cross validation using L1 and L2 regularizations, and the dropout regularization with probabilities 0.2 and 0.4, to prevent overfitting.

The ConvBLSTM model used 3D data, imputed with a generalized GAIN model. The first layers of the model applied convolutional and pooling operations [7], with a kernel that convolutes only operations across different dates for each of the features. Theoretically, this approach could have resulted in a loss of time information but this did not apply in our case because we had a different set of features for different visit codes. The approach also guaranteed that information was preserved for the most recent

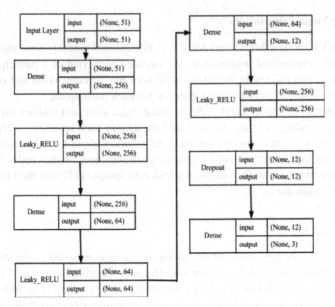

Fig. 2. MLP1: multilayer perceptron model 1.

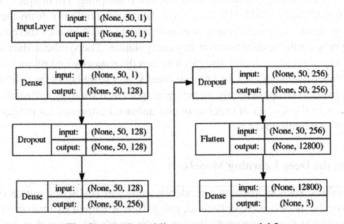

Fig. 3. MLP2: multilayer perceptron model 2.

visit codes with a large proportion of missing values for most persons. Following the convolutional operations, we applied a bidirectional LSTM layer [8, 13] because feature distribution helps preserve time information. After that, we applied a flatten operation to the data and sent the result to the final dense layer with softmax activation function, which gave us the probabilities for each target class (DEM, MCI, and CN). Figure 4 shows the topology of the ConvBLSTM model.

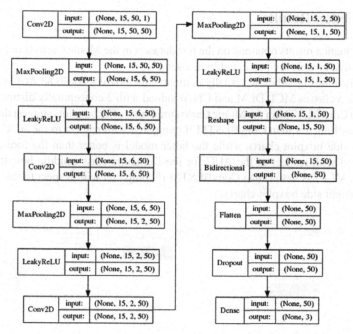

Fig. 4. ConvBLSTM: the Convolutional Bidirectional Long Short-Term Memory model

2.7 Monte Carlo Simulation and Performance Metrics

The stability of well-performing models MLP1, MLP2 and ConvBLSTM was investigated using a Monte Carlo simulation consisting of 100 iterations of the models' tuning and evaluation. We remind that the models' tuning consisted of performing a grid search in a 10-fold cross-validation on the training set (75% of the data) for identifying the best hyperparameters' values. The following performance metrics were evaluated on the test set (25% of the data), and recorded with each Monte Carlo iteration: accuracy, Cohen's kappa statistic, sensitivity (with respect to each class) and specificity (with respect to each class). In addition, for each iteration we recorded the area under the curve (AUC) for each class versus the rest.

2.8 Hardware and Software

The MLP1 model was implemented using Python on Tesla k80 GPU, together with a number of packages, including TensorFlow, Keras, hyperopt, scikit-learn, imbalanced-learn, numpy, pandas. The MLP2 model and the ConvBLSTM model were implemented with Python on Nvidia GTX 1070 Ti GPU under CUDA API. The same Python packages were used as for the MLP1 model, with the exception of Keras, which was used as part of the embedded tf.keras API from TensorFlow.

3 Prediction Results

The performance results obtained on the test datasets in the Monte Carlo simulations for our predictive models are presented in this section.

In particular, Fig. 5 below visually compares the boxplots of ROC AUC values for each of the 3 classes MCI, DEM and CN, obtained with 2 conceptually distinct models MLP2 and ConvBLSTM in the Monte Carlo simulation. The boxplots suggest that MLP2 is significantly better than ConvBLSTM if compared with respect to the AUC for MCI class (left side boxplot chart), while the latter model is better than the former model if compared with respect to the AUC for the Dementia class (middle boxplot chart). Moreover, MLP2 is better than ConvBLSTM if compared with respect to the AUC for CN class (right side boxplot chart).

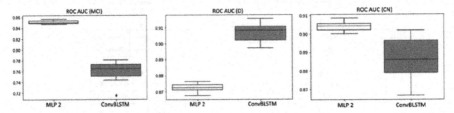

Fig. 5. Monte Carlo: ROC AUC boxplots for mild cognitive impairment (left), dementia (centre) and cognitive normal (right).

Numeric results for the performance metrics in the Monte Carlo simulations for the MLP models MLP1 and MLP2, are presented in Tables 2 and 3, and those for the ConvBLSTM model are presented in Tables 4 and 5. In particular the class-specific performances of the models for dementia, such as AUC, sensitivity and specificity, as well as the non-specific performances accuracy and kappa, are reflected in Tables 2 and 4.

Table 2. Monte Carlo: accuracy and kappa statistic for all 3 classes (DEM, MCI, CN), and AUC, sensitivity and specificity of dementia vs all other classes – MLP1 and MLP2

Model	ACC	KAPPA	AUC$_{DEM}$	SENS$_{DEM}$	SPEC$_{DEM}$
MLP 1	0.86	0.78	0.92	0.87	0.97
	SD 0.01	SD 0.01	SD 0.01	SD 0.03	SD 0.01
MLP 2	0.85	0.76	0.87	0.77	0.98
	SD 0.01	SD 0.01	SD 0.01	SD 0.01	SD 0.01

Table 3. Monte Carlo: sensitivity and specificity of MCI vs all other classes, and of CN vs all other classes respectively – for MLP1 and MLP2 models.

Model	SENS$_{MCI}$	SPEC$_{MCI}$	SENS$_{CN}$	SPEC$_{CN}$
MLP 1	0.86	0.87	0.94	0.87
	SD 0.03	SD 0.03	SD 0.02	SD 0.04
MLP 2	0.88	0.83	0.87	0.94
	SD 0.01	SD 0.01	SD 0.01	SD 0.01

Table 4. Monte Carlo: accuracy and kappa statistic for all 3 classes (DEM, MCI, CN), and AUC, sensitivity and specificity of dementia vs all other classes – for the ConvBLSTM model

Model	ACC	KAPPA	AUC$_{DEM}$	SENS$_{DEM}$	SPEC$_{DEM}$
Conv-BLSTM (GAIN)	0.82	0.72	0.90	0.86	0.94
	SD 0.01	SD 0.01	SD 0.01	SD 0.01	SD 0.01
Conv-BLSTM (SMOTE)	0.82	0.71	0.91	0.88	0.93
	SD 0.02	SD 0.03	SD 0.01	SD 0.01	SD 0.01

Table 5. Monte Carlo: sensitivity and specificity of MCI vs all other classes, and of CN vs all other classes respectively – for the ConvBLSTM model

Model	SENS$_{MCI}$	SPEC$_{MCI}$	SENS$_{CN}$	SPEC$_{CN}$
Conv-BLSTM (GAIN)	0.69	0.86	0.84	0.94
	SD 0.04	SD 0.01	SD 0.03	SD 0.02
Conv-BLSTM (SMOTE)	0.66	0.88	0.87	0.91
	SD 0.04	SD 0.01	SD 0.03	SD 0.03

4   Discussion and Conclusion

Deep Learning requires more sophisticated data pre-processing and feature engineering, as well as much more computational time for training a single model in comparison to statistical implementations. The three Deep Learning models explored here were based on two different Deep Feedforward Networks which are Multi-Layer Perceptron (MLP) models, and one Convolutional Bidirectional Long Short-Term Memory (ConvBLSTM) model.

The ReliefF algorithm was the method of choice here used for selecting the most informative features and was a better technique, in this case, in comparison to Information Gain; SMOTE was used to mitigate the large class imbalance in the data.

The ConvBLSTM model was chosen in order to preserve time information in the data. However, the performance of this model was impacted by the lack of visit code data. The GAIN imputation used to mitigate this was suboptimal as the training algorithm made random imputations of the mask layers and projected an initial data distribution on the matrix of data for each person, without considering any inherent differences between various datasets in the ADNI repository. A proposed enhancement for further work is to analyse missing value groups between different visit codes instances and to train the GAIN model for meaningfully different ADNI sub-project data groups, and not for the whole distribution.

The architecture of the neural network based models may be subject of further improvements in forthcoming work. In the case of the ConvBLSTM model, we experimented with a different shape for convolutional operations, so that it would transform participant matrices feature-wise and not time-wise, but this did not lead to a good convergence. An improvement could be to use the convolution operation separately, and apply the transfer learning technique from our already trained model with different convolution operations while freezing the original layers. This would result in a model graph with two different convolution branches and two different bidirectional LSTM layers that would be multiplied before the final dense layer [14].

The magnetic resonance imaging, positron emission tomography and genetic data in the ADNI dataset were only available in numeric and categorical format, and were not as raw images in this study. This meant that the potential power of Convolutional Neural Networks (CNNs) to classify image data was not explored here.

All of the models were able to recognize patterns differentiating the three classes DEM (Dementia), MCI (Minor Cognitive Impairment) and CN (Cognitive Normal), which indicates that each of these machine learning approaches has the capacity to accurately predict dementia and mild cognitive impairment.

Potential directions for further exploration include: (1) improving the modelling of time-based associations, with GAIN imputation to mitigate the sparseness in data for recent visits; (2) exploring other neural network model architectures, such as separating the convolution operation; and (3) applying CNNs to raw image data as these are likely to provide better performance for image classification. With these enhancements it is envisaged that individualized dementia-risk scores will be created and time information will be modelled more effectively.

References

1. World Health Organisation. Dementia. www.who.int/en/news-room/fact-sheets/detail/dementia
2. Prince, M., Wimo, A., Guerchet, M., Ali, G., Wu, Y., Prina, M.: The Global Impact of Dementia, An analysis of prevalence, incidence, cost and trends. World Alzheimer Report (2015)
3. Barrett, E., Burns, A.: Dementia Revealed. What Primary Care Needs to Know. Department of Health UK (2014)
4. Panch, T., Szolovits, P., Atun, R.: Artificial intelligence, machine learning and health systems. J. Glob. Health 8(2), 020303 (2018)
5. Yu, K., Beam, A., Kohane, I.: Artificial intelligence in healthcare. Nat. Biomed. Eng. 2(10), 719–731 (2018)
6. ADNIMERGE, Alzheimer's Disease Neuroimaging Initiative. ADNIMERGE: Alzheimer's Disease Neuroimaging Initiative. R package version 0.0.1

7. Lo, S.-C.B., Chan, H.-P., Lin, J.-S., Li, H., Freedman, M.T., Mun, S.K.: Artificial convolution neural network for medical image pattern recognition. Neural Netw. **8**(7–8), 1201–1214 (1995)

8. Huang, Z., Xu, W., Yu, K.: Bidirectional LSTM-CRF Models for Sequence Tagging, preprint https://arxiv.org/abs/1508.01991 (2015)

9. Spolaôr, N., Cherman, E.A., Monard, M.C., Lee, H.D.: ReliefF for multi-label feature selection. In: Brazilian Conference on Intelligent Systems (2013)

10. Morgan, C., Ashcroft, D.M., Kontopantelis, E., Stamate, D., Reeves, D.: Can dementia risk be predicted using routine electronic health records? In: Society for Academic Primary Care Conference (SAPC ASM) (2019)

11. Yoon, J., Jordon, J., van Der Schaar, M.: GAIN: missing data imputation using generative adversarial nets, preprint https://arxiv.org/abs/1806.02920 (2018)

12. Qazi, N., Raza, K.: Effect of feature selection, SMOTE and under sampling on class imbalance classification. In: 14th UKSim, pp. 145–150 (2012)

13. Gers, F.A., Schmidhuber, J., Cummins, F.: Learning to forget: continual prediction with LSTM. In: 9th International Conference on Artificial Neural Networks, pp. 850–855 (1999)

14. Yildirim, Ö.: A novel wavelet sequence based on deep bidirectional LSTM network model for ECG signal classification. Comput. Biol. Med. **96**, 189–202 (2018)

15. Spuhler, K.D., Gardus III, J., Gao, Y., DeLorenzo, C., Parsey, R., Huang, C.: Synthesis of patient-specific transmission data for PET attenuation correction for PET/MRI neuroimaging using a convolutional neural network. J. Nuclear Med. **60**, 555–560 (2019)

16. Stamate, D., Kim, M., Lovestone, S., Legido-Quigley, C., et al.: A metabolite-based machine learning approach to diagnose Alzheimer's-type dementia in blood: results from the European Medical Information Framework for Alzheimer's Disease biomarker discovery cohort. Alzheimer's Dement.: Transl. Res. Clin. Interv. **5**, 933–938 (2019)

17. Alzheimer's Disease Neuroimaging Initiative. http://adni.loni.usc.edu/

18. Kuhn, M., Johnson, K.: Applied Predictive Modeling. Springer, New York (2013). https://doi.org/10.1007/978-1-4614-6849-3

19. Robinson, L., Trenell, M.: Dementia: risk reduction. Newcastle Institute of Ageing (2016)

20. Amoroso, N., La Rocca, M., Bruno, S., Maggipinto, T., Monaco, A., Bellotti, R., et al.: Brain structural connectivity atrophy in Alzheimer's disease, preprint arXiv:1709.02369 (2017)

21. Walters, K., et al.: Predicting dementia risk in primary care: development and validation of the Dementia Risk Score using routinely collected data. BMC Med. **14**(1), 6 (2016). https://doi.org/10.1186/s12916-016-0549-y

22. Brauser, D.: 'Machine Learning' Tool May Predict Dementia Development Up to 10 Years Later. Medscape (2016)

Bridging the Gap Between AI and Healthcare Sides: Towards Developing Clinically Relevant AI-Powered Diagnosis Systems

Changhee Han[1,2,3]($\boxtimes$) iD, Leonardo Rundo[4,5] iD, Kohei Murao[3],
Takafumi Nemoto[6], and Hideki Nakayama[1]

[1] Graduate School of Information Science and Technology,
The University of Tokyo, Tokyo, Japan
[2] LPIXEL Inc., Tokyo, Japan
han@lpixel.net
[3] Research Center for Medical Big Data,
National Institute of Informatics, Tokyo, Japan
[4] Department of Radiology, University of Cambridge, Cambridge, UK
[5] Cancer Research UK Cambridge Centre, University of Cambridge, Cambridge, UK
[6] Department of Radiology, Keio University School of Medicine, Tokyo, Japan

Abstract. Despite the success of Convolutional Neural Network-based Computer-Aided Diagnosis research, its clinical applications remain challenging. Accordingly, developing medical Artificial Intelligence (AI) fitting into a clinical environment requires identifying/bridging the gap between AI and Healthcare sides. Since the biggest problem in Medical Imaging lies in data paucity, confirming the clinical relevance for diagnosis of research-proven image augmentation techniques is essential. Therefore, we hold a clinically valuable AI-envisioning workshop among Japanese Medical Imaging experts, physicians, and generalists in Healthcare/Informatics. Then, a questionnaire survey for physicians evaluates our pathology-aware Generative Adversarial Network (GAN)-based image augmentation projects in terms of Data Augmentation and physician training. The workshop reveals the intrinsic gap between AI/Healthcare sides and solutions on *Why* (i.e., clinical significance/interpretation) and *How* (i.e., data acquisition, commercial deployment, and safety/feeling safe). This analysis confirms our pathology-aware GANs' clinical relevance as a clinical decision support system and non-expert physician training tool. Our findings would play a key role in connecting inter-disciplinary research and clinical applications, not limited to the Japanese medical context and pathology-aware GANs.

Keywords: Translational research · Computer-aided diagnosis · Generative adversarial networks · Data augmentation · Physician training

© IFIP International Federation for Information Processing 2020
Published by Springer Nature Switzerland AG 2020
I. Maglogiannis et al. (Eds.): AIAI 2020, IFIP AICT 584, pp. 320–333, 2020.
https://doi.org/10.1007/978-3-030-49186-4_27

1 Introduction

Convolutional Neural Networks (CNNs) have enabled accurate/reliable Computer-Aided Diagnosis (CAD), occasionally outperforming expert physicians [1–3]. However, such research results cannot be easily transferred to a clinical environment. Artificial Intelligence (AI) and Healthcare sides have a huge gap around technology, funding, and people [4]. In Japan, the biggest challenge lies in medical data sharing because each hospital has different ethical codes and tends to enclose collected data without annotating them for AI research. This differs from the US, where National Cancer Institute provides annotated medical images [5].

Therefore, a Research Center for Medical Big Data was launched in November 2017: collaborating with 6 Japanese medical societies and 6 institutes of informatics, we collected large-scale annotated medical images for CAD research. Using over 60 million available images, we achieved prominent research results, presented at major Computer Vision [6] and Medical Imaging conferences [7]. Moreover, we published 6 papers [8–13] on Generative Adversarial Network (GAN)-based medical image augmentation [14]. Since the GANs can generate realistic samples with desired pathological features *via* many-to-many mappings, they could mitigate the medical data paucity *via* Data Augmentation (DA) and physician training.

Aiming to further identify/bridge the gap between AI and Healthcare sides in Japan towards developing medical AI fitting into a clinical environment in five years, we hold a workshop for 7 Japanese people with various AI and/or Healthcare background. Moreover, to confirm the clinical relevance for diagnosis of the pathology-aware GAN methods, we conduct a questionnaire survey for 9 Japanese physicians who interpret Computed Tomography (CT) and Magnetic Resonance (MR) images in daily practice. Figure 1 outlines our investigation.

Contributions. Our main contributions are as follows:

- **AI and Healthcare Workshop:** We firstly hold a clinically valuable AI-envisioning workshop among Medical Imaging experts, physicians, and Healthcare/Informatics generalists to bridge the gap between AI/Healthcare sides.
- **Questionnaire Survey for Physicians:** We firstly present both qualitative/quantitative questionnaire evaluation results for many physicians about research-proven medical AI.
- **Information Conversion:** Clinical relevance discussions imply that our pathology-aware GAN-based interpolation and extrapolation could overcome medical data paucity *via* DA and physician training.

2 Pathology-Aware GAN-Based Image Augmentation

In terms of interpolation, GAN-based medical image augmentation is reliable because acquisition modalites (e.g., X-ray, CT, MR) can display the human body's strong anatomical consistency at fixed position while clearly reflecting inter-subject variability [15,16]. This is different from natural images, where various objects can appear at any position; accordingly, to tackle large inter-subject/pathology/modality variability, we proposed to use noise-to-image GANs (e.g., random noise samples to diverse pathological images) for (*i*) medical DA and (*ii*) physician training [8]. While the noise-to-image GAN training is much more difficult than training image-to-image GANs [17] (e.g., a benign image to a malignant one), it can increase image diversity for further performance boost.

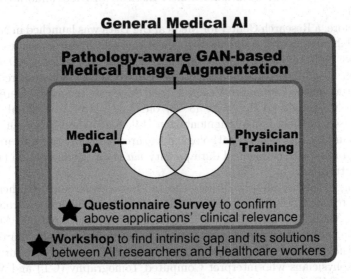

Fig. 1. Overview of our discussions towards developing clinically relevant AI-powered diagnosis systems: (*i*) A workshop for 7 Japanese people with various AI and/or Healthcare background to develop medical AI fitting into a clinical environment in five years; (*ii*) A questionnaire survey for 9 Japanese physicians to confirm our pathology-aware GAN-based realistic/diverse image augmentation's clinical relevance—the medical DA requires high diversity whereas the physician training requires high realism.

Regarding the DA, the GAN-generated images can improve CAD based on supervised learning [18–20]. For the physician training, the GANs can display novel desired pathological images and help train medical trainees despite infrastructural/legal constraints [21]. However, we have to devise effective loss functions and training schemes for such applications. Diversity matters more for the DA to sufficiently fill the real image distribution, whereas realism matters more for the physician training not to confuse the trainees.

So, how can we perform GAN-based DA/physician training with only limited annotated images? Always in collaboration with physicians, for improving 2D classification, we combined the noise-to-image and image-to-image GANs [9,10].

Nevertheless, further DA applications require pathology localization for detection and advanced physician training needs the generation of images with abnormalities, respectively. To meet both clinical demands, we proposed 2D/3D bounding box-based GANs conditioned on pathology position/size/appearance. Indeed, the bounding box-based detection requires much less physicians' annotation effort than segmentation [22].

In terms of extrapolation, the pathology-aware GANs are promising because common and/or desired medical priors can play a key role in the conditioning—theoretically, infinite conditioning instances, external to the training data, exist and enforcing such constraints have an extrapolation effect *via* model reduction [23]. For improving 2D detection, we proposed Conditional Progressive Growing of GANs that incorporates rough bounding box conditions incrementally into a noise-to-image GAN (i.e., Progressive Growing of GANs [24]) to place realistic/diverse brain metastases at desired positions/sizes on 256×256 MR images [11]. Since the human body is 3D, for improving 3D detection, we proposed 3D Multi-Conditional GAN that translates noise boxes into realistic/diverse $32 \times 32 \times 32$ lung nodules [25] placed at desired position/size/attenuation on CT scans [12]. Interestingly, inputting the noise box with the surrounding tissues has the effect of combining the noise-to-image and image-to-image GANs.

We succeeded to (*i*) generate images even realistic for physicians and (*ii*) improve detection using synthetic training images, respectively; they require different loss functions and training schemes. However, to exploit our pathology-aware GANs as a (*i*) non-expert physician training tool and (*ii*) clinical decision support system, we need to confirm their clinical relevance for diagnosis—such information conversion [26] techniques to overcome the data paucity, not limited to our pathology-aware GANs, would become a clinical breakthrough.

3 Methods

3.1 AI and Healthcare Workshop

- **Subjects:** 2 Medical Imaging experts (i.e., a Medical Imaging researcher and a medical AI startup entrepreneur), 2 physicians (i.e., a radiologist and a psychiatrist), and 3 Healthcare/Informatics generalists (i.e., a nurse and researcher in medical information standardization, a general practitioner and researcher in medical communication, and a medical technology manufacturer's owner and researcher in health disparities).
- **Experiments:** As its program shows (Table 1), during the workshop, we conduct 2 activities: (*Learning*) Know the overview of Medical Image Analysis, including state-of-the-art research, well-known challenges/solutions, and the summary of our pathology-aware GAN projects; (*Thinking*) Find the intrinsic gap and its solutions between AI researchers and Healthcare workers after sharing their common and different thinking/working styles. This workshop was held on March 17th, 2019 at Nakayama Future Factory, Open Studio, The University of Tokyo, Tokyo, Japan.

3.2   Questionnaire Survey for Physicians

- **Subjects:** 3 physicians (i.e., a radiologist, a psychiatrist, and a physiatrist) committed to (at least one of) our pathology-aware GAN projects and 6 project non-related radiologists without much AI background. This paper's authors are surely not included.
- **Experiments:** Physicians are asked to answer the following questionnaire within 2 weeks from December 6th, 2019 after reading 10 summary slides written in Japanese about general Medical Image Analysis and our pathology-aware GAN projects along with example synthesized images. We conduct both qualitative (i.e., free comments) and quantitative (i.e., five-point Likert scale [27]) evaluation: Likert scale 1 = very negative, 2 = negative, 3 = neutral, 4 = positive, 5 = very positive.

- **Question 1:** Are you keen to exploit medical AI in general when it achieves accurate and reliable performance in the near future? (five-point Likert scale) Please tell us your expectations, wishes, and worries (free comments).

Table 1. Workshop program to (*i*) know the overview of Medical Image Analysis and (*ii*) find the intrinsic gap and its solutions between AI researchers/Healthcare workers. *Indicates activities given by a facilitator (i.e., the first author), such as lectures.

Time (mins)	Activity
	Introduction
10	1. Explanation of the workshop's purpose and flow*
10	2. Self-introduction and explanation of motivation for participation
5	3. Grouping into two groups based on background*
	Learning: Knowing Medical Image Analysis
15	1. TED speech video watching: *Artificial Intelligence Can Change the future of Medical Diagnosis**
35	2. Lecture: Overview of Medical Image Analysis including state-of-the-art research, well-known challenges/solutions, and the summary of our pathology-aware GAN projects* (its video in Japanese: https://youtu.be/rTQLknPvnqs)
10	3. Sharing expectations, wishes, and worries about Medical Image Analysis (its video in Japanese: https://youtu.be/ILPEGga-hkY)
10	Intermission
	Thinking: Finding How to Develop Robust Medical AI
25	1. Identifying the intrinsic gap between AI/Healthcare sides after sharing their common and different thinking/working styles
60	2. Finding how to develop gap-bridging medical AI fitting into a clinical environment in five years
10	Intermission
	Summary
25	1. Presentation
10	2. Sharing workshop impressions and ideas to apply obtained knowledge (its video in Japanese: https://youtu.be/F31tPR3m8hs)
5	3. Answering a questionnaire about satisfaction and further comments
5	4. Closing remarks*

- **Question 2:** What do you think about using GAN-generated images for DA? (five-point Likert scale).
- **Question 3:** What do you think about using GAN-generated images for physician training? (five-point Likert scale).
- **Question 4:** Any comments/suggestions about our projects towards developing clinically relevant AI-powered systems based on your experience?

4 Results

4.1 Workshop Results

We show the clinically-relevant findings from this Japanese workshop. **Gap Between AI and Healthcare Sides Gap 1:** AI, including Deep Learning, provides unclear decision criteria, does it make physicians reluctant to use it for diagnosis in a clinical environment?

- **Healthcare side:** We rather expect applications other than diagnosis. If we use AI for diagnosis, instead of replacing physicians, we suppose a *reliable second opinion*, such as alert to avoid misdiagnosis, based on various clinical data not limited to images—every single diagnostician is anxious about their diagnosis. AI only provides minimum explanation, such as a heatmap showing attention, which makes persuading not only the physicians but also patients difficult. Thus, the physicians' intervention is essential for intuitive explanation. Methodological safety and feeling safe are different. In this sense, pursuing explainable AI generally decreases AI's diagnostic accuracy [28], so physicians should still serve as mediators by engaging in high-level conversation or interaction with patients. Moreover, according to the medical law in most countries including Japan, only doctors can make the final decision. The first autonomous AI-based diagnosis without a physician was cleared by the Food and Drug Administration in 2018 [29], but such a case is exceptional.
- **AI side:** Compared with other systems or physicians, Deep Learning's explanation is not particularly poor, so we require too severe standards for AI; the word *AI* is excessively promoting anxiety and perfection. If we could thoroughly verify the reliability of its diagnosis against physicians by exploring uncertainty measures [30], such intuitive explanation would be optional.

Gap 2: Are there any benefits to actually introducing medical AI?

- **Healthcare side:** After all, even if AI can achieve high accuracy and convenient operation, hospitals would not introduce it without any commercial benefits. Moreover, small clinics, where physicians are desperately needed, often do not have CT or MR scanners [31].
- **AI side:** The commercial deployment of medical AI is strongly tied to diagnostic accuracy; so, if it can achieve significantly outstanding accuracy at various tasks in the near future, patients would not visit hospitals/clinics without AI. Accordingly, introducing medical AI would become profitable in five years.

Gap 3: Is medical AI's diagnostic accuracy reliable?

- **Healthcare side:** To evaluate AI's diagnostic performance, we should consider many metrics, such as sensitivity and specificity. Moreover, its generalization ability for medical data highly relies on inter-scanner/inter-individual variability [32]. How can we evaluate whether it is suitable as a clinically applicable system?
- **AI side:** Generally, alleviating the risk of overlooking the diagnosis is the most important, so sensitivity matters more than specificity unless their balance is highly disturbed. Recently, such research on medical AI that is robust to different datasets is active [33].

How to Develop Medical AI Fitting into a Clinical Environment in Five Years
Why: Clinical significance/interpretation

- **Challenges:** We need to clarify which clinical situations actually require AI introduction. Moreover, AI's early diagnosis might not be always beneficial for patients.
- **Solutions:** Due to nearly endless disease types and frequent misdiagnosis coming from physicians' fatigue, we should use it as alert to avoid misdiagnosis [34] (e.g., reliable second opinion), instead of replacing physicians. It should help prevent oversight in diagnostic tests not only with CT and MR, but also with blood data, chest X-ray, and mammography before taking CT and MR [35]. It could be also applied to segmentation for radiation therapy [36], neurosurgery navigation [37], and pressure ulcers' echo evaluation. Along with improving the diagnosis, it would also make the physicians' workflow easier, such as by denoising [38]. Patients should decide whether they accept AI-based diagnosis under informed consent.

How: Data acquisition

- **Challenges:** Ethical screening in Japan is exceptionally strict, so acquiring and sharing large-scale medical data/annotation are challenging—it also applies to Europe due to General Data Protection Regulation [39]. Considering the speed of technological advances in AI, adopting it for medical devices is difficult in Japan, unlike in medical AI-ready countries, such as the US, where the ethical screening is relatively loose in return for the responsibility of monitoring system stability. Moreover, whenever diagnostic criteria changes, we need further reviews and software modifications. For example, the Tumor-lymph Node-Metastasis (TNM) classification [40] criteria changed for oropharyngeal cancer in 2018 and for lung cancer in 2017, respectively. Diagnostic equipment/target changes also require large-scale data/annotation acquisition again.
- **Solutions:** For Japan to keep pace, the ethical screening should be adequate to the other leading countries. Currently, overseas research and clinical trials are proceeding much faster, so it seems better to collaborate with overseas

companies than to do it in Japan alone. Moreover, complete medical checkup, which is extremely costly, is unique in East Asia, thus Japan could be superior in individuals' multiple medical data—Japan is the only country, where most workers aged 40 or over are required to have medical checkups once a year regardless of their health conditions by the Industrial Safety and Health Act [41]. To handle changes in diagnostic criteria/equipment and overcome dataset/task dependency, it is necessary to establish a common database creation workflow [42] by regularly entering electronic medical records into the database. For reducing data acquisition/annotation cost, AI techniques, such as GAN-based DA [12] and domain adaptation [43], would be effective.

How: Commercial deployment

- **Challenges:** Hospitals currently do not have commercial benefits to actually introduce medical AI.
- **Solutions:** For example, it would be possible to build AI-powered hospitals [44] operated with less staff. Medical manufacturers could also standardize data format [45], such as for X-ray, and provide some AI services. Many IT giants like Google are now working on medical AI to collect massive biomedical datasets [46], so they could help rural areas and developing countries, where physician shortage is severe [31], at relatively low cost.

How: Safety and feeling safe

- **Challenges:** Considering multiple metrics, such as sensitivity and specificity [47], and dataset/task dependency [48], accuracy could be unreliable, so ensuring safety is challenging. Moreover, reassuring physicians and patients is important to actually use AI in a clinical environment [49].
- **Solutions:** We should integrate various clinical data, such as blood test biomarkers and multiomics, with images [35]. Moreover, developing bias-robust technology is important since confounding factors are inevitable [50]. To prevent oversight, prioritizing sensitivity over specificity is essential while maintaining a balance [51]. We should also devise education for medical AI users, such as result interpretation, to reassure patients [52].

4.2   Questionnaire Survey Results

We show the questions and Japanese physicians' response summaries. Concerning the following **Questions 1, 2, 3**, Fig. 2 visually summarizes the expectation scores on medical AI (i.e., general medical AI, GANs for DA, and GANs for physician training) from both 3 project-related physicians and 6 project non-related radiologists.

Question 1: Are you keen to exploit medical AI in general when it achieves accurate and reliable performance in the near future?

- **Response summary:** As expected, the project-related physicians are AI-enthusiastic while the project non-related radiologists are also generally very positive about the medical AI. Many of them appeal the necessity of AI-based diagnosis for more reliable diagnosis because of the lack of physicians. Meanwhile, other physicians worry about its cost and reliability. We may be able to persuade them by showing expected profitability (e.g., currently CT scanners have an earning rate 16% and CT scans require 2–20 min for interpretation in Japan). Similarly, we can explain how experts annotate medical images and AI diagnoses disease based on them (e.g., multiple physicians, not a single one, can annotate the images *via* discussion).

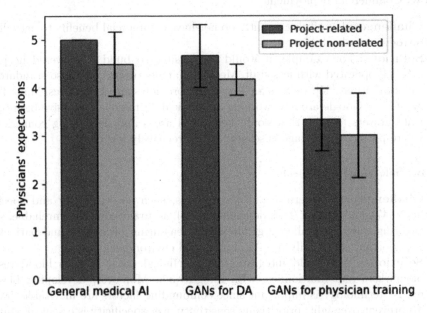

Fig. 2. Bar chart of the expectations on medical AI, expressed by five-point Likert scale scores, from 9 Japanese physicians: 3 project-related physicians and 6 project non-related radiologists, respectively. The vertical rectangles and error bars denote the average scores with 95% confidence intervals.

Question 2: What do you think about using GAN-generated images for DA?

- **Response summary:** As expected, the project-related physicians are very positive about the GAN-based DA while the project non-related radiologists are also positive. Many of them are satisfied with its achieved accuracy/sensitivity improvement when available annotated images is limited. However, similarly to their opinions on general Medical Image Analysis, some physicians question its reliability.

Question 3: What do you think about using GAN-generated images for physician training?

- **Response summary:** We generally receive neutral feedback because we do not provide a concrete physician training tool, but instead general pathology-aware generation ideas with example synthesized images—thus, some physicians are positive, and some are not. A physician provides a key idea about a pathology-coverage rate for medical student/expert physician training, respectively. For extensive physician training by GAN-generated atypical images, along with pathology-aware GAN-based extrapolation, further GAN-based extrapolation would be valuable.

Question 4: Any comments/suggestions about our projects towards developing clinically relevant AI-powered systems based on your experience?

- **Response summary:** Most physicians look excited about our pathology-aware GAN-based image augmentation projects and express their clinically relevant requests. The next steps lie in performing further GAN-based extrapolation, developing reliable and clinician-friendly systems with new practice guidelines, and overcoming legal/financial constraints.

5   Conclusion

Our first clinically valuable AI-envisioning workshop between people with various AI and/or Healthcare background reveals the intrinsic gap between both sides and its preliminary solutions. Regarding clinical significance/interpretation, medical AI could play a key role in supporting physicians with diagnosis, therapy, and surgery. For data acquisition, countries should utilize their unique medical environment, such as complete medical checkup for Japan. Commercial deployment could come as AI-powered hospitals and medical manufacturers' AI service. To assure safety and feeling safe, we should integrate various clinical data and devise education for medical AI users. We believe that such solutions on *Why* and *How* would play a crucial role in connecting inter-disciplinary research and clinical applications.

Through a questionnaire survey for physicians, we confirm our pathology-aware GANs' clinical relevance for diagnosis as a clinical decision support system and non-expert physician training tool: many physicians admit the urgent necessity of general AI-based diagnosis while welcoming our GAN-based DA to handle the lack of medical images. Thus, GAN-powered physician training is promising only under careful tool designing.

We find that better DA and expert physician training require further generation of images with abnormalities. Therefore, for better GAN-based extrapolation, we plan to conduct (*i*) generation by parts with coordinate conditions [53], (*ii*) generation with both image and radiogenomic conditions [54], and (*iii*) transfer learning among different body parts and disease types. Whereas this paper only explores the Japanese medical context and pathology-aware GANs, our findings are more generally applicable and can provide insights into the clinical practice in other countries.

References

1. Hwang, E.J., Park, S., Jin, K., et al.: Development and validation of a deep learning-based automatic detection algorithm for active pulmonary tuberculosis on chest radiographs. Clin. Infect. Dis. **69**(5), 739–747 (2018)
2. Wu, N., Phang, J., Park, J., et al.: Deep neural networks improve radiologists' performance in breast cancer screening. In: Proceedings of the International Conference on Medical Imaging with Deep Learning (MIDL). arXiv:1907.08612 (2019)
3. McKinney, S.M., Sieniek, M., Godbole, V., et al.: International evaluation of an AI system for breast cancer screening. Nature **577**(7788), 89–94 (2020)
4. Allen Jr., B., Seltzer, S.E., Langlotz, C.P., et al.: A road map for translational research on artificial intelligence in medical imaging: from the 2018 National Institutes of Health/RSNA/ACR/The Academy Workshop. J. Am. Coll. Radiol. **16**(9, Part A), 1179–1189 (2019)
5. Clark, K., Vendt, B., Smith, K., et al.: The Cancer Imaging Archive (TCIA): maintaining and operating a public information repository. J. Digit. Imaging **26**, 1045–1057 (2013). https://doi.org/10.1007/s10278-013-9622-7
6. Tokunaga, H., Teramoto, Y., Yoshizawa, A., Bise, R.: Adaptive weighting multi-field-of-view CNN for semantic segmentation in pathology. In: Proceedings of the Conference on Computer Vision and Pattern Recognition (CVPR), pp. 12597–12606 (2019)
7. Kanayama, T., et al.: Gastric cancer detection from endoscopic images using synthesis by GAN. In: Shen, D., et al. (eds.) MICCAI 2019. LNCS, vol. 11768, pp. 530–538. Springer, Cham (2019). https://doi.org/10.1007/978-3-030-32254-0_59
8. Han, C., Hayashi, H., Rundo, L., et al.: GAN-based synthetic brain MR image generation. In: Proceedings of the IEEE International Symposium on Biomedical Imaging (ISBI), pp. 734–738 (2018)
9. Han, C., et al.: Infinite brain MR images: PGGAN-based data augmentation for tumor detection. In: Esposito, A., Faundez-Zanuy, M., Morabito, F.C., Pasero, E. (eds.) Neural Approaches to Dynamics of Signal Exchanges. SIST, vol. 151, pp. 291–303. Springer, Singapore (2020). https://doi.org/10.1007/978-981-13-8950-4_27
10. Han, C., Rundo, L., Araki, R., et al.: Combining noise-to-image and image-to-image GANs: brain MR image augmentation for tumor detection. IEEE Access **7**, 156966–156977 (2019)
11. Han, C., Murao, K., Noguchi, T., et al.: Learning more with less: conditional PGGAN-based data augmentation for brain metastases detection using highly-rough annotation on MR images. In: Proceedings of the ACM International Conference on Information and Knowledge Management (CIKM), pp. 119–127 (2019)
12. Han, C., Kitamura, Y., Kudo, A., et al.: Synthesizing diverse lung nodules wherever massively: 3D multi-conditional GAN-based CT image augmentation for object detection. In: Proceedings of the International Conference on 3D Vision (3DV), pp. 729–737 (2019)
13. Han, C., Murao, K., Satoh, S., Nakayama, H.: Learning more with less: GAN-based medical image augmentation. Med. Imaging Tech. **37**(3), 137–142 (2019)
14. Goodfellow, I., Pouget-Abadie, J., Mirza, M., et al.: Generative adversarial nets. In: Proceedings of the Advances in Neural Information Processing Systems (NIPS), pp. 2672–2680 (2014)
15. Hsieh, J.: Computed Tomography: Principles, Design, Artifacts, and Recent Advances. SPIE, Bellingham (2009)

16. Brown, R.W., Cheng, Y.N., Haacke, E.M., et al.: Magnetic Resonance Imaging: Physical Principles and Sequence Design. Wiley, Hoboken (2014)

17. Tmenova, O., Martin, R., Duong, L.: CycleGAN for style transfer in X-ray angiography. Int. J. Comput. Assist. Radiol. Surg. **14**(10), 1785–1794 (2019). https://doi.org/10.1007/s11548-019-02022-z

18. Frid-Adar, M., Diamant, I., Klang, E., et al.: GAN-based synthetic medical image augmentation for increased CNN performance in liver lesion classification. Neurocomputing **321**, 321–331 (2018)

19. Madani, A., Moradi, M., Karargyris, A., Syeda-Mahmood, T.: Chest X-ray generation and data augmentation for cardiovascular abnormality classification. In: Proceedings of the Medical Imaging: Image Processing, vol. 10574, 105741M (2018)

20. Konidaris, F., Tagaris, T., Sdraka, M., Stafylopatis, A.: Generative adversarial networks as an advanced data augmentation technique for MRI data. In: Proceedings of the International Joint Conference on Computer Vision, Imaging and Computer Graphics Theory and Applications (VISIGRAPP), pp. 48–59 (2019)

21. Finlayson, S.G., Lee, H., Kohane, I.S., Oakden-Rayner, L.: Towards generative adversarial networks as a new paradigm for radiology education. In: Proceedings of the Machine Learning for Health (ML4H) Workshop. arXiv:1812.01547 (2018)

22. Rundo, L., Militello, C., Russo, G., Vitabile, S., Gilardi, M.C., Mauri, G.: GTVCUT for neuro-radiosurgery treatment planning: an MRI brain cancer seeded image segmentation method based on a cellular automata model. Nat. Comput. **17**(3), 521–536 (2017). https://doi.org/10.1007/s11047-017-9636-z

23. Stinis, P., Hagge, T., Tartakovsky, A.M., Yeung, E.: Enforcing constraints for interpolation and extrapolation in generative adversarial networks. J. Comput. Phys. **397**, 108844 (2019)

24. Karras, T., Aila, T., Laine, S., Lehtinen, J.: Progressive growing of GANs for improved quality, stability, and variation. In: Proceedings of the International Conference on Learning Representations (ICLR). arXiv:1710.10196v3 (2018)

25. Al-Shabi, M., Lan, B.L., Chan, W.Y., et al.: Lung nodule classification using deep local-global networks. Int. J. Comput. Assist. Radiol. Surg. **14**(10), 1815–1819 (2019). https://doi.org/10.1007/s11548-019-01981-7

26. Honda, T., Matsubara, Y., Neyama, R., et al.: Multi-aspect mining of complex sensor sequences. In: Proceedings of the IEEE International Conference on Data Mining (ICDM) (2019, in press)

27. Allen, I.E., Seaman, C.A.: Likert scales and data analyses. Qual. Prog. **40**(7), 64–65 (2007)

28. Adadi, A., Berrada, M.: Peeking inside the black-box: a survey on Explainable Artificial Intelligence (XAI). IEEE Access **6**, 52138–52160 (2018)

29. Abràmoff, M.D., Lavin, P.T., Birch, M., et al.: Pivotal trial of an autonomous AI-based diagnostic system for detection of diabetic retinopathy in primary care offices. NPJ Digit. Med. **1**(1), 39 (2018)

30. Nair, T., Precup, D., Arnold, D.L., Arbel, T.: Exploring uncertainty measures in deep networks for multiple sclerosis lesion detection and segmentation. Med. Image Anal. **59**, 101557 (2020)

31. Jankharia, G.R.: Commentary-radiology in India: the next decade. Indian J. Radiol. Imaging **18**(3), 189 (2008)

32. O'Connor, D., Potler, N.V., Kovacs, M., et al.: The healthy brain network serial scanning initiative: a resource for evaluating inter-individual differences and their reliabilities across scan conditions and sessions. Gigascience **6**(2), giw011 (2017)

33. Rundo, L., Han, C., Nagano, Y., et al.: USE-Net: incorporating squeeze-and-excitation blocks into U-Net for prostate zonal segmentation of multi-institutional MRI datasets. Neurocomputing **365**, 31–43 (2019)

34. Vandenberghe, M.E., Scott, M.L.J., Scorer, P.W., et al.: Relevance of deep learning to facilitate the diagnosis of HER2 status in breast cancer. Sci. Rep. **7**, 45938 (2017)

35. Li, X., Wang, Y., Li, D.: Medical data stream distribution pattern association rule mining algorithm based on density estimation. IEEE Access **7**, 141319–141329 (2019)

36. Agn, M., Law, I., af Rosenschöld, P.M., Van Leemput, K.: A generative model for segmentation of tumor and organs-at-risk for radiation therapy planning of glioblastoma patients. In: Proceedings of the Medical Imaging: Image Processing, vol. 9784, p. 97841D (2016)

37. Abi-Aad, K.R., Anderies, B.J., Welz, M.E., Bendok, B.R.: Machine learning as a potential solution for shift during stereotactic brain surgery. Neurosurgery **82**(5), E102–E103 (2018)

38. Yang, Q., Yan, P., Zhang, Y., et al.: Low-dose CT image denoising using a generative adversarial network with Wasserstein distance and perceptual loss. IEEE Trans. Med. Imaging **37**(6), 1348–1357 (2018)

39. Rumbold, J.M.M., Pierscionek, B.: The effect of the general data protection regulation on medical research. J. Med. Internet Res. **19**(2), e47 (2017)

40. Sobin, L.H., Gospodarowicz, M.K., Wittekind, C.: TNM Classification of Malignant Tumours, 7th edn. Wiley, Hoboken (2011)

41. Nawata, K., Matsumoto, A., Kajihara, R., Kimura, M.: Evaluation of the distribution and factors affecting blood pressure using medical checkup data in Japan. Health **9**(1), 124–137 (2016)

42. Mansour, R.P.: Visual charting method for creating electronic medical documents. US Patent 10,262,106, 16 April 2019

43. Ghafoorian, M., et al.: Transfer learning for domain adaptation in MRI: application in brain lesion segmentation. In: Descoteaux, M., Maier-Hein, L., Franz, A., Jannin, P., Collins, D.L., Duchesne, S. (eds.) MICCAI 2017. LNCS, vol. 10435, pp. 516–524. Springer, Cham (2017). https://doi.org/10.1007/978-3-319-66179-7_59

44. Chen, A., Zhang, Z., Li, Q., et al.: Feasibility study for implementation of the AI-powered Internet+ Primary Care Model (AiPCM) across hospitals and clinics in Gongcheng county, Guangxi, China. Lancet **394**, S44 (2019)

45. Laplante-Lévesque, A., Abrams, H., Bülow, M., et al.: Hearing device manufacturers call for interoperability and standardization of internet and audiology. Am. J. Audiol. **25**(3S), 260–263 (2016)

46. Morley, J., Taddeo, M., Floridi, L.: Google Health and the NHS: overcoming the trust deficit. Lancet Digit. Health **1**(8), e389 (2019)

47. Rossini, G., Parrini, S., Castroflorio, T., et al.: Diagnostic accuracy and measurement sensitivity of digital models for orthodontic purposes: a systematic review. Am. J. Orthod. Dentofacial Orthop. **149**(2), 161–170 (2016)

48. Huang, K., Cheng, H., Zhang, Y., et al.: Medical knowledge constrained semantic breast ultrasound image segmentation. In: Proceedings of the International Conference on Pattern Recognition (ICPR), pp. 1193–1198 (2018)

49. Krittanawong, C.: The rise of artificial intelligence and the uncertain future for physicians. Eur. J. Intern. Med. **48**, e13–e14 (2018)

50. Li, H., Jiang, G., Zhang, J., et al.: Fully convolutional network ensembles for white matter hyperintensities segmentation in MR images. NeuroImage **183**, 650–665 (2018)

51. Jain, A., Ratnoo, S., Kumar, D.: Addressing class imbalance problem in medical diagnosis: a genetic algorithm approach. In: Proceedings of the International Conference on Information, Communication, Instrumentation and Control (ICICIC), pp. 1–8 (2017)
52. Wartman, S.A., Combs, C.D.: Reimagining medical education in the age of AI. AMA J. Ethics **21**(2), 146–152 (2019)
53. Lin, C.H., Chang, C., Chen, Y., et al.: COCO-GAN: generation by parts via conditional coordinating. In: Proceedings of the International Conference on Computer Vision (ICCV), pp. 4512–4521 (2019)
54. Xu, Z., Wang, X., Shin, H., et al.: Correlation via synthesis: end-to-end nodule image generation and radiogenomic map learning based on generative adversarial network. arXiv:1907.03728 (2019)

Know Yourself: An Adaptive Causal Network Model for Therapeutic Intervention for Regaining Cognitive Control

Nimat Ullah[✉] [iD] and Jan Treur[iD]

Social AI Group, Vrije Universiteit Amsterdam, Amsterdam, The Netherlands
nimatullah09@gmail.com, j.treur@vu.nl

Abstract. Long term stress often causes depression and neuronal atrophies that in turn can lead to a variety of health problems. As a result of these cellular changes, also molecular changes occur. These changes, that include increase of glucocorticoids and decrease of the brain-derived neurotrophic factor, have the unfortunate effect that they decrease the cognitive abilities needed for the individual to solve the stressful situation. Such cognitive abilities like reappraisal and their adaptation mechanisms turn out to be substantially impaired while they are needed for regulation of the negative emotions. However, antidepressant treatments and some other therapies have proved to be quite effective for the strengthening of such cognitive abilities. This study introduces an adaptive causal network model for this phenomenon where a subject loses his or her cognitive abilities (negative metaplasticity) due to long-term stress and re-improve these cognitive abilities (positive metaplasticity) through mindfulness-based cognitive therapy (MBCT). Simulation results have been reported for demonstration of the phenomenon.

Keywords: Adaptive causal modeling · Negative metaplasticity · Positive metaplasticity · Mindfulness · Cognition · Therapy · Reappraisal

1 Introduction

Potentially, there can be various reasons and forms of cognitive alteration such as decline in cognitive abilities with age where the rate of decline varies from person to person (Verhaeghen 2011). The decline can be due to long-term stress (Garcia 2002) or due to reduction of flexibility as the person grows older (Charles 2010). This paper, specifically focuses on the consequences caused by long-term stress. Stress has become one of the most common negative emotion these days. Various studies have found that long-term stress can have very severe consequences (Garcia 2002; Mazure et al. 2002; Tennant 2001). These studies highlight the cognitive decline as one of the main consequences of long-term stressors. Decrease in the brain-derived neurotrophic factor (BDNF) as a result of increase in the glucocorticoids was found to be the main reason of cell loss and hence the alteration in synaptic plasticity at a cellular level (Fuchs and Gould 2000; Sapolsky 1999).

© IFIP International Federation for Information Processing 2020
Published by Springer Nature Switzerland AG 2020
I. Maglogiannis et al. (Eds.): AIAI 2020, IFIP AICT 584, pp. 334–346, 2020.
https://doi.org/10.1007/978-3-030-49186-4_28

However, there are a number of studies like (Garland et al. 2009; Garland et al. 2011) who put forward that mindfulness-based cognitive therapy (MBCT) combining different techniques from therapies and training helps in strengthening cognitive abilities, specifically those impaired by long-term stress. It works by focusing on present moment, gaining awareness of one's self and accepting the reality. This study considers MBCT instead of standard CBT, being one of the most studied therapies, due to the fact that some studies like (Troy et al. 2013) have compared MBCT and CBT in terms of their performance and found that MBCT is more effective.

Moreover, this study considers an adaptive causal network modeling approach (Treur 2020) to model the above mentioned phenomena because stressful emotions and their effects form an adaptive and cyclic process which this approach particularly handles quite effectively as demonstrated, for example, in (Ullah and Treur 2020; Ullah and Treur 2019). This modeling approach can be considered as a branch in the causal modeling area which has a long tradition in AI; e.g., see (Kuipers 1984; Kuipers and Kassirer 1983; Pearl 2009). It distinguishes itself by a dynamic perspective on causal relations, according to which causal relations exert causal effects over time, and in addition these causal relations themselves can change over time as well. The basic type of network model used is called a *multilevel adaptive temporal-causal network model*. By adding dynamics an adaptation to causal modeling, applications become possible that otherwise would be out of reach of causal modeling. This provides a useful approach to translate (supported by a dedicated modeling environment) qualitative processes as known from empirical literature into adaptive causal network models that can be used for simulation.

In the rest of the paper, Sect. 2 gives brief account of the literature on the subject, Sect. 3 presents the adaptive cognitive network model which is explained by simulation results in detail in Sect. 4. Finally, the paper is concluded in Sect. 5.

2 Background Literature

Research on stress, being one of the common negative emotions, has come up with vital long lasting negative consequences of it (Mazure et al. 2002; Tennant 2001). Studies like (Garcia 2002) have shown its contribution to depression and the adaptive casual way it affects one's cognitive abilities. It's worth noting here that the structural and functional changes brought about by stress are similar to those of depression. At a cellular level, stress and depression cause cell and neuronal losses. By (Sapolsky 1999) such cellular changes in the hippocampus are linked to the increase in plasma levels of glucocorticoid hormones like cortisol.

At a molecular level, the cellular deficiencies were found to take place at the hippocampus. This happens most of the time due to the decrease in the expression of the brain-derived neurotrophic factor (BDNF) associated with elevation of glucocorticoids (Fuchs and Gould 2000; Sapolsky 1999). Similarly, (Smith et al. 1995) also support the notion that high levels of glucocorticoids induced by stress and its administration are considered to be down-regulating the hippocampal expression. (Mocchetti et al. 1996) considers the reduction of BDNF, which supports neuronal survival and function in the hippocampus, as the potential mediating action of glucocorticoid on hippocampus. The effect of the boost of glucocorticoids is referred to as negative metaplasticity as it down-regulates adaptivity of the hippocampal synaptic connectivity. In contrast, the boost in the expression of BDNF is referred to as positive metaplasticity as it strengthens adaptivity of the connectivity in the hippocampus. As a result of these changes, the subject loses his control of cognitive abilities and is unable to efficiently regulate his emotions in an adaptive manner. On the other hand, the same mechanisms can reversibly be used via treatment as well (Garcia 2002), i.e., through the increase in the expression of BDNF. Synapses are responsible for the processing and transmission of neural information with some efficacy and their alteration is referred to as 'synaptic plasticity', which is a form of (first-order) adaptation. The mechanisms described above indicate how synaptic plasticity itself also can change, which is a form of second-order adaptation usually called metaplasticity. When metaplasticity enhances the adaptive cognitive function, by Garcia (2002) it is called positive metaplasticity and if this change brings some impairment to the adaptive cognitive function, then it's referred to as negative metaplasticity this indeed has been observed in case of long-term stress (Foster 1999; Kim and Yoon 1998); similar cognitive impairment has been observed in various other studies in both humans (Lupien et al. 1997) and animals (Mizoguchi et al. 2000).

Cognitive reappraisal, as a strategy, is considered to be a very adaptive strategy which contributes to positive psychological health by reducing negative effects and it is associated to improved memory and interpersonal functioning in contrast to, for example, suppression (Butler et al. 2003; Gross and John 2003; John and Gross 2004).

To deal with the cognitive decline caused by long-term stressors, we consider Mindfulness-based cognitive therapy (MBCT) as found quite effective by various researchers in this domain (Garland et al. 2009; Garland et al. 2011). The purpose of MBCT is to improve psychological health by increasing mindfulness. This therapy combines Kabat-Zinn's (Kabat-Zinn 1990) mindfulness-based stress reduction program with the techniques used in Cognitive Behavior Therapy (CBT). Kabat-Zinn's Mindfulness-Based stress reduction program involves daily meditation and self-awareness exercises. The MBCT training program aims at promoting acceptance of thoughts/feelings without being judgmental, focusing on the present moment and awareness of one's self (Coffey et al. 2010). According to (Allen et al. 2006) practicing of acceptance is meant to develop the ability to distinguish oneself from the contents of negative thoughts and recognize that emotions are non-permanent events. This is achieved by bringing a person in a decentered metacognitive state which may help in increasing cognitive flexibility through disengagement from the initial negative thoughts/appraisal, as has been considered essential by (Garland et al. 2009). After disengagement comes the ability to successfully switch from negative appraisal to positive appraisal and previous research credits this ability to mindfulness training (Jha et al. 2007). Similarly, MBCT encourages individuals to focus on the present moment as it is expected to give insight into one's own (wide range of) stimuli related physical sensations, feelings and thoughts. Otherwise, chances of failure increase in the reappraisal/re-interpretation of the thoughts without knowing about it.

3   Multilevel Adaptive Cognitive Modeling

This section of the paper gives overview of the multilevel adaptive causal network modeling approach (Treur 2016, 2020) that has been used for development and simulation of the adaptive causal network model. Table 1 summarizes the main conceptual and numerical representations for this adaptive causal modeling approach. The Network-oriented modeling approach provides a library of over 36 combination functions. Apart from the available combination functions, self-defined combination functions can also be added to the library which makes the approach quite flexible and supports a wide application range.

Table 1. Conceptual and numerical representations for the adaptive causal modeling approach

Concept	Conceptual representation	Explanation
States and connections	$X, Y, X \to Y$	Describes the nodes (representing states) and links (representing causal connections between states) of the network structure
Connection weight	$\omega_{X,Y}$	The *connection weight* $\omega_{X,Y} \in [-1, 1]$ represents the strength of the causal impact of state X on state Y through connection $X \to Y$
Aggregating multiple impacts on a state	$\mathbf{c}_Y(..)$	For each state Y (a reference to) a *combination function* $\mathbf{c}_Y(..)$ is chosen to combine the causal impacts of other states on state Y
Timing of the effect of causal impact	η_Y	For each state Y a *speed factor* $\eta_Y \geq 0$ is used to represent how fast a state is changing upon causal impact
Concept	Numerical representation	Explanation
State values over time t	$Y(t)$	At each time point t each state Y in the model has a real number value in $[0, 1]$
Single causal impact	$\mathbf{impact}_{X,Y}(t)$ $= \omega_{X,Y} X(t)$	At t state X with a connection to state Y has an impact on Y, using connection weight $\omega_{X,Y}$
Aggregating multiple causal impacts	$\mathbf{aggimpact}_Y(t)$ $= \mathbf{c}_Y\left(\mathbf{impact}_{X1,Y}(t), \ldots, \mathbf{impact}_{Xk,Y}(t)\right)$ $= \mathbf{c}_Y\left(\omega_{X1,Y} X_1(t), \ldots, \omega_{Xk,Y} X_k(t)\right)$	The aggregated causal impact of multiple states X_i on Y at t, is determined using combination function $\mathbf{c}_Y(..)$
Timing of the causal effect	$Y(t + \Delta t) = Y(t) +$ $\eta_Y\left[\mathbf{aggimpact}_Y(t) - Y(t)\right]\Delta t$ $= Y(t) + \eta_Y[\mathbf{c}_Y(\omega_{X1,Y} X_1(t), \ldots,$ $\omega_{Xk,Y} X_k(t)) - Y(t)]\Delta t$	The causal impact on Y is exerted over time gradually, using speed factor η_Y; here the X_i are all states with outgoing connections to state Y

Table 2 provides explanation of the various states of the adaptive causal network model proposed in this paper. In the table and the figure, the background colors differentiate between the (adaptation) levels of the model and are part of the meaning; they are

Table 2. Overview of the states of the multi-level network model in Fig. 1

	State	Explanation	Level
X_1	ws_s	World state for stimulus 's'	Base level
X_2	ss_s	Sensor state for stimulus 's'	
X_3	srs_s	Sensory representation state for stimulus 's'	
X_4	gs_b	Goal state for body 'b'	
X_5	ps_g	Preparation state for goal 'g'	
X_6	es_g	Execution state for goal 'g'	
X_7	ss_b	Sensor state for body state b	
X_8	srs_b	Sensory representation state for body sate b	
X_9	fs_b	Feeling state for body state b	
X_{10}	ps_b	Preparation state for body state b	
X_{11}	es_b	Expression execution state for body state b	
X_{12}	bs_-	Belief state for negative belief -	
X_{13}	bs_+	Belief state for positive belief +	
X_{14}	$accp_b$	Acceptance state for body	
X_{15}	aw_b	Awareness state for body	
X_{16}	pr_b	Present moment state of body	
X_{17}	cs_{reapp}	Control state for cognitive reappraisal	
X_{18}	$\mathbf{W}_{fs_b,cs_{reapp}}$	Reified representation state for connection weight $\omega_{fs_b,cs_{reapp}}$	First Reification Level
X_{19}	$\mathbf{M}_{W_{fs_b,cs_{reapp}}}$	Reified representation state for persistence factor μ for $\mathbf{W}_{fs_b,cs_{reapp}}$	Second Reification Level
X_{20}	$\mathbf{H}_{W_{fs_b,cs_{reapp}}}$	Reified representation state for speed factor η for $\mathbf{W}_{fs_b,cs_{reapp}}$	

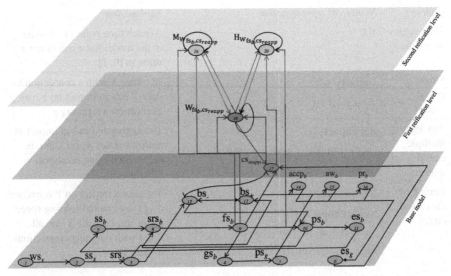

Fig. 1. Adaptive causal network model for therapeutic intervention for long-term stress

called reification levels as they explicitly represent by their states some of the network characteristics of the level below. For instance, the base model has 17 states, the first reification level modeling synaptic plasticity (first-order adaptation) has 1 state and the second reification level modeling metaplasticity (second-order adaptation) has 2 states. The modeled synaptic plasticity at the first reification level is addressed by the (adaptive) connection weight $\omega_{fs_b, cs_{reapp}}$ by representing it by reification state $\mathbf{W}_{fs_b, cs_{reapp}}$. The metaplasticity concerning adaptation of the dynamics of the adaptation of this connection weight is addressed by the second-level reification states $\mathbf{M}_{\mathbf{W}_{fs_b, cs_{reapp}}}$ for the persistence of the learnt effects and $\mathbf{H}_{\mathbf{W}_{fs_b, cs_{reapp}}}$ for the speed of the adaptation process.

The connectivity picture of the model shown in Fig. 1 demonstrates the phenomenon where long-term stress causes negative metaplasticity of cognitive abilities, i.e. loss in the ability to use and adapt cognitive reappraisal for regulation of negative emotions. In contrast, mindfulness-based cognitive therapy has been used as an intervention to obtain positive metaplasticity through which upregulation of the cognitive reappraisal ability can take place. In Fig. 1, the bottom level represents the base model which performs the basic function of regulation of emotions through reappraisal. Cognitive reappraisal changes one's perception/belief about the stimulus: from a negative to a positive interpretation in this model. Generally, reappraisal can also be from positive believe to negative belief depending upon the demand of the situation, but in this model, reappraisal refers to the reinterpretation to positive belief from negative. The first reification level modeling first-order adaptation, represents Hebbian learning for the connection from base state fs_b to base state cs_{reapp}. The dynamics of this state $\mathbf{W}_{fs_b, cs_{reapp}}$ represents the learning taking place for this connection at the base level. Learning doesn't only refer to increase in the strength of the connection, it can also involve decrease in the strength of the connection over time. In this model, as demonstrated in Figs. 2 and 3, initially, this connection from fs_b, to cs_{reapp} gets weaker due to long-term stress. When this connection gets weaker, the negative (body) feeling state fs_b gets higher which activates the goal state gs_b to address the problem, as also demonstrated in (Mohammadi Ziabari and Treur 2019) for such kind of interventions. This goal state represents the intention of the person to undergo some kind of therapy for getting back on track. In the scenario used for this model, goal activation refers to the intention to undergo MBCT (therapy) but in general it can be any kind of intervention, for example, antidepressant treatment as suggested in (Garcia 2002) etc.

The second-order reification states $\mathbf{M}_{\mathbf{W}_{fs_b, cs_{reapp}}}$ and $\mathbf{H}_{\mathbf{W}_{fs_b, cs_{reapp}}}$ represent the persistence factor μ and the speed factor η for $\mathbf{W}_{fs_b, cs_{reapp}}$ respectively. The $\mathbf{M}$-state refers to how long or short the $\mathbf{W}$-state retains its learned value. Similarly, the $\mathbf{H}$-state controls the speed factor of the $\mathbf{W}$-state managing how fast or slow the $\mathbf{W}$-state learns of forgets the value. Change in the speed and persistence factor is referred to as metaplasticity. The upregulation is called positive while the downregulation is called negative metaplasticity.

The two boxes below give insight into the different dynamics of the network. The colors and the boxes itself are part of the meaning and standard defined by network oriented modeling approach in (Treur 2020). In the boxes below, Box 1 represent the role matrices $\mathbf{mb}$ and $\mathbf{mcw}$. In role matrix $\mathbf{mb}$, each state has a row which shows all the incoming connections to that state. It's worth mentioning here that in matrix $\mathbf{mb}$ each state has only the incoming connections which are either at the same level or come from a lower level. The downward connections are indicated in the other role matrices instead. For instance, in matrix **mcw** the state number X_{18} (i.e., the **W**-state) represents the adaptive connection from X_9 to X_{17} at the base level, the causal effect of which is modeled by the downward connection from X_{18}. Moreover, the values between -1 and 1 in role matrix **mcw** represent the connection weights of the incoming connections to that specific state. The specific values in the role matrices allow to get the simulation pattern shown in Fig. 2 and Fig. 3. The values were chosen such that the simulation results exhibit a pattern similar to the ones found in the literature.

In Box 2, role matrices **mcfw**, **mcfp** and **ms** are given. Matrix **mcfw** indicates the combination function used at a state X_i for the aggregation of incoming causal impact from other states to state X_i. For instance, state X_1 uses the identify function, state X_{18} uses the Hebbian combination function and state X_{19} uses **alogistic(..)** as combination function, defined by

$$\textbf{alogistic}_{\sigma,\tau}(V_1, \ldots, V_k) = \left[\frac{1}{1 + e^{-\sigma(V_1 + \cdots V_k - \tau)}} + \frac{1}{1 + e^{\sigma\tau}}\right](1 + e^{-\sigma\tau}) \quad (1)$$

Moreover, the first-order adaptation state X_{18} uses the Hebbian learning combination function **hebb(..)** defined by

$$\textbf{hebb}_{\mu}(V_1, V_2, W) = V_1 V_2 (1 - W) + \mu W \quad (2)$$

mb connectivity:		1	2	3	4
base connectivity					
X_1	ws_s	X_1			
X_2	ss_s	X_1			
X_3	srs_s	X_2			
X_4	gs_b	X_9			
X_5	ps_g	X_4			
X_6	es_g	X_{14}	X_{15}	X_{16}	
X_7	ss_b	X_{11}			
X_8	srs_b	X_7	X_{10}		
X_9	fs_b	X_8			
X_{10}	ps_b	X_9	X_{12}		
X_{11}	es_b	X_{10}			
X_{12}	bs_-	X_3	X_{13}	X_{17}	
X_{13}	bs_+	X_3	X_{12}		
X_{14}	$accp_b$	X_5	X_9		
X_{15}	aw_b	X_5	X_9		
X_{16}	pr_b	X_5	X_9		
X_{17}	cs_{reapp}	X_6	X_9	X_{12}	
X_{18}	$W_{fs_b \cdot cs_{reapp}}$	X_9	X_{17}	X_{18}	
X_{19}	$M_{W_{fs_b \cdot cs_{reapp}}}$	X_9	X_{17}	X_{18}	X_{19}
X_{20}	$H_{W_{fs_b \cdot cs_{reapp}}}$	X_9	X_{17}	X_{18}	X_{20}

mcw connectivity:		1	2	3	4
connection weights					
X_1	ws_s	1			
X_2	ss_s	1			
X_3	srs_s	1			
X_4	gs_b	0.8			
X_5	ps_g	1			
X_6	es_g	0.5	0.5	0.58	
X_7	ss_b	1			
X_8	srs_b	0.27	0.3		
X_9	fs_b	1			
X_{10}	ps_b	0.1	0.45		
X_{11}	es_b	1			
X_{12}	bs_-	0.6	-0.4	-0.21	
X_{13}	bs_+	0.4	-0.35		
X_{14}	$accp_b$	0.9	0.2		
X_{15}	aw_b	0.9	0.2		
X_{16}	pr_b	0.9	0.2		
X_{17}	cs_{reapp}	1	X_{18}	0.2	
X_{18}	$W_{fs_b \cdot cs_{reapp}}$	1	1	1	
X_{19}	$M_{W_{fs_b \cdot cs_{reapp}}}$	-1	1	1	1
X_{20}	$H_{W_{fs_b \cdot cs_{reapp}}}$	-1	1	1	1

Box 1 Role matrices for connectivity

Role matrix **mcfp** provides the parameter values for each of the combination function used as indicated in **mcfw**. In the row for X_{18}, in column 2 (for Hebb) of **mcfp**, X_{19} in the red cell indicates the downward causal connection from the second-order reification state to X_{18}. In other words, X_{19} is responsible for the persistence factor of X_{18}. Similarly, the X_{20} in the red cell at the row for X_{18} in role matrix **ms** indicates the downward causal connection from the second-order reification level to the first-order reification level. In other words, X_{20} is responsible for the speed factor of X_{18}. These downward causal connections are essential to causally effectuate the adaptation processes modeled as dynamics at the higher levels.

4   Simulation Results

This section of the paper explains the simulation results given in Fig. 2 and Fig. 3 obtained from the adaptive causal network model with connectivity depicted in Fig. 1. These results were obtained using the values in **Box 1**, **Box 2**, and the initial values serving as inputs to the states as give in Table 3. Moreover, initial values of the states are given below in Table 2. Figure 2 demonstrates the basic phenomenon as discussed in the literature section, where long-term stressors decrease the adaptive cognitive abilities of the subject. The subject decides to undergo MBCT and regains his cognitive abilities. Figure 2 shows the patterns for all the states at the base level. In the figure, it can be seen that initially the control state cs_{reapp} for reappraisal gets activated as soon as negative belief bs_- gets higher. The control state cs_{reapp} changes the beliefs representing the interpretation of the person and, therefore, bs_- decreases while bs_+ increases. But during this time, gradual decrease in the activation level of cs_{reapp} can be observed and that's the time when bs_- also stays higher and bs_+ stays lower.

Table 3. Initial values serving as inputs to the states

State	ws_s	All other base states	$\mathbf{W}_{fs_b,\,cs_{reapp}}$	$\mathbf{HW}_{fs_b,\,cs_{reapp}}$	$\mathbf{MW}_{fs_b,\,cs_{reapp}}$
Initial value	1	0	0.3	0.5	0.9

mcfw aggregation — Combination function weights

		1 alogistic	2 hebb	3 Id
X_1	ws_s			1
X_2	ss_s			1
X_3	srs_s			1
X_4	gs_b	1		
X_5	ps_g			1
X_6	es_g	1		
X_7	ss_b			1
X_8	srs_b	1		
X_9	fs_b			1
X_{10}	ps_b	1		
X_{11}	es_b			1
X_{12}	bs_-	1		
X_{13}	bs_+	1		
X_{14}	$accp_b$	1		
X_{15}	aw_b	1		
X_{16}	pr_b	1		
X_{17}	cs_{reapp}	1		
X_{18}	$\mathbf{W}_{fs_b,cs_{reapp}}$		1	
X_{19}	$\mathbf{M}_{\mathbf{W}_{fs_b,cs_{reapp}}}$	1		
X_{20}	$\mathbf{H}_{\mathbf{W}_{fs_b,cs_{reapp}}}$	1		

mcfp aggregation — Combination function parameters

		1 Alogistic σ	1 Alogistic τ	2 Hebb μ	3 id
X_1	ws_s				1
X_2	ss_s				1
X_3	srs_s				1
X_4	gs_b	18	0.62		
X_5	ps_g				1
X_6	es_g	10	0.8		
X_7	ss_b				1
X_8	srs_b	10	0.3		
X_9	fs_b				1
X_{10}	ps_b	10	0.3		
X_{11}	es_b				1
X_{12}	bs_-	8	0.2		
X_{13}	bs_+	10	0.15		
X_{14}	$accp_b$	10	0.55		
X_{15}	aw_b	10	0.55		
X_{16}	pr_b	10	0.55		
X_{17}	cs_{reapp}	10	0.44		
X_{18}	$\mathbf{W}_{fs_b,cs_{reapp}}$			X_{19}	
X_{19}	$\mathbf{M}_{\mathbf{W}_{fs_b,cs_{reapp}}}$	10	0.85		
X_{20}	$\mathbf{H}_{\mathbf{W}_{fs_b,cs_{reapp}}}$	10	0.8		

ms timing — Speed factors

		1
X_1	ws_s	0
X_2	ss_s	1
X_3	srs_s	1
X_4	gs_b	0.01
X_5	ps_g	0.01
X_6	es_g	0.01
X_7	ss_b	0.2
X_8	srs_b	0.2
X_9	fs_b	0.2
X_{10}	ps_b	0.2
X_{11}	es_b	0.2
X_{12}	bs_-	0.2
X_{13}	bs_+	0.2
X_{14}	$accp_b$	0.01
X_{15}	aw_b	0.01
X_{16}	pr_b	0.01
X_{17}	cs_{reapp}	0.28
X_{18}	$\mathbf{W}_{fs_b,cs_{reapp}}$	X_{20}
X_{19}	$\mathbf{M}_{\mathbf{W}_{fs_b,cs_{reapp}}}$	0.01
X_{20}	$\mathbf{H}_{\mathbf{W}_{fs_b,cs_{reapp}}}$	0.01

Box 2 Role matrices for aggregation and timing

In the meanwhile, gradual increase in the goal state gs_b takes place, since as a result of decrease in the activation level of cs_{reapp} the intensity of negative feelings fs_b increases which in turn activates the goal gs_b to undergo some therapy to regain cognitive control. Goal state gs_b activates the states that lay on the casual pathway describing undergoing MBCT therapy. Finally, when the execution state for goal (therapy) es_g gets enough momentum, as a result cs_{reapp} again gets activated. This means that the person has re-gained his cognitive reappraisal ability during execution of the therapy; however, at this stage the person still depends on the therapy for this. While the person is again becoming able to reappraise his or her emotions with a little help from the therapy, due to the achieved lower levels of fs_b, the negative metaplasticity at the second reification level turns into positive metaplasticity by increasing the too low values of the two second-order states $\mathbf{M}_{\mathbf{W}_{fs_b,cs_{reapp}}}$ and $\mathbf{H}_{\mathbf{W}_{fs_b,cs_{reapp}}}$ to higher values. Therefore, now a learning process takes place (that was blocked before by the negative metaplasticity due to low values of $\mathbf{M}_{\mathbf{W}_{fs_b,cs_{reapp}}}$ and $\mathbf{H}_{\mathbf{W}_{fs_b,cs_{reapp}}}$), strengthening the connection from fs_b to cs_{reapp}. Due to the successful reappraisal, fs_b goes down, so it deactivates the goal gs_b to undergo therapy. This deactivates the therapy but the person is still able to reappraise his emotions effectively as before. It means that overall the connection has re-strengthened as a result of MBCT. This figure represents two phases of the subject's life. First phase is before his cognitive abilities are impaired by long-term stress where he regulates his emotions multiple times during that phase. The second phase is after the therapy when he's again able to use his cognitive abilities. The fluctuation in the graphs represent the cyclic nature of the stressful emotions that takes place repeatedly in cycles and are regulated accordingly.

This pattern is exactly as various researchers from cognitive and neuro-sciences have described in their articles as summarized in the background study in Sect. 2, for instance (Garcia 2002).

Figure 3 gives insight into the first and second-level reification states. The states represented in the figure are $\mathbf{W}_{fs_b,cs_{reapp}}$, $\mathbf{M}_{\mathbf{W}_{fs_b,cs_{reapp}}}$, $\mathbf{H}_{\mathbf{W}_{fs_b,cs_{reapp}}}$ which represent the adaptation of the indicated connection weight based on Hebbian learning, and the persistence factor and the speed factor for this adaptation, respectively. First it is shown that all these three states decrease because of the long-term stressors, but after activation of the therapy in the base model all three states again get higher values. It's worth noting here that all the states stay high even after the therapy gets deactivated in the base model. This shows that the person's wellbeing does not depend on the therapy anymore.

As already mentioned, $\mathbf{W}$ represent the Hebbian learning taking place at the connection indicated by fs_b, cs_{reapp} in the base model. The speed factor and persistence of the $\mathbf{W}$ is handled by the $\mathbf{H}$ and $\mathbf{M}$-states respectively. In the Fig. 3 initially negative metaplasticity occurs (lower values of the second-order states) because of long-term stressors, so the learning gets blocked in a sense. Due to this, the connection fs_b, cs_{reapp} gets weaker. But as the person undergoes MBCT, due to the positive effects that has on the second-order states $\mathbf{M}_{\mathbf{W}_{fs_b,cs_{reapp}}}$ and $\mathbf{H}_{\mathbf{W}_{fs_b,cs_{reapp}}}$, the connection fs_b, cs_{reapp} again gets stronger and the person is again able to reappraise his emotions even after there is no more therapy. These results in Fig. 2 and Fig. 3 are completely in line with the findings from cognitive and neurosciences.

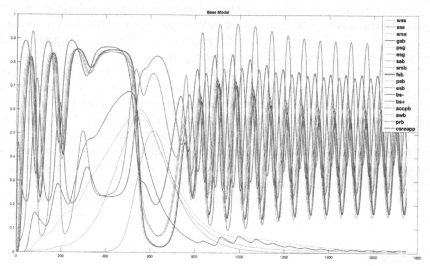

Fig. 2. Loss and gain of adaptive cognitive abilities as a result of long-term stressor (base level)

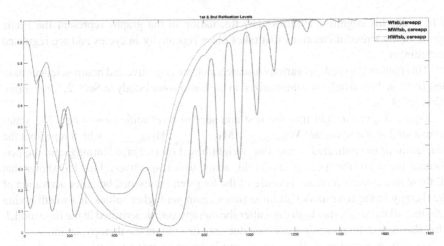

Fig. 3. First and second-order reification states for plasticity and metaplasticity over time

5 Conclusion

In this paper, the ideas on negative and positive metaplasticity from the (empirical) literature have been modeled and given a concrete shape in the form of simulations. On one hand, the multi-order adaptive causal network model presented in this paper gives a concrete shape to the abstract ideas, on the other hand it also highlights the consequences of long-term stressors along with the possible treatment for the psychological problems they cause. Findings from various neurological and psychological studies were brought together and presented in the form of a concrete computational model. Various therapies have already been modeled in work like (Ziabari and Treur 2019), but this paper considers therapy as a permanent treatment to cope with negative metaplasticity for the first time.

The paper also shows how the applicability scope of causal modeling, which has a long tradition in AI (e.g., see (Kuipers 1984; Kuipers and Kassirer 1983; Pearl 2009)), can substantially be widened by adding dynamics and (multi-order) adaptation to causal modeling. Doing this, causal modeling becomes applicable to modeling of dynamic and adaptive processes that otherwise would be out of reach for causal modeling.

The introduced second-order adaptive network model provides a good basis to develop a patient model (based on a virtual agent) for virtual training of therapists. In future, the author aims at extending the model by incorporating more neurological states for both the negative and positive metaplasticity and the neurological states involved in MBCT.

References

Allen, N.B., Blashki, G., Gullone, E.: Mindfulness-based psychotherapies: a review of conceptual foundations, empirical evidence and practical considerations. Aust. N. Z. J. Psychiatry **40**(2), 285–294 (2006). https://doi.org/10.1111/j.1440-1614.2006.01794.x

Butler, E.A., Egloff, B., Wlhelm, F.H., Smith, N.C., Erickson, E.A., Gross, J.J.: The social consequences of expressive suppression. Emotion 3(1), 48–67 (2003). https://doi.org/10.1037/1528-3542.3.1.48

Charles, S.T.: Strength and vulnerability integration: a model of emotional well-being across adulthood. Psychol. Bull. 136(6), 1068–1091 (2010). https://doi.org/10.1037/a0021232

Coffey, K.A., Hartman, M., Fredrickson, B.L.: Deconstructing mindfulness and constructing mental health: understanding mindfulness and its mechanisms of action. Mindfulness 1(4), 235–253 (2010). https://doi.org/10.1007/s12671-010-0033-2

Foster, T.C.: Involvement of hippocampal synaptic plasticity in age-related memory decline. Brain Res. Rev. 30(3), 236–249 (1999). https://doi.org/10.1016/s0165-0173(99)00017-x

Fuchs, E., Gould, E.: In vivo neurogenesis in the adult brain: regulation and functional implications. Eur. J. Neurosci. 12(7), 2211–2214 (2000)

Garcia, R.: Stress, metaplasticity, and antidepressants. Curr. Mol. Med. 2(7), 629–638 (2002). https://doi.org/10.2174/1566524023362023

Garland, E., Gaylord, S., Park, J.: The role of mindfulness in positive reappraisal. Explore 5(1), 37–44 (2009). https://doi.org/10.1016/j.explore.2008.10.001

Garland, E.L., Gaylord, S.A., Fredrickson, B.L.: Positive reappraisal mediates the stress-reductive effects of mindfulness: an upward spiral process. Mindfulness 2(1), 59–67 (2011). https://doi.org/10.1007/s12671-011-0043-8

Gross, J.J., John, O.P.: Individual differences in two emotion regulation processes: implications for affect, relationships, and well-being. J. Pers. Soc. Psychol. 85(2), 348–362 (2003). https://doi.org/10.1037/0022-3514.85.2.348

Jha, A.P., Krompinger, J., Baime, M.J.: Mindfulness training modifies subsystems of attention. Cogn. Affect. Behav. Neurosci. 7(2), 109–119 (2007)

John, O.P., Gross, J.J.: Healthy and unhealthy emotion regulation: personality processes, individual differences, and life span development. J. Pers. 72(6), 1301–1334 (2004). https://doi.org/10.1111/j.1467-6494.2004.00298.x

Kabat-Zinn, J.: Full Catastrophe Living: Using the Wisdom of Your Body and Mind to Face Stress, Pain and Illness, 1st edn. Delta, New York (1990)

Kim, J.J., Yoon, K.S.: Stress: metaplastic effects in the hippocampus. Trends Neurosci. 21(12), 505–509 (1998). https://doi.org/10.1016/s0166-2236(98)01322-8

Kuipers, B.: Commonsense reasoning about causality: deriving behavior from structure. Artif. Intell. 24, 169–203 (1984). https://doi.org/10.1016/0004-3702(84)90039-0

Kuipers, B., Kassirer, J.P.: How to discover a knowledge representation for casual reasoning by studying an expert pysician. In: Proceedings of the IJCAI 1983 (1983)

Lupien, S.J., et al.: Stress-induced declarative memory impairment in healthy elderly subjects: relationship to cortisol reactivity. J. Clin. Endocrinol. Metab. 82(7), 2070–2075 (1997). https://doi.org/10.1210/jcem.82.7.4075

Mazure, C.M., Maciejewski, P.K., Jacobs, S.C., Bruce, M.L.: Stressful life events interacting with cognitive/personality styles to predict late-onset major depression. Am. J. Geriatr. Psychiatry 10(3), 297–304 (2002)

Mizoguchi, K., Yuzurihara, M., Ishige, A., Sasaki, H., Chui, D.H., Tabira, T.: Chronic stress induces impairment of spatial working memory because of prefrontal dopaminergic dysfunction. J. Neurosci. 20(4), 1568–1574 (2000)

Mocchetti, I., Spiga, G., Hayes, V., Isackson, P., Colangelo, A.: Glucocorticoids differentially increase nerve growth factor and basic fibroblast growth factor expression in the rat brain. J. Neurosci. 16(6), 2141–2148 (1996). https://doi.org/10.1523/JNEUROSCI.16-06-02141.1996

Mohammadi Ziabari, S.S., Treur, J.: Cognitive modeling of mindfulness therapy by autogenic training. In: Satapathy, S.C., Bhateja, V., Somanah, R., Yang, X.-S., Senkerik, R. (eds.) Information Systems Design and Intelligent Applications. AISC, vol. 863, pp. 53–66. Springer, Singapore (2019). https://doi.org/10.1007/978-981-13-3338-5_6

Pearl, J.: Causality. Cambridge University Press, Cambridge (2009)

Sapolsky, R.M.: Glucocorticoids, stress, and their adverse neurological effects: relevance to aging. Exp. Gerontol. **34**(6), 721–732 (1999). https://doi.org/10.1016/S0531-5565(99)00047-9

Smith, M.A., Makino, S., Kvetnansky, R., Post, R.M.: Stress and glucocorticoids affect the expression of brain-derived neurotrophic factor and neurotrophin-3 mRNAs in the hippocampus. J. Neurosci. **15**(3), 1768–1777 (1995)

Tennant, C.: Work-related stress and depressive disorders. J. Psychosom. Res. **51**(5), 697–704 (2001). https://doi.org/10.1016/S0022-3999(01)00255-0

Treur, J.: Network-Oriented Modeling: Addressing Complexity of Cognitive, Affective and Social Interactions. Springer, Cham (2016). https://doi.org/10.1007/978-3-319-45213-5

Treur, J.: Network-Oriented Modeling for Adaptive Networks: Designing Higher-Order Adaptive Biological. Mental and Social Network Models. Springer, Cham (2020). https://doi.org/10.1007/978-3-030-31445-3

Troy, A.S., Shallcross, A.J., Davis, T.S., Mauss, I.B.: History of mindfulness-based cognitive therapy is associated with increased cognitive reappraisal ability. Mindfulness **4**(3), 213–222 (2013). https://doi.org/10.1007/s12671-012-0114-5

Ullah, N., Treur, J.: The choice between bad and worse: a cognitive agent model for desire regulation under stress. In: Baldoni, M., Dastani, M., Liao, B., Sakurai, Y., Zalila Wenkstern, R. (eds.) PRIMA 2019. LNCS (LNAI), vol. 11873, pp. 496–504. Springer, Cham (2019). https://doi.org/10.1007/978-3-030-33792-6_34

Ullah, N., Treur, J.: Better late than never: a multilayer network model using metaplasticity for emotion regulation strategies. In: Cherifi, H., Gaito, S., Mendes, J.F., Moro, E., Rocha, L.M. (eds.) COMPLEX NETWORKS 2019. SCI, vol. 882, pp. 697–708. Springer, Cham (2020). https://doi.org/10.1007/978-3-030-36683-4_56

Verhaeghen, P.: Cognitive processes and ageing. In: Stuart-Hamilton, I. (ed.) An Introduction to Gerontology, pp. 159–193. Cambridge University Press, Cambridge (2011). https://doi.org/10.1017/CBO9780511973697.006

Multi-omics Data and Analytics Integration in Ovarian Cancer

Archana Bhardwaj[(✉)] and Kristel Van Steen

GIGA-R Centre, BIO3 – Medical Genomics, University of Liège, Liège, Belgium
a.bhardwaj@uliege.be

Abstract. Cancer, which involves the dysregulation of genes via multiple mechanisms, is unlikely to be fully explained by a single data type. By combining different "omes", researchers can increase the discovery of novel bio-molecular associations with disease-related phenotypes. Investigation of functional relations among genes associated with the same disease condition may further help to develop more accurate disease-relevant prediction models. In this work, we present an integrative framework called Data & Analytic Integrator (DAI), to explore the relationship between different omics via different mathematical formulations and algorithms. In particular, we investigate the combinatorial use of molecular knowledge identified from omics integration methods netDx, iDRW and SSL, by fusing the derived aggregated similarity matrices and by exploiting these in a semi-supervised learner. The analysis workflows were applied to real-life data for ovarian cancer and underlined the benefits of joint data and analytic integration.

Keywords: Multi-omics integration · Semi-supervised learning · Network medicine

1 Introduction

Worldwide, ovarian cancer has the worst prognosis and highest mortality rate [1]. Coupling biomarker discovery to survival traits can increase our understanding about relevant tumor mechanisms and may provide insights into early detection strategies and/or preventive actions. The abundance of data due to advancements in high throughput sequencing technologies and carefully established data repositories are essential in this context. Cancer biology is complex and requires systems views to unravel the complexity. One of the Big Data cancer repositories are made available via The Cancer Genome Atlas Program (TCGA - https://www.cancer.gov/about-nci/organization/ccg/research/structural-genomics/tcga). It comprises multiple omics collections such as transcriptome, methylation and copy number variant (CNV) data. A transcriptome refers to the full range of messenger RNA that is produced in a particular cell or tissue type. A methylome, giving rise to methylation data, comprises the set of all nucleic acid methylation modifications in the genome of an organism or in a particular cell. CNVs are a specific type of DNA variation referring to copies of sections of the genome, the number of which varying

© IFIP International Federation for Information Processing 2020
Published by Springer Nature Switzerland AG 2020
I. Maglogiannis et al. (Eds.): AIAI 2020, IFIP AICT 584, pp. 347–357, 2020.
https://doi.org/10.1007/978-3-030-49186-4_29

between individuals. Even though non-omics data should not be ignored, in general, adopting multi-omics integrative strategies in cancers, like ovarian cancer, are believed to be the road to travel by, irrespective of whether subtyping or (survival) prediction is the aim (e.g., [2, 3]). With the vast amount of data to be mined, it is not surprising that machine learning tools have become indispensable in the data integration field, including multi-view methods for joint clustering of multiple data types [4], auto-encoder architectures based on omics and clinical data to study a variety of cancer-relevant traits [5], and deep-learners for robust cancer survival prediction [3].

While performing multi-omics integration, several challenges exist, such as validating the added value of multi-omics data integrative methods over single-omics analyses, assessing at which stage to perform the integration (e.g., early – data integration before analytic modelling, late – integration of modelling results), and how to deal with concordant and discordant relationships between multi-omics datasets in cancers. Here, we explore the performance of a novel combined omics data and analytics integrator (DAI) and compare it to state-of-the-art multi-omics data integrative approaches. We define performance in terms of optimized prediction or classification of ovarian cancer patients into short-term (less than 3 years) or long-term survival (at least 3 years). The categorization based on the threshold of 3 years of survival was inspired by [6]. We consider 3 omics data types: genomic (CNVs), epigenomic (methylation) and transcriptomic (gene expression). Notably, epigenomics refers to "epi"-genetic ("epi" from Greek: on top of) modifications that affect gene expression regulation but does not change the genomic sequence itself.

The paper is organized as follows. In Sect. 2, data overview and preparation steps are outlined. Analytical workflows are detailed in Sect. 3. Results are presented in Sect. 4. A discussion and closing remarks are given in Sect. 5.

2 Data Overview and Preparation

CNV, methylation and gene expression data for ovarian cancer were retrieved from the TCGA data portal via TCGA2STAT [7] software. In particular, we first discarded patients who did not have the 3 omics data types available. We then used the OMICSBind function TCGA2STAT to merge the available data and subsequently performed sample filtration following [8]. OMICSBind returns a combined data matrix for samples that are common to two types of molecular input data. Thereafter, we discarded patients having "vital status" as "dead" and "days to death" as "non-positive" or "NA", and we discarded patients having "vital status" as "alive" and "days to last follow-up" as "non-positive" or "NA" [8]. Next, we created two groups: ST (<3 years of survival) and LT (≥3 years of survival). Based on the above filtration criteria, LT/ST status was available for all patients included in this study (i.e. no missing labels). For each data type, we eliminated genes with a missing rate across all samples >20%. Remaining missing omics data entries were imputed with the kNNImpute function in R. In particular, each missing feature for a sample was replaced by a weighted average of the corresponding features from k nearest neighbors of that sample, weighted by the distance of the neighbors [9]. The resulting dataset for integrative analyses comprised 100 ST and 130 LT survivors, with information available on 22618 CNVs, 12644 methylation and 12043 gene expression features.

3 Analytical Workflows

As the aim is to optimize classification/prediction of LT/ST survival status and to exploit the integrated information of 3 omics datasets, we used the following promising integrative approaches as starting point: iDRW [10], netDx [11], and SSL [6], with default options, unless specified otherwise. Each of these methods adopt different paths towards generating omics features, that is the basis to assess similarities between patients. Apart from applying the original work-flows, patient-similarity matrices obtained from each approach were fused (when applicable) to create a single matrix per method, which was submitted to a graph-based learning method as in [6], so as to classify patients into LT/ST survival groups (Fig. 1). More details are given in the following paragraphs.

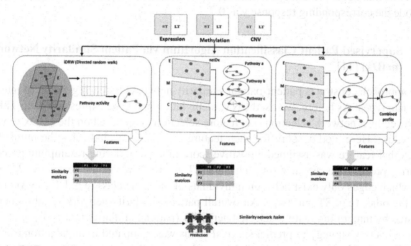

Fig. 1. DAI workflow to create data and analytics integrated patient similarity networks using adaptions of the machine learning approaches iDRW, netDx and SSL (see text). We employed molecular information on expression (E), methylation (M) and CNV (C) data and created new features on the basis of which to assess patient similarity. For iDRW an integrated omics network was used to derive pathway activity scores as new features. In netDx, features were pathway-genes and omics-specific patient similarity networks were derived for each pathway (4 are shown). Linked to SSL, original gene measurements (data-driven) and specific knowledge-based gene sets (knowledge-driven) features that carry information about disease relevance and protein networks were used. Developed patient similarity networks were combined into a method-specific single network. Principles of similarity network fusion were used to generate an analytics integrated patient similarity network, which served as input to a semi-supervised learning method to predict LT/ST survival state.

3.1 Integrative Directed Random Walk-Based Workflow (IDRW) [10]

In this approach gene-gene networks are built for each omics dataset, supported by KEGG pathway information (https://www.genome.jp/kegg/), from which an integrated directed network is derived. Then a random walk is performed on this integrated network. Significant genes in the integrated gene-gene networks and their weights from the DRW method contribute to integrated pathway-activity scores [10]. For our purposes we used a customized version from the authors to handle >2 omics. Out of 327, only the six significant pathways (T-test of pathway-activity across LT/ST survivor classes) were kept. Next, in line with the original iDRW workflow, a regression model in R was applied that classified the samples into ST and LT classes. As in the current study the focus lies on integration and not on variations of prediction model paradigms, we replaced the logistic regression model by a graph-based semi-supervised learner that can be applied with missing classes and with multiple input data types (i.e., iDRW + SSL). We thus converted the patient similarity matrix W_{iDRW} to a Laplacian L_{iDRW} and obtained final class predictions by solving $(I + \mu L_{iDRW})^{-1}y$, with y encoded as $(-1, 1)$ corresponding to (ST survival, LT survival) and μ a trade-off parameter, following the single-graph based semi-supervised learner of [6]. Note that in the presence of a missing survival status for a patient, it would suffice to encode the corresponding response y as 0.

3.2 Supervised Patient Classification Algorithm via Patient Similarity Networks (netDx) [11]

The approach constructs patient-patient similarity networks for each gene set of interest per data types. As before, we used CNV, methylation and gene expression data. This is followed by a network selection (i.e. feature selection) step based on the netDx scoring procedure. Here, netDx score for each feature (i.e. pathway) indicates the number of times that feature was assigned a positive score in a query during resampling process. Scoring process was repeated for each class (ST and LT). At end, the best network is one for which edges only exist between individuals of the same class (e.g. LT survivor) and not the other (e.g. ST survivor). An overall patient similarity network is subsequently created by integrating feature-selected networks (patient similarity matrix W_{netDx}). The original netDx strategy to predict survival status was compared to an adaption (netDX + SSL) using the semi-supervised learner as before with predicted classes obtained by solving $(I + \mu L_{netDx})^{-1}y$ (L_{netDx}: the graph Laplacian linked to W_{netDx}).

3.3 Graph-Based Semi-supervised Learning (SSL) [6]

Also here, the approach is based on creating patient similarity matrices for each omics data type separately. However, the features used to assess patient-to-patient similarity is different from the previous approaches. In particular, pre-defined gene sets as "genomic knowledge" were downloaded from the Molecular Signatures Database (MSigDB 7.0).32 [12]: chemical and genetic perturbations and canonical pathways (C2), motif (C3) and cancer gene sets (C4), gene ontology (C5), and immunological signatures (C7), involving 5501, 831, 858, 9996 and 4872 gene sets, respectively. We also collected a list of 2067 "seed genes" from the OCGene database, appended with genes from Papp et al. [13], leading to a unique seed gene list of 2072 genes. These were submitted to ToppGenet [14] to prioritize neighboring genes of the seeds based on functional similarity to the seeds or topological features in a protein-protein interaction network. The top 1% prioritized genes (1600 genes) were used to refine the MSigDB-derived "genomic knowledge" gene sets (number of genes in C2: 3132, C3: 568, C4: 449, C5: 5593, C7: 2711). Gene measurements per patient were subsequently averaged within each genomic knowledge gene set and were used to create "knowledge-driven" patient similarity matrices W_{c2_exp}, W_{c2_cnv}, W_{c2_meth}, W_{c3_exp}, W_{c3_cnv}, W_{c3_meth}, W_{c4_exp}, W_{c4_cnv}, W_{c4_meth}, W_{c5_exp}, W_{c5_cnv}, W_{c5_meth}, W_{c7_exp}, W_{c7_cnv}, and W_{c2_meth}. This is in contrast to using all original gene measurements, which would lead to "data-driven" similarity matrices W_{DD_E}, W_{DD_CNV}, W_{DD_METH}, for unfiltered measurements of gene expression, CNV and methylation, respectively. For each gene set of interest, the weights α for these matrices were estimated so as to optimize LT/ST survival class prediction as in the minimization problem $\min_{\alpha} y^T \left(I + \sum_{k=1}^{K} \alpha_k L_k \right)^{-1} y$, $\sum_k \alpha_k \leq \mu$ (K: number of graphs $= 3$; L_k: Laplacian corresponding to graph k; y: class response vector). The final class predictions were obtained by $\left(I + \sum_{k=1}^{K=3} \alpha_k L_k \right)^{-1} y$.

3.4 Data and Analytics Integrator (DAI)

We started by adapting netDx as follows. We obtained a single similarity matrix by fusing ST and LT specific similarity matrices. In particular, multiple pathway profiles for ST patients were integrated while adopting Similarity Network Fusion (SNF) analytics [15], leading to $W_{netDx-ST}$, and similar for LT patients, leading to an aggregated patient similarity matrix $W_{netDx-LT}$. The fused matrix was denoted by W_{netDx_S}, where the underscore "S" now refers to the pooled LT and ST survivors. This matrix was subsequently converted to a Laplacian L_{netDx_S} for use in the semi-supervised learner of [6], as explained before.

Then SSL was adapted to generate a single patient similarity matrix by first retrieving the software's weights α_i for each data type i (expression, methylation, and CNV). Second, we normalized the retained weights (i.e. new weights sum to 1) to form an integrated patient similarity matrix $W_{SSL\_adapted} = \sum_i \alpha_i M_i$, with Mi denoting the patient similarity matrix derived from omics data type i. Lastly, we built a shell around adaptations of iDRW, netDx and SSL integrating all three matrices W_{iDRW}, W_{netDx_S} and $W_{SSL\_adapted}$. In particular, we SNF-fused the three matrices and converted the resulting data and analytics aggregated patient similarity matrix W_{DAI} to a Laplacian L_{DAI}. Class predictions were obtained by solving $(I + \mu L_{DAI})^{-1} y$, with y encoded as $(-1, 1, 0)$ corresponding to (ST survivor, LT survivor, unknown survival state), and with μ a trade-off parameter between predictions close to the given label and predictions not too different from those for graph-adjacent nodes. DAI (Fig. 1) also allows the option to apply the multi-graph based SSL model of [6] on L_{iDRW}, L_{netDx} and L_{SSL} directly, instead of first fusing patient similarity matrices and second applying a single-graph semi-supervised learner. Key (dis-)similarities between DAI and iRDW, netDx and SSL in their original forms are summarized in Table 1.

Table 1. Highlighted (dis-)similarities between DAI and original implementations of iDRW, netDx and SSL.

Highlight	SSL	netDx	iDRW	DAI
Input data (Omics based)	Data-driven or knowledge-driven	Data driven	Data driven	Knowledge-driven
Weighted data accommodated	Yes	No	No	Yes
Feature selection	No additional feature selection	Identify omics-specific features leading to pathway-specific profiles	Identify genes contributing to pathway activity score	No additional feature selection
Output includes patient similarity matrix	Yes	Yes	No	Yes
Prediction model via multi/single graph-based semi-supervised learning	Yes	No	No	Yes

4 Results

4.1 Single Omics Analyses

We first analyzed each omics data type separately, in a data-driven and knowledge-drive fashion, as explained in Sect. 3 (SSL). For the ovarian samples, W_{c2_cnv}, W_{c3_cnv}, W_{c4_cnv}, W_{c5_cnv} and W_{c7_cnv}, typically gave rise to the highest AUC values (Fig. 2). Overall, knowledge-based SSL outperformed data-driven SSL. Single graph-based SSL based on patient similarity for C2, C3, C4, C5, and C7 typically increased AUC estimates compared to data-driven approaches.

4.2 Multi Omics Integration

Next, based on 3-omics integration with iDRW, we identified numerous significant genes. Using the original workflow of iDRW, a total number of 1145, 2544, and 1846 genes from CNV, methylation, and expression omics, respectively, were found to be uniquely significant (unadjusted p value < 0.05). We mapped all these genes on their respective chromosomes (Fig. 3A: circular plot). Only 32 were common (*ABCA2, ACSL3, AKAP1, ALPI, AP3B2, APTX, ARPC2, CD79B, CLTC, COLEC12, COX11, CSHL1, CYB561, DDX42, EXOSC9, FAm83E, HOXB9, INTS9, LEPROTL1, mPHOSPH10, NFE2L1, NR2F2, OSBPL7, PmL, RPP25, SNF8, SNRPG, TIA1, TJP1, TUBD1, UBAC1, ZNF652*). Characteristic for these common genes was that they appeared to be highly co-expressed (Fig. 3B). iDRW's multi-omics view highlighted 6 statistically significant pathways,

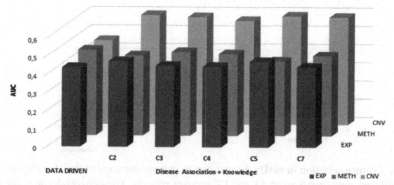

Fig. 2. Single-omics prediction performance. By data type AUC estimates for data-driven (W_{DD_exp}, W_{DD_cnv}, W_{DD_meth}) and knowledge-driven models (W_{c2_exp}, W_{c2_cnv}, W_{c2_meth}, W_{c3_exp}, W_{c3_cnv}, W_{c3_meth}, W_{c4_exp}, W_{c4_cnv}, W_{c4_meth}, W_{c5_exp}, W_{c5_cnv}, W_{c5_meth}, W_{c7_exp}, W_{c7_cnv}, and W_{c7_meth}).

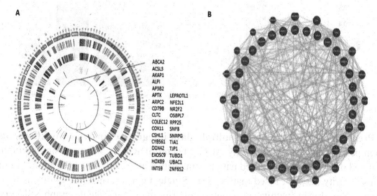

Fig. 3. iDRW identified significant genes. A. Circos plot: Chromosomal distribution of significant genes unique to either CNV, methylation, and expression omics. The outer circle represents the chromosomal bands. First inner circles represent the distribution of omics specific (inward direction: CNV, methylation, and expression) genes. The red circle (i.e. fourth layer) encapsulates significant genes common across three omics after integration; B. Gene-expression network of 32 significant genes, common to all considered ovarian omics data types. (Color figure online)

implying that their corresponding pathway scores were significantly different between LT/ST survivors. The AUC of the original iDRW using a logistic regression model was estimated to be 0.32, which is lower than AUC = 0.51 with our adapted version using W_{iDRW} and predictions based on single graph semi-supervised learning.

Application of netDx to ovarian patient samples showed a higher number of KEGG pathways crossing the netDx threshold criterion for LT/ST survival prediction (Fig. 4). These pathways profiles were converted into patient specific similarity matrices to derive group specific W_{netDx_ST} and W_{netDx_LT} weight matrices (see also Sect. 3).

Furthermore, the original netDx implementation gave AUC = 0.50 (Fig. 5). The adapted version with SNF fused similarity matrix W_{netDx_S} submitted to single-graph

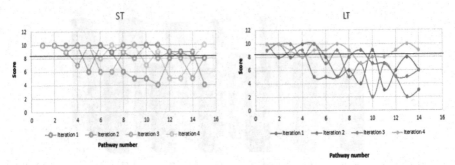

Fig. 4. Feature selection in netDx. Scatter plot to indicate the score of the 15 pathways across four different runs (iteration) in ST and LT survivor patients. The horizontal line is the netDx proposed threshold.

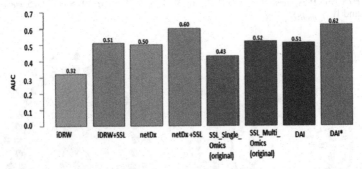

Fig. 5. Prediction performance of multiple data integrative analysis workflows and DAI. Prediction performance is measured by AUC. Legend: iDRW + SSL uses single-graph SSL to the integrated pathway-activity based patient similarity matrix; netDx + SSL applies multi-graph SSL to a similarity network fused matrix; SSL_Single_Omics (original) exploits a single-omics based knowledge-driven (pathways) patient similarity matrix and single-graph SSL, giving rise to an omics-specific AUC; AUCs are averaged across multi-omics;; SSL_Multi_Omics (original) employs knowledge-driven (pathways) patient similarity matrices across multiple omics combined with multi-graph SSL; DAI combines fusion of W_{iDRW}, W_{netDx_S} and $W_{SSL\_adapted}$ with single graph-based SSL. DAI* differs from DAI in that W_{iDRW}, W_{netDx_S} and $W_{SSL\_adapted}$ are not fused but combined with multip-graph based SSL.

SSL [6] improved the performance (AUC = 0.60). Multi-omics profiles were integrated across biological knowledge to increase prediction with SSL over a single omics approach. By integration of W_{c2_exp}, W_{c2_cnv}, and W_{c2_meth} we achieved an AUC of 0.52. The predication accuracies were quite similar for other sources of biological knowledge. In particular, integration of W_{c3_exp}, W_{c3_cnv}, W_{c3_meth}, led to AUC = 0.51; integration of W_{c4_exp}, W_{c4_cnv}, W_{c4_meth}, W_{c5_exp}, W_{c5_cnv}, W_{c5_meth}, W_{c7_exp}, W_{c7_cnv}, an W_{c7_meth} resulted in AUC prediction accuracies of 0.51, 0.55, and 0.53, respectively.

The current test version of DAI can be seen as a simple wrapper approach around multiple data integrative analytics to increase class prediction. Rather than submitting W_{iDRW}, W_{netDx_S}, $W_{SSL\_adapted}$ to a multiple graph-based semi-supervised learner, we primarily focused on obtaining a fused similarity network and single graph-based learning. With this setting, DAI's estimated AUC of 0.51 clearly outperformed the original iDRW, yet showed comparable performance to the original implementations of netDx and SSL. Among the original implementations of iDRW, netDx and SSL multi-omics prediction strategies, iDRW was the worst performer (AUC = 0.32). Interestingly, our adapted version of netDx (i.e. adapting the prediction model itself) outperformed all other considered strategies, including iDRW + SSL and DAI that performed similarly to the original SSL multi-omics integrative method (Fig. 5). To investigate whether there was an added value of multiple over single graph-based learning, combined with netDX and learning over 19 graphs, a smaller AUC was obtained compared to netDx + SSL (not shown), but DAI's performance (involving 3 graphs) increased to give the highest AUC (0.62) among all considered approaches (Fig. 5 – DAI*).

5 Discussion and Final Remarks

We introduced a Data and Analytic Integrator (DAI) that attempts to improve disease class prediction accuracy by integrating multi-omics data and analytics in various ways. The current workflow integrates 3 types of omics data, being CNV, methylation and gene expression data, and 3 analytic frameworks, represented by iDRW, netDx and SSL. Each of the analytics approaches derives information from multi-omics in a unique way and thus maximize their potential of providing complementary information towards class predication. In DAI, extracted information from each approach is translated into a single patient similarity matrix. The matrices for each of the analytic approaches are then combined. The current implementation of DAI uses Similarity Network Fusion to create an aggregated matrix but also allows using the individual matrices directly into a graph-based semi-supervised learner to predict class membership. The latter seems to be advantageous in terms of AUC performance, especially when the number of graphs for learning is relatively small. More work is needed though to investigate the impact of aggregating highly heterogeneous analytics.

As disease-associated genes are helpful in generating hypotheses about disease mechanisms, we investigated the utility of filtered gene sets, by making explicit use of earlier reported disease-gene associations. Little added value was achieved by doing so, compared to using unfiltered gene sets, except for giving rise to reduced computation times. One explanation may lie in the fact that association models and prediction models have different aims and evaluation criteria. Pathways highlighted by DAI (in particular RANBP2 pathways; SMARCA4 pathway; NOL7 pathways; diabetes pathways) were found to be implicated in ovarian, breast, cervical, and neuroblastoma cancer types [16–18].

In summary, our pilot results have shown that the exploitation of knowledge-based gene sets can substantially increase prediction performance. Furthermore, letting the data speak for themselves, in that the contribution of multiple omics data types in prediction models is estimated from the data, seems to boost prediction performance, but cannot receive all the credits. For instance, simply changing a logistic regression prediction model for a predictor based on a single aggregated patient similarity matrix was sufficient to create a top performer. Also, including a poor performer in similarity network fusion of three patient similarity matrices, based on multi-omics view from 3 analytic approaches, did not work decremental. Hence, future work will include the further exploitation of knowledge-driven data in DAI in combination with more elaborate non-linear aggregation of method-specific patient similarity matrices that estimate the relative contribution of each such matrix with the objective to maximize prediction accuracy.

Acknowledgement. We thank So Yeon Kim for sharing extended code of iDRW that allows the integration of 3 omics data types. A.B, and K.V.S acknowledge funding by Télévie 2015 "PDAC-xome: Exome sequencing in PDAC" (convention n° 7.4629.15), Télévie 2016 "Drivers and markers in pancreatic cancer" (convention n° 7.4502.16), and FRS-FNRS – CDR 2017 "SysMedPC" (convention n° J.0061.17).

References

1. Momenimovahed, Z., Tiznobaik, A., Taheri, S., Salehiniya, H.: Ovarian cancer in the world: epidemiology and risk factors. Int. J. Women's health **11**, 287–299 (2019)
2. Shen, R., Mo, Q., Schultz, N., Seshan, V.E., Olshen, A.B., Huse, J., et al.: Integrative subtype discovery in glioblastoma using iCluster. PLoS ONE **7**(4), e35236 (2012)
3. Chaudhary, K., Poirion, O.B., Lu, L., Garmire, L.X.: Deep learning–based multi-omics integration robustly predicts survival in liver cancer. Clin. Cancer Res. **24**(6), 1248–1259 (2008)
4. Rappoport, N., Shamir, R.: Multi-omic and multi-view clustering algorithms: review and cancer benchmark. Nucleic Acids Res. **46**(20), 10546–10562 (2018)
5. Simidjievski, N., Bodnar, C., Tariq, I., Scherer, P., Terre, H.A., Shams, Z., et al.: Variational autoencoders for cancer data integration: design principles and computational practice. Front. Genet. **10**, 1205 (2019)
6. Kim, D., Joung, J.-G., Sohn, K.-A., Shin, H., Park, Y.R., Ritchie, M.D., et al.: Knowledge boosting: a graph-based integration approach with multi-omics data and genomic knowledge for cancer clinical outcome prediction. J. Am. Med. Inform. Assoc. **22**(1), 109–120 (2014)
7. Wan, Y.W., Allen, G.I., Liu, Z.: TCGA2STAT: simple TCGA data access for integrated statistical analysis in R. Bioinformatics **32**(6), 952–954 (2016)
8. Dereli, O., Oğuz, C., Gönen, M.: Path2Surv: pathway/gene set-based survival analysis using multiple kernel learning. Bioinformatics **35**(24), 5137–5145 (2019)
9. Xu, A., Chen, J., Peng, H., Han, G., Cai, H.: Simultaneous interrogation of cancer omics to identify subtypes with significant clinical differences. Front. Genetics **10**, 236 (2019)
10. Kim, S.Y., Jeong, H.H., Kim, J., Moon, J.H., Sohn, K.A.: Robust pathway-based multi-omics data integration using directed random walks for survival prediction in multiple cancer studies. Biol. Dir. **14**(1), 8 (2019)
11. Pai, S., Hui, S., Isserlin, R., Shah, M.A., Kaka, H., Bader, G.D.: netDx: interpretable patient classification using integrated patient similarity networks. Mol. Syst. Biol. **15**(3), e8497 (2019)
12. Liberzon, A., Subramanian, A., Pinchback, R., Thorvaldsdóttir, H., Tamayo, P., Mesirov, J.P.: Molecular signatures database (MSigDB) 3.0. Bioinformatics **27**(12), 1739–1740 (2011)
13. Papp, E., Hallberg, D., Konecny, G.E., Bruhm, D.C., Adleff, V., Noë, M., et al.: Integrated genomic, epigenomic, and expression analyses of ovarian cancer cell lines. Cell Rep. **25**(9), 2617–2633 (2018)

14. Chen, J., Bardes, E.E., Aronow, B.J., Jegga, A.G.: ToppGene Suite for gene list enrichment analysis and candidate gene prioritization. Nucleic Acids Res. **37**(Suppl. 2), W305–W311 (2009)
15. Wang, B., Mezlini, A.M., Demir, F., Fiume, M., Tu, Z., Brudno, M., et al.: Similarity network fusion for aggregating data types on a genomic scale. Nat. Methods **11**(3), 333 (2014)
16. Connor, Y.D., Miao, D., Lin, D.I., Hayne, C., Howitt, B.E., Dalrymple, J.L., et al.: Germline mutations of SMARCA4 in small cell carcinoma of the ovary, hypercalcemic type and in SMARCA4-deficient undifferentiated uterine sarcoma: clinical features of a single family and comparison of large cohorts. Gynecol. Oncol. **157**(1), 106–114 (2020)
17. Doçi, C.L., Mankame, T.P., Langerman, A., Ostler, K.R., Kanteti, R., Best, T., et al.: Characterization of NOL7 gene point mutations, promoter methylation, and protein expression in cervical cancer. Int. J. Gynecol. Pathol.: off. J. Int. Soc. Gynecol. Pathol. **31**(1), 15–24 (2012)
18. Blanco Jr., L.Z., Kuhn, E., Morrison, J.C., Bahadirli-Talbott, A., Smith-Sehdev, A., Kurman, R.J.: Steroid hormone synthesis by the ovarian stroma surrounding epithelial ovarian tumors: a potential mechanism in ovarian tumorigenesis. Mod. Pathol. **30**(4), 563–576 (2017)

Overlap-Based Undersampling Method for Classification of Imbalanced Medical Datasets

Pattaramon Vuttipittayamongkol$^{(\boxtimes)}$ and Eyad Elyan

Robert Gordon University, Aberdeen, UK
{p.vuttipittayamongkol,e.elyan}@rgu.ac.uk

Abstract. Early diagnosis of some life-threatening diseases such as cancers and heart is crucial for effective treatments. Supervised machine learning has proved to be a very useful tool to serve this purpose. Historical data of patients including clinical and demographic information is used for training learning algorithms. This builds predictive models that provide initial diagnoses. However, in the medical domain, it is common to have the positive class under-represented in a dataset. In such a scenario, a typical learning algorithm tends to be biased towards the negative class, which is the majority class, and misclassify positive cases. This is known as the class imbalance problem. In this paper, a framework for predictive diagnostics of diseases with imbalanced records is presented. To reduce the classification bias, we propose the usage of an overlap-based undersampling method to improve the visibility of minority class samples in the region where the two classes overlap. This is achieved by detecting and removing negative class instances from the overlapping region. This will improve class separability in the data space. Experimental results show achievement of high accuracy in the positive class, which is highly preferable in the medical domain, while good trade-offs between sensitivity and specificity were obtained. Results also show that the method often outperformed other state-of-the-art and well-established techniques.

Keywords: Imbalanced data · Medical diagnosis · Medical prediction · Class overlap · Classification · Undersampling · Nearest neighbour · Machine learning

1 Introduction

In the past decade, machine learning has been widely used to aid medical diagnosis. Same as in other domains, hidden knowledge can be discovered based on previous information. This is often too complicated to be done by hand or through simple statistical techniques, especially when there are many related features and the data is large. In the medical domain, it is important that prevention and early diagnosis are carried out to avoid further complications and achieve better treatment outcomes [2]. Hence, detecting possible existence or

© IFIP International Federation for Information Processing 2020
Published by Springer Nature Switzerland AG 2020
I. Maglogiannis et al. (Eds.): AIAI 2020, IFIP AICT 584, pp. 358–369, 2020.
https://doi.org/10.1007/978-3-030-49186-4_30

occurrence of diseases is of high interest in supervised learning. This is achieved by training classification models to predict the patients' conditions based on the given symptoms and personal information.

It is common in the medical domain that a dataset has an uneven class distribution. In many situations, the class of interest rarely occurs, hence its samples are limited compared to the other classes. However, traditional learning algorithms are generally designed to maximise the overall prediction accuracy. Thus, on imbalanced datasets, they tend to be biased in classification towards the majority class and fail to detect anomaly cases. A number of solutions have been proposed to handle classification of datasets with skewed class distributions, so-called imbalanced datasets. Many of them focused on a medical dataset of a specific disease [1,15,23] while others proved their performance on several medical-related datasets [10,24].

Learning from imbalanced medical datasets are seen in a wide range of problems. Besides classification of well-known public datasets such as breast cancer Wisconsin and Pima Indian diabetes, other types of classification tasks have also been carried out. These include classification of electrodiogram (ECG) heartbeats [13], image classification of breast cancer [15] and video classification of bowel cancer [23]. Regardless of problem types, a common objective is to achieve high prediction accuracy, especially on the positive class, which is under-represented.

Rebalancing class distributions seems to be a typical approach to handle imbalanced medical datasets. However, it was shown in the literature that solutions based on improving the visibility of positive samples in the overlapping region could produce significantly higher positive class accuracy (sensitivity) [4,19,20].

In this paper, we propose a framework for improving classification of imbalanced medical datasets. Aiming at high sensitivity on the diagnosis, an overlap-based undersampling method is used. Recursive searching of neighbouring instances is employed to identify instances in the overlapping region. Then, overlapped negative instances are removed to maximise the presence of positive instances to the learning algorithm.

The rest of this paper is organised as follows. In Sect. 2, we discuss existing methods for handling imbalanced medical datasets. Section 3 gives the details of the proposed framework. Section 4 contains experimental setup including brief descriptions of real-world medical datasets used in the experiment. In Sect. 5, results and discussion are provided. Finally, in Sect. 6, we conclude the paper and discuss potential future directions.

2   Related Literature

Despite high interests in classification of medical data, the common issue of imbalanced class distributions is not often addressed [14]. This is evidenced by a review paper discussing existing methods used for medical datasets classification [14]. Only 1 out of 71 proposed solutions considered the class imbalanced issue.

To tackle class imbalance, long-established methods such as random under-sampling, SMOTE [6], ENN [22] and ADASYN [12] were still used in many recent studies [2,7]. Although improvements in results were reported, they have been constantly outperformed by newer methods.

Novel methods for handling imbalanced medical datasets have also been proposed. In [10], the authors selectively chosen minority class instances for oversampling based on their nearest neighbours. Minority class instances were defined as noise, unstable or boundary samples. Then, noisy instances were removed and only boundary instances were oversampled using linear interpolation techniques. The method showed improvement over SMOTE and an extension of SMOTE. However, it has disadvantages of high parameters dependency and the risk of losing important information in eliminating minority class instances.

In [18], a new technique for determining the final outputs for medical datasets with multiple minority classes was used. Unlike the traditional majority voting approach, classes were assigned based on the highest weighted combination of accuracy, sensitivity, specificity and AUC. Results showed trivial improvement over the traditional method and the improvement might not be attributed to increases in the minority class accuracy, which is highly desirable in the medical domain.

Wan et al. designed a scoring function that assigned ranking to differentiate between minority class and majority class instances [21]. Boosting was adopted to carried out automatic scoring. The method could improved sensitivity on medical datasets further than a cost-sensitive approach and other well-known ensemble-based methods. Moreover, it has the benefit of no prior costs required, which is often unknown and hard to estimate.

One of the latest techniques, Generative Adversarial Net (GAN), was employed in [24] to synthesise minority class instances. It was combined with a multilayer extreme learning machine (ELM) algorithm and showed superior performance to other techniques used with ELM such as weighting and SMOTE. The method also consumed low computational time.

Rather than using a method to broadly handle datasets of multiple diseases, many studies focused on a specific disease such as cancers [7,15,23], polyps [3] and osteoporosis [2]. For instance, Yuan et al. proposed an ensemble-based deep learning approach for detecting bowel cancer [23]. They modified the loss function to penalise the classifier when missclassifying samples that have been correctly classified in the previous iteration. However, results showed that the method was comparable to a long-established ensemble, RUSBoost, in terms of sensitivity and computational time. Other methods for classification of cancer datasets were also proposed [15,16]. In [15], an evolutionary algorithm was used as an undersampling approach to select the most significant samples, then combined with Boosting. The method improved the classification of a breast cancer dataset over other ensemble-based techniques. Similarly, in [16], a cost-sensitive ensemble integrated with a genetic algorithm was proposed to handle an imbalanced breast thermogram dataset. The method provided higher sensitivity than other existing ones. Even so, a common drawback of these ensemble-based solutions is high computational costs.

ECG datasets of heartbeats are also of high interest, and they are generally highly imbalanced, where most heartbeats are normal. With complicated components and morphology of ECG, deep convolutional neural networks (CNN) are often employed for classification tasks [1,13]. CNN is used in combination with many other techniques to enhance results. These include Borderline-SMOTE, feature selection and two-phase training presented in [11]. Two-phase training, introduced by Havaei et al., is known as a promising training technique for imbalanced data. In the first stage, a balanced portion of the data is used to train so that CNN can distinguish different classes. Then in the second stage, the original imbalanced data is fed to fine-tune the output layer parameters.

3 The Proposed Framework

The proposed framework for improving prediction on imbalanced medical datasets is presented in Fig. 1. Firstly, the training data is preprocessed using normalisation and the overlap-based undersampling technique. Then, the preprocessed data is used to train a learning algorithm to build a predictive model. Finally, the model is evaluated with the testing data.

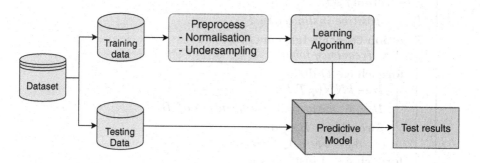

Fig. 1. The proposed framework for classification of imbalanced medical datasets

In the data preprocessing step, we aim at maximising the presence of minority class instances in the overlapping region. The undersampling method based on recursive neigbourhood searching [19] is used. For convenience, we refer to it as URNS. The method will perform a challenging task of identifying overlapped negative instances by considering their k nearest neighbours. These instances are then removed to reduce the complexity of the learning task and reduce the bias classification towards the negative class. Since URNS employs a distance-based technique, its sensitivity to noise has to be concerned. To address the issue, we propose that the data is normalised before URNS is applied. Here, we used standard scores (z-scores) as the normalisation method. The detailed discussion on URNS is provided as follows.

3.1   The URNS Method

The main objective of URNS is to maximise the visibility of the minority class to the learning algorithm. This is achieved by eliminating majority class instances from the overlapping region. To identify potential overlapped instances, the k-Nearest Neighbours algorithm (kNN) is used. The local surroundings of each minority class instances are carefully explored. Majority class instances that are in close proximity, which are highly likely to weaken the appearance of minority class ones, are to be removed. However, to prevent excessive elimination, we consider removing only instances that each impacts more than one minority class sample. On the other hand, sufficient elimination is also ensured by the recursive searching. That is the search is carried out twice, where the output of the first round becomes the input of the second round of searching.

Algorithm 1: Recursive Neighbour Search Undersampling

Data: training set, k

Result: undersampled training set

begin

    $T \leftarrow training\ set$;

    $T_{pos} \leftarrow positive\ instances\ in\ T$;

    Function CommonNeighbour (T, Q, k):

        $A \leftarrow frequency\ table$;

        foreach $q \in Q$ **do**

            $B \leftarrow kNN(q, T, k)$;

            $B_{neg} \leftarrow majority\ class\ members\ of\ B$;

            foreach $y \in B_{neg}$ **do**

                $A_y.freq \leftarrow A_y.freq + 1$;

        foreach $q \in A.instance$ **do**

            if $A_x.freq > 1$ **then**

                $X \leftarrow X \cup \{x\}$;

        return X;

    $R_1 \leftarrow CommonNeighbour(T, T_{pos}, k)$;

    $R_2 \leftarrow CommonNeighbour(T, R_1, k)$;

    $\hat{T} \leftarrow T - \{R_1 \cup R_2\}$;

    return $(\hat{T})$

Algorithm 1 describes the process of the recursive neighbour search undersampling method, and Fig. 2 illustrates the detection of potential overlapped instances. The method begins with searching for k nearest neighbours of all minority class instances (queries). A majority class neighbour that any two queries have in common is considered as an overlapped instance. This is shown

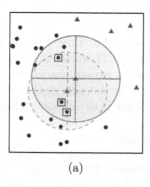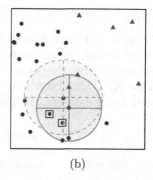

(a)   (b)

Fig. 2. Recursive neighbour Searching involves (a) detecting common majority class neighbours of positive instances followed by (b) searching further for the nearest neighbours they have in common. All majority class common neighbours (in blue boxes) from both steps are then eliminated [19]. (Color Figure Online)

in Fig. 2a, where the common neighbours are marked with blue boxes. Then, to ensure thorough detection, the search is repeated by using the common neighbours detected in the first step as the queries. Their common neighbours are then searched for as depicted in Fig. 2b. Finally, all common neighbours in the majority class from both steps are removed from the training data.

To allow generalisation of the method across any datasets, we present the use of an adaptive k value as shown in Eq. 1 in the neighbour searching algorithm. A rule of thumb where k is related to the square root of the data size (N) was considered. Then, the value of k was adjusted so that it is at the same time proportional to the imbalance ratio (IR). This add-on will also help enhance the discovery of overlapped majority class instances.

$$k = \sqrt{N} + \sqrt{IR} \tag{1}$$

4   Experiment

4.1   Setup

We carried out an experiment using five real-world binary-class datasets. Each dataset was partitioned into training and testing data at 70:30. Random Forest (RF) was chosen as the learning algorithm as it is one of the most-used classifiers for imbalanced datasets [8]. Also, it showed promising results on sensitivity with a better trade-off between sensitivity and specificity than other algorithms [2]. The performance of our method was compared against well-established and state-of-the-art algorithms. These were SMOTE [6], BLSMOTE [9], DBSMOTE [5] and k-means undersampling [17]. The parameters of these methods were set as in the original works. The methods were evaluated in terms of sensitivity, specificity, G-mean and F1-score. Except for KDD's breast cancer, where sufficient data was available, 10-fold cross-validation was used in the training phase for the purpose of model selection.

4.2   Datasets

Five datasets used in the experiment are presented in Table 1 in ascending order of IR with their general information. Wisconsin, Thoracic, Cleveland and Thyroid were obtained from the UCI repository[1]. Breast cancer was given as a challenge in the KDD Cup 2008[2]. We cleaned the datasets so that there were no missing values. In all datasets, the positive class is the minority class.

Table 1. Datasets

Dataset	Instances	Features	IR	% neg
Wisconsin	683	9	1.86	65.00
Thoracic	470	17	5.71	85.11
Cleveland	173	13	12.31	92.49
Thyroid	7200	21	12.48	92.58
Breast cancer	102294	117	163.20	99.39

Wisconsin Breast Cancer. The Wisconsin breast cancer dataset, widely-known as Wisconsin, was collected at the University of Wisconsin Hospitals, USA during 1989–1991. The class labels are diagnoses of malignant (positive) or benign (negative) breast mass. Other given information is cells characteristics.

Thoracic Surgery. The data was collected from patients who underwent major lung resections for primary lung cancer at Wroclaw Thoracic Surgery Centre, Poland during 2007–2011. The prediction labels are one-year survival period, which are died (positive) and survived (negative). Model training will be based on patients' personal information, conditions, behaviour and symptoms.

Cleveland Heart Disease. The dataset consists of databases obtained from patients in different regions, namely Cleveland, Long Beach, Hungary and Switzerland. Patients with the presence of heart disease (positive) are to be distinguished from those with absence (negative).

Thyroid. The records were collected by the Garavan Institute of Sydney, Australia. The problem is to determine whether a patient referred to the clinic is hypothyroid. The original dataset contains 3 classes: normal, hyperfunction and subnormal function. The normal cases (negative) occupies over 92% of the dataset and the last two classes are the minority groups. In our experiment, we combined hyperfunction and subnormal function and recognised both cases as hypothyroid (positive).

---

[1] https://archive.ics.uci.edu/ml/index.php.

[2] https://www.kdd.org/kdd-cup/view/kdd-cup-2008.

Breast Cancer. The dataset is composed of features computed from X-ray images of breasts for early detection of breast cancer. Each sample is labelled with malignant (positive) or benign (negative). This dataset is very large and extremely imbalanced with positive instances of less than 1%.

5   Results and Discussion

Experimental results show that our proposed framework were effective in handling classification of imbalanced medical datasets. URNS showed better results than the well-established and state-of-the-art methods by achieving the highest sensitivity and the highest G-mean on most datasets. Across all datasets, sensitivity and G-mean were significantly improved over the baseline (RF with no resampling). These results are presented in Tables 2, 3, 4, 5 and 6, where the highest value in each evaluation metric is highlighted in **bold**.

Table 2. Results on Wisconsin

Method	Sensitivity	Specificity	G-mean	F1-score
Baseline	94.37	**96.97**	95.66	**94.37**
URNS	**98.59**	93.18	**95.85**	93.33
SMOTE	94.37^a	96.97^a	95.66^a	**94.37**^{a}
BLSMOTE	94.37^a	96.97^a	95.66^a	**94.37**^{a}
DBSMOTE	94.37^a	96.97^a	95.66^a	**94.37**^{a}
kmUnder	95.77^b	95.45^b	95.61^b	93.79^b

^{a} No changes in the results after applying the method
^{b} Results obtained with modified parameter setting

Table 2 shows the results on Wisconsin breast cancer dataset. Our URNS method provided the highest sensitivity of 98.59% and the highest G-mean of 95.85%. These were achieved with high specificity and F1-score. It should be noted that the other methods failed to work on this dataset. In particular, the SMOTE-based methods, i.e., SMOTE, BLSMOTE and DBSMOTE, had no effects on the classification results. This could have been because insufficient positive samples were synthesised, which was due to their objective to rebalance data. As a result, the presence of the positive class, especially around the boundary regions, could not be improved. As opposed, our method does not factor the imbalance ratio and the removal only depends on the amount of class overlap. Lastly, kmUnder could not be carried out using the k value proposed in the original work since there were fewer distinct samples than k. Thus, we replaced it with $k = N_{minority}/2$. However, it did not give better results than URNS.

As shown in Table 3, URNS achieved the best sensitivity and F1-score on Thoracic surgery dataset. It is worth pointing out that this dataset is very hard

Table 3. Results on Thoracic

Method	Sensitivity	Specificity	G-mean	F1-score
Baseline	0	**99.17**	0	0
URNS	**95.24**	5.83	23.57	**25.97**
SMOTE	9.52	89.17	29.14	11.11
BLSMOTE	9.52	87.5	28.87	10.53
DBSMOTE	9.52	97.5	30.47	15.38
kmUnder	80.95	20.83	**41.07**	25.56

to classify. This can be seen from the baseline results that none of the positive test cases were correctly identified. Moreover, none of the methods could produce high sensitivity and high specificity at the same time. This high trade-off between the accuracy of the two classes indicates that the dataset is likely to suffer from severe class overlap. Due to this trade-off, even though URNS achieved very high sensitivity of 95.24%, it had the lowest specificity. Thus, URNS is preferable when it is required that nearly all death cases are correctly predicted, otherwise an alternative method providing a more compromised result needs to be explored.

Table 4. Results on Cleveland

Method	Sensitivity	Specificity	G-mean	F1-score
Baseline	33.33	**100**	57.74	50
URNS	**100**	93.75	96.82	66.67
SMOTE	**100**	97.92	98.95	85.71
BLSMOTE	**100**	91.67	95.74	60
DBSMOTE	**100**	**100**	**100**	**100**
kmUnder	**100**	39.58	62.92	17.14

From Table 4, our method perfectly classified the positive test cases on the Cleveland heart disease dataset. Its specificity and G-mean were high and comparable to SMOTE, BLSMOTE and DBSMOTE. Due to the high class imbalance nature of the dataset, F1-score of URNS was much lower than those of SMOTE and DBSMOTE even though their specificity values were not far different. This is because F1-score considers true positives and false positives. Thus, in a highly class imbalanced situation, F1-score will be strongly negatively affected by high false positives, which could be misleading when considering the metric alone. Compared to kmUnder, our method provided a substantially higher trade-off between sensitivity and specificity. This could be attributed to less information loss of the URNS method.

Table 5. Results on Thyroid

Method	Sensitivity	Specificity	G-mean	F1-score
Baseline	98.74	99.75	99.24	**97.82**
URNS	**100**	99.2	**99.6**	95.21
SMOTE	98.74[a]	99.75[a]	99.24[a]	**97.82**[a]
BLSMOTE	98.11	98.15	98.13	88.64
DBSMOTE	98.74[a]	99.75[a]	99.24[a]	**97.82**[a]
kmUnder	0	**100**	0	0

[a] No changes in the results after applying the method

As can be seen from Table 5, our URNS method provided the best trade-off between sensitivity and specificity on the Thyroid dataset. This is evidenced by the highest G-mean of 99.60%. With the highest sensitivity of 100% achieved, it also yielded high values of specificity and F1-score, which were competitive with the other methods except kmUnder. Note that SMOTE and DBSMOTE led to no changes in the classification results. BLSMOTE had lower performance than the baseline. Lastly, kmUnder completely failed to handle the dataset.

Table 6. Results on breast cancer

Method	Sensitivity	Specificity	G-mean	F1-score
Baseline	29.57	**99.98**	54.37	44.72
URNS	74.73	93.49	**83.59**	12.03
SMOTE	45.16	99.75	67.12	**48.55**
BLSMOTE	33.33	99.89	57.7	44.13
DBSMOTE	36.02	99.84	59.97	44.37
kmUnder	**93.01**	40.27	61.2	1.86

Finally, results on the large and extremely imbalanced dataset of breast cancer are presented in Table 6. Our method achieved the second highest sensitivity, which was lower than kmUnder but significantly higher than the other methods. Essentially higher specificity, G-mean and F1-score indicate that URNS had a better trade-off than kmUnder. URNS showed high specificity and the highest G-mean of 83.59%. Its low F1-score was due to the bias caused by very high class imbalance as discussed above.

6 Conclusions

In this paper, we handled imbalanced medical datasets using an overlap-based undersampling method. By recursively exploring the neighbourhood of instances, majority class instances potentially in the overlapping region were identified. Then, removal of these instances led to better acknowledgement of minority class instances. Results on real-world datasets showed that the URNS method provided high sensitivity, which is highly desirable in the medical domain, while offering good trade-offs between the accuracy rates of the positive class and the negative class. Moreover, these results were competitive with those of other state-of-the-art and well-established solutions. This can be attributed to some advantages of URNS over other methods. First, the resampling rate is independent of class imbalance and based on the amount of class overlap. Second, the method specifically addresses the problem of class overlap, which often causes errors in classification. Furthermore, this method was implemented with an adaptive k value and no parameter setting is needed. These enable generalisation of the method across any medical datasets. A potential future direction will include improving the framework. A method for setting k value to also be adaptive to the local surroundings of instances such as data density and regional class distribution may improve identification of overlapped instances and hence classification results. To allow wider applicability on real-world medical problems, a framework for multi-class datasets will be developed.

References

1. Acharya, U.R., et al.: A deep convolutional neural network model to classify heartbeats. Comput. Biol. Med. **89**, 389–396 (2017)
2. Bach, M., Werner, A., Żywiec, J., Pluskiewicz, W.: The study of under- and over-sampling methods' utility in analysis of highly imbalanced data on osteoporosis. Inf. Sci. **384**, 174–190 (2017)
3. Bae, S.H., Yoon, K.J.: Polyp detection via imbalanced learning and discriminative feature learning. IEEE Trans. Med. Imaging **34**(11), 2379–2393 (2015)
4. Bunkhumpornpat, C., Sinapiromsaran, K.: DBMUTE: density-based majority under-sampling technique. Knowl. Inf. Syst. **50**(3), 827–850 (2016). https://doi.org/10.1007/s10115-016-0957-5
5. Bunkhumpornpat, C., Sinapiromsaran, K., Lursinsap, C.: DBSMOTE: density-based synthetic minority over-sampling technique. Appl. Intell. **36**(3), 664–684 (2012). https://doi.org/10.1007/s10489-011-0287-y
6. Chawla, N.V., Bowyer, K.W., Hall, L.O., Kegelmeyer, W.P.: SMOTE: synthetic minority over-sampling technique. J. Artif. Intell. Res. **16**, 321–357 (2002)
7. Fotouhi, S., Asadi, S., Kattan, M.W.: A comprehensive data level analysis for cancer diagnosis on imbalanced data. J. Biomed. Inform. **90**, 103089 (2019)
8. Haixiang, G., Yijing, L., Shang, J., Mingyun, G., Yuanyue, H., Bing, G.: Learning from class-imbalanced data: review of methods and applications. Expert Syst. Appl. **73**, 220–239 (2017)
9. Han, H., Wang, W.-Y., Mao, B.-H.: Borderline-SMOTE: a new over-sampling method in imbalanced data sets learning. In: Huang, D.-S., Zhang, X.-P., Huang, G.-B. (eds.) ICIC 2005. LNCS, vol. 3644, pp. 878–887. Springer, Heidelberg (2005). https://doi.org/10.1007/11538059_91
10. Han, W., Huang, Z., Li, S., Jia, Y.: Distribution-sensitive unbalanced data over-sampling method for medical diagnosis. J. Med. Syst. **43**(2), 39 (2019). https://doi.org/10.1007/s10916-018-1154-8

11. Havaei, M., et al.: Brain tumor segmentation with deep neural networks. Med. Image Anal. **35**, 18–31 (2017)
12. He, H., Bai, Y., Garcia, E.A., Li, S.: ADASYN: adaptive synthetic sampling approach for imbalanced learning. In: 2008 IEEE International Joint Conference on Neural Networks (IJCNN 2008). IEEE World Congress on Computational Intelligence, pp. 1322–1328. IEEE (2008)
13. Jiang, J., Zhang, H., Pi, D., Dai, C.: A novel multi-module neural network system for imbalanced heartbeats classification. Expert Syst. Appl.: X **1**, 100003 (2019)
14. Kalantari, A., Kamsin, A., Shamshirband, S., Gani, A., Alinejad-Rokny, H., Chronopoulos, A.T.: Computational intelligence approaches for classification of medical data: state-of-the-art, future challenges and research directions. Neurocomputing **276**, 2–22 (2018)
15. Krawczyk, B., Galar, M., Jeleń, Ł., Herrera, F.: Evolutionary undersampling boosting for imbalanced classification of breast cancer malignancy. Appl. Soft Comput. **38**, 714–726 (2016)
16. Krawczyk, B., Schaefer, G., Woźniak, M.: A hybrid cost-sensitive ensemble for imbalanced breast thermogram classification. Artif. Intell. Med. **65**(3), 219–227 (2015)
17. Lin, W.C., Tsai, C.F., Hu, Y.H., Jhang, J.S.: Clustering-based undersampling in class-imbalanced data. Inf. Sci. **409**, 17–26 (2017)
18. Shilaskar, S., Ghatol, A.: Diagnosis system for imbalanced multi-minority medical dataset. Soft. Comput. **23**(13), 4789–4799 (2018). https://doi.org/10.1007/s00500-018-3133-x
19. Vuttipittayamongkol, P., Elyan, E.: Neighbourhood-based undersampling approach for handling imbalanced and overlapped data. Inf. Sci. **509**, 47–70 (2020)
20. Vuttipittayamongkol, P., Elyan, E., Petrovski, A., Jayne, C.: Overlap-based undersampling for improving imbalanced data classification. In: Yin, H., Camacho, D., Novais, P., Tallón-Ballesteros, A.J. (eds.) IDEAL 2018. LNCS, vol. 11314, pp. 689–697. Springer, Cham (2018). https://doi.org/10.1007/978-3-030-03493-1_72
21. Wan, X., Liu, J., Cheung, W.K., Tong, T.: Learning to improve medical decision making from imbalanced data without a priori cost. BMC Med. Inform. Decis. Making **14**(1), 111 (2014). https://doi.org/10.1186/s12911-014-0111-9
22. Wilson, D.L.: Asymptotic properties of nearest neighbor rules using edited data. IEEE Trans. Syst. Man Cybern. **3**, 408–421 (1972)
23. Yuan, X., Xie, L., Abouelenien, M.: A regularized ensemble framework of deep learning for cancer detection from multi-class, imbalanced training data. Pattern Recogn. **77**, 160–172 (2018)
24. Zhang, L., Yang, H., Jiang, Z.: Imbalanced biomedical data classification using self-adaptive multilayer ELM combined with dynamic GAN. Biomed. Eng. Online **17**(1), 181 (2018). https://doi.org/10.1186/s12938-018-0604-3

Natural Language

An Overview of Chatbot Technology

Eleni Adamopoulou$^{(\boxtimes)}$ ⓘ and Lefteris Moussiades ⓘ

Department of Computer Science, International Hellenic University, Agios Loukas,
65404 Kavala, Greece
{eladamo,lmous}@cs.ihu.gr

Abstract. The use of chatbots evolved rapidly in numerous fields in recent years, including Marketing, Supporting Systems, Education, Health Care, Cultural Heritage, and Entertainment. In this paper, we first present a historical overview of the evolution of the international community's interest in chatbots. Next, we discuss the motivations that drive the use of chatbots, and we clarify chatbots' usefulness in a variety of areas. Moreover, we highlight the impact of social stereotypes on chatbots design. After clarifying necessary technological concepts, we move on to a chatbot classification based on various criteria, such as the area of knowledge they refer to, the need they serve and others. Furthermore, we present the general architecture of modern chatbots while also mentioning the main platforms for their creation. Our engagement with the subject so far, reassures us of the prospects of chatbots and encourages us to study them in greater extent and depth.

Keywords: Chatbot · Chatbot architecture · Artificial Intelligence · Machine learning · NLU

1 Introduction

Artificial Intelligence (AI) increasingly integrates our daily lives with the creation and analysis of intelligent software and hardware, called intelligent agents. Intelligent agents can do a variety of tasks ranging from labor work to sophisticated operations. A chatbot is a typical example of an AI system and one of the most elementary and widespread examples of intelligent Human-Computer Interaction (HCI) [1]. It is a computer program, which responds like a smart entity when conversed with through text or voice and understands one or more human languages by Natural Language Processing (NLP) [2]. In the lexicon, a chatbot is defined as "A computer program designed to simulate conversation with human users, especially over the Internet" [3]. Chatbots are also known as smart bots, interactive agents, digital assistants, or artificial conversation entities.

Chatbots can mimic human conversation and entertain users but they are not built only for this. They are useful in applications such as education, information retrieval, business, and e-commerce [4]. They became so popular because there are many advantages of chatbots for users and developers too. Most implementations are platform-independent and instantly available to users without needed installations. Contact to the chatbot is spread through a user's social graph without leaving the messaging app the chatbot lives

I. Maglogiannis et al. (Eds.): AIAI 2020, IFIP AICT 584, pp. 373–383, 2020.
https://doi.org/10.1007/978-3-030-49186-4_31

in, which provides and guarantees the user's identity. Moreover, payment services are integrated into the messaging system and can be used safely and reliably and a notification system re-engages inactive users. Chatbots are integrated with group conversations or shared just like any other contact, while multiple conversations can be carried forward in parallel. Knowledge in the use of one chatbot is easily transferred to the usage of other chatbots, and there are limited data requirements. Communication reliability, fast and uncomplicated development iterations, lack of version fragmentation, and limited design efforts for the interface are some of the advantages for developers too [5].

The rest of the paper is organized as follows. In Sect. 2, we briefly present the history of chatbots and highlight the growing interest of the research community. In Sect. 3, some issues about the association with chatbots are discussed, while in Sect. 4, essential concepts relevant to chatbot technology are described. Next, in Sect. 5, we present a classification of existing chatbots while in Sect. 6, we present the underlying chatbot architecture and the leading platforms for their development. Finally, Sect. 7 reports conclusions and highlights further research topics.

2   History

Alan Turing in 1950 proposed the Turing Test ("Can machines think?"), and it was at that time that the idea of a chatbot was popularized [6]. The first known chatbot was Eliza, developed in 1966, whose purpose was to act as a psychotherapist returning the user utterances in a question form [7]. It used simple pattern matching [8] and a template-based response mechanism. Its conversational ability was not good, but it was enough to confuse people at a time when they were not used to interacting with computers and give them the impetus to start developing other chatbots [5]. An improvement over ELIZA was a chatbot with a personality named PARRY developed in 1972 [9]. In 1995, the chatbot ALICE was developed which won the Loebner Prize, an annual Turing Test, in years 2000, 2001, and 2004. It was the first computer to gain the rank of the "most human computer" [10]. ALICE relies on a simple pattern-matching algorithm with the underlying intelligence based on the Artificial Intelligence Markup Language (AIML) [11], which makes it possible for developers to define the building blocks of the chatbot knowledge [10]. Chatbots, like SmarterChild [12] in 2001, were developed and became available through messenger applications. The next step was the creation of virtual personal assistants like Apple Siri [13], Microsoft Cortana [14], Amazon Alexa [15], Google Assistant [16] and IBM Watson [17].

As shown in Fig. 1 according to Scopus [18], there was a rapid growth of interest in chatbots especially after the year 2016. Many chatbots were developed for industrial solutions while there is a wide range of less famous chatbots relevant to research and their applications [19].

Documents by year

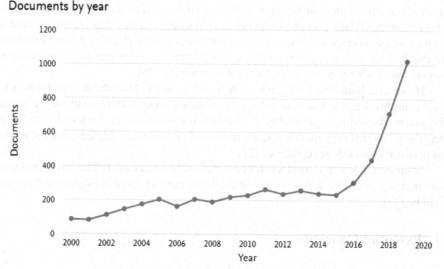

Fig. 1. Search results in Scopus by year for "chatbot" or "conversation agent" or "conversational interface" as keywords from 2000 to 2019.

3 Associate with Chatbots

Why do users use chatbots? Chatbots seem to hold tremendous promise for providing users with quick and convenient support responding specifically to their questions. The most frequent motivation for chatbot users is considered to be productivity, while other motives are entertainment, social factors, and contact with novelty. However, to balance the motivations mentioned above, a chatbot should be built in a way that acts as a tool, a toy, and a friend at the same time [8].

The reduction in customer service costs and the ability to handle many users at a time are some of the reasons why chatbots have become so popular in business groups [20]. Chatbots are no longer seen as mere assistants, and their way of interacting brings them closer to users as friendly companions [21]. According to a study, social media user requests on chatbots for customer service are emotional and informational, with the first category rate being more than 40% and with users not intending to take specific information [22]. Machine learning is what gives the capability to customer service chatbots for sentiment detection and also the ability to relate to customers emotionally as human operators do [23].

Concerning the user's trust in chatbots, it depends on factors relative to the chatbot itself, like how much it responds like a human, how it is self-presented, and how much professional its appearance is. Nevertheless, it depends also on factors relative to its service contexts, like the brand of the chatbot host, privacy and security in the chatbot, and other risk issues about the topic of the request [10]. Human-likeness can be suggested by using human figures (visual cues), human-associated names, or identity (identity cues) and mimicking of human languages (conversational cues) [24]. It has already been studied the influence of personification and interactivity in people's disclosures around sensitive topics, such as psychological stressors [25]. Important to mention is that chatbots still lack empathy understanding meaning and that they are not as capable as humans of understanding conversational undertones. Though progress has been made in this field, and soon machines will not only be able to understand what somebody is saying but also what is the feeling of what he is saying [26].

However, a biased view of gender is revealed, as most of the chatbots perform tasks that echo historically feminine roles and articulate these features with stereotypical behaviors. Accordingly, general or specialized chatbots automate work that is coded as female, given that they mainly operate in service or assistance related contexts, acting as personal assistants or secretaries [21].

Soon we will live in a world where conversational partners will be humans or chatbots, and in many cases, we will not know and will not care what our conversational partner will be [27].

4   Essential Concepts

Below are some fundamental concepts related to chatbot technology.

Pattern Matching is predicated on representative stimulus-response blocks. A sentence (stimuli) is entered, and output (response) is created consistent with the user input [11]. Eliza and ALICE were the first chatbots developed using pattern recognition algorithms. The disadvantage of this approach is that the responses are entirely predictable, repetitive, and lack the human touch. Also, there is no storage of past responses, which can lead to looping conversations [28].

The **Artificial Intelligence Markup Language** (AIML) was created from 1995 to 2000, and it is based on the concepts of Pattern Recognition or Pattern Matching technique. It is applied to natural language modeling for the dialogue between humans and chatbots that follow the stimulus-response approach. It is an XML-based markup language and it is tag-based. As shown in Fig. 2, AIML is based on basic units of dialogue called categories (tag <category>) which are formed by user input patterns (tag <pattern>) and chatbot responses (tag <template>) [11].

```
<aiml version="1.0.1" encoding="UTF-8"?>
   <category>
      <pattern>  My   name   is   *   and   I   am   *   years   old  </pattern>
      <template>  Hello  <star/>.  I  am  also  <star index="2"/>  years  old!</template>
   </category>
</aiml>
```

Fig. 2. Example of AIML code

Latent Semantic Analysis (LSA) may be used together with AIML for the development of chatbots. It is used to discover likenesses between words as vector representation [29]. Template-based questions like greetings and general questions can be answered using AIML while other unanswered questions use LSA to give replies [30].

Chatscript, being the successor of the AIML language, is an expert system, which consists of an open-source scripting language and the engine that runs it. It is comprised of rules which are associated with topics, finding the best item that matches the user query string and executing a rule in that topic. Chatscript also includes long-term memory in the form of \$ variables which can be used to store specific user information like the name or age of the user. It is also case-sensitive, widening the possible responses that can be given to the same user input based on the intended emotion, as uppercase is typically used in conversations to indicate emphasis [28].

RiveScript is a plain text, line-based scripting language for the development of chatbots and other conversational entities. It is open-source with available interfaces for Go, Java, JavaScript, Perl, and Python [31].

Natural Language Processing (NLP), an area of artificial intelligence, explores the manipulation of natural language text or speech by computers. Knowledge of the understanding and use of human language is gathered to develop techniques that will make computers understand and manipulate natural expressions to perform desired tasks [32]. Most NLP techniques are based on machine learning.

Natural Language Understanding (NLU) is at the core of any NLP task. It is a technique to implement natural user interfaces such as a chatbot. NLU aims to extract context and meanings from natural language user inputs, which may be unstructured and respond appropriately according to user intention [32]. It identifies user intent and extracts domain-specific entities. More specifically, an **intent** represents a mapping between what a user says and what action should be taken by the chatbot. Actions correspond to the steps the chatbot will take when specific intents are triggered by user inputs and may have parameters for specifying detailed information about it [28]. Intent detection is typically formulated as sentence classification in which single or multiple intent labels are predicted for each sentence [32].

An **entity** is a tool for extracting parameter values from natural language inputs. For example, consider the sentence "What is the weather in Greece?". The user intent is to learn the weather forecast. The entity value is Greece. Therefore, the user asks for the weather forecast in Greece [33]. Entities can be either system-defined or developer-defined. For example, the system entity @sys.date corresponds to standard date references like 10 August 2019 or the 10th of August [28]. Domain entity extraction usually referred to as a slot-filling problem, is formulated as a sequential tagging problem where parts of a sentence are extracted and tagged with domain entities [32].

Finally, **contexts** are strings that store the context of the object the user is referring to or talking about. For example, a user might refer to a previously defined object in his following sentence. A user may input "Switch on the fan." Here the context to be saved is the fan so that when a user says, "Switch it off" as the next input, the intent "switch off" may be invoked on the context "fan" [28].

5   Types of Chatbots

Chatbots can be classified using different parameters: the knowledge domain, the service provided, the goals, the input processing and response generation method, the human-aid, and the build method.

Classification based on the **knowledge domain** considers the knowledge a chatbot can access or the amount of data it is trained upon. **Open domain** chatbots can talk about general topics and respond appropriately, while **closed domain** chatbots are focused on a particular knowledge domain and might fail to respond to other questions [34].

Classification based on the **service provided** considers the sentimental proximity of the chatbot to the user, the amount of intimate interaction that takes place, and it is also dependent upon the task the chatbot is performing. **Interpersonal** chatbots lie in the domain of communication and provide services such as Restaurant booking, Flight booking, and FAQ bots. They are not companions of the user, but they get information and pass them on to the user. They can have a personality, can be friendly, and will probably remember information about the user, but they are not obliged or expected to do so. **Intrapersonal** chatbots exist within the personal domain of the user, such as chat apps like Messenger, Slack, and WhatsApp. They are companions to the user and understand the user like a human does. **Inter-agent** chatbots become omnipresent while all chatbots will require some inter-chatbot communication possibilities. The need for protocols for inter-chatbot communication has already emerged. Alexa-Cortana integration is an example of inter-agent communication [34].

Classification based on the **goals** considers the primary goal chatbots aim to achieve. **Informative** chatbots are designed to provide the user with information that is stored beforehand or is available from a fixed source, like FAQ chatbots. **Chat-based/Conversational** chatbots talk to the user, like another human being, and their goal is to respond correctly to the sentence they have been given. **Task-based** chatbots perform a specific task such as booking a flight or helping somebody. These chatbots are intelligent in the context of asking for information and understanding the user's input. Restaurant booking bots and FAQ chatbots are examples of Task-based chatbots [34, 35].

Classification based on the **input processing and response generation method** takes into account the method of processing inputs and generating responses. There are three models used to produce the appropriate responses: **rule-based** model, **retrieval-based** model, and **generative** model [36].

Rule-based model chatbots are the type of architecture which most of the first chatbots have been built with, like numerous online chatbots. They choose the system response based on a fixed predefined set of rules, based on recognizing the lexical form of the input text without creating any new text answers. The knowledge used in the chatbot is humanly hand-coded and is organized and presented with conversational patterns [28]. A more comprehensive rule database allows the chatbot to reply to more types of user input. However, this type of model is not robust to spelling and grammatical mistakes in user input. Most existing research on rule-based chatbots studies response selection for single-turn conversation, which only considers the last input message. In more human-like chatbots, multi-turn response selection takes into consideration previous parts of the conversation to select a response relevant to the whole conversation context [37].

A little different from the rule-based model is the **retrieval-based** model, which offers more flexibility as it queries and analyzes available resources using APIs [36]. A retrieval-based chatbot retrieves some response candidates from an index before it applies the matching approach to the response selection [37].

The **generative** model generates answers in a better way than the other three models, based on current and previous user messages. These chatbots are more human-like and use machine learning algorithms and deep learning techniques. However, there are difficulties in building and training them [36].

Another classification for chatbots considers the amount of **human-aid** in their components. **Human-aided** chatbots utilize human computation in at least one element from the chatbot. Crowd workers, freelancers, or full-time employees can embody their intelligence in the chatbot logic to fill the gaps caused by limitations of fully automated chatbots. While human computation, compared to rule-based algorithms and machine learning, provides more flexibility and robustness, still, it cannot process a given piece of information as fast as a machine, which makes it hard to scale to more user requests [35].

Chatbots can also be classified according to the permissions provided by their development platform. Development platforms can be of open-source, such as RASA, or can be of proprietary code such as development platforms typically offered by large companies such as Google or IBM. **Open-source platforms** provide the chatbot designer with the ability to intervene in most aspects of implementation. **Closed platforms**, typically act as black boxes, which may be a significant disadvantage depending on the project requirements. However, access to state-of-the-art technologies may be considered more immediate for large companies. Moreover, one may assume that chatbots developed based on large companies' platforms may be benefited by a large amount of data that these companies collect.

Of course, chatbots do not exclusively belong to one category or another, but these categories exist in each chatbot in varying proportions.

6 Design and Development

The design and development of a chatbot involve a variety of techniques [29]. Understanding what the chatbot will offer and what category falls into helps developers pick the algorithms or platforms and tools to build it. At the same time, it also helps the end-users understand what to expect [34].

The requirements for designing a chatbot include accurate knowledge representation, an answer generation strategy, and a set of predefined neutral answers to reply when user utterance is not understood [38]. The first step in designing any system is to divide it into constituent parts according to a standard so that a modular development approach can be followed [28]. In Fig. 3, a general chatbot architecture is introduced.

The process starts with a user's request, for example, "What is the meaning of environment?", to the chatbot using a messenger app like Facebook, Slack, WhatsApp, WeChat or Skype, or an app using text or speech input like Amazon Echo [39].

After the chatbot receives the user request, the Language Understanding Component parses it to infer the user's intention and the associated information (intent: "translate," entities: [word: "environment"]) [35].

Once a chatbot reaches the best interpretation it can, it must determine how to proceed [40]. It can act upon the new information directly, remember whatever it has understood and wait to see what happens next, require more context information or ask for clarification.

When the request is understood, action execution and information retrieval take place. The chatbot performs the requested actions or retrieves the data of interest from

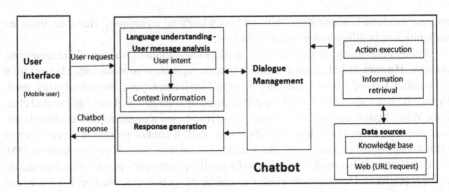

Fig. 3. General chatbot architecture

its data sources, which may be a database, known as the Knowledge Base of the chatbot, or external resources that are accessed through an API call [35].

Upon retrieval, the Response Generation Component uses Natural Language Generation (NLG) [41] to prepare a natural language human-like response to the user based on the intent and context information returned from the user message analysis component. The appropriate responses are produced by one of the three models mentioned in Sect. 5 of the paper: rule-based, retrieval based, and generative model [36].

A Dialogue Management Component keeps and updates the context of a conversation which is the current intent, identified entities, or missing entities required to fulfill user requests. Moreover, it requests missing information, processes clarifications by users, and asks follow-up questions. For example, the chatbot may respond: "Would you like to tell me as well an example sentence with the word environment?" [35].

Many commercial and open-source options are available for the development of a chatbot. The number of chatbot-related technologies is already overwhelming and growing each day [42]. Chatbots are developed in two ways: using any programming language like Java, Clojure, Python, C++, PHP, Ruby, and Lisp or using state-of-the-art platforms. At this time, we are distinguishing six leading NLU cloud platforms that developers can use to create applications able to understand natural languages: Google's DialogFlow [43], Facebook's wit.ai [44], Microsoft LUIS [45], IBM Watson Conversation [17], Amazon Lex [46], and SAP Conversation AI [47]. All these platforms are supported by machine learning. They share some standard functionality (they are cloud-based, they support various programming and natural languages) but differ significantly in other aspects [33]. Other known chatbot development platforms are RASA [48], Botsify [49], Chatfuel [50], Manychat [51], Flow XO [52], Chatterbot [53], Pandorabots [54], Botkit [55], and Botlytics [56].

7 Conclusions

Minimal human interference in the use of devices is the goal of our world of technology. Chatbots can reach out to a broad audience on messaging apps and be more effective than humans are. At the same time, they may develop into a capable information-gathering tool. They provide significant savings in the operation of customer service departments. With further development of AI and machine learning, somebody may not be capable of understanding whether he talks to a chatbot or a real-life agent.

We consider that this research provides useful information about the basic principles of chatbots. Users and developers can have a more precise understanding of chatbots and get the ability to use and create them appropriately for the purpose they aim to operate.

Further work of this research would be exploring in detail existing chatbot platforms and compare them. It would also be interesting to examine the degree of ingenuity and functionality of current chatbots. Some ethical issues relative to chatbots would be worth studying like abuse and deception, as people, on some occasions, believe they talk to real humans while they are talking to chatbots.

Acknowledgments. This work is partially supported by the MPhil program "Advanced Technologies in Informatics and Computers", hosted by the Department of Computer Science, International Hellenic University.

References

1. Bansal, H., Khan, R.: A review paper on human computer interaction. Int. J. Adv. Res. Comput. Sci. Softw. Eng. **8**, 53 (2018). https://doi.org/10.23956/ijarcsse.v8i4.630
2. Khanna, A., Pandey, B., Vashishta, K., Kalia, K., Bhale, P., Das, T.: A study of today's A.I. through chatbots and rediscovery of machine intelligence. Int. J. u- e-Serv. Sci. Technol. **8**, 277–284 (2015). https://doi.org/10.14257/ijunesst.2015.8.7.28
3. chatbot I Definition of chatbot in English by Lexico Dictionaries. https://www.lexico.com/en/definition/chatbot
4. Abu Shawar, B.A., Atwell, E.S.: Chatbots: are they really useful? J. Lang. Technol. Comput. Linguist. **22**, 29–49 (2007)
5. Klopfenstein, L., Delpriori, S., Malatini, S., Bogliolo, A.: The rise of bots: a survey of conversational interfaces, patterns, and paradigms. In: Proceedings of the 2017 Conference on Designing Interactive Systems, pp. 555–565. Association for Computing Machinery (2017)
6. Turing, A.M.: Computing machinery and intelligence. Mind **59**, 433–460 (1950). https://doi.org/10.1093/mind/LIX.236.433
7. Weizenbaum, J.: ELIZA—a computer program for the study of natural language communication between man and machine. Commun. ACM **9**, 36–45 (1966). https://doi.org/10.1145/365153.365168
8. Brandtzaeg, P.B., Følstad, A.: Why people use chatbots. In: Kompatsiaris, I., et al. (eds.) Internet Science, pp. 377–392. Springer, Cham (2017). https://doi.org/10.1007/978-3-319-70284-1_30
9. Colby, K.M., Weber, S., Hilf, F.D.: Artificial paranoia. Artif. Intell. **2**, 1–25 (1971). https://doi.org/10.1016/0004-3702(71)90002-6
10. Wallace, R.S.: The anatomy of A.L.I.C.E. In: Epstein, R., Roberts, G., Beber, G. (eds.) Parsing the Turing Test: Philosophical and Methodological Issues in the Quest for the Thinking Computer, pp. 181–210. Springer, Cham (2009). https://doi.org/10.1007/978-1-4020-6710-5_13
11. Marietto, M., et al.: Artificial intelligence markup language: a brief tutorial. Int. J. Comput. Sci. Eng. Surv. **4** (2013). https://doi.org/10.5121/ijcses.2013.4301
12. Molnár, G., Zoltán, S.: The role of chatbots in formal education. Presented at the 15 September 2018

13. Siri. https://www.apple.com/siri/
14. Personal Digital Assistant - Cortana Home Assistant – Microsoft. https://www.microsoft.com/en-us/cortana
15. What exactly is Alexa? Where does she come from? And how does she work? https://www.digitaltrends.com/home/what-is-amazons-alexa-and-what-can-it-do/
16. Google Assistant, your own personal Google. https://assistant.google.com/
17. IBM Watson. https://www.ibm.com/watson
18. Scopus - Document search. https://www.scopus.com/search/form.uri?display=basic
19. Colace, F., De Santo, M., Lombardi, M., Pascale, F., Pietrosanto, A., Lemma, S.: Chatbot for e-learning: a case of study. Int. J. Mech. Eng. Robot. Res. **7**, 528–533 (2018). https://doi.org/10.18178/ijmerr.7.5.528-533
20. Ranoliya, B.R., Raghuwanshi, N., Singh, S.: Chatbot for university related FAQs. In: 2017 International Conference on Advances in Computing, Communications and Informatics (ICACCI), Udupi, pp. 1525–1530 (2017)
21. da Costa, P.C.F.: Conversing with personal digital assistants: on gender and artificial intelligence. J. Sci. Technol. Arts **10**, 59–72 (2018). https://doi.org/10.7559/citarj.v10i3.563
22. Xu, A., Liu, Z., Guo, Y., Sinha, V., Akkiraju, R.: A new chatbot for customer service on social media. In: Proceedings of the 2017 CHI Conference on Human Factors in Computing Systems, pp. 3506–3510. ACM, New York (2017)
23. Følstad, A., Nordheim, C.B., Bjørkli, C.A.: What makes users trust a chatbot for customer service? An exploratory interview study. In: Bodrunova, S.S. (ed.) INSCI 2018. LNCS, vol. 11193, pp. 194–208. Springer, Cham (2018). https://doi.org/10.1007/978-3-030-01437-7_16
24. Go, E., Sundar, S.S.: Humanizing chatbots: the effects of visual, identity and conversational cues on humanness perceptions. Comput. Hum. Behav. **97**, 304–316 (2019). https://doi.org/10.1016/j.chb.2019.01.020
25. Sannon, S., Stoll, B., DiFranzo, D., Jung, M., Bazarova, N.N.: How personification and interactivity influence stress-related disclosures to conversational agents. In: Companion of the 2018 ACM Conference on Computer Supported Cooperative Work and Social Computing, pp. 285–288. ACM, New York (2018)
26. Fernandes, A.: NLP, NLU, NLG and how Chatbots work. https://chatbotslife.com/nlp-nlu-nlg-and-how-chatbots-work-dd7861dfc9df
27. Dale, R.: The return of the chatbots. Nat. Lang. Eng. **22**, 811–817 (2016). https://doi.org/10.1017/S1351324916000243
28. Ramesh, K., Ravishankaran, S., Joshi, A., Chandrasekaran, K.: A survey of design techniques for conversational agents. In: Kaushik, S., Gupta, D., Kharb, L., Chahal, D. (eds.) ICICCT 2017. CCIS, vol. 750, pp. 336–350. Springer, Singapore (2017). https://doi.org/10.1007/978-981-10-6544-6_31
29. Akma, N., Hafiz, M., Zainal, A., Fairuz, M., Adnan, Z.: Review of chatbots design techniques. Int. J. Comput. Appl. **181**, 7–10 (2018). https://doi.org/10.5120/ijca2018917606
30. An e-business chatbot using AIML and LSA - Semantic Scholar. https://www.semanticscholar.org/paper/An-e-business-chatbot-using-AIML-and-LSA-Thomas/906c91ca389a29b47a0ec072d54e23ddaa757c88
31. Artificial Intelligence Scripting Language - RiveScript.com. https://www.rivescript.com/
32. Jung, S.: Semantic vector learning for natural language understanding. Comput. Speech Lang. **56**, 130–145 (2019). https://doi.org/10.1016/j.csl.2018.12.008
33. Canonico, M., Russis, L.D.: A comparison and critique of natural language understanding tools. Presented at the (2018)
34. Nimavat, K., Champaneria, T.: Chatbots: an overview types, architecture, tools and future possibilities. Int. J. Sci. Res. Dev. **5**, 1019–1024 (2017)
35. Kucherbaev, P., Bozzon, A., Houben, G.-J.: Human-aided bots. IEEE Internet Comput. **22**, 36–43 (2018). https://doi.org/10.1109/MIC.2018.252095348

36. Hien, H.T., Cuong, P.-N., Nam, L.N.H., Nhung, H.L.T.K., Thang, L.D.: Intelligent assistants in higher-education environments: the FIT-EBot, a chatbot for administrative and learning support. In: Proceedings of the Ninth International Symposium on Information and Communication Technology, pp. 69–76. ACM, New York (2018)

37. Wu, Y., Wu, W., Xing, C., Zhou, M., Li, Z.: Sequential Matching Network: A New Architecture for Multi-turn Response Selection in Retrieval-based Chatbots. arXiv:1612.01627 [cs] (2016)

38. Augello, A., Gentile, M., Dignum, F.: An overview of open-source chatbots social skills. In: Diplaris, S., Satsiou, A., Følstad, A., Vafopoulos, M., Vilarinho, T. (eds.) INSCI 2017. LNCS, vol. 10750, pp. 236–248. Springer, Cham (2018). https://doi.org/10.1007/978-3-319-77547-0_18

39. Zumstein, D., Hundertmark, S.: Chatbots – an interactive technology for personalized communication, transactions and services. IADIS Int. J. WWW/Internet **15**, 96–109 (2017)

40. Fern, A., et al.: The Best Open Source Chatbot Platforms in 2019 (2019). https://blog.verloop.io/the-best-open-source-chatbot-platforms-in-2019/

41. Singh, S., Darbari, H., Bhattacharjee, K., Verma, S.: Open source NLG systems: a survey with a vision to design a true NLG system. **9**, 4409–4421 (2016)

42. Nayyar, D.A.: Chatbots and the Open Source Tools You Can Use to Develop Them (2019). https://opensourceforu.com/2019/01/chatbots-and-the-open-source-tools-you-can-use-to-develop-them/

43. Dialogflow. https://dialogflow.com/

44. Wit.ai. https://wit.ai/

45. LUIS (Language Understanding) – Cognitive Services – Microsoft Azure. https://www.luis.ai/home

46. Amazon Lex – Build Conversation Bots. https://aws.amazon.com/lex/

47. SAP Conversational AI | Automate Customer Service With AI Chatbots. https://cai.tools.sap

48. Rasa: Open source conversational AI. https://rasa.com/

49. Botsify: Botsify - Create Automated Chatbots Online for Facebook Messenger or Website. https://botsify.com

50. Chatfuel. https://chatfuel.com/

51. ManyChat – Chat Marketing Made Easy. https://manychat.com/

52. AI Online Chatbot Software, Live Chat on Websites. https://flowxo.com/

53. About ChatterBot — ChatterBot 1.0.2 documentation. https://chatterbot.readthedocs.io/en/stable/

54. Pandorabots: Home. https://home.pandorabots.com/home.html

55. Botkit: Building Blocks for Building Bots. https://botkit.ai/

56. Botlytics. https://www.botlytics.co/

Applying an Intelligent Personal Agent on a Smart Home Using a Novel Dialogue Generator

Anastasios Alexiadis[✉], Alexandros Nizamis, Ioannis Koskinas, Dimosthenis Ioannidis, Konstantinos Votis, and Dimitrios Tzovaras

Centre for Research and Technology Hellas, Information Technologies Institute (CERTH/ITI), 6km Charilaou-Thermi, Thessaloniki, Greece
{talex,alnizami,jkosk,djoannid,kvotis,Dimitrios.Tzovaras}@iti.gr

Abstract. Nowadays, Intelligent Personal Agents include Natural Language Understanding (NLU) modules, that utilize Machine Learning (ML), which can be included in different kind of applications in order to enable the translation of users' input into different kinds of actions, as well as ML modules that handle dialogue. This translation is attained by the matching of a user's sentence with an intent contained in an Agent. This paper introduces the first generation of the CERTH Intelligent Personal Agent (CIPA) which is based on the RASA (https://rasa.com/) framework and utilizes two machine learning models for NLU and dialogue flow classification. Besides the architecture of CIPA—Generation A, a novel dialogue-story generator that is based on the idea of adjacency pairs is introduced. By utilizing on this novel-generator, the agent is able to create all the possible dialog trees in order to handle conversations without training on existing data in contrast with the majority of the current alternative solutions. CIPA supports multiple intents and it is capable of classifying complex sentences consisting of two user's intents into two automatic operations from the part of the agent. The introduced CIPA—Generation A has been deployed and tested in a real-world scenario at Centre's of Research & Technology Hellas (CERTH) nZEB Smart Home (https://smarthome.iti.gr/) in two different domains, energy and health domain.

Keywords: Intelligent agents · Natural Language Understanding · Multiple intents · Smart home

1 Introduction

Fast growth in natural language understanding capabilities of intelligent virtual agents enables its wide use on different types of consumers devices. This type of personal agents have been the focus of research for a long time with plenty of significant results. ELIZA [1] and IBM Shoebox [2] are considered as two of the most representative research's first outcomes. Nowadays, there is a wide variety of Intelligent Personal Assistants (IPAs) such as Google Assistant, Amazon

© IFIP International Federation for Information Processing 2020
Published by Springer Nature Switzerland AG 2020
I. Maglogiannis et al. (Eds.): AIAI 2020, IFIP AICT 584, pp. 384–395, 2020.
https://doi.org/10.1007/978-3-030-49186-4_32

Alexa and Apple Siri that attempt to assist humans either by providing them with information or by performing specific tasks for them as a results of their communication with humans using natural language. These systems aim to solve two different but related problems: (a) the natural language understanding and (b) the management of the flow of the conversation (dialogue). Based on these two aforementioned problems, different frameworks have been designed. These frameworks provide to researchers and developers systems that are capable of handling the above problems in order to enable them to develop Virtual Assistants and chat-bots for specific domains. These IPAs frameworks include Google Dialogflow, Amazon Lex, Facebook WIT.AI and RASA. Most of these frameworks are accessed through web Application Programming Interfaces (APIs) and the developers do not have access neither to their source code or to the machine learning models that power them.

Leveraging on the above described research results and frameworks, the first version of CERTH Intelligent Personal Agent is designed and introduced in this work. The main key innovative aspect of the proposed agent is the adoption of a novel dialogue-story generator that is based on the idea of adjacency pairs and is able to cover all the possible states in the communication between the humans and the agent. Furthermore, CIPA Generation A is able to support applications of different domains and it is available in a multi-intent model.

The paper is structured as follows. Following the Introduction, a brief literature review is presented. The proposed CIPA–gen A's architecture is analyzed in Sect. 3. In the same section, the dialogue-story generator that is based on the idea of adjacency pairs is presented as well. Section 4 consists the evaluation section and provides a brief description of the use cases of both domains, energy and health. Finally, the conclusions and future work are drawn in Sect. 5.

2   Related Works

Modern conversational agents relies on natural language processing, supported by AI and ML techniques and execute tasks based on verbal interaction with end-users. As the proposed work is driven by results of research and development in the fields of task-based agents, conversational agents, and dialogue generation, a brief analysis of related works in these fields is documented in this section.

The proposed work is strongly related with dialogue generation in general. Modeling the future direction of a dialogue is crucial to generate coherent dialogues and execute tasks. To this aim, data-driven conversation modeling mostly build upon recurrent neural networks (RNNs) [13] that enable an agent to learn how to map rules between input messages and responses from a massive amount of training data. StarSpace [3] is a general-purpose neural embedding model that can solve a wide variety of problems including text classification. The StarSpace model consists of learning entities that are described by a set of discrete features. Its main contributions are (a) the generalization across diverse problems and (b) the comparison between embeddings of different types. Industry leaders such as Google Assistant, Amazon Alexa and Apple Siri are based on ML models for dialogue generation. Recently, Google introduces Meena[1], an end-to-end, neural conversational model that learns to respond sensibly to a given conversational context. Meena has a single evolved transformer encoder block for processing the conversation context block and 13 evolved transformer decoder blocks to formulate an actual response. The Encoder-Decoder Model for dialogue generation is a popular technique that is adopted from research community as well. Serban et al. [11] introduces an encoder-decoder model for generating dialogues. It is based in a neural network-based architecture, with hierarchical latent variables that capture dependencies over an extended conversation history. To the same direction, a variational hierarchical conversation RNNs model is introduced in [12]. Other recent approaches are based on Reinforcement Learning techniques. Li et al. [14] train two models, a generative model to produce response sequences, and a discriminator to distinguish between the human-generated dialogues and the machine-generated ones. The generative model receives the output of the discriminator as a reward in order to push the complete system to generate dialogues that mostly resemble human dialogues. In [15] deep reinforcement learning for dialogue generation method was lately proposed. The method optimizes long-term rewards and uses a encoder-decoder architecture. It simulates the conversation between two virtual agents to discover possible actions while learns to maximize expected reward. Furthermore, plenty approaches related to task-oriented RL methods have been proposed [16].

Besides the general works related to dialogue generation, smart home automation applications that were led by task oriented dialogue systems have been introduced. Recently, Park et al. [17], introduced a framework for development of task-oriented dialogue systems in a Smart Home environment. Domain knowledge is required for the proposed task-oriented dialogue system. The framework ontologically expresses the required knowledge and is able to build a dialogue system by editing the dialogue knowledge. In addition to this, a more intelligent conversation is enabled by providing a hierarchical argument structure to manage the various argument representations that are included in natural language sentences. The dialogue management system of the introduced framework is based on a rule-based systems. Rule-based dialogue management systems are defining the possible states of a dialogue and the behavior of the system for the given state. Both finite-state [8] and information-state [9] based systems are used. Especially, the finite-state methodology is highly-correlated to the adjacency pairs methodology, that is introduced from the presented work, as a concept for dialogue generation. However both finite-state and information-state techniques are vulnerable to errors in NLU, because they are entirely dependent on the result of the NLU to recognize the intent of the user. A smart home automation system, named Cassandra, has been introduced in [10]. A voice assisting system based on automatic speech recognition and NLU is proposed in order to enable user to control smart home appliances. Opposite to the presented work in this paper, Cassandra uses a knowledge base of predefined scenarios to handle the dialog with the user. After analysis and parameters' extraction based on user's

---

[1] https://ai.googleblog.com/2020/01/towards-conversational-agent-that-can.html

voice command, a predifined scenario is selected. Furthermore, the end-user is able to alter Cassandra's configuration by creating scenarios or editing existing ones.

In contrast with the above described related works, that are based on dialogue generation for existing knowledge and data, the proposed CIPA Gen—A introduces a novel dialogue generator that does not require the availability of historical data. It is based on adjacency pairs and enables the generation of all the possible dialogue trees. Adjacency pairs were first introduced by Schegloff and Sacks [18] as the basic foundation of conversational structure can be used for the evaluation of dialogue stories. Midgley et al. [7] proposed a new method of dialogue segmentation in order to leverage the gathered information from more than one previous utterances identifying adjacency pair labels that occur with high frequency.

3   CIPA Gen—A Architecture

In this section the design approaches of CERTH Intelligent Personal Agent Gen–A architecture are presented. CIPA–gen A is designed in order to support multiple task domains. This capability is enabled by auto-generated Python scripts that sub-class RASA Core's Action classes, one class for each action included in a CIPA domain, and by providing a call-back function that handles the actions for that domain. Moreover, CIPA–gen A offers an API that supports exchange of messages in JSON format. Furthermore, it includes an SQL database for the storage of user's messages and an action history. CIPA–gen A is provided with a multiple intent model. The novel dialogue generator from CIPA–gen A and the supported multi intent model are explained in details in the following subsections. The technical descriptions come alongside with examples related to the supported domains by the Agent and their datasets in order to describe better the application of the proposed agent and be easier for the reader to understand and follow the concepts in this document.

3.1   Supported Domains and Datasets

CIPA–gen A has been deployed in CERTH nZEB Smart Home in two different domains, energy and health. The CERTH nZEB Smart Home is a rapid prototyping and a novel technologies demonstration infrastructure resembling a real domestic building where occupants can experience actual living scenarios while exploring various innovating smart IoT-based technologies with provided Energy, Health, Big Data, Robotics and Artificial Intelligence (AI) services.

CERTH nZEB Smart Home - Energy Domain: Smart Home is equipped with energy domain related IoT devices that monitor the energy consumption and production, and the conditions of the entire building while various algorithms can support automation and energy efficiency scenarios. A dataset has been created from the testers' (Smart Home occupants) inputs, thereafter referred to as the *gold* set of the energy domain. This set is continuously extended for nZEB Smart Home's energy domain. Due to its size (1608 samples) it is solely used as a test set. A common pattern that has been found in the data was the omission of information crucial for an operation to be performed. For example a user could utter "Turn on the lights" without providing the room where he/she wanted the action to be performed. Moreover, there were observations such as "Turn the lights" which did not contain the fully specified action information, such as if the user wanted the lights on or off. As the user could request a change for the state of the lights of another room this is crucial information.

CERTH nZEB Smart Home - Health Domain: Health related IoT devices monitors a variety of physiological attributes, and enabling the extraction of valuable data through intelligent processing towards preventing situations that could lead to harmful outcomes. A dataset, based on the testers' inputs, has been created. Thereafter, it is referred as the *gold* set of the health domain. This set is continuously extended and updated for nZEB Smart Home's health domain. Due to its size (1972 samples) it is solely used as a test set.

3.2   Novel Dialogue Generator

A novel dialogue-story generator that is based on the idea of adjacency pairs has been designed and developed for CIPA–gen A. The proposed story generator models dialogues that consist of a subset of adjacency pairs based on the following two assumptions: (a) the user may omit information required by an action and thus the Agent have to ask for that information by interacting with the user in a dialogue and (b) the user may not cooperate by following up the conversation. For example, instead of replying to a question posed by the Agent, the user may request another action.

For every action requested, the story generator produces conversation flows with various pieces of the missing but required information and defines the conversation flows that interact with the user to obtain this missing information. This produces all valid conversation flows for each action. For example, consider the "Turn on hvac" action from the nZEB Smart Home energy domain. The introduced generator will produce the following valid stories for this action:

- The user gives the room and the on/off switch.
- The user gives only the room and the agent asks for the on/off operation.
- The user gives only the on/off operation and the agent asks for the room.
- The user doesn't give the required information and the agent first asks for the room and after getting a correct reply asks for the on/off option.

In addition, the generator produces dialogues for handling valid multi-intents by the user. For example in the `turn-hvac+change-hvac-mode` multi intent, the generator will produce all valid adjacency pair flows for the union of their slots. At the end of each story it will call each action sequentially. If a slot is common to both intents and the user has not provided it, the CIPA–gen A will ask it once.

Algorithm 1 Write Story

1: **procedure** WRITE STORY($intent, intenToProcess, included,$
   $excluded, runRepeatedAct, mappings, c, repeat$)
2:     $s \leftarrow 0$
3:     **if** $repeat > -1$ **then**
4:         *write* repeat signature number c
5:     **else**$multiIntent(intent)$
6:         *write* multi intent signature number c
7:         *write* single intent signature number c
8:     **end if**
9:     $WriteStoryII(intent, included, mappings)$
10:    **for** $i \leftarrow 0$ to $excluded.length$ **do**
11:        *write* ' -' $slots\_to\_act\_map[excluded[i]]$
12:        **if** $repeat = i$ and $runRepeatedAct = False$ and $s = 0$ **then**
13:            $WriteStoryII(intentToProcess, [\,], mappings)$
14:            *write* ' - $utter\_repeat$'
15:            **break**
16:        **else**$repeat = i$ and $s = 0$
17:            $WriteStoryII(intentToProcess, [excluded[i]], mappings)$
18:            $s \leftarrow 1$
19:            **continue**
20:        **end if**
21:        *write* '* inform{' $(excluded[i], slot\_example\_fun(excluded[i]))$ '}'
22:    **end for**
23:    **if** $!multiIntent(intent)$ and $(repeat = -1$ or $runRepeatedAct = True)$ **then**
24:        *write* ' -' $intent\_to\_act\_map[intent]$
25:    **else**$repeat = -1$ or $runRepeatedAct = True$
26:        **for** each intent $inte\_$ of multi intent $intent$ **do**
27:            *write* ' -' $intent\_to\_act\_map[inte\_]$
28:        **end for**
29:    **end if**
30:    *write* ' - $action\_restarted$'
31: **end procedure**

Moreover, it produces invalid conversation flows in which the user does not cooperate by following up the conversation with the supported intents but he/she follows up the conversation with a different intent. The invalid conversation flows restart the conversation after the agent tells the user that it did not understand his/her intentions.

The proposed algorithm that generates stories (Algorithm 1 to 3), takes as inputs an intent to action mapping, a function that maps intents to a set of slots (that are needed for the intent), a slot example relation that maps slots to random slot instances, a slot to action mapping (that specifies which action requests each slot) and a mapping of valid multi intents. The algorithm automatically generates the story file (in the RASA format) for training the dialogue models. It operates by including a tree-expansion phase to produce all possible story combinations that have slot information missing in the original user message, as well as all invalid variations of them in which the user's reply is classified to a different intent from the expected one.

Algorithm 2 Generate all stories of an intent

1: **procedure** GEN INTENT(*intentToProcess, rc, mappings*)
2:     **if** *multIntent(intentToProcess)* **then**
3:         *Split intentToProcess to individual intents*
4:         *slots ← union of slots of the individual intents*
5:     **else**
6:         *slots ← slots of intentToProcess*
7:     **end if**
8:     *c* = 1
9:     *slotsComb = SlotsCombinations(slots)*
10:    **for** *include, exluded* in *slotsComb* **do**
11:        *WriteStory(intentToProcess, intentToProcess, included,*
       excluded, runRepeated = False, mappings, c, −1)
12:        *c ← c + 1*
13:        **for** *excludedItem[i]* in *excluded* **do**
14:            **for** *intent* in *intent_to_act_map* **do**
15:                **if** *intent ≠ intentToProcess* **then**
16:                    *WriteStory(intentToProcess, intent, included,*
       excluded, runRepeatedAct = False, mappings, rc, i)
17:                **else**
18:                    *WriteStory(intenToProcess, intent, included,*
       excluded, runRepeatedAct = True, mappings, rc, i)
19:                **end if**
20:                *rc ← rc + 1*
21:            **end for**
22:        **end for**
23:    **end for**
       return *rc*
24: **end procedure**

3.3   Single and Multi Intent Models

Both single and multi intent models are supported by the CIPA–gen A. In the single intent model the Chatito, a third-party open-source natural language generator is used in order to create a training set for the NLU module. For the NLU model we did not select Rasa NLU's default model that uses the SpaCy Natural Language Processing system but opted to use a RASA NLU Model based on StarSpace for intent classification. Afterwards, we defined in Chatito's DSL (Domain Specific Language) examples of intents for the nZEB Smart Home domain. These do not correspond to an exact one to one mapping, as we defined some extra intents. The two most important of them are the *inform* and *nounderstand* intents. The first should match all user messages that inform the agent about a slot (a parameter) of an action, such as the location of an action, in the case it was not included in the original message. The second one, the nounderstand intent, should match all user messages that resemble messages that correspond to other intents but that do not make sense. For example consider the user message "turn on the lights" and the message "turn on the door", the first message should be classified to the *turn the lights* intent, whereas the second should be classified to the *nounderstand* intent. By specifying these intents and text patterns in the Chatito DSL, we used the generator to generate our training set for the NLU Model.

For the dialogue Model, we used the default Rasa Core's LSTM and memorization models. Furthermore, we added a fallback policy for both the intent classification and dialogue classification models. Therefore, the Agent will be able to reply that he does not understand anything in the cases of low classification confidence.

Algorithm 3 Story Gen - Main function of the algorithm

1: **procedure** STORY GEN(*domain_mappings*)
2:     Write a story of no understand classified to utter default
3:     $rc \leftarrow 1$
4:     **for** *intent* in *intent_to_action_map* **do**
5:         $rc \leftarrow GenIntent(intent, rc, domain\_mappings)$
6:     **end for**
7:     **for** $(inte1 \rightarrow intents)$ in multi_intent_map **do**
8:         **for** *inte2* in *intents* **do**
9:             $rc \leftarrow GenIntent(inte1' +' inte2, rc, domain\_mappings)$
10:        **end for**
11:    **end for**
12:    **for** *slot* in *slots_to_act_map* **do**
13:        Write a story consisting of a single inform and a slot
    classified to no-understand
14:    **end for**
15:    Write a story consisting of a single inform classified to no-understand
16: **end procedure**

For the multi intent model we developed an algorithm that takes as input a Chatito DSL file alongside with a mapping of valid multi intents and then, it produces an extended Chatito DSL file that includes patterns of these multi intents by combining the patterns of their single intents counterparts. Moreover, it uses a random variable to consider omitted duplicated words when combining messages from two intents. For example, in the case of combining "please turn on the hvac in the kitchen" with "turn the hvac mode on", the second "hvac" word will probably be omitted in the training example for the multi-itent.

Thereafter, we used the output of this algorithm to generate our NLU training examples using Chatito. The output of this algorithm produced a multitude of multi intent data that surpassed the single-intent ones in the data. To balance the classes we post-processed this output by tripling to quadrupling the single-intent observations in the data.

Furthermore, we extended RASA NLU's adaption of the StarSpace method, by adding more hidden layers to the NLU model and training an NLU Model for multi-intent classification. We used a $512 \times 384 \times 256 \times 128$ configuration for network A and a 384×128 configuration for network B.

4   Evaluation

In this section the evaluation results for the multi intent models and for both domains, energy and health, are presented. Prior to introducing the results of the

evaluation, an assumption for miss-classification should be taken into consideration. Most of the miss-classification were of the type `no-understand` to `inform`.
This miss-classification type is not a problem as it is handled by the dialogue
model. We can classify orphan `inform`s (that is `inform`s that do not occur in the
middle of the dialogue) as `no-understand`s, whereas incorrect `inform`s in the
middle of a dialogue that do no match any of the valid dialogue flows generated
by the generator presented in this paper are handled by the produced stories for
invalid scenarios. So, if we consider this type of miss-classifications as correct
our accuracy is increased (column *Accuracy2* in Table 1).

Table 1. Evaluation metrics

Evaluation	Accuracy	Accuracy 2	Lowest conf
M-Intent (Energy)	98.39%	99%	0.53
M-Intent (Health)	99.14%	99.75%	0.47

Table 1 presents the results of the multi-intent NLU model on the gold sets
of the two domains. We have included various text patterns for these intents in
the Chatito DSL and generated our training data for the NLU Models as we did
not have access to enough real world user messages for these domains. Training
our NLU Models on real-world data should provide a stronger model for these
domains.

To evaluate the dialogue generator we present examples of valid use cases. A
user can phrase a command for the agent in various ways. Moreover, the agent
can converse with the user if it recognizes an intent that information is missing.
For example if the user commands the agent to "turn the lights", the agent
will reply with either "where?" or "in which room?". If the user replies with
a valid room (i.e., "In the kitchen") the agent will reply "Do you want them
on or off?". If the user replies with either "on" or "off" the agent will execute
the *turn-lights action*. A representative example of the generative process of the

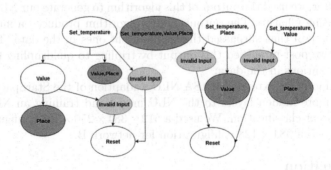

Fig. 1. Finite automata of story generator example

algorithm through the conversational flow between the agent and a user, as far as set temperature-single intent command are concerned:

- Active node = stands on an active state that has not fulfilled all requirements and is able to generate new scenarios
- Final node = stands on a final state and cannot generate new scenarios

A case example was used as part of the evaluation process. The example aims to demonstrate that the Novel Dialogue Generator of CIPA–gen A is able to produce all the possible use case scenarios in its conversation with a user by adopting the concept of adjacency pairs. In particular, in the presented example the agent requires information about the value of the temperature and the place (room). There are 4 initial cases: (a) the user defines the intent accompanied by the value or (b) the place or (c) both of them or (d) the case that the user inputs only the intent. In the next step of this discourse, the user can either provide valid information that fulfills a slot or provide an invalid input that is categorized in one of the 14 invalid operations in the sector of energy (lights, hvac, etc) including no-understand operation in the case of no relevant input. Every single invalid input in the Fig. 1 diagram is translated to multiple alternative rejected scenarios. The node that contains value and place is a final node. On the contrary, the rest of the nodes need at least one more slot to be fulfilled in order to end up in a final state. In the same logic, every active node generates more complex scenarios until a final node is born. In this example, the *set_temperature* intent generates 64 unique scenarios that cover all possible contingencies and they are illustrated from the colourful nodes in the Fig. 1. Each node represents the given input from the user. Calculation of all possible generated scenarios in a multi intent command is a process identical to the single intent, differentiating only in the number of slots. Multi intents commands have as slots the union of the slots of two intents. So, for example in the *set_temperature_lights* intent that is responsible to set the temperature in a room and turn the lights, the available slots comprise the union of (value, place) from *set_temperature* and *lights_on_off* from light. The total number of generated scenarios are 281.

The evaluation of CIPA story generator was an experimental process that conducted through a comparison between the enumeration of all possible scenarios that can be generated from a dialogue among an agent and a user as illustrated in the Fig. 1 and the actual scenarios that are generated from CIPA (Parsed data from the generated file and calculated the sum of generated scenarios for every intent). CIPA is evaluated as a competent agent that behaves properly covering all contingencies.

5   Conclusions and Future Work

In conclusion, this paper introduces a general purpose task-based agent, equipped with a novel dialogue generator, and it was applied to two domains of the CERTH *nZEB Smart Home*, energy and health domain. The developed CIPA–gen A is based on the RASA framework and supports multi-intent for the aforementioned domains capable of recognizing up to two intents per user command. Towards to a domain agnostic agent, the code-base was engineered so as to be easily applied to different domains. CIPA–gen A uses an embedding method for its NLU model that is not based on pre-trained word vectors of a specific language so as it can be generalized to *any* natural language. The training data should have to be generated for that language, using the tools selected and the algorithms developed, while omitting a dependency on SpaCy for slot extraction. This consists the only limitation on training a model for a different language. In addition, messages from real-world users should be collected in order to build a real usage data set that can be used for both training the model to achieve greater accuracy and for evaluation purposes. Furthermore, a novel dialogue generator, based on the idea of adjacency pairs, has been implemented in order to enable the proposed agent to generate all the possible scenarios in a conversation between the agent and a user.

Future research and development related to CIPA agent will be focused on the extension of the agent's capabilities. A first enhancement would be the replacement of the LSTM of the dialogue model with a differential neural computer (DNC) [4]. Next releases of the agent will support complex objectives in its action model, i.e., the agent will be able to form a complex plan of action based on its conversation with the user and execute it and by utilizing AI Planning research [6]. The new system will be the CERTH Advanced Multi-Domain Reasoning and Planning System (CAMDRaPS). This idea has been considered before by MIT researchers [5] but it has not been developed yet. In general, it is planned to add an open-source classical planner, more advanced conversion model from intents/entities to planning actions/variables, machine learning techniques to improve classification of the domain, Probabilistic Planning and Probabilistic Reasoning over Time.

Acknowledgements. This project has received funding from the European Union's Horizon 2020 research and innovation programme under grant agreements No 643607 (myAirCoach) & No 732679 (ACTIVAGE).

References

1. Weizenbaum, J.: ELIZA a computer program for the study of natural language communication between man and machine. Commun. ACM **9**, 36–45 (1996). https://doi.org/10.1145/365153.365168
2. Dersch, W.C.: Shoebox - a voice responsive machine. Datamation **8**(6), 47–50 (1962)
3. Wu, L., Fisch, A., Chopra, S., Adams, K., Bordes, A., Weston, J.: StarSpace: embed all the things!. In: Proceedings of the Thirty-Second AAAI Conference on Artificial Intelligence (2018)
4. Graves, A., et al.: Hybrid computing using a neural network with dynamic external memory. Nature **538**(7626), 471–476 (2016)

5. Yu, P., Shen, J., Yeh, P.Z., Williams, B.: Towards personal assistants that can help users plan. In: Traum, D., Swartout, W., Khooshabeh, P., Kopp, S., Scherer, S., Leuski, A. (eds.) IVA 2016. LNCS (LNAI), vol. 10011, pp. 424–428. Springer, Cham (2016). https://doi.org/10.1007/978-3-319-47665-0_47

6. Nau, D., Ghallab, M., Traverso, P.: Automated Planning: Theory and Practice. Morgan Kaufmann Publishers Inc., San Francisco (2004)

7. Midgley, T.D., Harrison, S., MacNish, C.: Empirical verification of adjacency pairs using dialogue segmentation. In: Proceedings of the 7th SIGdial Workshop on Discourse and Dialogue, pp. 104–108. Association for Computational Linguistics (2006)

8. Goddeau, D., Meng, H., Polifroni, J., Seneff, S., Busayapongchai, S.: A form-based dialogue manager for spoken language applications. In: Proceedings of the International Conference on Spoken Language Processing, Philadelphia, PA, USA, 3–6 October 1996, pp. 701–704 (1996)

9. Traum, D.R., Larsson, S.: The information state approach to dialogue management. In: van Kuppevelt, J., Smith, R.W. (eds.) Current and New Directions in Discourse and Dialogue. Text, Speech and Language Technology, vol. 22. Springer, Dordrecht (2003). https://doi.org/10.1007/978-94-010-0019-2_15

10. Dumitrescu, S.D.: Cassandra smart-home system description. In: 2017 International Conference on Speech Technology and Human-Computer Dialogue (SpeD), Bucharest, pp. 1–6 (2017)

11. Serban, I.V., Sordoni, A., Lowe, R., Charlin, L., Pineau, J., Courville, A.C, Bengio, Y.: A hierarchical latent variable encoder-decoder model for generating dialogues. In: AAAI, pp. 3295–3301 (2017)

12. Park, Y., Cho, J., Kim, G.: A hierarchical latent structure for variational conversation modeling. ArXiv, abs/1804.03424 (2018)

13. Vinyals, O., Le, Q.: A neural conversational model. In: Proceedings of ICML Deep Learning Workshop (2015)

14. Li, J., Galley, M., Brockett, C., Spithourakis, G.P., Gao, J., Dolan, B.: A persona-based neural conversation model. In: Proceedings of the 54th Annual Meeting of the Association for Computational Linguistics (Volume 1: Long Papers), Berlin, Germany, pp. 994–1003 (2016)

15. Li, J., Monroe, W., Ritter, A., Galley, M., Gao, J., Jurafsky, D.: Deep reinforcement learning for dialogue generation (2016)

16. Schatzmann, J., Weilhammer, K., Stuttle, M., Young, S.: A survey of statistical user simulation techniques for reinforcement-learning of dialogue management strategies. Knowl. Eng. Rev. 21(02), 97–126 (2006)

17. Park, Y., Kang, S., Seo, J.: An efficient framework for development of task-oriented dialog systems in a smart home environment. Sensors (Basel) 18(5), 1581 (2018). https://doi.org/10.3390/s18051581

18. Schegloff, E.A., Sacks, H.: Opening up closings. Semiotica 8(4), 289–327 (1973)

Exploring NLP and Information Extraction to Jointly Address Question Generation and Answering

Pedro Azevedo⬡, Bernardo Leite[(✉)]⬡, Henrique Lopes Cardoso⬡,
Daniel Castro Silva⬡, and Luís Paulo Reis⬡

Artificial Intelligence and Computer Science Lab (LIACC), Department of
Informatics Engineering (DEI), Faculty of Engineering, University of Porto (FEUP),
Rua Dr. Roberto Frias, s/n, 4200-465 Porto, Portugal
{pedro.jazevedo,bernardo.leite,hlc,dcs,lpreis}@fe.up.pt

Abstract. Question Answering (QA) and Question Generation (QG) have been subjects of an intensive study in recent years and much progress has been made in both areas. However, works on combining these two topics mainly focus on how QG can be used to improve QA results. Through existing Natural Language Processing (NLP) techniques, we have implemented a tool that addresses these two topics separately. We further use them jointly in a pipeline. Thus, our goal is to understand how these modules can help each other. For QG, our methodology employs a detailed analysis of the relevant content of a sentence through Part-of-speech (POS) tagging and Named Entity Recognition (NER). Ensuring loose coupling with the QA task, in the latter we use Information Retrieval to rank sentences that might contain relevant information regarding a certain question, together with Open Information Retrieval to analyse the sentences. In its current version, the QG tool takes a sentence to formulate a simple question. By connecting QG with the QA component, we provide a means to effortlessly generate a test set for QA. While our current QA approach shows promising results, when enhancing the QG component we will, in the future, provide questions for which a more elaborated QA will be needed. The generated QA datasets contribute to QA evaluation, while QA proves to be an important technique for assessing the ambiguity of the questions.

Keywords: Question Generation · Question Answering · Named Entity Recognition · Part-of-speech tagging · Open Information Extraction

1 Introduction

Artificial intelligence has its impact on all areas of society, and Education is not left out. Posing questions is a central piece in the area of Education. From Professor-generated questions, it is possible not only to test the acquired knowledge but also to contribute to the student's continuous learning process. As the

I. Maglogiannis et al. (Eds.): AIAI 2020, IFIP AICT 584, pp. 396–407, 2020.
https://doi.org/10.1007/978-3-030-49186-4_33

amount of information available is increasing exponentially, accurately selecting relevant information has become an important goal. The task is a daily concern for teachers. The ability to automatically generate questions opens up several possibilities in this context. The prospect of generating questions by giving a text document as input, can provide substantial assistance for reading comprehension. Additionally, it is very important in the Education field since it allows the use of automatic mechanisms to find relevant information and create questions about those contents. It can also be useful since it teaches people how to search for information online correctly. Answering these questions does not always imply knowing where the answers are located. In fact, when developing Question Answering (QA) systems, we often need to rely on Information Retrieval to locate relevant information on the web [28].

With this in mind, we have developed a tool capable of generating factual questions and also answering them, regardless of the genre of textual content. Generated questions include factual questions such as *Who?, Where?, Which country?, When?, What?, How much?,* and *What organization?.* Factual questions allow us to question about specific facts. Usually, these facts may refer to information about people, places, dates, events, monetary values or even about certain organizations/institutions. Our tool uses Named Entity Recognition (NER) to extract entities from the text. In addition, we use POS tagging to find patterns in sentences and extract information that can be questioned. Questions are generated from well-defined rules and with the use of regular expressions to match certain patterns.

In tandem with the Question Generation (QG) component, we have developed an independent QA mechanism. Searching within a corpus, the QA mechanism is capable of ranking candidate sentences. These sentences are also evaluated by their content using Open Information Extraction mechanisms [12].

This combined approach is sensible because both tasks can help each other. QA can be used to assess if the generated question is correctly formulated, by retrieving relevant candidate sentences; or if the question is ambiguous, by finding more than one viable answer. On the other hand, QG can help evaluate the QA mechanism by creating a large number of questions. By manipulating the subject or the manner that questions are generated, it is possible to evaluate the robustness of the QA mechanism.

In Sect. 2, we present related work in the two topics of Question Answering and Question Generation. In Sect. 3, we explain the methodology used in our research, and the development of the question generation and answering components. Experimental results are presented in Sect. 4. Finally, in Sect. 5 we conclude and point out some observations about both current work and new ideas for future work.

2   State of the Art

Most existing works address either QG or QA. Hence, this section discusses the state of the art for QG and QA in a separate way. The works that do address QG together with QA focus only on creating QG models to improve QA [6,11,27,31].

This is done by augmenting QA datasets, or by training the QA model with the generated questions and fine-tuning on QA datasets.

2.1   Question Generation

Typically, QG can be divided into three distinct categories: syntax-based, semantic-based and template-based.

In a syntax-based approach [5], the goal is to convert declarative sentences into interrogative using several transformations. With a semantic approach [7], it is possible to obtain the semantic parse of a sentence using semantic role labeling (SRL). This approach may provide a deeper level of analysis, when compared to the syntax-based one. It also applies the necessary transformations. Finally, in a template-based approach [15] there are no transformation rules. This method extracts relevant content from the text and uses predefined question templates. Recent approaches make use of neural networks [8,13] in order to automatically generate questions from large datasets.

Some specific algorithms have been developed regarding automatic question generation. Topic modeling and noun phrase extraction have been used to create questions from different text passages with a holistic approach [17]. The use of an agent for generation of factual questions has been proposed [24] in order to assess the knowledge of learners and verify their understandings. Factoid gap-filling questions are usually employed [1]: the system extracts informational sentences from the paragraphs in order to generate a gap or a set of gaps that will be hidden from the sentences. These elements will have to be filled out. Through semantic analysis it has been possible to extract important features such as semantic roles and then work with sentence patterns [7]. The generation of factoid questions with Recurrent Neural Networks (RNN) [22] uses a novel neural network approach in order to convert texts or facts into Natural Language questions. Then, the generated questions and their answers may be evaluated by both evaluation metrics and humans.

QG has been used to support several areas of study such as language learning [25], history [20], vocabulary and grammar [10], science [4] and technologies [30].

Our approach aims to bridge the studies made for generating factual questions in English through a syntactic analysis combined with Named Entity Recognition. We also generate factual questions from Dependency Analysis.

2.2   Question Answering

A base structure, for question answering, needs to be defined in order to find answers, so a generic pipeline is created with three major modules [21]: Question Analysis, Passage Retrieval and Answer Extraction. In Question Analysis, the main goal is to analyze the question. The module receives an input with unstructured text and identifies semantic and syntactic elements that define the question. This information will be encoded in a structured way to be used in the remainder modules. Passage Retrieval can be based on a search engine. Using the given query, the most similar passages or sentences are retrieved. Queries are formulated based on the information extracted in the Question Analysis stage, and are used to find information suitable for answering the posed question. Different candidates are then evaluated, for which dynamic sources such as the Web and online databases can be explored. Using the appropriate representation of the question and each candidate passage, candidate answers are extracted from the passages and ranked in terms of probable correctness in the Answer Extraction module. An answer can be formulated based on this information.

There are different ways to approach the analysis of a question. Most known approaches perform stop-word removal, conversion of the inflected term to its canonical form, query term expansion, and syntactic query generation. For example, Pakray et al. [19] uses the Stanford Dependency parser to recognize query and target result types. IBM Watson uses rule- and classification-based methods to analyze different parts of the input query [9]. Phrase-level dependency graph was adopted by Xu et al. [29] to determine question structure and the domain dataset was used to instantiate created patterns.

The Information Retrieval phase of any QA system is always a challenge, considering that sources, in some cases, are not the same. Problems arise on the credibility of the sources or how the information is structured. Thus, an IR system needs to be generic in order to extract information from different types of texts: structured and unstructured. Therefore, to deduce the intention of the query different approaches are used, such as a syntactic analysis to extract information from the query [26] as well as the use of structured information such as knowledge graphs able to disambiguate information [23].

Widely well-known techniques on Answer Extraction (AE) are n-grams, patterns, named entities and syntactic structures. A very important AE technique introduced a merging score strategy based on relevant terms [14].

3   System Overview

In our approach, we produced a tool capable of automatically generating questions from any text source, which are then sent to an independent module that analyzes the questions and tries to answer them based on a provided corpus. Figure 1 shows a system overview in which the red blocks represent QG operations, and blue ones compose the QA module. The dashed line aims to highlight the input and output elements that together intersect the QG and the QA modules, enabling the evaluation of the latter. As visible in the diagram, the same documents serve as input for both mechanisms. The key generated in QG is matched against the answer generated in QA in the evaluation step. "SPO (*subject, predicate, object*) Tuples" is the information acquired from the Question Analysis, to be further explained in Sect. 3.2.

3.1   Question Generation

NLTK[1] was used to tokenize the obtained sentences. We used spaCy[2] for POS tagging and NER, supporting the identification of the following entity labels:

---

[1] http://www.nltk.org/.

[2] https://spacy.io/models/pt.

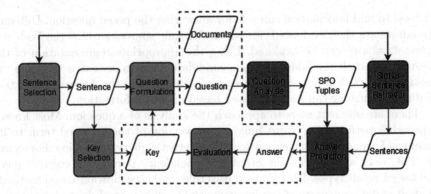

Fig. 1. System Overview, including QG steps (in red) and QA steps (in blue). (Color figure online)

PER (People, including fictional), DATE (Absolute or relative dates or periods), GPE (Countries, cities, states), LOC (Non-GPE locations, mountain ranges, bodies of water), ORG (Companies, agencies, institutions), MONEY (Monetary values, including unit) and EVENT (Named hurricanes, battles, wars, sports events). Additionally, spaCy also identifies context-specific token vectors, POS tags, dependency parse and named entities.

The first step for generating questions is directly related to the selection of sentences with one or more questionable facts. To address this phase, we use POS tagging combined with NER. Through the information obtained by these two tasks, it is possible to understand the structure of a sentence with some depth. Through POS we can know the grammatical class of each of the words and by identifying the entities we can find out if a phrase is expressing information about people, places, dates, countries, cities, states, money, or events. For instance, when a sentence begins with an entity type person followed by a verb, it is likely that the sentence is related to a certain person performing a particular action; we can thus ask who is the person performing that specific action.

Once the candidate sentence is selected, it is necessary to find out if its structure corresponds to at least one of the established rules. These rules allow us to check for the existence of questionable facts and to find patterns. For that, we use regular expressions. If there are matches on the established rules, we assume that a question can be generated from that sentence.

The next phase is to select the word or set of words that will be the answer (key selection) to the generated question. These words are the questionable facts and may be referring to people (*Who? What people?*), locations (*Where?, Which country?*), dates (*When?*), amounts (*How much?*), organizations (*Which organization?*) and events (*Which event?*).

The last phase is responsible for making the necessary transformations from the original sentence in order to create the interrogative sentence. Transformation rules take into account the position of pronouns, auxiliary and main verbs as well as phrases that are in active and passive form. The detail of the entire process is shown in Table 1.

Table 1. Steps from sentence handling to question generation.

Sentence pattern before POS and NER	**The French Revolution** was a period of intense political and social upheaval in France
Sentence pattern after POS	\<DET\> \<PROPN\> \<PROPN\> \<AUX\> \<DET\> \<NOUN\> \<ADP\> \<ADJ\> \<ADJ\> \<CCONJ\> \<ADJ\> \<NOUN\> \<ADP\> \<PROPN\> \<PUNCT\>
Sentence pattern after POS and NER	**\<EVENT\>** \<AUX\> \<DET\> \<NOUN\> \<ADP\> \<ADJ\> \<ADJ\> \<CCONJ\> \<ADJ\> \<NOUN\> \<ADP\> \<PROPN\> \<PUNCT\>
Expression used as rule in QG	**\<EVENT\>** \<(?:AUXorVERB)\>.*? \<PUNCT\>
Generated question	**Which event** was a period of intense political and social upheaval in France?

We have considered several particular cases where the sentences needed certain adjustments. These cases are listed as follows:

- The phrase is in the passive form. The system will transform it to active form;
- The phrase has the main verb and the auxiliary verb. The system will create the question in a way that both verbs stay in the correct positions;
- The phrase has more than one entity. The system will detect the given entities and, if it finds a relationship between them, it will question accordingly;
- The main verb is in the past and, for that case, it changes the verb tense.

Table 2 showcases several examples of questions generated according to the type of entities, number of entities and word position.

3.2   Question Answering

When designing the Question Answering module, it is necessary to not create assumptions, allowing this module to be completely independent of the Question Generation one. Three major steps identified in Subsect. 2.2 were followed to develop this module.

As the generated questions are of the *Wh* type, an abstraction of the analysis of the question was made. Thus, NER is performed to analyze the question. Each question may or may not have an entity. Thus, the extraction of all subjects and all objects are done. In order to do this, the Stanford CoreNLP [16] was used to obtain a Dependency Parsing Tree and also POS tagging. The returned entities, subjects and objects will be referenced as keywords.

Table 2. Original sentences and generated questions

Entity/Entities	Sentence and question
PER = Paul	S: **Paul** was the son of Henry of Burgundy and Teresa, the illegitimate daughter of King Alfonso VI of León and Castile Q: **Who** was the son of Henry of Burgundy and Teresa?
PER = Anne PER = Henry	S: **Henry and Anne** reigned jointly as count and countess of Portugal Q: **What people** reigned jointly as count and countess of Portugal?
GPE = Portugal	S: **Portugal** was conquered by Afonso I Q: **Which country** was conquered by Afonso I?
ORG = The Congress of Manastir	S: **The Congress of Manastir** had chosen the Latin script as the one to be used to write the language Q: **Which organization** had chosen the Latin script as the one to be used to write the language?
MONEY = 50 thousand dollars	S: One bedroom apartment costs **50 thousand dollars** Q: **How much** costs the one bedroom apartment?
PER = Henry	S: A car is cleaned by **Henry** Q: **Who** did clean a car?
DATE = 1109	S: Paul was born in **1109** Q: **When** was Paul born?

After analyzing the question, we retrieve sentences that may contain the information that can answer the question. The sentences are extracted from a predefined set of documents regarding a knowledge topic relevant to the question. To achieve this purpose, a metric of similarity is created between the keywords extracted in the previous step and the sentences that are more likely to contain the necessary content to answer the question. This metric consists in the statistical measure of the Term Frequency–Inverse Document Frequency (TF-IDF) followed by a cosine-similarity. The score has a range between 0 and 1. The sentences retrieved are the x most relevant sentences that scored a minimum y of the similarity metric. The variables x and y are hyper-parameters and need to be fine-tuned for different datasets. In this case, they were set as $x = 3$ and $y = 0.75$, extracting a sufficient number of sentences that would not overload the AE system and maintaining the most relevant sentences.

To extract the answer, we process the obtained sentences by extracting triples, consisting of subject (S), predicate (P) and object (O) [3]. To perform this extraction we use the Stanford OpenIE [2].

As an example, from the text *José Saramago was born in Portugal. Bernardo Azevedo wrote this sentence.* the following triples are extracted:

- 'subject': 'José Saramago', **predicate**: 'was', **object**: 'born'
- 'subject': 'José Saramago', **predicate**: 'was born in', **object**: 'Portugal'
- 'subject': 'Bernardo Azevedo', **predicate**: 'wrote', **object**: 'sentence'

With these extracted triples, we perform lemmatization to every predicate. Since at this stage the objective is to obtain the answer, it is necessary to verify which predicate in the triple matches the one in the question. All triples in which the sentence relationship is not present are discarded. After this selection, we check if the question contains the subject or the object. If the question contains both, the triple is discarded. If the question has only the subject/object, the object/subject will be retrieved as a possible answer. In the end, there will be a list of all predicted answers.

An example is presented as follows, which shows the extracted triples that were used to predict the answer.

- **Question:** What people reigned jointly as count and countess of Portugal?
- **Most relevant sentence:** Henry and Anne reigned jointly as count and countess of Portugal.
  - **Triple 1:** 'subject': 'Anne', 'predicate': 'reigned jointly as', 'object': 'count of Portugal',
  - **Triple 2:** 'subject': 'Henry', 'predicate': 'reigned jointly as', 'object': 'countess of Portugal'
  - **Predicted Answer:** Anne Henry
  - **Correct Answer:** Henry and Anne

4   Experimental Evaluation

We used the *Wikipedia Sentences 2*[3] dataset, available on *Kaggle*. It consists of a collection of 7.8 million sentences from August 2018 English Wikipedia dump. From this dataset, we created 50 different documents, each containing 20 randomly selected sentences. A small control group (handwritten sentences) of 3 documents was also created. This enabled us to analyze specific cases in order to improve the process of question generation.

Bearing in mind that there is no standard method for evaluating the quality of the generated questions, have developed a pilot test with five English teachers. Our survey contained 10 generated questions from the same text source (sentences from our dataset) and each question is evaluated according to the following criteria:

- **Objectivity of the Question** - Do you consider the question objective? (1 - Nothing objective, 5 - Very objective)
- **Question Extension** - How do you characterize question extension? (1 - Not long, 5 - Too long)
- **Grammatically** - Do you consider the question to be grammatically correct? (1 - Very Incorrect, 5 - Totally Correct)
- **Answerability** - How many answers do you think this question might have? (No answer, One, Two or more)

The results of the first three metrics can be seen in Table 3.

---

[3] https://www.kaggle.com/mikeortman/wikipedia-sentences.

Table 3. Averages scores for Objectivity, Grammatically and Question Extension

Metric	Avg. score
Objectivity (1–5)	3,64
Extension (1–5)	3,14
Grammaticality (1–5)	3,42

Regarding answerability, little consensus was achieved regarding the number of answers given a question. We assume this happens because there are several interpretations that can be caused by the presence of multiple entities in the sentence or external knowledge (in addition to what is written in the sentence).

Overall, the teachers considered the questions to be objective. Some ambiguities aroused when there were multiple entities identified for the same sentence. In other cases, generic questions have also introduced ambiguity. Regarding grammatically, we conclude that the questionable term (used at the beginning of the question) may not be the most appropriate in some cases. Also, the main inconsistencies are due to verb conjugation. Question extension is adequate most of the times but it needs treatment, mainly to remove unnecessary parts. The appropriateness of the question length may not always be the most suitable. To improve that, we would have to better understand the context in which the question is asked, that is, how long the question needs to be.

Assessing whether the answer found is the right one is not enough to evaluate the QA mechanism. Understanding the question is important and, for example, predicting "Paul" when the correct answer is "John" should not be considered as bad as "London". With this in mind, a metric of similarity was created. This consists of a model of word2vec embeddings [18], trained with the dataset available in the *Gensim* API called *text8*. This allows us to neutralize the score if the predicted and correct answers have a similar type or meaning.

To evaluate the QA mechanism, a pipeline was created to join the different tasks. Thus, all the documents generated from the *wikisentences* were read, followed by the generation of the questions and performing a search for their answers. Results are shown in Table 4. These results are divided on five groups: Wiki Documents, Controlled Documents, Entity Questions, Entity Questions without the generating question with the entity ORG and Dependency Questions. Evaluation metrics include: Short Answer (using the similarity metric), Correct Triple (whether the triple contains the answer) and Correct Sentence (as compared to the one used by the QG module to generate the question). The type of questions generated from Wiki Documents and Controlled Documents contained Entity and Dependency questions.

Overall, the results were good. Thus, we find that by adding different types of questions that include new entities, the QA system can generalize well, which demonstrates its robustness. This is supported by the fact that, when adding questions about the *ORG* entity, results are similar. However, results were not as good for questions generated through dependency analysis. This is probably

Table 4. Results of the different question sets and evaluated in 3 metrics

Dataset	Question generated	Short answer	Correct triple	Correct sentence
Wiki documents	311	78,5%	87,4%	96,4%
Controlled documents	43	89,3%	95,2%	100%
Entity questions w/o ORG	301	80,9%	88,5%	98,8%
Entity questions	334	81,2%	89,8%	98,1%
Dependency questions	20	32,7%	80,0%	100%

due to the fact that an extracted triple does not contain the necessary information. It is possible to state that creating different types of questions can help to
evaluate the robustness of the QA module. These results also revealed a decrease
in the performance of the QA mechanism for questions generated from the Wiki
Documents when compared to the Controlled Documents. This is due to the
fact that, in some cases, there are very broad questions and some questions are
not being asked in the best possible way, reveling ambiguity problems like the
appearance of different answer possibilities. After a closer look into the system,
ambiguities have been found. For example, in the question, from the wiki documents, *What did Maria have?* the answer *New Pet* was predicted, although the
correct answer could be *A Motorcycle*. This happened because there were two
sentences that contained two reasonable answers: *Maria has a motorcycle.* and
Maria and Bob have adopted a new pet.

5 Conclusions and Future Work

Question Generation and Question Answering are two independent yet highly
related tasks. If it is true that the amount of available data has increased exponentially, it is also true that there is a need for discernment and responsibility to
select important and valid information. This is important to filter information
that can be questionable.

From a learning perspective, access to trustworthy information provides a
variety of contents for the teachers. For the students, the available content can
help them to enhance their knowledge. Bearing this in mind, the developed tool
has a clear advantage: provide content such as generated questions and their
answers. This helps the teacher to automatically generate questions and make
slight changes if needed. For the student, it allows to test their skills with different
questions from different contexts.

The results of the presented tool are promising. Even so, we are aware that
there is a lot to improve in each of the modules. Our main goal is to present a
direction of research on the possible bidirectionality of the QG and QA tasks.
The analyzed results show a possible way to evaluate the robustness of the QA
mechanism based on the ambiguity of the generated questions.

We intend to expand our tool with other types of approaches that allow a
more extensive analysis of the sentences. In addition, we intend to decrease the
number of grammatical errors that can be verified in the process of generating
questions. Generating factual questions for the English language has a lot of

possibilities and some of the suggestions can be: generating questions from a combination of more than one sentence, the ability to handle more complex text and to have standard evaluation techniques. The possession of gold-standard test data is also very important.

The most important conclusion that we can draw from our study is the possibility of generating datasets with question and answer pairs in a completely automatic manner. This way, less human intervention will be necessary to create this type of content. In this automatic generation process (both for questions and answers) it is necessary to guarantee the quality of the generated content. We see a promising future for these tasks using modern Machine Learning techniques.

References

1. Agarwal, M., Mannem, P.: Automatic gap-fill question generation from text books. In: Proceedings of the 6th Workshop on Innovative Use of NLP for Building Educational Applications, IUNLPBEA 2011, pp. 56–64. Association for Computational Linguistics, USA (2011)
2. Angeli, G., Premkumar, M.J.J., Manning, C.D.: Leveraging linguistic structure for open domain information extraction. In: Proceedings of the 53rd Annual Meeting of the Association for Computational Linguistics and the 7th International Joint Conference on Natural Language Processing (Volume 1: Long Papers), pp. 344–354 (2015)
3. Banko, M., Cafarella, M.J., Soderland, S., Broadhead, M., Etzioni, O.: Open information extraction from the web. IJCAI 7, 2670–2676 (2007)
4. Conejo, R., Guzmán, E., Trella, M.: The SIETTE automatic assessment environment. Int. J. Artif. Intell. Educ. 26, 270–292 (2015)
5. Danon, G., Last, M.: A syntactic approach to domain-specific automatic question generation. CoRR (2017)
6. Duan, N., Tang, D., Chen, P., Zhou, M.: Question generation for question answering. In: Proceedings of the 2017 Conference on Empirical Methods in Natural Language Processing, pp. 866–874 (2017)
7. Flor, M., Riordan, B.: A semantic role-based approach to open-domain automatic question generation. In: Proceedings of the Thirteenth Workshop on Innovative Use of NLP for Building Educational Applications, pp. 254–263. Association for Computational Linguistics, New Orleans, June 2018
8. Harrison, V., Walker, M.A.: Neural generation of diverse questions using answer focus, contextual and linguistic features. CoRR abs/1809.02637 (2018)
9. High, R.: The Era of Cognitive Systems: An Inside Look at IBM Watson and How it Works. IBM Corporation, Redbooks (2012)
10. Hoshino, A., Nakagawa, H.: Predicting the difficulty of multiple-choice close questions for computer-adaptive testing. Nat. Lang. Process. Appl. 279 (2010)
11. Hu, S., Zou, L., Zhu, Z.: How question generation can help question answering over knowledge base. In: Tang, J., Kan, M.-Y., Zhao, D., Li, S., Zan, H. (eds.) NLPCC 2019. LNCS (LNAI), vol. 11838, pp. 80–92. Springer, Cham (2019). https://doi.org/10.1007/978-3-030-32233-5_7
12. Khot, T., Sabharwal, A., Clark, P.: Answering complex questions using open information extraction. arXiv preprint arXiv:1704.05572 (2017)
13. Kumar, V., Ramakrishnan, G., Li, Y.F.: A framework for automatic question generation from text using deep reinforcement learning. arXiv abs/1808.04961 (2018)

14. Le, J., Zhang, C., Niu, Z.: Answer extraction based on merging score strategy of hot terms. Chin. J. Electron. **25**(4), 614–620 (2016)
15. Le, N.-T., Pinkwart, N.: Evaluation of a question generation approach using semantic web for supporting argumentation. Res. Pract. Technol. Enhanc. Learn. **10**(1), 1–19 (2015). https://doi.org/10.1007/s41039-015-0003-3
16. Manning, C.D., Surdeanu, M., Bauer, J., Finkel, J., Bethard, S.J., McClosky, D.: The Stanford CoreNLP natural language processing toolkit. In: Association for Computational Linguistics (ACL) System Demonstrations, pp. 55–60 (2014)
17. Mazidi, K.: Automatic question generation from passages. In: Gelbukh, A. (ed.) CICLing 2017. LNCS, vol. 10762, pp. 655–665. Springer, Cham (2018). https://doi.org/10.1007/978-3-319-77116-8_49
18. Mikolov, T., Chen, K., Corrado, G., Dean, J.: Efficient estimation of word representations in vector space. arXiv preprint arXiv:1301.3781 (2013)
19. Pakray, P., Bhaskar, P., Banerjee, S., Pal, B.C., Bandyopadhyay, S., Gelbukh, A.F.: A hybrid question answering system based on information retrieval and answer validation. In: CLEF (Notebook Papers/Labs/Workshop) (2011)
20. Papasalouros, A., Kanaris, K., Kotis, K.: Automatic generation of multiple choice questions from domain ontologies. In: Proceedings of the IADIS International Conference e-Learning 2008, vol. 1, pp. 427–434 (2008)
21. Prager, J., Chu-Carroll, J., Brown, E.W., Czuba, K.: Question answering by predictive annotation. In: Strzalkowski, T., Harabagiu, S.M. (eds.) Advances in Open Domain Question Answering. TLTB, vol. 32, pp. 307–347. Springer, Dordrecht (2008). https://doi.org/10.1007/978-1-4020-4746-6_10
22. Serban, I.V., et al.: Generating factoid questions with recurrent neural networks: the 30m factoid question-answer corpus. CoRR abs/1603.06807 (2016)
23. Shekarpour, S., Marx, E., Ngomo, A.C.N., Sina, S.: Semantic interpretation of user queries for question answering on interlinked data. Elsevier-Web Semantics (2015)
24. Stancheva, N.S., Popchev, I., Stoyanova-Doycheva, A., Stoyanov, S.: Automatic generation of test questions by software agents using ontologies. In: 2016 IEEE 8th International Conference on Intelligent Systems (IS), pp. 741–746, September 2016
25. Susanti, Y., Tokunaga, T., Nishikawa, H., Obari, H.: Evaluation of automatically generated English vocabulary questions. Res. Pract. Technol. Enhanc. Learn. **12**, Article no. 11 (2017)
26. Unger, C., Bühmann, L., Lehmann, J., Ngonga Ngomo, A.C., Gerber, D., Cimiano, P.: Template-based question answering over RDF data. In: Proceedings of the 21st International Conference on World Wide Web, pp. 639–648. ACM (2012)
27. Wang, T., Yuan, X., Trischler, A.: A joint model for question answering and question generation. arXiv preprint arXiv:1706.01450 (2017)
28. Wu, P., Zhang, X., Feng, Z.: A survey of question answering over knowledge base. In: Zhu, X., Qin, B., Zhu, X., Liu, M., Qian, L. (eds.) CCKS 2019. CCIS, vol. 1134, pp. 86–97. Springer, Singapore (2019). https://doi.org/10.1007/978-981-15-1956-7_8
29. Xu, K., Zhang, S., Feng, Y., Zhao, D.: Answering natural language questions via phrasal semantic parsing. In: Zong, C., Nie, J.-Y., Zhao, D., Feng, Y. (eds.) NLPCC 2014. CCIS, vol. 496, pp. 333–344. Springer, Heidelberg (2014). https://doi.org/10.1007/978-3-662-45924-9_30
30. Zampirolli, F., Batista, V., Quilici-Gonzalez, J.A.: An automatic generator and corrector of multiple choice tests with random answer keys. In: 2016 IEEE Frontiers in Education Conference (FIE), pp. 1–8, October 2016
31. Zhang, S., Bansal, M.: Addressing semantic drift in question generation for semi-supervised question answering. arXiv preprint arXiv:1909.06356 (2019)

Hong Kong Protests: Using Natural Language Processing for Fake News Detection on Twitter

Alexandros Zervopoulos[1]([⊠]), Aikaterini Georgia Alvanou[1],
Konstantinos Bezas[1], Asterios Papamichail[1], Manolis Maragoudakis[2],
and Katia Kermanidis[1]

[1] Department of Informatics, Ionian University, Corfu, Greece
{c19zerv,c19alva,c19beza,c19papa,kerman}@ionio.gr
[2] Department of Information and Communication Systems Engineering,
University of the Aegean, Samos, Greece
mmarag@aegean.gr

Abstract. The automation of fake news detection is the focus of a great deal of scientific research. With the rise of social media over the years, there has been a strong preference for users to be informed using their social media account, leading to a proliferation of fake news through them. This paper evaluates the veracity of politically-oriented news and in particular the tweets about the recent event of Hong Kong protests, with the aid of a dataset recently published by Twitter. From this dataset, Chinese tweets are translated into English, which are kept along with originally English tweets. By utilizing a language-independent filtering process, relevant tweets are identified. To complete the dataset, tweets originating from valid sources are used as the real portion, with journalists rather than news agencies being considered, which constitutes a novel aspect of the methodology. Well-known Machine Learning algorithms are used to classify tweets, which are represented by a feature value vector that is extracted, selected and preprocessed from the datasets and mainly revolves around language use, with word entropy being a novel feature. The results derived from these algorithms highlight morphological, lexical and vocabulary differences between tweets spreading fake and real news, which are for the most part in accordance with past related work.

Keywords: Fake news detection · Natural Language Processing · Machine Learning · Twitter · Hong Kong protests

1 Introduction

Social media, popular or not, allow both borderless communication and a plethora of information to be spread at a dizzying speed around the world,

I. Maglogiannis et al. (Eds.): AIAI 2020, IFIP AICT 584, pp. 408–419, 2020.
https://doi.org/10.1007/978-3-030-49186-4_34

justifying the choice of the largest percentage of them to keep up to date with domestic and global news events, via Facebook, Twitter and so on. However, the validity of the news is not guaranteed, as it may be hampered by conspiracy, political expediencies and interests. By extension, the spread of fake news contributes to a common and everyday phenomenon, which can undermine values and ideals and, thus, needs addressing.

The spread of fake news is particularly prevalent in politically oriented content, especially so on Twitter, where it has been found that the rate of dissemination of fake news is higher than that of real news [12,20]. In this context, computer science can also be used as the primary asset and tool for false detection in news releases from Twitter user accounts, helping to counteract and eliminate this phenomenon. Since the process of falsehood detection by conventional methods, such as the involvement of certified journalists, is a costly process, due to the financial costs and the lengthy periods of time required to complete it, technological approaches, starring Artificial Intelligence, have gained ground [11], instead. The recent (June 2019) events of the Hong Kong protests related to political controversy have been of great concern to the public because of the violent turn and the high turnout of citizens inside and outside China's borders [14]. As a result, a plethora of tweets were triggered, raising the question of the validity of their content. It is therefore important to study the extent of the fake news spread on Twitter about this event.

In this paper, the problem of automatically distinguishing between tweets spreading fake and real news is tackled through the use of Machine Learning (ML) algorithms. In order to accomplish this, an initial dataset published by Twitter[1] regarding the Hong Kong protests is used to represent the fake portion of the data used for classification. This dataset contains tweets in a multitude of languages, though only English and Chinese tweets are utilized, with the aid of machine translation. Relevant tweets are pinpointed through a filtering process that is language-independent, making selective use of machine translation. A collection of tweets is gathered to represent the real portion of the dataset, which are considered trustworthy based on the account posting the tweet. News agency and journalist accounts are considered trustworthy sources for the purposes of this study. The assembled dataset is publicly available for research purposes in Humanistic and Social Informatics Laboratory's website[2]. From the assembled dataset, a plethora of linguistic features are extracted, preprocessed and selected to be used as inputs for a variety of well-established ML algorithms, namely Naive Bayes, Support Vector Machines (SVMs), C4.5 and Random Forest. Twitter text has idiosyncrasies that render its linguistic processing quite interesting and that have been tackled in various contexts, the TraMOOC system being one of them [18]. The derived models indicate significant differences in morphological, lexical and vocabulary features between tweets spreading fake and real news. In contrast to previous studies, journalists are investigated

---

[1] https://transparency.twitter.com/en/information-operations.html.
[2] https://hilab.di.ionio.gr/index.php/en/datasets/.

regarding trustworthiness, rather than just news agencies, and word entropy is used as a novel feature, which plays an important role in classification.

The rest of this paper is organized as follows. An overview of related literature is presented in Sect. 2, while the applied methodology is described in Sect. 3. Section 4 specifies the produced results and finally, conclusions are drawn in Sect. 5.

2   Related Work

The detection of fake news, and, in particular, those that are spread through social media, has been extensively researched by the scientific community. Specifically, in [11,17], the technical challenges in automating fake news detection using Natural Language Processing (NLP) tactics are presented, while a comparison between the used datasets, features, models and their respective performances is provided, with the aim of facilitating future studies.

On the other hand, Ahmed et al. [1] approach the issue with the help of text analysis with N-gram attributes (up to 4-gram size) and by using 6 different machine learning techniques for classification. The model that reached a standing out performance is the linear SVM with the use of unigram attributes. Conroy et al. [5] focused on the detection of fake news, with the aim of presenting a hybrid approach based on a combination of linguistic and network-analysis techniques. For both categories, machine learning tools that ultimately contribute to successful detection are described.

In addition, Buntain and Golbeck [3] are occupied with automating the detection of fake news on Twitter via 2 existing datasets to analyze the structure and behavior of potentially fake Twitter threads, assessing their proximity to the thread with the help of the Buzzfeed dataset[3]. The aim is to determine the appropriate characteristics of training capable models for predicting falsehood.

While multiple well-established datasets exist, experimentation focusing on specific events regularly takes place. This poses an array of challenges, primarily due to the fact that expertly-annotated data is hard to come by. As such, attempts have been made to circumvent the need for experts' opinions by utilizing data-driven techniques. One such example is the work by Helmstetter [7], who consider the credibility of a tweet's source as a proxy for the trustworthiness of the tweet itself, achieving high prediction scores.

In [19], authors deliberate the classification of fake and verified news and the promotion of 4 categories of fake news: propaganda, satire, hoaxes and clickbait. The analysis and experimentation is based on Twitter and, in fact, the data collection takes place over a period of 2 weeks, during the terrorist attacks in Brussels in 2016. For classification, linguistically-infused neural network models are created, based on the content of tweets and social network interactions. Finally, they conclude that morphological and grammatical features are not efficient.

---

[3] https://github.com/BuzzFeedNews/2016-10-facebook-fact-check.

3   Methodology

3.1   Fake News Dataset

The initial fake news dataset is retrieved from Twitter's Election Integrity Hub[4], where three sets were disclosed in August and September 2019. In greater detail, this dataset consists of 13,856,454 tweets in total and includes 31 fields, which represent tweet-related features about both the tweet's text and the user. In the present study, Twitter is regarded as a reliable source and they have deemed these accounts to be "deliberately and specifically attempting to sow political discord in Hong Kong,"[5]; thus, this dataset is considered as ground truth with respect to the fake news portion of the assembled dataset.

However, as per Twitter's description of the dataset, the accounts involved tend to be fake, post spam and act in a coordinated manner, which has also been investigated in the literature [17]. Hence, not all tweets are relevant to the spread of fake news related to the Hong Kong protests, deeming mandatory an initial preprocessing step. Furthermore, due to the specificities of the events, which take place in China, it is assumed that most of the relevant tweets would be worded either in Chinese or English, as the latter is more prevalent in the Twitter platform.

To better visualize and understand the contents of the dataset, the tweets' text is preprocessed, from which word clouds are constructed. Namely, hashtags, mentions and URLs are removed from the texts in English and in Chinese. Those in Chinese are also translated into English, using Google's Translation API available through the Google Cloud[6] platform. Afterwards, a frequency word cloud is created for each language, with the aid of Python's word cloud[7] module, containing at most 400 words. It should be noted that these word clouds are derived only from the first two out of the three of Twitter's datasets, as described at the beginning of this subsection. The resulting word clouds are depicted in Fig. 1a, Fig. 1b. Chinese tweets are evidently more relevant to the events than their English counterparts.

To precisely identify the tweets spreading false information regarding Hong Kong protests, the ensuing filtering methodology is followed, which is largely based on the assumption that a tweet's hashtags also indicate the content of a tweet's text. It is worth pointing out that this process is language-independent, which is particularly advantageous in this case, as it is impractical to translate millions of Chinese tweets into English. Moreover, the presented filtering process is overall fairly efficient, requiring a short amount of time, typically a few minutes using commodity hardware.

The methodology can be broken down in the subsequent manner. First of all, a list of hashtags related to Hong Kong protests is manually constructed, comprising both English and Chinese hashtags. Afterwards, hashtags appearing

---

[4] https://transparency.twitter.com/en/information-operations.html.
[5] https://tinyurl.com/y3ffrblt.
[6] https://cloud.google.com/translate/.
[7] https://github.com/amueller/word_cloud.

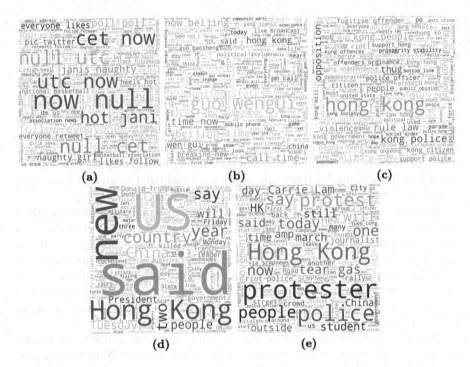

Fig. 1. Frequency word clouds formed from tweets: (a) In the fake news dataset worded in English. (b) In the fake news dataset worded in Chinese. (c) Resulting from the filtering process. (d) In the news agency dataset. (e) In the journalist dataset.

in tweets along with at least one of the previously mentioned hashtags are kept track of, and their co-occurrence counts are calculated across the entire dataset. Finally, tweets are deemed relevant if they contain a hashtag with a co-occurrence count higher than an arbitrary threshold (in this case, 100). It is evident that, due to this approach, tweets without hashtags cannot be considered relevant.

Upon the completion of the filtering steps, only 3,908 tweets, worded in English or Chinese, are considered to be relevant to the spread of fake news related to Hong Kong protests. The word cloud constructed by the collection of these relevant tweets is depicted in Fig. 1c.

3.2   Real News Dataset

As previously mentioned, the dataset corresponding to real news has not been retrieved in the same manner as the fake news, i.e. no real news dataset has been made publicly available by Twitter. Additionally, well-established datasets are not appropriate for this study, due to the fact that they do not specifically deal with events surrounding the Hong Kong protests. To address these challenges, the real news dataset has to be constructed from the ground up here. In more detail, since expert labeling is a costly process, news posts about Hong Kong protests appearing in generally well-regarded news agencies are often assumed to be reliable sources[8]. Therefore, tweets from the accounts of news agencies are retrieved, rather than articles.

Moreover, one would expect news agencies' tweets to differ in style from those of personal accounts, e.g. more formal speech, fewer replies to other users, more references to the agency's articles, etc. As such, in this study, tweets from journalists of well-regarded news agencies are also considered as real news. The search and retrieval of relevant twitter accounts of news agencies and journalists employed by them takes place manually and, in this case, the following news agencies are considered: BBC News, Reuters, Bloomberg, BuzzFeed, Channel News Asia, CNN, Agence France-Presse, South China Morning Post, Wall Street Journal, The New York Times, The Associated Press, The Washington Post and Quartz. Having completed the search and retrieval of the aforementioned twitter accounts, 13 accounts of news agencies and 107 accounts of journalists are gathered. Using Twitter's user timeline API, 41,996 tweets from news agencies' accounts and 103,359 tweets from journalists' accounts are collected.

The retrieved data seem to be supporting the assumption that the tweets contained in the fake news dataset are more similar to those of journalists than those of news agencies. A few notable statistics derived from the unfiltered data and the first two fake news datasets are listed to showcase some differences and the aforementioned notion of 'similarity': On average, a tweet contains approximately 0.22 hashtags in the fake news dataset, 0.23 hashtags when posted by a journalist and 0.1 hashtags when posted by a news agency. Additionally, the mean number of urls in a tweet is 0.3 in the fake news dataset, 0.35 when posted by a journalist, and 0.82 when posted by a news agency. Lastly, on average, each of the accounts posting tweets have 4.1 followers in the fake news dataset, 15.14 in the journalist dataset, and 11,509.08 in the news agency dataset.

Frequency word clouds are also derived from relevant tweets found in these two datasets and are shown in Fig. 1d, Fig. 1e. While both are evidently relevant to Hong Kong events, the more objective, news-based narrative of the news agency dataset differs from the journalist and fake news dataset. Thus, it becomes clear that the tweets collected from journalist accounts are more similar to those in the fake news dataset, when considering both tweet content and account characteristics.

Using the filtering process described previously, 5,388 and 666 of the tweets posted by journalists and news agencies, respectively, are considered relevant, with the latter being fewer due to the fact that the filtering process relies purely on hashtags, which news agencies don't use as often, as was already showcased. Due to both the low number of tweets and dissimilarity to the fake news dataset, the news agency dataset is entirely dropped and not further studied. All in all, the assembled dataset consists of 3,908 and 5,388 tweets spreading fake and real news, respectively.

---

[8] https://www.4imn.com/news-agencies/.

3.3 Features

The selected features are purely linguistic in nature and they represent a single
tweet. While the literature indicates that network-related features are worth
investigating, most of the information about the fake news dataset has been
made unavailable by Twitter and is no longer accessible on the platform, as the
accounts involved in the disclosed datasets have been banned. Regarding the
derivation of features, various preprocessing stages are necessary in some cases,
which are mentioned when appropriate. The features add up to 38 in total,
including the class label, and their Pearson correlation heatmap is depicted in
Fig. 2. Even at this early stage, the features of tweet entropy, tweet length and
type to token ratio (TTR) are highly correlated with the class label, so they are
likely to be important in classification.

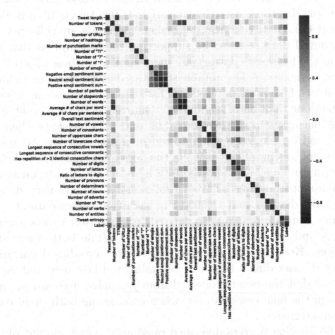

Fig. 2. Pearson correlation heatmap of features.

The TTR is calculated after the text has been tokenized as an indication
of the tweet's richness in vocabulary. A plethora of features are used in the
morphological level. These include a large assortment of counts regarding the
tweet's: length in characters, tokens, included URLs, hashtags, certain punctuation marks ('?', '!', '?!') and total punctuation marks, emojis, periods, stopwords,
words, vowels, consonants, upper and lower case characters, digits and letters.
Since these counts are expected to be correlated with the tweet's length, they are
normalized by dividing them with the latter. Furthermore, the average number

of characters per word and average number of words per period are calculated, along with the length of the longest sequence of vowels and consonants in a word. Finally, if a character's consecutive appearance takes place more than thrice, it is marked in the form of a boolean value. Last but not least, the entropy of a tweet is calculated through the equation: $S = -\sum_i P_i \log P_i$, where P_i is the probability of word i, which has been stemmed and converted to lowercase, appearing in the dataset. Entropy has been used in this case as an indicator of word importance for a tweet, but other similar metrics, such as word weighted frequency could be considered instead.

A number of Part of Speech (PoS) features are included, which keep track of the corresponding occurrences in the tweet: verbs, entities, pronouns, determiners, adverbs, as well as the proposition "to". These features are calculated with the aid of the NLTK module's [8] recommended PoS tagger and, much like the morphological features representing counts, they are normalized by the tweet's length.

A wide array of semantic features are used to provide higher-level information about the tweet. These include: positive, neutral and negative sentiment derived from text and emojis. The sentiment of emojis is calculated based on the list provided by [10] and summed up for each emoji found in the tweet text. Similarly, the text sentiment is calculated according to the AFINN word list [9], which is also available as a Python package[9].

3.4 Algorithms

In the sequel, the ML algorithms, feature preprocessing and selection methods are considered. Literature has deemed effective the use of Naive Bayes, SVMs and Decision Trees for predicting the veracity of news. As such, four different algorithms are used for the training and evaluation of classification models: Naive Bayes, SVM, C4.5 and Random Forests [2] of C4.5. The rather popular Scikit-Learn Python module [13] implements these algorithms and is being used for the purposes of this study. Regarding SVMs, the Radial Basis Function kernel is made use of and the tweaking of parameters gamma and C is optimized through the use of the grid search hyper parameter tuning technique.

Furthermore, certain algorithms require feature preprocessing or selection to become more effective. In all cases, feature selection can significantly reduce training time, and aids in the prevention of overfitting. As such, for all algorithms except Random Forest, all features are ranked according to their mutual information and the top 10 of those are selected. Due to the fact that decision trees are harder to understand if preprocessing is applied to the data, it is avoided in this study. However, in the case of Naive Bayes and SVMs, all feature values v have been normalized to values v_{norm} lying in the $[0, 1]$ range through the transformation $v_{norm} = (v - v_{min})/(v_{max} - v_{min})$, where v_{min}, v_{max} are the minimum and maximum values across all values of a given feature, respectively.

---

[9] https://pypi.org/project/afinn/.

4 Results

4.1 Feature Selection

Having completed the feature selection process for the collected datasets, the top 10 most significant features according to the mutual information metric are: tweet length, TTR, number of punctuation marks, number of periods, average number of characters per sentence, number of adverbs, number of "to", number of verbs, number of entities and tweet entropy.

Since the features are mostly linguistic, the goal is to identify patterns in language use between tweets spreading fake and real news. Thus, one would perhaps expect language to be more formal and structurally sound in tweets disseminating real news than those disseminating fake ones. The most significant features derived from the feature selection process seem to be in accordance with that theory. For instance, low TTR indicates repetition of words, which may be inversely associated with conceptual variance, while adverbs constitute a hard-to-interpret source of information [4].

4.2 Machine Learning Models

The ML algorithms utilized for the classification of tweets spreading fake and real news are trained using the collected datasets and the corresponding evaluation results are presented. The dataset consists of 3,910 and 5,388 tweets spreading fake and real news, respectively, with the majority class baseline being 57.9%. In all cases, 5-fold cross validation is used to increase reliability of results.

Table 1. Evaluation results of the ML algorithms used in the classification process.

Algorithm	Class	Precision	Recall	F1 Score
Naive Bayes	Fake	90.1%	85.4%	87.6%
	Real	89.7%	92.8%	91.2%
	Average	89.9%	89.1%	89.4%
SVM	Fake	96.0%	84.0%	89.6%
	Real	89.4%	97.5%	93.3%
	Average	92.7%	90.8%	91.4%
C4.5	Fake	94.7%	84.7%	89.3%
	Real	89.8%	96.6%	93.0%
	Average	92.3%	90.6%	91.2%
Random Forest	Fake	97.5%	84.3%	90.3%
	Real	89.7%	98.4%	93.8%
	Average	93.6%	91.3%	92.1%

As can be observed from Table 1, although all algorithms perform fairly well, the most highly performing algorithm is Random Forest, achieving a macro average F1 Score of 92.4%. A noteworthy observation from these results is that all algorithms tend to score higher in precision in the Fake rather than the Real class, whereas the opposite is true for recall; recall scores are noticeably lower for the Fake class when compared to Real. Since the impact of a tweet spreading fake news could be considered significant, one could argue that models achieving higher recall scores would be preferred, even if precision scores suffered somewhat. Nevertheless, both scores are adequately high, such that no additional performance concerns are raised.

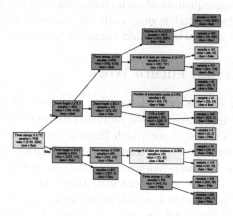

Fig. 3. A decision tree resulting from the C4.5 algorithm with maximum depth set to four.

In order to gain a better understanding of the factors that affect classification, an indicative decision tree resulting from C4.5 is depicted in Fig. 3. It is evident that tweet entropy contributes to an exceedingly high degree in the classification, as 75.7% and 96.7% of the tweets belonging to the Fake and Real class, respectively, are correctly identified within the first split. This would indicate that tweets in the Fake class contain more infrequent words than those in the Real class, which could be an aftereffect of the tweets being translated. Other notable features include tweet length, number of punctuation marks and a few of the morphological features, such as number of verbs. All things considered, according to such a model, a tweet spreading fake news would exhibit the following traits: unconventional vocabulary, longer length, fewer punctuation marks and shorter sentences.

The obtained results are compared to those found in the literature. Granik and Mesyura [6] scored an accuracy of 74% employing Naive Bayes. Ahmed et al. [1] scored an accuracy of 92% utilizing Linear SVM. Rubin et al. [16] achieved 87% F-score using SVM model. Sentiment features do not seem to be very important, as is also highlighted by literature [15]. Unlike the results presented here, Volkova et al. [19] conclude that morphological and grammar features are not important, although they do find that their results contradict previous work.

The overall high classification scores can be attributed to a number of factors resulting from the methodology. For one, even though journalists' tweets may be more similar to the ones in the fake dataset than those of news agencies, they may still differ somewhat noticeably, making classification easier. Furthermore, the employed methodology relies purely on linguistic traits, with the ones in the fake news dataset being mostly translated to English from Chinese. Therefore, there is significant risk that the model may distinguish traits originating from the translation, which could explain the exceedingly high importance of the tweet's length and entropy. This is further accentuated if one considers the intricacies of the Chinese language, such as the fact that a single symbol corresponds to an English word. The consequences of such intricacies are evident in the translated dataset: a Chinese tweet that is originally 125 characters long is translated to 585 characters in English, which would normally be too long for a tweet.

5 Conclusions and Future Work

This study focused on detecting fake news in tweets related to Hong Kong protests, using ML algorithms. It used an initial dataset with fake news that has been publicized by Twitter, from which English and Chinese tweets are taken into account, with the latter being translated into English. Relevant tweets were identified through a language-independent process, utilizing hashtag information, enabling the selective use of machine translation. A whole new dataset with real news was built as well, which comes from Twitter accounts of worldwide news agencies. Interesting (in comparison to other studies) is the fact that real news were also retrieved from the personal Twitter accounts of the journalistic team of these agencies. The assembled datasets were used to train and evaluate ML algorithms, once the necessary feature extraction, preprocessing and selection was accomplished, to represent each tweet as a feature value vector. The obtained results achieved high classification scores and indicate that tweets spreading fake news and real news differ noticeably in linguistic features and most notably in tweet length, vocabulary. Word entropy was also deemed very important in classification, a feature not commonly used in similar studies.

Even though the obtained results are promising, there is still much room for improvement and experimentation in future work: (i) study the impact translation has on the results; (ii) include and compare different kinds of features, especially ones related to user and network characteristics; and (iii) utilize more modern ML algorithms, such as deep neural networks.

Acknowledgments. This project has received funding from the GSRT for the European Union's Horizon 2020 research and innovation programme under grant agreement No 644333.

References

1. Ahmed, H., Traore, I., Saad, S.: Detection of online fake news using N-gram analysis and machine learning techniques. In: Traore, I., Woungang, I., Awad, A. (eds.) ISDDC 2017. LNCS, vol. 10618, pp. 127–138. Springer, Cham (2017). https://doi.org/10.1007/978-3-319-69155-8_9

2. Breiman, L.: Random forests. Mach. Learn. **45**(1), 5–32 (2001)
3. Buntain, C., Golbeck, J.: Automatically identifying fake news in popular Twitter threads. In: 2017 IEEE International Conference on Smart Cloud (SmartCloud), pp. 208–215. IEEE (2017)
4. Conlon, S.P.N., Evens, M.: Can computers handle adverbs? In: The 15th International Conference on Computational Linguistics, COLING 1992, vol. 4 (1992)
5. Conroy, N.J., Rubin, V.L., Chen, Y.: Automatic deception detection: methods for finding fake news. Proc. Assoc. Inf. Sci. Technol. **52**(1), 1–4 (2015)
6. Granik, M., Mesyura, V.: Fake news detection using Naive Bayes classifier. In: 2017 IEEE First Ukraine Conference on Electrical and Computer Engineering (UKRCON), pp. 900–903, May 2017
7. Helmstetter, S., Paulheim, H.: Weakly supervised learning for fake news detection on Twitter. In: 2018 IEEE/ACM International Conference on Advances in Social Networks Analysis and Mining (ASONAM), pp. 274–277, August 2018
8. Loper, E., Bird, S.: NLTK: the natural language toolkit. arXiv preprint cs/0205028 (2002)
9. Nielsen, F.Å.: A new ANEW: evaluation of a word list for sentiment analysis in microblogs. arXiv preprint arXiv:1103.2903 (2011)
10. Novak, P.K., Smailović, J., Sluban, B., Mozetič, I.: Sentiment of emojis. PloS One **10**(12), e0144296 (2015)
11. Oshikawa, R., Qian, J., Wang, W.Y.: A survey on natural language processing for fake news detection. arXiv preprint arXiv:1811.00770 (2018)
12. Parmelee, J.H., Bichard, S.L.: Politics and the Twitter revolution: how tweets influence the relationship between political leaders and the public. Lexington Books (2011)
13. Pedregosa, F., et al.: Scikit-learn: machine learning in python. J. Mach. Learn. Res. **12**(Oct), 2825–2830 (2011)
14. Purbrick, M.: A report of the 2019 Hong Kong protests. Asian Aff. **50**(4), 465–487 (2019)
15. Rashkin, H., Choi, E., Jang, J.Y., Volkova, S., Choi, Y.: Truth of varying shades: analyzing language in fake news and political fact-checking. In: Proceedings of the 2017 Conference on Empirical Methods in Natural Language Processing, pp. 2931–2937 (2017)
16. Rubin, V.L., Conroy, N., Chen, Y., Cornwell, S.: Fake news or truth? Using satirical cues to detect potentially misleading news. In: Proceedings of the Second Workshop on Computational Approaches to Deception Detection, pp. 7–17 (2016)
17. Shu, K., Sliva, A., Wang, S., Tang, J., Liu, H.: Fake news detection on social media: a data mining perspective. SIGKDD Explor. Newsl. **19**(1), 22–36 (2017)
18. Sosoni, V., et al.: Translation crowdsourcing: creating a multilingual corpus of online educational content. In: Proceedings of the Eleventh International Conference on Language Resources and Evaluation (LREC 2018) (2018)
19. Volkova, S., Shaffer, K., Jang, J.Y., Hodas, N.: Separating facts from fiction: linguistic models to classify suspicious and trusted news posts on Twitter. In: Proceedings of the 55th Annual Meeting of the Association for Computational Linguistics (Volume 2: Short Papers), pp. 647–653 (2017)
20. Vosoughi, S., Roy, D., Aral, S.: The spread of true and false news online. Science **359**(6380), 1146–1151 (2018)

On the Learnability of Concepts
With Applications to Comparing Word Embedding Algorithms

Adam Sutton[✉] and Nello Cristianini

University of Bristol, Bristol BS8 1UB, UK
adam.sutton@bristol.ac.uk

Abstract. Word Embeddings are used widely in multiple Natural Language Processing (NLP) applications. They are coordinates associated with each word in a dictionary, inferred from statistical properties of these words in a large corpus. In this paper we introduce the notion of "concept" as a list of words that have shared semantic content. We use this notion to analyse the learnability of certain concepts, defined as the capability of a classifier to recognise unseen members of a concept after training on a random subset of it. We first use this method to measure the learnability of concepts on pretrained word embeddings. We then develop a statistical analysis of concept learnability, based on hypothesis testing and ROC curves, in order to compare the relative merits of various embedding algorithms using a fixed corpora and hyper parameters. We find that all embedding methods capture the semantic content of those word lists, but fastText performs better than the others.

Keywords: Word embedding · Linear classifier · Concepts

1 Introduction

Word embedding is a technique used in Natural Language Processing (NLP) to map a word to a numeric vector, in a way that semantic similarity between two words is reflected in geometric proximity in the embedding space. This allows NLP algorithms to keep in consideration some aspects of meaning, when processing words. Typically word embeddings are inferred by algorithms from large corpora based on statistical information. These are unsupervised algorithms, in the sense no explicit information about the meaning of words is given to the algorithm. Word embeddings are used as input to multiple downstream systems such as text classifiers [19] or machine translations [2].

An important problem in designing word embeddings is that of evaluating their quality, since a measure of quality can be used to compare the merits of different algorithms, different training sets, and different parameter settings.

© IFIP International Federation for Information Processing 2020
Published by Springer Nature Switzerland AG 2020
I. Maglogiannis et al. (Eds.): AIAI 2020, IFIP AICT 584, pp. 420–432, 2020.
https://doi.org/10.1007/978-3-030-49186-4_35

Importantly, it can also be used as an objective function to design new and more effective procedures to learn embeddings from data. Currently, most word embedding methods are trained based on statistical co-occurrence information and are then assessed based on criteria that are different than the training ones.

Cosine similarity and euclidean distances have shown the ability to represent semantic relationships between words such as in GloVe where the vector representations for the words man, woman, king and queen are such that [13]:

$$king - queen \approx man - woman \tag{1}$$

Schnabel et al. [16] identifies two families of criteria: intrinsic and extrinsic, the first family assessing properties that a good embedding should have (eg: analogy, similarity, etc.), the second assessing their contribution as part of a software pipeline (eg. in machine translation).

We propose a criterion of quality for word embeddings, and then we present a statistical methodology to compare different embeddings. The criterion would fall under the intrinsic class of methods in the classification of Schnabel et al. [16], and has similarities with both their coherence criterion and with their categorization and relatedness criteria. However it makes use of the notion of "concept learnability" based on statistical learning ideas. We make use of extensional definitions of concepts, as they have been defined by [1]. Intuitively, a concept is a subset of the universe, and it is learnable if it is possible for an algorithm to recognise further members after learning a random subset of its members.

The key part in this study is that of a "concept". If the set of all words in a corpus is called a vocabulary (which can be seen as the universe), we define any subset of the vocabulary as a concept. We call a concept learnable if it is possible for a learning algorithm to be trained on a random subset of its words, and then recognise the remaining words. We argue that concept learnability captures the essence of semantic structure, and if the list of words has been carefully selected, vetted and validated by rigorous studies, it can provide an objective way to measure the quality of the embedding.

In the first experiment we will measure the learnability of Linguistic Inquiry and Word Count (LIWC) lists. We compare LIWC lists to randomly generated word lists for popular pretrained word embeddings of three different algorithms (GloVe [13], word2vec [10], and fastText [9]). We show that LIWC concepts are represented in all embeddings through statistical testing.

In our second experiment we compare the learnability of different types of embedding algorithms and settings, using a linear classifier. We compare three of these embedding methods (GloVe [13], word2vec [10], and fastText [9]) to each other. We use the same method as previous, however for this experiment we train with the same hyper parameters and corpus across all three word embeddings [20]. We show that from this experiment fastText performs the best, performing significantly better than both word2vec and GloVe.

This study is a statistical analysis of how a given word embedding affects the learnability of a set of concepts, and therefore how well it captures their meaning. We report on the statistical significance of how learnable various concepts are under different types of embedding, demonstrating a protocol for the comparison of different settings, data sets, algorithms. At the same time this also provides a method to measure the semantic consistency of a given set of words, such as those routinely used in Social Psychology, eg. in the LIWC technique.

2   Related Work

Word embedding algorithms can be generated taking advantage of the statistical co-occurrence of words, assuming that words that appear together often have a semantic relationship. Three such algorithms that take advantage of this assumption are fastText [9], word2vec [8], and GloVe [13].

There has been a lot of work focused on providing evaluation and understanding for word embeddings. Schnabel et al. have looked at two schools of evaluation; intrinsic and extrinsic [16]. Extrinsic evaluations alone are unable to define the general quality of a word embedding. The work also shows the impact of word frequency on results, particularly with the cosine similarity measure that is commonly used. Intrinsic methods have also had criticisms, with Faruqui et al. calling word similarity and word analogy tasks unsustainable and showing issues with the method [5].

Nematzadeh et al. showed that GloVe and word2vec have similar constraints when compared to earlier work on geometric models [11]. For example, a human defined triangle inequality such as "asteroid" being similar to "belt" and "belt" being similar to "buckle" are not well represented within these geometric models.

Schwarzenberg et al. have defined "Neural Vector Conceptualization" as a method to interpret what samples from a word vector space belong to a certain concept [17]. The method was able to better identify meaningful concepts related to words using non linear relations (when compared to cosine similarity). This method uses a multi class classifier with the Microsoft Concept Graph as a knowledge base providing the labels for training.

Sommerauer and Fokkens have looked at understanding the semantic information that has been captured by word embedding vectors [18] using concepts provided by [3] and training binary classifiers for these concepts. Their proposed method shows that using a pretrained word2vec model some properties of words are represented within the embeddings, while others are not. For example, functions of a word and how they interact are represented (e.g. having wheels and being dangerous), however appearance (e.g. size and colour) are not as well represented.

3   Methods and Resources

3.1   Embeddings

A corpus **C** is a collection of documents from sources such as news articles, or Wikipedia. From **C** we can extract a set of words to be a vocabulary **V**. Each document in **C** is a string of words (in which the ordering of words within the document is used as part of the embedding algorithm). With a vocabulary $\mathbf{V}$ and a corpus a function $\mathbf{\Phi}$ to be defined such that $\mathbf{\Phi} : \mathbf{V} \to \mathbb{R}^d$, which is mapping every word in the vocabulary to a d dimensional vector. Word vectors from a word embedding are commonly formalised as w.

Using an embedding method $\mathbf{\Phi}$, we will now define the action of going from words in a vocabulary to an embedded space as: $\mathbf{\Phi}(word_j \in V) = w_j \in \mathbb{R}^d$. A word vector for a given word will now be defined as w. Word vectors are generally normalised to unit length for measurement in word analogy or word similarity tasks:

$$\hat{w} = \frac{w}{||w||} \tag{2}$$

3.2   Concepts

In this paper we make use of the notion of a 'concept' defined as any subset of the vocabulary, that is a set of words. Sometimes we will use the expression "list of words", for consistency with the literature in social psychology, but we will never make use of the order in that list, so that we effectively use "list" as another expression for "set", in this article. We define this as a set of words $\mathbf{L} \subseteq \mathbf{V}$ (or for an embedding a set of points in $\mathbb{R}^d$ such that $\Phi(\mathbf{L}) \subseteq \Phi(\mathbf{V})$).

We use the word vectors from a word list to define this concept in an embedding. In general, a concept is defined as any subset of a set (or a "universe"). We would normally define a concept as an unordered list of words that have been created, validated, and understood by humans that should be learnable by machines. However for the purpose of this paper a concept can be defined as any subset of words from $\mathbf{V}$. This use is consistent with the Extensional Definition of a concept used in logic, and the same definition of concept as used in the probably approximately correct model of machine learning [1].

3.3   Linear Classification

A classifier is a function that maps elements of an input space (a universe, in our case a vocabulary) to a classification space. A binary linear classifier is a function that classifies vectors of a vector space R^d into two classes, as follows:

$$f : \mathbf{R}^d \to \{0, 1\}, f(x) = \sigma(\langle x, w \rangle + b) \tag{3}$$

We will learn linear classifiers from data, using the Perceptron Algorithm on a set of labeled data, which is a set of vectors labeled as belonging to class 1 or class 0. As we will learn concepts formed by words, and linear classifiers only operate on vectors, we will apply them to the vector space generated by the word embedding, as follows.

A linear classifier is a simple supervised machine learning model used to classify membership of an input. We will use a single layer perceptron with embeddings as input to see if it is possible for a perceptron to predict half of a word list, while being trained on its other half.

Given a word list $\mathbf{L}$ such that $\Phi(\mathbf{L}) \subseteq \Phi(\mathbf{V}) \subseteq \mathbb{R}^d$ we will define the words from this list as $\mathbf{L}^c = \mathbf{V} \setminus \mathbf{L}$. We will use $\mathbf{L}$ and $\mathbf{L}^c$ to define a train set and test set for our perceptron. We will first uniformly sample half of the words of $\mathbf{L}$, we will then sample in equal amount from $\mathbf{L}^c$. We will then append these two word lists to make $\mathbf{L_{train}}$. To produce a test set $\mathbf{L_{test}}$ we will take the remaining words that haven't been sampled from $\mathbf{L}$, and sample the same number of words again from $\mathbf{L}^c$.

A member of the training set can be defined as $l_i \in \Phi(\mathbf{L_{train}})$. We define our prediction function $\hat{y}$ as:

$$\hat{y} = \sigma((\sum_i^d \theta_i l_i) + b) \tag{4}$$

where θ and b are the training parameters of the classifier and σ is the sigmoid function. We will then train the perceptron using the cross entropy loss function:

$$J = -\frac{1}{|\mathbf{L}|} \sum_i^{|\mathbf{L}|} y_i \log \hat{y}_i + (1 - y_i) \log(1 - \hat{y}_i) \tag{5}$$

where y_i is the correct class of the training sample.

3.4   Linguistic Inquiry and Word Count

This study uses lists of words generated by the LIWC project [12], a long-running effort in social psychology to handcraft, vet and validate lists of words of clinical value to psychologists. They typically aim at capturing concerns, interests, emotions, topics, of psychological significance. LIWC lists are well suited to an experiment of this kind as the words within them are common and relevant to any cross-domain corpus.

Table 1. Sample words from the LIWC word lists used in experiments

Full name	Sample words	List name
Positive emotions	happy, pretty, good	posemo
Negative emotions	hate, worthless, enemy	negemo
Anger processes	hate, kill, pissed	anger
Biological processes	eat, blood, pain	bio
Relativity	area, bend, exit	relative
Affective processes	happy, ugly, bitter	affect
Social processes	talk, us, friend	social
Work concerns	work, class, boss	work
Family concerns	mom, brother, cousin	family
Health concerns	weak, heal, blind	health

Table 1 shows samples of the ten word lists used in this study as well their full names, and what they will be described as when used in the context of this study. Most word lists used have hundreds of words in them. Family is the smallest word list with a total of 54 words being used. These word samples will used to extensionally define word lists as concepts within the embedding.

4 Measuring Performance of Linear Classifiers

We will measure the performance of a linear classifier by using the receiver operating characteristic (ROC) curve, a quantity defined as the performance of a binary classifier as its prediction threshold is changed between the lowest probable prediction and its highest probable prediction. This curve plots the True Positive Rate (also known as the Recall) and the False Positive Rate (also known as the fall-out) at each classification threshold possible. We also show the accuracy of the classifier, and the precision.

Our first experiment will look at the three word embedding algorithms of GloVe, word2vec, and fastText with regards to how they perform using pre-trained word embeddings readily available online. Our second experiment will compare all three algorithms performance under identical conditions, with the same training corpus and hyper parameters.

We will take the input as the embedding representations for words, and the output being a binary classification if the word belongs to that LIWC word set (L) or not. For the training set L_{train}, we will uniformly random sample half of the words from the list L we are experimenting on. We will then sample an equal number of words from L^c. For the test set L_{test} we take the remaining words from L, and again sample another equal set of negative test samples from L^c.

We repeat this method 1,000 times, and for each iteration of this test we generate new word lists L_{train} and L_{test} each time. This method of a linear classifier has been defined in Eq. 4 and Eq. 5. This experiment is performed for the 10 LIWC word lists listed in Table 1. We take their average across all 1,000 iterations of the experiment we performed.

4.1 Learning Concepts from GloVe, Word2vec, and FastText

In this section we will look at the ability of three different word embedding algorithms to capture information in word lists that reflect real world concepts.

To ensure that these metrics are statistically significant, we have created a null-hypothesis of making random concepts based on random word lists (L_{random}) and performing the same classification task on the random concept. We repeat this test 1000 times and take the best performance for each of the metrics we look at for these random lists (which will be defined as $L_{random(max)}$). Of 1000 tests, we hypothesise no concept defined by a random word list outperforms any of the word lists we test on.

Table 2. Average Performance of a Linear Classifiers using LIWC word lists on GloVe word embeddings to identify members of its own set. Random lists are also tested to obtain a p-value and compare performances. These embeddings perform better than random word lists resulting in a p-value of < 0.001

L	Size	Accuracy	Recall	FPR	Prec	AUC
L_{posemo}	392	0.915	0.902	0.079	0.919	0.964
L_{negemo}	492	0.913	0.913	0.085	0.915	0.965
L_{anger}	184	0.888	0.880	0.103	0.896	0.950
L_{bio}	558	0.895	0.871	0.087	0.909	0.954
$L_{relative}$	632	0.937	0.935	0.059	0.940	0.979
L_{affect}	908	0.910	0.906	0.085	0.914	0.962
L_{social}	396	0.906	0.887	0.075	0.922	0.962
L_{work}	322	0.899	0.880	0.081	0.916	0.959
L_{family}	54	0.884	0.893	0.125	0.881	0.956
L_{health}	232	0.895	0.880	0.105	0.893	0.953
$L_{random(max)}$	400	0.547	0.32	0.115	0.617	0.574
$L_{random(avg)}$	400	0.500	0.198	0.198	0.502	0.501

GloVe. We will set GloVe to be our embedding algorithm (Φ), with the corpus **C** being a collection of Wikipedia and Gigaword 5 news articles. These embeddings are pretrained and available online on the GloVe web-page [6]. These word embeddings are open for anyone to use, and can be used to repeat these experiments.

Table 2 shows the performance and statistics of ten different word lists from LIWC. $L_{random(avg)}$ shows the average performance of concepts defined from random word lists. $L_{random(max)}$ shows the best performing random word list for each test statistic.

An accuracy of approximately 0.9 shows a high general performance. The precision and recall show that these word lists are able to accurately discern remaining members of its list and words that are not a part of the concept. After a thousand iterations of random word lists the best performing random lists (shown in $L_{random(max)}$) were performing worse than each LIWC word list, giving a p-val of < 0.001 for each word list.

word2vec. We will use word2vec as our embedding algorithm (Φ), with the corpus **C** being a dump of Wikipedia from April 2018 [22] using the conventional skip-gram model. These embeddings are available online on the Wikipedia2Vec web-page [22]. These word embeddings are open for anyone to use, and can be used to repeat these experiments.

Table 3 shows the performance and statistics of ten different word lists from LIWC while using the word2vec embedding algorithm. $L_{random(avg)}$ and $L_{random(max)}$ again show the average and best performances of random word lists.

Table 3. Average Performance of Linear Classifiers using LIWC word lists on word2vec embeddings to identify members of its own set. Random lists are also tested to obtain a p-value and compare performances. These embeddings perform better than all random word lists resulting in a p-value of < 0.001

L	Size	Accuracy	Recall	FPR	Prec	AUC
L_{posemo}	392	0.904	0.914	0.115	0.888	0.959
L_{negemo}	492	0.923	0.920	0.081	0.919	0.970
L_{anger}	184	0.890	0.906	0.126	0.879	0.953
L_{bio}	558	0.890	0.901	0.120	0.882	0.954
$L_{relative}$	632	0.911	0.952	0.135	0.876	0.963
L_{affect}	908	0.886	0.947	0.177	0.842	0.950
L_{social}	396	0.893	0.911	0.123	0.881	0.957
L_{work}	322	0.877	0.910	0.154	0.855	0.947
L_{family}	54	0.874	0.912	0.164	0.853	0.953
L_{health}	232	0.893	0.899	0.113	0.889	0.959
$L_{random(max)}$	400	0.545	0.27	0.055	0.68	0.576
$L_{random(avg)}$	400	0.498	0.128	0.130	0.494	0.500

An accuracy of approximately 0.9 shows a high general performance, although it performs slightly worse than GloVe's pre-trained embeddings. This shows that the word2vec embedding algorithm Φ applied to the corpus **C** yields word vectors that represent the real world meaning of words. The AUC is extracted from the scores of the sigmoid within the classifier. Overall word2vec performs slightly worse than GloVe embeddings in most metrics. However while the source corpora is very similar, GloVe has additional sources of information. The p-values for these word lists in comparison to random word lists is again <0.001 showing that these word lists that have a real world representation are represented accurately within the embedding.

fastText. The third and final word embedding algorithm (Φ) we will test is fast-Text [7]. The corpus **C** is a collection of Wikipedia, "UMBC WebBase corpus" and statmt.org news [9]. These embeddings are also pretrained word embeddings that are available from the fastText website.

Table 4 shows the performance statistics of the fastText word embeddings using our proposed method to evaluate word embeddings. $L_{random(avg)}$ and $L_{random(max)}$ show the random performance, while the other lists are LIWC word lists and their respective performances.

A precision of 1 in the best performing random word lists are insignificant as the recall is shown to be poor, due to predicting most samples to be negative. The p-val of all of the word lists defined by LIWC is <0.001 as after one thousand iterations no random list outperformed any of LIWC lists. This again means that

Table 4. Average Performance of Linear Classifiers using LIWC word lists on fastText embeddings to identify members of its own set. Random lists are also tested to obtain a p-value and compare performances. These embeddings perform better than random word lists resulting in a p-value of < 0.001

L	Size	Accuracy	Recall	FPR	Prec	AUC
L_{posemo}	392	0.928	0.925	0.068	0.931	0.977
L_{negemo}	492	0.937	0.934	0.067	0.932	0.978
L_{anger}	184	0.940	0.965	0.084	0.919	0.981
L_{bio}	558	0.917	0.933	0.098	0.905	0.970
$L_{relative}$	632	0.933	0.966	0.099	0.907	0.977
L_{affect}	908	0.886	0.947	0.177	0.842	0.950
L_{social}	396	0.927	0.920	0.074	0.925	0.973
L_{work}	322	0.918	0.914	0.077	0.922	0.970
L_{family}	54	0.966	0.975	0.041	0.960	0.995
L_{health}	232	0.931	0.940	0.078	0.924	0.980
$L_{random(max)}$	400	0.51	0.04	0.0	1.0	0.562
$L_{random(avg)}$	400	0.500	0.007	0.006	0.427	0.505

these word lists represent a real world concept, and that the embeddings are able to capture this information of this concept by using members of the set within the embedding to define it.

4.2   Comparing Embeddings

In this section we will be comparing the performance of the three word embedding algorithms used in the previous experiment. However, for this experiment the hyper parameters and the corpora trained will be fixed for the purpose of direct comparison. All embeddings have been generated by ourselves using the three word embedding algorithms word2vec (skip-gram), GloVe, and fastText.

The AUC metric we have previously shown can be viewed as a measure of the learnability of an embedded concept. This compares the true positive rate (also known as the recall) and the false positive rate and shows the performance at each threshold that is possible within the classifier on for a given word lists test set.

This AUC could be seen as the performance of that binary classifier, and also as a measure of the quality of each embedding and a measure of the quality of each word list. The better the performance of an embedding, the higher perceived quality of that embedding. The better a list performs on all embeddings, the higher the quality of that list.

To accurately compare the performance of the embedding algorithms, we perform the same test as shown in Sect. 4.1. However we ensure that a number of parameters are kept the same for each embedding, to maintain fairness. For this test, we will ensure that the corpus used to train will be identical between all embeddings. The corpus (C) used for all three embedding algorithms will be a dump from the English Wikipedia taken from the first of July, 2019 [20]. The embedding dimension d will be set to 300. A word must appear a minimum of five times to be embedded, and the context window of all words is five.

In Table 5 we show the AUC performance of all three embedding algorithms used in the paper. The fastText embedding algorithm is shown to have the highest performing embedding for 8 of the 10 lists that have been tested. Glove performs best on two lists, and generally performs better than word2vec overall. These performances are consistent with previous comparisons of these word embeddings [9,13]. The word list $L_{relative}$ is shown to have the best overall performance across all three non-random embeddings, demonstrating the quality of that list.

Table 5. AUC performance of word lists for each embedding algorithm used in these experiments, along with the average AUC for an embedding across all lists. Bold denotes the embedding algorithm that performs best for a given word list. Italic denotes the best performing list for each embedding algorithm.

L	GloVe	word2vec	fastText
L_{posemo}	0.961	0.929	**0.965**
L_{negemo}	0.965	0.945	**0.973**
L_{anger}	0.957	0.928	**0.970**
L_{bio}	0.960	0.935	**0.974**
$L_{relative}$	*0.971*	*0.927*	*0.961*
L_{affect}	**0.960**	0.944	0.958
L_{social}	0.960	0.925	**0.973**
L_{work}	0.947	0.909	**0.970**
L_{family}	0.948	0.864	**0.963**
L_{health}	0.952	0.923	**0.975**
Mean	0.958	0.922	**0.968**
Median	0.960	0.927	**0.970**

We tested the statistical significance of the performance differences observed between GloVe and fastText. To this purpose we performed a Wilcoxon signed-rank test, using the median of the AUCs from each embedding as the test statistic [21]. We use the Wilcoxon signed-rank test as the fastText mean AUCs shown in Table 5 do not represent a normal distribution.

We propose a null hypothesis that the median difference of fastText and GloVe AUCs (as shown in Table 5) are 0. We use a sample size of 10 as the difference of no pairs are equal to zero. We set our alpha to 0.01 for a one sided (right) tail test, where the test statistic W_{crit} is 5. We find our resulting W_{test} to be 3, which leads us to reject the null hypothesis and show that fastText outperforming GloVe is statistically significant, for the word lists that we are testing. This gives us a p-value of 0.0088.

5   Conclusion

In this paper, we have shown that word embeddings are able to capture the meaning of human defined word lists. We have shown the ability of embedding algorithms in learning concepts from word lists. In particular we have shown this quality in word2vec, GloVe and fastText. We have shown that learning embeddings from real data can represent real world concepts defined extensionally, utilising word lists provided by LIWC.

We have also shown the relative performance of GloVe, fastText, and word2vec when using LIWC word lists to form concepts using similar corpora that derive most of their corpora from Wikipedia. fastText performs better in the majority of situations for all word lists we have tested from LIWC, while GloVe outperforms word2vec generally. However as all algorithms use slightly different corpora, this result may change depending on the corpora used.

This measure of performance of word embeddings can be used in the future as a measure of "quality" of word embeddings. While there are other methods that look at the performance of word embeddings by evaluating their performance in a specific task [15], our method differs in that it looks at an embeddings general ability to understand human defined concepts. There has also been criticism of evaluating word embeddings using only word similarity tasks [5]. This method can also be used in another way as a measure of the quality of word lists and their ability to accurately describe a concept, providing an assumption or proof that an embedding is performing suitably to the users needs.

Future work with this method would involve extensive testing of the method using with varying differing hyper parameters to see the optimal performance of these embedding algorithms. An example of this is the impact of embedding dimension on performance. Another experiment could be looking at the performance of this test on deep contextualized embeddings such as ELMo [14] and BERT [4]. These embeddings have been shown to have better performance on many tasks that employ word embeddings. While these embeddings are optimized for their specific end tasks, they train embeddings before that tuning process takes place. There is potential to compare these embeddings by testing the extracted embedding with a linear classifier, or fine tuning their full model to our task. However a key benefit for sentence embeddings is the context of words around them, which our task will not benefit from.

Further work could be focused on the performance of different word lists and concepts within word embeddings. The benefit of this could be to validate word lists that are not as carefully curated as LIWC word lists. These word lists may come from different fields, as LIWC is focused on clinical psychology other word lists may perform differently. Different source corpora may also change the performance of these word lists due to the meaning of some words changing from domain to domain.

References

1. Anthony, M., Biggs, N.: Computational Learning Theory, vol. 30. Cambridge University Press, Cambridge (1997)
2. Cho, K., et al.: Learning phrase representations using RNN encoder-decoder for statistical machine translation. arXiv preprint arXiv:1406.1078 (2014)
3. Devereux, B.J., Tyler, L.K., Geertzen, J., Randall, B.: The centre for speech, language and the brain (CSLB) concept property norms. Behav. Res. Methods **46**(4), 1119–1127 (2014)
4. Devlin, J., Chang, M.W., Lee, K., Toutanova, K.: Bert: Pre-training of deep bidirectional transformers for language understanding. arXiv preprint arXiv:1810.04805 (2018)
5. Faruqui, M., Tsvetkov, Y., Rastogi, P., Dyer, C.: Problems with evaluation of word embeddings using word similarity tasks. arXiv preprint arXiv:1605.02276 (2016)
6. Pennington, J., Socher, R., Manning, C.D.: Wikipedia 2014 + Gigaword 5 pretrained word embeddings. http://nlp.stanford.edu/data/glove.6B.zip, Accessed 07 Oct 2019
7. Joulin, A., Grave, E., Bojanowski, P., Mikolov, T.: Bag of tricks for efficient text classification. arXiv preprint arXiv:1607.01759 (2016)
8. Mikolov, T., Chen, K., Corrado, G., Dean, J.: Efficient estimation of word representations in vector space. arXiv preprint arXiv:1301.3781 (2013)
9. Mikolov, T., Grave, E., Bojanowski, P., Puhrsch, C., Joulin, A.: Advances in pretraining distributed word representations. In: Proceedings of the International Conference on Language Resources and Evaluation, LREC 2018 (2018)
10. Mikolov, T., Sutskever, I., Chen, K., Corrado, G.S., Dean, J.: Distributed representations of words and phrases and their compositionality. In: Advances in neural information processing systems, pp. 3111–3119 (2013)
11. Nematzadeh, A., Meylan, S.C., Griffiths, T.L.: Evaluating vector-space models of word representation, or, the unreasonable effectiveness of counting words near other words. In: CogSci (2017)
12. Pennebaker, J.W., Francis, M.E., Booth, R.J.: Linguistic inquiry and word count: Liwc 2007, Mahway: Lawrence Erlbaum Associates 71 (2001)
13. Pennington, J., Socher, R., Manning, C.: Glove: Global vectors for word representation. In: Proceedings of the 2014 conference on empirical methods in natural language processing, EMNLP, pp. 1532–1543 (2014)
14. Peters, M.E., et al.: Deep contextualized word representations. arXiv preprint arXiv:1802.05365 (2018)
15. Rajpurkar, P., Zhang, J., Lopyrev, K., Liang, P.: Squad: 100,000+ questions for machine comprehension of text. arXiv preprint arXiv:1606.05250 (2016)
16. Schnabel, T., Labutov, I., Mimno, D., Joachims, T.: Evaluation methods for unsupervised word embeddings. In: Proceedings of the 2015 Conference on Empirical Methods in Natural Language Processing, pp. 298–307 (2015)
17. Schwarzenberg, R., Raithel, L., Harbecke, D.: Neural vector conceptualization for word vector space interpretation. arXiv preprint arXiv:1904.01500 (2019)
18. Sommerauer, P., Fokkens, A.: Firearms and tigers are dangerous, kitchen knives and zebras are not: testing whether word embeddings can tell. arXiv preprint arXiv:1809.01375 (2018)

19. Tang, D., Wei, F., Yang, N., Zhou, M., Liu, T., Qin, B.: Learning sentiment-specific word embedding for twitter sentiment classification. In: Proceedings of the 52nd Annual Meeting of the Association for Computational Linguistics (Volume 1: Long Papers). vol. 1, pp. 1555–1565 (2014)
20. Wikimedia: enwiki dump on 20190701. https://dumps.wikimedia.org/enwiki/20190701/. Accessed 07 Jul 2019
21. Wilcoxon, F.: Individual comparisons by ranking methods. In: Kotz, S., Johnson, N.L. (eds.) Breakthroughs in Statistics. Springer Series in Statistics (Perspectives in Statistics). Springer, New York (1992). https://doi.org/10.1007/978-1-4612-4380-9_16
22. Yamada, I., Asai, A., Shindo, H., Takeda, H., Takefuji, Y.: Wikipedia2vec: an optimized tool for learning embeddings of words and entities from wikipedia. arXiv preprint 1812.06280 (2018)

Towards Fashion Recommendation: An AI System for Clothing Data Retrieval and Analysis

Maria Th. Kotouza[✉] [iD], Sotirios–Filippos Tsarouchis,
Alexandros-Charalampos Kyprianidis, Antonios C. Chrysopoulos,
and Pericles A. Mitkas

Electrical and Computer Engineering, Aristotle University of Thessaloniki,
54124 Thessaloniki, Greece
maria.kotouza@issel.ee.auth.com

Abstract. Nowadays, the fashion industry is moving towards fast fashion, offering a large selection of garment products in a quicker and cheaper manner. To this end, the fashion designers are required to come up with a wide and diverse amount of fashion products in a short time frame. At the same time, the fashion retailers are oriented towards using technology, in order to design and provide products tailored to their consumers' needs, in sync with the newest fashion trends. In this paper, we propose an artificial intelligence system which operates as a personal assistant to a fashion product designer. The system's architecture and all its components are presented, with emphasis on the data collection and data clustering subsystems. In our use case scenario, datasets of garment products are retrieved from two different sources and are transformed into a specific format by making use of Natural Language Processes. The two datasets are clustered separately using different mixed-type clustering algorithms and comparative results are provided, highlighting the usefulness of the clustering procedure in the clothing product recommendation problem.

Keywords: Clothing data · Web-crawling · Meta-data analysis · Mixed-type clustering

1 Introduction

The fashion clothing industry is moving towards fast fashion, enforcing the retail markets to design products at a quicker pace, while following the fashion trends and their consumer's needs. Thus, artificial intelligence (AI) techniques are introduced to a company's entire supply chain, in order to help the development of innovative methods, solve the problem of balancing supply and demand, increase the customer service quality, aid the designers, and improve overall efficiency [1]. Recently, an increasing number of projects in the fashion industry make use of AI techniques, including projects run by Google and Amazon.

The use of AI techniques was not possible before the adoption of e-commerce sites and information and communications technology (ICT) systems from the traditional

I. Maglogiannis et al. (Eds.): AIAI 2020, IFIP AICT 584, pp. 433–444, 2020.
https://doi.org/10.1007/978-3-030-49186-4_36

fashion industry, due to data deficiency. Nowadays, the overflowing amount of data deriving from the daily use of e-commerce sites and the data collected by fashion companies enable solutions related to the fashion design process using AI techniques. Popular fashion houses have provided remarkable AI-driven solutions, such as the Hugo Boss AI Capsule Collection[1], in which a new collection is developed entirely by an AI system, as well as the Reimagine Retail[2] from the collaboration of Tommy Hilfiger, IBM and Fashion Institute of Technology, which aims to identify future industry trends and to improve the design process.

This work focuses on the creative part of the fashion industry, the fashion designing process. To this end, an intelligent and semi-autonomous decision support system for fashion designers is proposed. This system can act as a personal assistant, by retrieving, organizing and combining data from many sources, and, finally, suggesting clothing products taking into account the designer's preferences. The system combines natural language processing (NLP) techniques to analyze the information accompanying the clothing images, computer vision algorithms to extract characteristics from the images and enrich their meta-data, and machine learning techniques to analyze the raw data and to train models that can facilitate the decision-making process.

Several research works have been presented in the field of clothing data analysis, most of them involving clothing classification and feature extraction based on images, dataset creation, as well as product recommendation. In the work of [2], the DeepFashion dataset was created consisting of 800,000 images characterized by many features and labels. In the work of [3], a sequence of steps is outlined in order to learn the features of a clothing image, which includes the following: a) image description retrieval, b) feature learning for the top and bottom part of the human body, c) feature extraction using deep learning, d) usage of pose estimation techniques, and e) hierarchical feature representation learning using deep learning. Other related efforts [4–6] present how to train models using image processing and machine learning techniques for feature extraction.

However, little work has been done in analyzing the meta-data accompanying clothing images. In this work, apart from proposing an AI system which involves many subsystems as part of the clothing design process that can be combined together in order to help the designers with the decision-making process, we emphasize on the data collection, meta-data analysis and clustering techniques that can be applied to improve recommendations.

2   Methods

In this section, we present the proposed decision support system for the designers' creative processes. The system is developed in such a way to be able to model the designer's preferences automatically and be user-friendly at the same, in order to be easily handled by individuals without knowledge of the action planning research field. The system is composed of two interconnected components:

---

[1] https://www.hugoboss.com/fashionstories/digitalisation-is-and-remains-a-big-trend-which-has-already-been-embraced-by-hugo-boss/fs-story-1e6xd6hk2kr8e.html.

[2] https://www.ibm.com/blogs/think/2018/01/tommyhilfiger-ai/.

1. **Offline component**: This component performs (a) data collection from internal and external sources, (b) data storage and management to Databases, and (c) data analysis processes that produce the artificial models which provide personalized recommendations to the end-users.
2. **Online component**: This component comprises mainly the user interface (UI). The users, who are usually fashion designers with limited technical experience, are able to easily set their parameters via the graphical UI, visualize their results and provide feedback on the system results.

The overall system architecture is depicted in Fig. 1, whereas the major subsystems/processes are further analyzed in the following subsections.

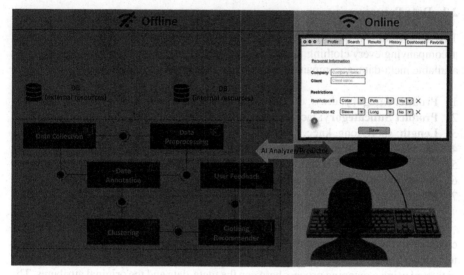

Fig. 1. The proposed system architecture.

2.1 Data Collection

There are two different sources used for training, as well as for the recommendation process: the internal and external data sources.

Internal Data. Each company has its own production line, rules and designing styles that are influenced by the fashion trends. The creativity team usually use an inspiration or starting point based on clothes coming from the company's previous collections and adapt them to the new fashion trends. The internal data are usually organized in relational databases and can be reached by the Data Collection subsystem.

External Data. The most common designers' source for new ideas is browsing on the collections of other popular online stores. To this end, the system includes a web crawler, the e-shops crawler, which is able to retrieve clothing information, i.e. clothing images accompanied by their meta-data. The online shops that are supported so far are Asos, Shtterstock, Zalando and s.Oliver.

Another important inspiration source for the designers are social media platforms, especially Pinterest and Instagram. To this end, a second web crawler, the social-media crawler, was implemented, which is able to utilize existing APIs and retrieve information from the aforementioned platforms, including clothing images, titles of the post, descriptions and associated tags.

Both crawlers' infrastructure is extendable, so that they can be easily used for other online shops or social media platforms in the future.

2.2 Data Preprocessing

This subsystem is responsible for extracting the clothing attributes from the meta-data accompanying every clothing image. Some of the attributes that are extracted from the available meta-data, accompanied by some valid examples, are presented below:

1. **Product category**: dress, overall, pajamas, shorts, skirts.
2. **Product Subcategory**: jacket, coat, T-shirt, leggings.
3. **Length**: short, long, knee.
4. **Sleeve**: short, ¾ length, sleeveless.
5. **Collar Design**: shirt collar, peter pan, mao collar.
6. **Neck Design**: V-neck, square neck.
7. **Fit**: regular, slim.

For each attribute there is a dictionary, created by experienced fashion designers, that contains all the possible accepted values, including synonyms and abbreviations. NLP techniques are used for word-based preprocessing of all meta-data text. The attributes are extracted using a mapping process between the meta-data and the original attributes. The mapping is achieved by finding the occurrences of the words contained in the dictionaries to the meta-data. In the case of successful matching, the corresponding word is marked as a label to the respective attribute.

2.3 Data Annotation

The Data Annotation process complements the Data Collection and Data Preprocessing modules. It is used to enrich the extracted data with common clothing features that can be derived from images using computer vision techniques. Examples of clothing attributes that can be extracted from images include color, fabric and neck design.

It is widely known that color has the biggest impact on clothing, as it is related to location, occasion, season, and many other factors. Taking into consideration its importance, an intelligent computer vision component was implemented. This component has the capability to distinguish and extract the five most dominant colors of each clothing image. More specifically, the color of a clothing image is represented by the values of the RGB channels and its percentage, the color ranking specified by the percentage value and the most relevant general color label to the respective RGB value. The rest of the clothing attributes are extracted using deep learning techniques. Each attribute is represented by a single value from a set of predefined labels.

2.4 Clustering Based on Meta-Data

After the Data Collection and Annotation processes, all the data are available in a common format (row data) that can be analyzed using well-known state-of-the-art techniques. A common technique to organize data into groups of similar products is clustering. Clustering can speed up the recommendation process, by making the look-up subprocess quicker when it comes to significant amount of data. A practical example is a case where a user makes a search at the online phase: the system can limit the data used for product recommendation to those that are included in the clusters characterized by labels related to the user's search.

Several clustering algorithms can be used depending on the type of the data. Clothing data can be characterized by both numerical (i.e. product price) and categorical features (i.e. product category) in general. A detailed review of the algorithms used for mixed-type data clustering can be found in [7]. The algorithms can be divided in three major categories: a) partition-based algorithms, which build clusters and update centers based on partition, b) hierarchical clustering algorithms, which create a hierarchical structure that combines (agglomerative algorithms) or divides (division algorithms) the data elements into clusters, based on the elements' similarities, and c) model-based algorithms, which can either use neural network methods or statistical learning methods, choose a detailed model for every cluster and discover the most appropriate model. The algorithms that we use in this paper are as follows:

1. Kmodes[3]: A partition-based algorithm, which aims to partition the objects into k groups such that the distance from objects to the assigned cluster modes is minimized. The distance, i.e. the dissimilarity between two objects, is determined by counting the number of mismatches in all attributes. The number of clusters is set by the user.
2. Pam[4]: A partition-based clustering algorithm, which creates partitions of the data into k clusters around medoids. The similarities between the objects are obtained using the Gower's dissimilarity coefficient [8]. The goal is to find k medoids, i.e. representative objects, which minimize the sum of the dissimilarities of the objects to their closest representative object. The number of clusters is set by the user.
3. HAC[5]: A hierarchical agglomerative clustering algorithm, which is based on the pairwise object similarity matrix calculated using the Gower's dissimilarity coefficient. At the beginning of the process, each individual object forms its own cluster. Then, the clusters are merged iteratively until all the elements belong to one cluster. The clustering results are visualized as a dendrogram. The number of clusters is set by the user.

---

[3] https://www.rdocumentation.org/packages/klaR/versions/0.6-14/topics/kmodes.

[4] https://www.rdocumentation.org/packages/cluster/versions/2.1.0/topics/pam.

[5] https://www.rdocumentation.org/packages/stats/versions/3.6.2/topics/hclust.

4. FBHC[6]: A frequency-based hierarchical clustering algorithm [9], which utilizes the frequency of each label that occurs in each product feature to form the clusters. Instead of performing pairwise comparisons between all the elements of the dataset to determine objects' similarities, this algorithm builds a low dimensionality frequency matrix for the root cluster, which is split recursively as one goes down the hierarchy, overcoming limitations regarding memory usage and computational time. The number of clusters can be set by the user or by a branch breaking algorithm. This algorithm would iteratively compare the parent clusters with their children nodes, using evaluation metrics and user-selected thresholds.
5. VarSel[7]: A model-based algorithm, which performs the variable selection and the maximum likelihood estimation of the Latent class model. The variable selection is performed using the Bayesian information criterion. The number of clusters is determined by the model.

2.5 Clothing Recommender and User Feedback

The Clothing Recommender is the most important component of our system, since it combines all the aforementioned analysis results to create models that make personalized predictions and product recommendations. The internal and external data, the user's preferences, and the company's rules are all taken into consideration.

Moving on to the online component, the UI enables the designer to search for products using keywords. The extracted results can then be evaluated by the designer and the preferred products can be saved on their dashboard over time and for each product search. If the user is not satisfied by the recommendations, they have the ability either to renew their preferences or ask for new recommendations.

The offline and the online components are interconnected by a subsystem that is responsible for implementing the models feedback process. The user can approve or disapprove the proposed products based on their preferences, and this information is transmitted as input to a state-of-the-art Deep Reinforcement Learning algorithm, which assesses the end user's choices and re-trains the personalized user model. This is an additional learning mechanism evolving the original models over time, making the new search results more relevant and personalized.

3 Experimental Results

A real-life scenario is provided as a use case, in order to highlight the usefulness of the clustering procedure in the clothing product recommendation. Our team is collaborating with a fashion designer working for the Energiers Greek retail company, who is interested in designing the company's collection for the new season. She uses the garments designed and produced by the company in the previous season as a source of inspiration, combined with the Assos e-shop current collections.

---

[6] https://github.com/mariakotouza/FBHC.
[7] https://www.rdocumentation.org/packages/VarSelLCM/versions/2.0.1/topics/VarSelCluster.

3.1 Datasets

In this direction, the Company dataset was created by extracting the fashion products from the previous season from the company database, and the relevant E-shop dataset was retrieved using a web crawler. A total of 4674 images were collected by the e-shop crawler for the season winter 2020, by making queries involving different labels of the attributes *Product Category, Length, Sleeve, Collar* and *Fit*. The meta-data of the retrieved images and a pointer to the image location were stored in a relational database. The meta-data were tokenized and split into columns, by assigning values in the desired attributes, after preprocessing plain text using NLP techniques.

3.2 Data Clustering

In this section, the experimental results on the Company and E-shop datasets using the Kmodes, Pam, HAC, FBHC and VarSel algorithms are presented. The results are evaluated using four internal evaluation metrics:

a. **Entropy**, which quantifies the expected value of the information contained in the clusters.
b. **Silhouette**, which validates the consistency within the clusters.
c. **Within sum of square error (WSS)**, which is the total distance of data points from their respective cluster centroids and validates the consistency between the objects of each cluster.
d. **Identity** [9], which is expressed as the percentage of data contained in the cluster with an exact alignment regarding the feature's labels.

Lower values of Entropy and WSS, and higher values of Silhouette and Identity indicate better clustering results. The clustering results differ according to the applied clustering algorithms. Table 1 shows the normalized mutual information [10] of the algorithms that were tested, in a pairwise fashion. The values show some variance, with most of them being around 30%. It is worth mentioning that the Pam and FBHC algorithms share information that reaches 59.87%, which is something that can enhance their reliability. On the other hand, the least amount of information is shared between the clusters formed by VarSel and Kmodes, FBHC. The main reason seems to be that VarSel algorithm has automatically identified only 3 clusters, whereas the rest of them have formed 6 clusters. The number of clusters (k) for Kmodes, Pam, HAC and FBHC

Table 1. The normalized mutual information of the algorithms tested on the Company dataset.

	Kmodes	Pam	HAC	FBHC	VarSel
Kmodes	1	0.3855	0.3036	0.3535	0.0923
Pam	0.3855	1	0.3740	0.5987	0.2835
HAC	0.3036	0.3740	1	0.3082	0.3239
FBHC	0.3535	0.5987	0.3082	1	0.2682
VarSel	0.0923	0.2835	0.3239	0.2682	1

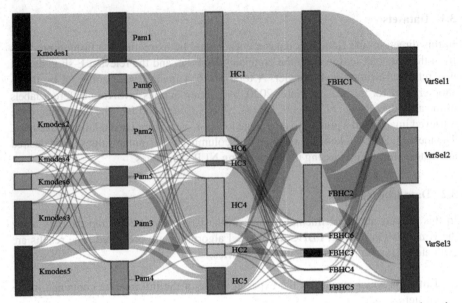

Fig. 2. The Sankey diagram of the clustering results achieved by the algorithms tested on the Company dataset.

was given as input parameter to the algorithms, after experimenting with varying values of k (2 to 12 clusters) and calculating the WSS and Silhouette metrics.

A graphical representation of the information shared across the clusters created by the different algorithms can be seen on the Sankey diagram depicted in Fig. 2. The figure makes clear that the Pam algorithm uniformly distributes the data objects across the 6 clusters, whereas Kmodes clustering results follow a normal distribution. The distributions of those two portioning algorithms seem to be close. The VarSel algorithm normally distributes the objects in a similar fashion, but in this case only 3 clusters are created. On the other hand, the hierarchical algorithms create two large size clusters, where the majority of the objects are assigned to, and four significantly smaller clusters.

Table 2 reports the comparison results of the clustering algorithms based on the values that they achieved at the evaluation metrics. The average values of the evaluation metrics are presented. The best results achieved by an algorithm are highlighted as boldface, whereas the second highest results are presented in italics. The table makes clear that there is not a unique best algorithm that achieves the best results in all the evaluation metrics, so the algorithm's selection depends on the application needs. The hierarchical algorithms achieved better results at the Entropy and Identity metrics, which means that the number of labels characterizing each feature in a cluster is small, whereas the partition-based algorithms outperform at the metrics that concern the distances between the objects of each cluster. Once again it is proved that the Pam algorithm uniformly distributes the data across clusters, and this is the reason why we select this algorithm for the rest of the analysis in this paper. A 2-dimensional representation of the distribution of the data into the six groups obtained by Pam can be seen in Fig. 3.

Table 2. The evaluation results for the Company dataset using various clustering methods.

	Kmodes	Pam	HAC	FBHC	VarSel
# Clusters	6	6	6	6	3
Entropy	0.3279	0.2500	*0.2050*	**0.1889**	0.5159
Silhouette	*0.3479*	**0.4660**	0.2093	0.0753	0.2478
WSS	*0.2800*	**0.2300**	0.4100	0.4900	0.4200
Identity	0.1296	0.1481	*0.1851*	**0.3888**	0.0740

The centroids of the Company dataset extracted by the Pam algorithm are depicted in Table 3 and Table 4 accordingly. The centroids are determined as the most frequent attribute values of the row data for each cluster. A more detailed representation of the groups' consistency for the attributes *Product Category* and *Gender* can be obtained using a heatmap (Fig. 4). By analyzing the consistency of each group and the distribution of the labels across the groups in the two datasets, one can observe that the Company dataset is characterized by six major categories, i.e. Set, Bermuda, Blouse for Men and Women, Dress, and Leggings. On the other hand, the E-shop dataset is characterized by Dress, Shirt, Trousers, Set, Romper, and Cardigan.

As for the rest of attributes, most of the products are characterized by Short Length in the Company dataset, whereas in the E-shop dataset the Medium and Knee Length are more frequent. The tables make clear that the Collar attribute has many missing values, so a good practice will be to recognize this attribute at the Data Annotation subsystem, using computer vision techniques. As for the Fit, the Regular Fit value is the most common in both datasets.

Therefore, when the fashion designer is interested in designing a red dress, she can set the parameters for the product category and the color through the UI of the system and press the search button. The system will then refer to the Company database and filter only the products that are included in the Group 4 created by the offline clustering procedure. The same procedure will be followed to filter only the products that belong to Group 1 in the e-shop's database. The two groups are then combined and the system can select only those products with the label "red" at the Color attribute. Next, this subset can be filtered even more according to the designer's additional preferences and the fashion trends to extract personalized recommendations. Finally, the designer can interact with the system to evaluate (grade) each recommended product, create her dashboard or even ask for new recommendations results if she is not satisfied at all.

4   Discussion

In this work, an intelligent system that automates the typical procedures followed by a fashion designer is described. The system can retrieve data from online sources and the designer's company database, transform plain text accompanying images into clothing features using dictionary mapping and NLP techniques, extract new features from the images using computer vision, and store all the information into a common format in a

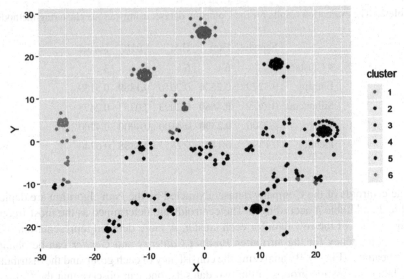

Fig. 3. A 2-dimensional representation of the distribution of the Company dataset into the six groups obtained by Pam.

Table 3. The clustering centroids using the Company dataset using Pam.

	Category	Gender	Length	Sleeve	Collar	Neck	Fit
C1	Set	Man	Short	Long	Shirt	Off shoulder	–
C2	Bermuda	Man	Short	Sleeveless	Flat Knitted Rib	Round	Regular
C3	Blouse	Man	Medium	Short	Shirt	Round	Regular
C4	Dress	Woman	Short	Sleeveless	Flat Knitted Rib	Round	–
C5	Blouse	Woman	Short	Short	–	Round	Regular
C6	Leggings	Woman	Capri	–	–	–	Slim

Table 4. The clustering centroids using the E-shop dataset from the web-crawler using Pam.

	Category	Gender	Length	Sleeve	Collar	Neck	Fit
C1	Dress	Woman	Medium	Raglan	–	Halterneck	Regular
C2	Shirt	Man	Medium	Long	Stand-up	Collar	Regular
C3	Trousers	Man	Medium	Flared	–	Collar	Cargo
C4	Set	Woman	Knee	Flared	–	Collar	Regular
C5	Romper	Woman	Knee	Flared	–	Off shoulder	Regular
C6	Gardigan	Woman	Knee	Long	–	V-neck	Regular

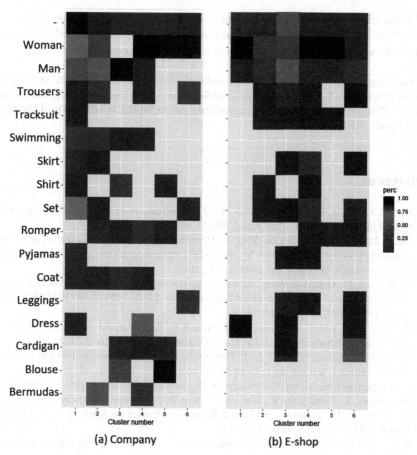

Fig. 4. The distribution of the features' labels across the six different clusters for (a) the Company and (b) the E-shop datasets using Pam.

relational database. The processed data can then be handled by state-of-the-art machine learning techniques including clustering, prediction models, and recommender systems. The paper focuses on presenting the system's architecture, emphasizing on the data collection and transformation processes, as well as the clustering procedures that can be used to organize the row data into groups. A real-life use case scenario was also presented, showing the usefulness of the clustering procedure in the product recommendation problem. Future work involves the augmentation of the Data Annotation process, enabling the extraction of new relevant attributes from non-annotated images. Additionally, the extended use of the products prices and the products' sales history can enrich the model creation process significantly, leading to more reasonable and personalized suggestions for the designers.

Additional steps can be taken in the direction of the improvement of the user-friendliness and the capabilities of the UI, which will be utilized by the designers to enter their preferences, search products, save the system's products recommendation and create dashboards. Finally, an extended set of experiments using new datasets and methods are needed, whereas testing and evaluation of the recommended products are going to be done by fashion designers in more real-life use case scenarios.

Acknowledgements. This research has been co-financed by the European Regional Development Fund of the European Union and Greek national funds through the Operational Program Competitiveness, Entrepreneurship and Innovation, under the call RESEARCH – CREATE – INNOVATE (project code: T1EDK-03464).

References

1. Guo, Z.X., Wong, W.K., Leung, S.Y.S., Li, M.: Applications of artificial intelligence in the apparel industry: a review. Text. Res. J. **81**(18), 1871–1892 (2011)
2. Liu, Z., Luo, P., Qiu, S., Wang, X., Tang, X.: DeepFashion: powering robust clothes recognition and retrieval with rich annotations. In: 2016 IEEE Conference on Computer Vision and Pattern Recognition (CVPR), no. 1, pp. 1096–1104 (2016)
3. Li, R., Feng, F., Ahmad, I., Wang, X.: Retrieving real world clothing images via multi-weight deep convolutional neural networks. Cluster Comput. **22**(3), 7123–7134 (2017). https://doi.org/10.1007/s10586-017-1052-8
4. Hidayati, S.C., You, C.W., Cheng, W.H., Hua, K.L.: Learning and recognition of clothing genres from full-body images. IEEE Trans. Cybern. **48**(5), 1647–1659 (2018)
5. Kalantidis, Y., Kennedy, L., Li, L.-J.: Getting the look: clothing recognition and segmentation for automatic product suggestions in everyday photos. In: Proceedings of the 3rd ACM Conference on International Conference on Multimedia Retrieval, pp. 105–112 (2013)
6. Vittayakorn, S., Yamaguchi, K., Berg, A.C., Berg, T.L.: Runway to realway: visual analysis of fashion. In: Proceedings - 2015 IEEE Winter Conference on Applications of Computer Vision, WACV 2015, pp. 951–958 (2015)
7. Balaji, K., Lavanya, K.: Clustering algorithms for mixed datasets: a review. Int. J. Pure Appl. Math. **18**(7), 547–556 (2018)
8. Gower, J.C.: A general coefficient of similarity and some of its properties. Biometrics **27**(4), 857–871 (1971)
9. Kotouza, M.Th., Psomopoulos, F.E., Mitkas, P.A.: A dockerized framework for hierarchical frequency-based document clustering on cloud computing infrastructures. J. Cloud Comput. **9**(1), 1–17 (2020)
10. Strehl, A., Ghosh, J.: Cluster ensembles—a knowledge reuse framework for combining multiple partitions. J. Mach. Learn. Res. **3**, 583–617 (2002)

Greek Lyrics Generation

Orestis Lampridis[✉], Athanasios Kefalas, and Petros Tzallas

School of Informatics, Aristotle University of Thessaloniki, University Campus,
54124 Thessaloniki, Greece
{lorestis,kefalasa,ptzallas}@csd.auth.gr

Abstract. This paper documents the efforts in implementing lyric generation machine learning models in the Greek language for the genre of Éntekhno music. To accomplish this, we used three different Long Short-Term Memory Recurrent Neural Network approaches. The first method utilizes word-level bi-directional network models, the second method expands on the first by learning the word embeddings on the initial layer of the network, while the last method is based on a char-level network model. Our experimental procedure, which utilized a high sample of human judges, shows that texts of lyrics generated by our models are of high quality and are not that easily distinguishable from actual lyrics.

Keywords: Lyrics generation · Natural language processing · Machine learning

1 Introduction

Natural Language Generation (NLG) is the process of generating text or speech with the use of structured data. Although precisely defining NLG has been proven difficult, a definition often used has been given in [20]. "NLG is a subfield of artificial intelligence and computational linguistics that is concerned with the construction of computer systems than can produce understandable texts in English or other human languages from some underlying non-linguistic representation of information".

NLG as a whole and more specifically the specialized task of creative writing is a task that humans can be quite effective at, while computational intelligence may find it rather difficult to generate creative and good quality text [4]. To this end, many attempts have been made in the literature to try and mimic human creativity, including automatic generation of poetry [5, 8, 16], metaphors [23], slogans [22] and others.

The domain of song lyrical writing has not been explored as much [17]. This problem can be more apparent due to the fact that song lyrics also have a musical aspect and may have a different style based on the genre of the song as well as secondary constraints of the task such as rhyming words and thematic tone definition.

O. Lampridis, A. Kefalas and P. Tzallas—Contributed equally to this work.

© IFIP International Federation for Information Processing 2020
Published by Springer Nature Switzerland AG 2020
I. Maglogiannis et al. (Eds.): AIAI 2020, IFIP AICT 584, pp. 445–454, 2020.
https://doi.org/10.1007/978-3-030-49186-4_37

To evaluate the performance of our models, we used human judges in order to evaluate the fidelity and credibility of our model when it comes to accurately imitating the lyrics of a real Éntekhno song. In other words, how likely it would be for the text outputted by our model to be the lyrics of an actual song.

This paper is structured as follows: In Sect. 2, we introduce various computational approaches which have been used to successfully generate either poetry or song lyrics. In Sect. 3, we demonstrate our dataset. In Sect. 4, we present our different approaches to solve this problem. In Sect. 5, we evaluate our different approaches with the help of human judges. Finally, in Sect. 6, we offer our conclusions and discuss several future directions for our work.

2 Related Work

While the study of text generation for creative writing purposes, such as poetry and song lyrics, is of great interest in academic areas such as linguistics and music, it is also of great importance for many subfields of computer science. Relevant literature can be found in research areas such as computational creativity, information extraction, natural language processing and machine learning.

In the work of Graves [9], Recurrent Neural Networks (RNNs) [21] were used for text generation and their high effectiveness was showcased. Graves used a variation of RNNs called Long Short-Term Memory (LSTM) [10] architecture to create a language model in the character level, which has a higher success at text generation than a regular RNN model. The results are impressive, as the network created was able to learn various grammatical rules, while also being able to accurately reproduce a considerable amount of vocabulary of words in the English language.

As already mentioned, automatic text generation for poetry and song lyrics, has also been explored in the literature. Language models have been used to generate poetic text, constrained by both a target style and a predefined form. These include Markov models as in [2] and models based on Support Vector Machines (SVMs) as in [5]. As far as rap lyrics are concerned, Wu et al. [1] present a system generating rap lyrics that outputs a single sequence of rap lyrics that are a response to a particular input. A different approach was given by Malmi et al. [14] by using a Deep Neural Network (DNN) and the RankSVM [11] algorithm. By utilizing full lines of lyrics from rap songs, they created 16-line verses.

The work which is most similar to ours is that of Potash et al. [19]. Given a sequence of rap lyrics and a specific rapper, an LSTM RNN model is trained and used to predict the next word. Their goal is to produce rap lyrics that are similar in style but not identical to already existing rap lyrics; a task known as ghost writing in the music industry.

3 Dataset

The task of generating song lyrics using machine learning algorithms requires the use of a proper dataset of song lyrics for training these algorithms. The dataset was downloaded mainly from three sources, stixoi.info, greeklyrics.gr and kithara.to by using a web scraper utility program. The scraper utility targeted the URL of the song details web page of each source by injecting an id that was generated from a randomly shuffled sequence of Long type numbers. The Long datatype was selected, because Long is an appropriate type for mapping a numeric primary key id in most widely used relational databases and therefore would have the highest probability of targeting an existing id in the source's database. The scrapper was setup to use a random back-off scheme with a 20 s minimum await time, to avoid interfering with the website's normal traffic as much as possible.

In cases where the scrapper requested a URL that was not a miss, the retrieved HTML document was queried using XPath to get the textual content of HTML DOM elements that contained the song title, lyrics, artist and other optional information such as the songs composer or lyricist. The raw textual data was then sanitized by removing special characters, duplicate whitespace characters, new line characters and symbols. Finally, after gathering an appropriately large amount of songs for the scope of our research, the dataset was exported as a csv.

The size of the raw dataset we have gathered for training the text generating language model, consisted of about 18000 entries of unprocessed songs of various artists and genres. Each song in the dataset consists of the song title and lyrics for each song and optionally in cases where it the data were available, the information of the artist, the composer, the music producer and lyricist. All entries in the raw dataset are strictly Greek songs and short poems or limericks that had been used in some traditional folk songs. The very first preprocessing step was to prune all entries in the dataset that had mixed language lyrics because the context switch between different languages along with the small number of samples with language context switches would impede the text generation model from effective learning and would negatively affect character-based models.

As different genres of music can have different styles of lyrics and specific phraseology, rhythm and even vocabulary, using the entirety of the dataset would only produce incoherent results regardless of the quality of the text generation model. For the purposes of our research, we selected songs of various artists from only the Greek Éntekhno genre. After dropping songs from all other genres the dataset was then pruned to a final size of 1150 songs.

The entries in the dataset were preprocessed to replace all whitespace characters in the lyrics of song with a space character. Furthermore, all remaining symbols and non-alphanumeric characters expect for punctuation were removed from all songs in the dataset. Additionally, alternate types of single and double quote characters were replaced by the respective standardized English counterpart. Finally, the last preprocessing step for the dataset was to correct all the disfluencies and typographical errors within the songs by hand and additionally expand words with apostrophes to their full variant form. Shortened word forms contained in the song lyrics as localisms and idioms were replaced by hand with the full variant or normalized form. Lastly, some explicit or offensive words were replaced in the corpus by alternates that had either a similar meaning or a similar phonetic rhythm or letter similarity to retain rhyme within the song.

4 Implementation

As seen in the literature, RNNs are appropriate for the modeling of natural language sequences as the RNN cells are able to retain contextual information about a sequence of tokens. As basic RNN units can present a variety of issues like the exploding or vanishing gradient problem [18], we settled on using LSTM units as the primary component of the neural networks. Sequences of word tokens may have long running dependencies in the context of a sentence that may be directed either from the beginning to the end or vice versa. The use of bidirectional LSTM units allows for the modeling of these forward and backward dependencies of a token in a given sequence. Deriving from the baseline architectural components succinctly described above we followed three different approaches that all utilize the bidirectional LSTM - RNN models and compared their results. For the first two models, we based our approach in [6, 7], while for the final model we drew influence from [15]. The code that was developed is available on our GitHub repository[1].

4.1 Word Level Bi-LSTM

The RNN LSTM model used in the first approach is based on the idea of training it with a large sequence of words and then trying to predict the next word by using in a one-hot vector representation of the sequences. As a first step, we had to read our lyrics in text form, convert to lowercase, to have fewer words and split the sentences into tokens. We chose to treat the newline as an individual word. The thought process behind this is that we are giving the LSTM the capability to decide when to start a new line. On this first approach, we didn't use any further pre-processing (i.e. punctuation removal), because we wanted to see if the network could also learn from these features and apply them when creating the new lyrics.

Before building the LSTM RNN model, we split our data into 98% for the training set and 2% for the test set. Several different architectures were explored for our network. After carefully examining the quality of the resulting lyrics and taking into account metrics such as accuracy and validation accuracy we ended up on using the following architecture: The first layer in the network consisted of a bidirectional RNN layer with 256 LSTM units. The second layer was a dropout layer with dropout = 0.2. Finally, there is a dense layer with softmax as activation function. The output of this layer is a vector of size equal to the number of words in the corpus which contains the probabilities for each available word in the corpus. The next word is then predicted by using the multinomial distribution to sample on these probabilities.

In order to fit the model, we had to use a data generator. We do this because in this approach we used one-hot representation of the sentences. One-hot vectorization results in vectors of 0 s with a single 1 in the column of the used word. Thus, each sentence is represented by an array of size the length of the sequences times the number of unique words in our corpus. Our total training set would have size equal to the number of sentences times the length of the sequences times the number of unique words. This number ends up being enormous. Using data generators, we feed the model

---

[1] https://github.com/orestislampridis/Greek-Lyrics-Generation.

with chunks of the training data, one for each step, instead of feeding everything at once. The generator function gets the list of sentences, the list of next words, and the size of the batch. For training the model, a shuffled set of the training sequences was used. The loss function used in this model is the categorical cross entropy function. For validation, we send another generator with the test data, so it gets evaluated every epoch. Finally, the optimizer used was adam [12].

4.2 Word Level Bi-LSTM with Trainable Embeddings

The model used in this approach uses a trainable Embedding layer before feeding the embeddings of a word sequence to a bidirectional LSTM. The raw text of the dataset is further preprocessed by removing all punctuation from the song lyrics and all tokens were converted to lowercase. By applying case folding to lowercase, we limit the size of the vocabulary the embedding layer will have to learn. After processing the raw text of a song, the lyrics were segmented in the word level and were consequently converted to sequences of tokens. Sequences with size smaller than the pre-configured sequence size were padded with a synthetic padding token. Along with the synthetic word used for padding an additional two synthetic words were introduced to the sequences, one marking the beginning of a song and one denoting the end of a song. The inputs of the embedding layer and the model as a whole are non one-hot encoded words, meaning that instead of vectors representing words the model accepts integers that map to a specific token in a predefined vocabulary that was extracted from the dataset.

The first layer in the model's neural network was a trainable embedding layer. The size of the output vector was set to 1024 dimensions. The next layer is a bidirectional RNN layer with 256 LSTM units. The output of this layer is a collection of logit values that are intercepted by a dropout layer. The final layer is a dense or regression layer with a softmax activation. The output of this layer is a vector with a number of dimensions equal to the size of the vocabulary. The next token is predicted by using the multinomial distribution to sample on the probabilities contained in the output vector.

For training the above model a shuffled set of all available sequences was used. In contrast with the word level approach, the loss function used in this model is the sparse categorical cross entropy function that is used on sparse categorical data. For validation, the entirety of available sequences were also used, in essence forcing the model to overfit on all of the sequences, while simultaneously increasing the dropout rate to 0.2. This method essentially forces the model to overfit on the dataset but still retain some generalization when predicting sequences. In later epochs of training, some of the data in the train set should be held out to increase the generalization of the model's predictions. The validation function used was sparse categorical accuracy while the optimizer used was adamax [12] for the first few epochs and later manually switched to adam. The use of adamax in the first few epochs is a minor optimization on training because that specific optimizer is the best suited for training models with embedding layers.

4.3 Character Level Bi-LSTM

In this approach, the data are left in their raw form. The vocabulary now consists of all the characters in the text data, including punctuation. Since we are working at character level there is no point in finding the most common words. Additionally, the data are not converted to lowercase in the final approach, because we noticed that the results were worse in this scenario. The only data processes are the mapping of the characters to integer numbers and the creation of sequences to feed the RNN model. The input value in each sequence is a 100-character string and the target value is simply the next character which is encoded by the One Hot Encoding method. This way we can give RNN a greater power to learn a probability for each possible target value. When a one hot encoding is used for the output variable, it may offer a finer set of predictions than a single label. The input is reshaped from a one-dimensional array to a two-dimensional array to feed the first layer of the RNN.

The model uses a trainable Embedding layer which works in the same fashion as the one in the Word level Bi-LSTM with Embeddings. It then feeds the embedding sequences of characters to the first layer of the RNN, which a bidirectional CuDNNLSTM layer with 512 nodes. The CuDNNLSTM is the same as a simple LSTM layer, the only difference is that it uses GPU for training, which makes it much faster. The hidden layer is another bidirectional CuDNNLSTM, again with 512 nodes. There is a Dropout layer set to 0.2 to avoid overfitting. Finally the output layer is a dense layer and the activation function is the softmax function as we want a distribution over the outputs. We use the adam optimize and the categorical cross-entropy loss.

Callbacks are made monitoring the train and validation accuracy. We train the model for 50 epochs. We use the validation split set to 0.2, which randomly splits up the data into a training set and test set and evaluates the model. The maximum validation accuracy achieved by this model is around 50%, which is an acceptable value for an NLG model. In the generation process we use both a random seed, sampling the sequences made before and a specific set seed. The seed is then mapped and reshaped and a prediction is made using the argmax function.

5 Evaluation

In evaluating the results of our approaches, we decided to use human judges for extrinsic evaluation of the three proposed neural language models. We were inspired by the work of Belz and Reiter [3] to set up our experimental procedure. We created a survey that contained 10 different texts in total. Out of these texts, 9 of them were from our 3 different models (3 texts for each model) and the last text was from an actual song. We did this to see how humans would rate an actual (i.e. not automatic generated) Éntekhno song. The participants didn't know that a text from an actual song would be shown. We expected this song to have the highest score and if this was the case, we would be certain about the validity of our survey. Then, we asked 76 volunteers to participate in the survey. A high percentage of them (96.1%) stated that they were familiar with the genre of Éntekhno.

The texts were presented in a random fashion and after each text, the volunteers were called to rate the accuracy of the generated lyrics on a scale of [1, 5] when it comes to imitating the lyrics of a real song. In other words, how likely it would be for the text they were reading to be the lyrics of an actual Éntekhno song. We call this accuracy the fidelity of our models. An example of the texts that were shown to the participants is shown in Table 1. To make the results more understandable to the international community, we tried to offer the best possible translation in the English language for the texts in Table 2. Note, that these translations do not accurately represent the capability of our models since each language has different ways to express creativity and our models were only trained using the Greek language. Lastly, average fidelity for each model according to the survey and after taking into account the three texts for each of our models is reported in Table 3.

Table 1. Sample texts shown to the participants during our experiment

	text shown
word_level	πως έγινε και αλλάξαμε πορεία και μένει το όνειρο ξανά στην ψυχή μου. παλίρροια, παλίρροια, μπήκες και η ζωή μου. πιες από τα μάτια μου φωτιά, και κάψε την ψυχή μου. σαν ξαφνική παλίρροια, μπήκες και τη ζωή μου. πιες από τα μάτια μου φωτιά, και κάψε την ψυχή μου. σαν ξαφνική παλίρροια, μπήκες μες τη ζωή μου
embeddings	να αράξω και τον καημό για να σε δω να σε αγκαλιάσω και θα μου πεις φοβάμαι μη σε χάσω δεν είσαι εδώ το όνειρο μου ξανά ζήσω και να φοβάμαι μη χαθείς όταν ξυπνήσω
char_level	Χρόνια γυρίζεις θεατής στου κόσμου τις κερκίδες μα πες μου τι αγκάλιασες σφιχτά από όσα είδες. Χρόνια καρδιές πολλές και σε χρόνια να παίζεις το παραθείο και στην καρδιά μου ποιημούρα αγάπη μου είπες μου πονάς τα μάτια σου κι εσύ το χρώμα της καρδιάς σου μια κορφή εκεί που πάει εσύ είσαι εγώ εγώ είσαι εσύ. Πάρε με μια καρδιά μου που σε αγαπώ σε αγαπώ το χρόνο να παίζει το φως μου και με παγιδεύεις
Actual	Από τα ίδια πάλι να γεμίσω το κεφάλι. Φέρε να πιω για να μου φύγει το βάρος γιατί απόψε θα πεθάνει ο χάρος. Μια γυναίκα με έχει κάνει το ποτό να μη με πιάνει. Θέλησε να με πληγώσει μα πικρά θα το πληρώσει. Φέρε να πιω

As we can see in Table 3, the word level model with pre-trained embeddings is the one with the highest fidelity, while the word level model is a close second. This is mostly because the embeddings model uses a more complex model than the word level model one, as it uses an additional embedding layer. The character level approach suffers from grammatical and syntax errors; therefore, it has been rated lower. Also, we noticed that the longer the lyrics generated the more difficult it is for each model to generate realistic lyrics.

Looking at the highest fidelity score in Table 4, it was interesting to see that the actual song scored a fidelity of 4.092, not that close to the perfect score of 5. Also, its score was not that far ahead from the second highest scoring text, which was generated by our word level embeddings model and had a fidelity score of 3.815. The third highest scoring text was also not that far behind with a score of 3.671.

Table 2. Sample texts shown to the participants translated in the English language

	Text shown
word_level	how it happened and we changed course and the dream stays in my soul again. tide, tide, you came in and my life. drink fire from my eyes, and burn my soul. like a sudden tide, you came into my life. drink fire from my eyes, and burn my soul. like a sudden tide, you came into my life
embeddings	to seize the misery to see you hug you and you will tell me i'm afraid i will lose you you're not here my dream live again and i'm afraid you get lost when i wake up
char_level	For years you have been watching the stands in the world, but tell me what you hugged tightly from what you saw. Years many hearts and in years to play the paratheo and in my heart my love poem you said you hurt my eyes and you the color of your heart a peak where it goes you are me I am you. Take me with a heart that I love you I love you time to play my light and you trap me
Actual	From the same again to fill the head. Bring me to drink so that my weight will go away because tonight the Reaper will die. A woman has made me stop drinking. She wanted to hurt me but she will pay dearly. Bring me a drink

Table 3. The average fidelity for each model

	word_level	embeddings	char_level
Fidelity	3.022	3.053	2.338

Table 4. The highest fidelity achieved by each model along with the fidelity of the actual lyrics

	word_level	embeddings	char_level	Actual
Fidelity	3.671	3.815	3.026	4.092

6 Conclusions and Future Work

With regards to future work, a larger dataset could be very beneficial. In our approach, we used a dataset of only 1150 songs, which is a compensatory number, but a larger dataset may produce better results. To this end, the Greek Music Dataset [13] could also be explored as it includes songs of the Éntekhno category. Also, even if we tried many different combinations of layers and tunings in our models, there are a lot more that can be used and bring a better output. Furthermore, an expansion to the whole process could be attempted. One idea could be to add a function to compute the style and tempo of each song and use that information to the generation process, this could produce more realistic lyrics imitating the style of actual songs. In addition, more complex features could be added to the RNN, such as the part of speech of each word, the phonetic representation of each word or the frequency of each word or character. In the character level approach, a function could be made to correct the syntax and grammatical errors or investigate if the words outputted from the network are correct by cross-checking them with a Greek dictionary of words. Finally, another approach to help with the rhyme of the song generated could be to create two RNN models, one that generates the song from start to finish and another one that does the opposite and takes as seed the end of the previous lyric.

Concluding, we are obligated to acknowledge the difficulty of the lyrics generation task. With all the progress that is made it can only be used for inspiration by a lyricist and not for immediate commercial use. Poetry and music are arts and have a high degree of creativity that emanate from each artist's soul and it is quite difficult to train a machine learning algorithm to copy them.

Acknowledgments. We would like to express our gratitude to prof. Grigorios Tsoumakas for assigning us this project and for his valuable help and guidance. We would also like to thank all 76 volunteers for the time and effort they took to participate in the survey used for evaluating our three proposed models along with the comments and suggestions they provided about possible improvements. This work is supported by the Data and Web Science MSc program of the School of Informatics at the Aristotle University of Thessaloniki.

References

1. Addanki, K., Wu, D.: Unsupervised rhyme scheme identification in hip hop lyrics using hidden Markov models. In: Dediu, A.-H., Martín-Vide, C., Mitkov, R., Truthe, B. (eds.) SLSP 2013. LNCS (LNAI), vol. 7978, pp. 39–50. Springer, Heidelberg (2013). https://doi.org/10.1007/978-3-642-39593-2_3

2. Barbieri, G., et al.: Markov constraints for generating lyrics with style. In: ECAI, vol. 242 (2012)

3. Belz, A., Reiter, E.: Comparing automatic and human evaluation of NLG systems. In: 11th Conference of the European Chapter of the Association for Computational Linguistics (2006)

4. Colton, S., Wiggins, G.A.: Computational creativity: the final frontier? In: ECAI, vol. 12 (2012)

5. Das, A., Gambäck, B.: Poetic machine: computational creativity for automatic poetry generation in Bengali. In: ICCC (2014)

6. Enrique, A.: Word-level LSTM text generator. Creating automatic song lyrics with Neural Networks, June 2018. https://medium.com/coinmonks/word-level-lstm-text-generator-creating-automatic-song-lyrics-with-neural-networks-b8a1617104fb. Accessed 27 Feb 2019

7. Enrique, A.: Automatic song lyrics generation with Word Embeddings, June 2018. https://medium.com/coinmonks/word-level-lstm-text-generator-creating-automatic-song-lyrics-with-neural-networks-b8a1617104fb. Accessed 27 Feb 2019

8. Gervás, P.: An expert system for the composition of formal Spanish poetry. In: Macintosh, A., Moulton, M., Coenen, F. (eds.) Applications and Innovations in Intelligent Systems VIII, pp. 19–32. Springer, London (2001). https://doi.org/10.1007/978-1-4471-0275-5_2

9. Graves, A.: Generating sequences with recurrent neural networks. arXiv preprint arXiv:1308.0850 (2013)

10. Hochreiter, S., Schmidhuber, J.: Long short-term memory. Neural Comput. **9**(8), 1735–1780 (1997)

11. Joachims, T.: Optimizing search engines using clickthrough data. In: Proceedings of the Eighth ACM SIGKDD International Conference on Knowledge Discovery and Data Mining (2002)
12. Kingma, D.P., Ba, J.: Adam: a method for stochastic optimization. arXiv preprint arXiv:1412.6980 (2014)
13. Makris, D., Kermanidis, K.L., Karydis, I.: The Greek audio dataset. In: Iliadis, L., Maglogiannis, I., Papadopoulos, H., Sioutas, S., Makris, C. (eds.) AIAI 2014. IAICT, vol. 437, pp. 165–173. Springer, Heidelberg (2014). https://doi.org/10.1007/978-3-662-44722-2_18
14. Malmi, E., et al.: Dopelearning: a computational approach to rap lyrics generation. In: Proceedings of the 22nd ACM SIGKDD International Conference on Knowledge Discovery and Data Mining. ACM (2016)
15. Ma'amari, M.: AI Generates Taylor Swift's Song Lyrics, September 2018. https://towardsdatascience.com/ai-generates-taylor-swifts-song-lyrics-6fd92a03ef7e. Accessed 27 Feb 2019
16. Oliveira, H.: Automatic generation of poetry: an overview. Universidade de Coimbra (2009)
17. Oliveira, H.G.: Tra-la-lyrics 2.0: automatic generation of song lyrics on a semantic domain. J. Artif. General Intell. 6(1), 87–110 (2015)
18. Pascanu, R., Mikolov, T., Bengio, Y.: Understanding the exploding gradient problem. CoRR, abs/1211.5063 2, 417 (2012)
19. Potash, P., Romanov, A., Rumshisky, A.: Ghostwriter: using an LSTM for automatic rap lyric generation. In: Proceedings of the 2015 Conference on Empirical Methods in Natural Language Processing (2015)
20. Reiter, E., Dale, R.: Building applied natural language generation systems. Nat. Lang. Eng. 3(1), 57–87 (1997)
21. Rumelhart, D.E., Hinton, G.E., Williams, R.J.: Learning representations by back-propagating errors. Nature 323(6088), 533–536 (1986)
22. Tomašic, P., Znidaršic, M., Papa, G.: Implementation of a slogan generator. In: Proceedings of 5th International Conference on Computational Creativity, Ljubljana, Slovenia, vol. 301 (2014)
23. Veale, T., Hao, Y.: A fluid knowledge representation for understanding and generating creative metaphors. In: Proceedings of the 22nd International Conference on Computational Linguistics (Coling 2008) (2008)

Correction to: An Intelligent Cloud-Based Platform for Effective Monitoring of Patients with Psychotic Disorders

Ilias Maglogiannis, Athanasia Zlatintsi, Andreas Menychtas,
Dennis Papadimatos, Panayiotis P. Filntisis, Niki Efthymiou,
George Retsinas, Panayiotis Tsanakas, and Petros Maragos

Correction to:
**Chapter "An Intelligent Cloud-Based Platform for Effective
Monitoring of Patients with Psychotic Disorders" in:
I. Maglogiannis et al. (Eds.):** *Artificial Intelligence Applications
and Innovations*, **IFIP AICT 584,
https://doi.org/10.1007/978-3-030-49186-4_25**

The original version of this chapter was revised. The first name of one of the authors inadvertantly contained a typo. The author's first name has been corrected to "Niki."

The updated version of this chapter can be found at
https://doi.org/10.1007/978-3-030-49186-4_25

I. Maglogiannis et al. (Eds.): AIAI 2020, IFIP AICT 584, p. C1, 2020.
https://doi.org/10.1007/978-3-030-49186-4_38

Author Index

Printed in the United States
by Baker & Taylor Publisher Services